Statistical Methods of Geophysical Data Processing

Statistical Methods of Geophysical Data Processing

Vladimir Troyan • Yurii Kiselev

Saint Petersburg State University, Russia

World Scientific

NEW JERSEY • LONDON • SINGAPORE • BEIJING • SHANGHAI • HONG KONG • TAIPEI • CHENNAI

Published by

World Scientific Publishing Co. Pte. Ltd.
5 Toh Tuck Link, Singapore 596224
USA office: 27 Warren Street, Suite 401-402, Hackensack, NJ 07601
UK office: 57 Shelton Street, Covent Garden, London WC2H 9HE

British Library Cataloguing-in-Publication Data
A catalogue record for this book is available from the British Library.

STATISTICAL METHODS OF GEOPHYSICAL DATA PROCESSING

ISBN-13 978-981-4293-74-7
ISBN-10 981-4293-74-1

Printed in Singapore by World Scientific Printers

Introduction

The study of the structure of the Earth and near-earth space requires simultaneous processing and interpretation of a large volume of experimental data. The presence of a random error, which one is inevitable at measurings, and the presence of a random noise in the wave fields, which distorts useful signals, leads to a necessity to use the probability-statistical methods for the analysis and interpretation of the geophysical information (Goltsman, 1971; Ryzhikov and Troyan, 1994; Troyan, 1982; Troyan and Sokolov, 1989; Fedorov, 1972; Sheriff and Geldart, 1995; Troyan and Kiselev, 2000). The digital recording of the geophysical fields and the computer-aided data processing with the subsequent interpretation now is widely used. The formalization of the interpretation process requires the links of a measured field with the parameters and a state of an investigated object, that is called frequently as the model of an experimental material. The similar model includes the scheme of the experiment together with the formalization of links of an observable geophysical field with parameters and the state of the medium, and also the presence of a random deviation between the experimental field and "idealized" one obtained as a result of a solution of the direct problem. The main goal of the processing and interpretation of the geophysical data consists in the restoration of desired parameters and the state of the medium. For the solution of this intricate problem it is necessary to use a maximum of a priori information about the medium, which can be obtained on the basis of the previous experiments. The systems of the processing and interpretation of the geophysical data in practice, as a rule, are on-line, that allows along with strict mathematical methods and routines as an unformalized device to include an intuition and an experience of the geophysicist.

The suggested textbook is based on the courses of lectures "Statistical methods of processing of the geophysical data" and "Inverse problems in geophysics", which are given by the authors for master's and post-graduate students of the Chair of Physics of the Earth at Physical Faculty of Saint Petersburg State University during last ten years.

In the first chapter the basic concepts of the probability theory are given. The space of elementary events and the relative-frequency, classic and geometrical definitions of probability are introduced. The formula of the total probability and the Bayes formula are given (Populis and Pillai, 2002; Cramer, 1946; Kolmogorov, 1956; Pugachev, 1984). The cumulative distribution function is introduced and its properties are considered (Brandt, 1998; Pugachev, 1984). Numerical characteristics of a distribution of probabilities are analyzed in details: an expectation, a variance, a coefficient of correlation, a median, the initial and central moments, an asymmetry coefficient and an excess. The characteristic functions and their properties are considered. The limit theorems of the probability theory are given (Populis and Pillai, 2002; Cramer, 1946). Various types of probability distributions are considered: binomial, Poisson, geometrical, normal, uniform, Student, Fisher, exponential, Laplace, Cauchy, logarithmic normal, χ^2–distribution etc. (Pugachev, 1984; Rao, 1972). The concept of the entropy and the information is introduced. The informations of Shannon and Fisher are considered; the possibilities of their use for an exposition of the interpretation quality of the geophysical data are analyzed (Rao, 1972). The properties of random functions are given. The autocorrelation and cross-correlation functions are introduced. The connection of the autocorrelation function with the power spectrum is considered (Pugachev, 1984).

The second chapter is devoted to an account of basic elements of the mathematical statistics. The basic concepts of the theory of decisions are introduced: structure of a decision space, a loss function, a resolution rule and sufficient statistics. The attention to the properties of estimates (consistency, bias, effectiveness, sufficiency, normality, robustness) are given. The examples of an estimation of the accuracy and reliability of the interpretation of geophysical fields are surveyed (Johnson and Lion, 1977; Goltsman, 1971; Cramer, 1946; Nikitin, 1986; Pugachev, 1984; Troyan and Sokolov, 1989; Fedorov, 1972).

In the third chapter the concept of the model of the measurement data is introduced. This model is a functional relationship between the observations and with the state and parameters of an investigated medium. The random noise is a very important part of the model. The distinctive feature of the statistical theory of interpretation is the assumption about a stochastic nature of an observed geophysical field. By depending on the statement of problem and purpose of interpretation, the models of an experimental material are subdivided into the quantitative interpretation, when the problem consists in a determination of the estimates of the desired parameters of a medium, the qualitative interpretation, when the problem consists in a choice of a state of the object (test of hypothesis) and the qualitative-quantitative interpretation, when the parameters and the state of the object are estimated simultaneously. The important points of a description of the model are the representation of properties of the random component and taking into account correctly of a priori information about properties of an investigated medium (Goltsman, 1971; Troyan, 1982; Troyan and Sokolov, 1989).

The fourth chapter is devoted to the description of the perfect relationship of the sounding signals (geophysical fields) with the parameters of a medium (examples of the solution of the direct geophysical problem). The elastic wave fields, which are used for the reconstruction of the Earth structure in the problems of the seismology and seismic exploration are surveyed in details. The basic equations of acoustics used at study the oceanic column and oceanic sedimentary cover are introduced. The mathematical model of propagation of the electromagnetic signals in an earth-crust and ionosphere is described. The transport equation for the problem of remote sensing of an atmosphere is introduced (Aki and Richards, 2002; Jensen *et al.*, 2000; Alekseev *et al.*, 1983; Kravtsov and Apresyan, 1996; Brekhovskikh and Godin, 1998; Brekhovskikh and Lysanov, 1991; Kravtsov, 2005; Petrashen *et al.*, 1985; Petrashen and Nakhamkin, 1973; Petrashen, 1978, 1980; Petrashen and Kashtan, 1984; Bleistein *et al.*, 2000).

In the fifth chapter the elements of the ray method, which is used widely for the solution of the problem of wave propagation is introduced. The geometrical optics method (Babic and Buldyrev, 1991; Babic *et al.*, 1999; Kravtsov, 2005; Bleistein *et al.*, 2000) is considered. This method is a short-wave asymptotics of the wave field in smoothly inhomogeneous, stationary and weakly conservative media (the reference heterogeneities are much greater than wavelength). The spatio-temporal approximation of the solution of the scalar wave equation is given (Babic *et al.*, 1999). The short-wave asymptotics for the Helmholtz homogeneous equation is introduced (WKBJ approximation). The ray method for propagation of elastic waves is considered in an assumption of the faster change of the characteristics of the wave process in a direction of a normal to the wave front in comparison with a change of characteristics of the medium (Kravtsov and Apresyan, 1996). For the description of the propagation of acoustic waves at ocean it is offered to use the ray approximation of the propagation in the almost stratified medium, i.e. in a medium with smoothly (in comparison with depth) varying velocity of propagation of a signal in a horizontal plane. The ray method for the description of the surface waves in a vertical inhomogeneous medium is introduced. The considered ray approximation of the propagation of electromagnetic fields is a basis for the description of the propagation processes in inhomogeneous media, and also has the significant methodological importance. On the basis of this description it is possible to get the transport equation and to establish connection of the statistical parameters of a medium with the parameters of the phenomenological theory of a transport of electromagnetic waves.

In the sixth chapter the methods of a parameter estimation of geophysical objects are described (Goltsman, 1971; Nikitin, 1986; Petrashen *et al.*, 1985; Ryzhikov and Troyan, 1994; Stratonovich, 1975; Troyan and Sokolov, 1989). The algorithms and examples of applying of the basic methods of the parameter estimation which have obtained widespread occurrence at the solution of the geophysical problems are introduced: the method of moments, the maximum likelihood method, the Newton-Le Cam method, various modifications of the least squares method, the Bayes criterion and method of the statistical regularization, criterion of the a posteriori probability ratio, the singular analysis, the least modules method, the robust methods (reparametrization method, Huber's method, Andrews's method), the method of Backus-Gilbert, the interval estimation method and the genetic algorithm. All methods, introduced in this chapter, can be used at problems of quantitative interpretation.

In the seventh chapter the statistical criteria for a choice of the model are given, which are meant for the solution of the problems of qualitative interpretation (Goltsman, 1971; Troyan and Sokolov, 1989). The problem of the test of parametric hypotheses and the a posteriori probabilities ratio are surveyed in details. The special attention is given to the signal resolution problem for the signals of a various geophysical nature. The information criterion of a choice of the model, which is basing on the maximum likelihood method and Shannon's information, is introduced. The iterative procedure of an estimation of parameters of interfering signals with simultaneous definition of a number of signals is represented (Akaike, 1974).

The eight chapter is devoted to the tasks of approximation of geophysical fields (Troyan and Sokolov, 1989). The methods of the spline approximation are introduced. An algorithm of one-dimensional approximation by cubic spline, periodic and parametric spline functions, two-dimensional spline, application of spline functions for a smoothing of histograms are considered. The algorithms of approximation of the seismic horizon and the velocity law by piecewise-polynomials together with the well observations are represented.

In the ninth chapter the mathematical notions in the terms of the functional analysis for a problem of the parameters estimation of geophysical objects are introduced (Ryzhikov and Troyan, 1994). The basic concepts and relations of the applied functional analysis (which are used in 10-th and 11 chapters) are briefly described. The definition of the ill-posed problems and the methods of their solution are considered. Some statistical criteria in the terms of the functional analysis are introduced. On the basis of the information approach the setting of the problem of the mathematical design of the geophysical experiment are tendered: a choice of the frequency and temporal intervals of measurings, choice of a spread the sources and receivers, and also their number.

The tenth chapter is devoted to a problem of a creation and interpretation of the tomography functionals, which make sense of the functions of the influence of various spatial domains of a medium on a separate measuring (Ryzhikov and Troyan, 1994). The norm of the tomography functional is determined by the intensity of the interaction of the incident and reversed fields. The examples of a build-up and the interpretation the the tomography functionals for the scalar wave equation, Lame equation, for the transport equation of a stationary sounding signal and for a diffusion equation are introduced. The build-up of the incident and reversed fields in a layered reference medium in the conformity with problems of the propagation of elastic, acoustic and electromagnetic waves are surveyed.

In the eleventh chapter the tomography methods of an image reconstruction of a medium (Ryzhikov and Troyan, 1994) are introduced. An algorithm of the reconstructive tomography is proposed on the basis of the statistical regularization method. The comparison of this method with the Backus–Hilbert method is held. The notion of the information sensitivity is considered. The measure of the information sensitivity of the observation field concerning a linear functional of parameters of a field can be the effective tool of a choice of the physically justified model. The original algorithm of the regularization for the problems of the three-dimensional ray tomography is introduced.

The twelfth chapter is devoted to the transforms and analysis of geophysical signals. The traditional transforms, such as the Fourier transform, Laplace transform, the Radon and Hilbert transforms with the reference to the analysis of seismograms, and rather new methods the cepstral and bispectra analysis are introduced. The traditional algorithms of the Wiener filtration, inverse filtration, dynamic filtration (Kalman filter) are surveyed. The original algorithm of the factor analysis is introduced, which can be successfully applied to searching latent periodicities and for a wide range of the interpretating problems.

In Appendix the tasks and computer exercises with a description of the required programs implemented under the MATLAB package are introduced. The realization of these computer exercises will allow the reader more deeply to understand a content and possibilities of the methods for the analysis and processing of the geophysical information which are introduced in the book.

The textbook is intended for the students, master students, post-graduate students of geophysical specialities as well engeneers and geoscientists. However it can be suit to the students and post-graduate students of other specialities, which are concerned to the analysis and handling of signals of any physical nature (radiophysics, optics, astrophysics, physical medicine, etc.).

We are deeply grateful to our teacher Prof. Fedor Goltsman. He is a founder of the statistical theory of interpretation (Goltsman, 1971).

We wish to express our sincere gratitude to Dr. Gennady Ryzhikov for for the discussions of the book content, including the materials received together and published in the monograph (Ryzhikov and Troyan, 1994) and many articles (Ryzhikov, 1976a,b; Ryzhikov and Troyan, 1985, 1986a,c,b, 1987, 1989; Ryzhikov *et al.*, 1989; Ryzhikov and Troyan, 1990).

Contents

Introduction v

1. Basic concepts of the probability theory 1

1.1 The Definition of Probability . . . 1
1.1.1 Set of elementary events . . . 1
1.1.2 Probability model with a finite number of outcomes . . . 4
1.1.3 Relative-frequency definition of probability . . . 5
1.1.4 Classical definition of probability . . . 5
1.1.5 Geometrical definition of probability . . . 6
1.1.6 Exercises . . . 6
1.2 Basic Properties of Probability . . . 7
1.2.1 Addition of probabilities . . . 7
1.2.2 Nonindependent and independent events . . . 9
1.2.3 The Bayes formula and complete probability . . . 10
1.2.4 Exercises . . . 11
1.3 Distribution Functions . . . 11
1.3.1 Random variables . . . 11
1.3.2 Distribution function . . . 12
1.3.3 The density function . . . 14
1.3.4 The distribution and density of function of one random argument . . . 15
1.3.5 Random vectors . . . 16
1.3.6 Marginal and conditional distributions . . . 17
1.3.7 The distributive law of two random variables . . . 18
1.3.8 Exercises . . . 21
1.4 The Numerical Characteristics of Probability Distributions . . . 23
1.4.1 Mathematical expectation . . . 23
1.4.2 Variance and correlation coefficients . . . 24
1.4.3 Quantiles . . . 26
1.4.4 Characteristics of a density function . . . 27

1.4.5 Exercises . . . 29
1.5 Characteristic and Generating Functions . . . 30
1.5.1 Moment generating function . . . 31
1.5.2 Probability generator . . . 32
1.5.3 Semi-invariants or cumulants . . . 33
1.5.4 Exercises . . . 33
1.6 The Limit Theorems . . . 34
1.6.1 Convergence in probability . . . 34
1.6.2 Chebyshev inequality . . . 34
1.6.3 The law of averages (Chebyshev's theorem) . . . 35
1.6.4 Generalised Chebyshev's theorem . . . 36
1.6.5 Markov's theorem . . . 36
1.6.6 Bernoulli theorem . . . 37
1.6.7 Poisson theorem . . . 37
1.6.8 The central limit theorem . . . 37
1.6.9 Exercises . . . 39
1.7 Discrete Distribution Functions . . . 39
1.7.1 Binomial distribution . . . 40
1.7.2 Poisson distribution . . . 41
1.7.3 Geometrical distribution . . . 42
1.7.4 Exercises . . . 43
1.8 Continuous Distributions . . . 45
1.8.1 Univariate normal distribution . . . 45
1.8.2 Multivariate normal distribution . . . 47
1.8.3 Uniform distribution . . . 51
1.8.4 χ^2–distribution . . . 51
1.8.5 Student's distribution (t-distribution) . . . 52
1.8.6 Fisher distribution and Z-distribution . . . 53
1.8.7 Triangular distribution . . . 55
1.8.8 Beta distribution . . . 55
1.8.9 Exponential distribution . . . 55
1.8.10 Laplace distribution . . . 56
1.8.11 Cauchy distribution . . . 57
1.8.12 Logarithmic normal distribution . . . 57
1.8.13 Significance of the normal distribution . . . 58
1.8.14 Confidence intervals . . . 59
1.8.15 Exercises . . . 60
1.9 Information and Entropy . . . 62
1.9.1 Entropy of the set of discrete states of system . . . 62
1.9.2 Entropy of the complex system . . . 63
1.9.3 Shannon information (discrete case) . . . 64

1.9.4 Entropy and information for systems with a continuous set of states 66
1.9.5 Fisher information 69
1.9.6 Exercises 72
1.10 Random Functions and its Properties 72
1.10.1 Properties of random functions 74
1.10.2 Properties of the correlation function 76
1.10.3 Action of the linear operator on a random function 77
1.10.4 Cross correlation function 77
1.10.5 Wiener–Khinchin theorem and power spectrum 78
1.10.6 Definition of estimations of the characteristics of random variables 81

2. Elements of mathematical statistics 83

2.1 The Basic Concepts of the Decision Theory 83
2.1.1 Distribution class of the statistical problem 85
2.1.2 The structure of the decision space and the loss function . 85
2.1.3 Decision rule 88
2.1.4 Sufficient statistic 90
2.2 Estimate Properties 91
2.2.1 Consistency 92
2.2.2 Bias 92
2.2.3 Rao–Cramer inequality. Efficiency 94
2.2.4 Sufficiency 97
2.2.5 Asymptotic normality 97
2.2.6 Robustness 98

3. Models of measurement data 99

3.1 Additive Model 99
3.2 Models of the Quantitative Interpretation 101
3.3 Regression Model 102
3.4 The Models of Qualitative Interpretation 104
3.5 The Models of Qualitative-Quantitative Interpretation 105
3.6 Random Components of Model and its Properties 106
3.7 Model with Random Parameters 111
3.8 A Priori Information 111

4. The functional relationships of sounding signal fields and parameters of the medium 115

4.1 Seismology and Seismic Prospecting 115
4.2 Acoustics of the Ocean 123

4.3 Wave Electromagnetic Fields in Geoelectrics and Ionospheric Sounding . . . 126
4.4 Atmospheric Sounding . . . 131

5. Ray theory of wave field propagation 135

5.1 Basis of the Ray Theory . . . 135
5.2 Ray Method for the Scalar Wave Equation . . . 138
5.3 Shortwave Asymptotic Form of the Solution of the One-Dimensional Helmholtz Equation (WKB Approximation) . . . 144
5.4 The Elements of Elastic Wave Ray Theory . . . 146
5.5 The Ray Description of Almost-Stratified Medium . . . 147
5.6 Surface Wave in Vertically Inhomogeneous Medium . . . 152
5.7 Ray Approximation of Electromagnetic Fields . . . 155
5.8 Statement of Problem of the Ray Kinematic Tomography . . . 159

6. Methods for parameter estimation of geophysical objects 163

6.1 The Method of Moments . . . 163
6.2 Maximum Likelihood Method . . . 164
6.3 The Newton–Le Cam Method . . . 165
6.4 The Least Squares Method . . . 167
6.5 LSM — Nonlinear Model . . . 169
6.6 LSM — Orthogonal Polynomials (Chebyshev Polynomials) . . . 170
6.7 LSM — Case of Linear Constraints . . . 172
6.8 Linear Estimation — Case of Nonstationary Model . . . 174
6.9 Bayes' Criterion and Method of Statistical Regularization . . . 175
6.10 Method of Maximum a Posteriori Probability . . . 176
6.11 The Recursion Algorithm of MAP . . . 178
6.12 Singular Analysis and Least Squares Method . . . 179
6.12.1 Resolution matrix . . . 184
6.13 The Method of Least Modulus . . . 184
6.14 Robust Methods of Estimation . . . 186
6.14.1 Reparametrization algorithm . . . 187
6.14.2 Huber robust method . . . 188
6.14.3 Andrews robust method . . . 189
6.15 Interval Estimation . . . 190
6.16 The Method of Backus and Gilbert of the Linear Inverse Problem Solution . . . 193
6.17 Genetic Algorithm . . . 196
6.17.1 Coding . . . 196
6.17.2 Selection . . . 197
6.17.3 Crossover . . . 197
6.17.4 Mutation . . . 197

6.17.5 Choice 199

7. Statistical criteria for choice of model 201

7.1 Test of Parametric Hypothesis 201
7.2 Criterion of a Posteriori Probability Ratio 202
7.3 The Signal Resolution Problem 206
7.4 Information Criterion for the Choice of the Model 208
7.5 The Method of the Separation of Interfering Signals 211

8. Algorithms of approximation of geophysical data 217

8.1 The Algorithm of Univariate Approximation by Cubic Splines . . 217
8.2 Periodic and Parametric Spline Functions 223
8.3 Application of the Spline Functions for Histogram Smoothing . . 226
8.4 Algorithms for Approximation of Seismic Horizon Subject to Borehole Observations 227
8.4.1 The Markovian type of correlation along the beds and no correlation between beds 229
8.4.2 Markovian correlation between the beds and no correlation along bed 230
8.4.3 Conformance inspection of seismic observation to borehole data concerning bed depth 230
8.4.4 Incorporation of random nature of depth measurement using borehole data 232
8.4.5 Application of a posteriori probability method to approximation of seismic horizon 234
8.4.6 Case of uncorrelated components of random vector 235
8.4.7 Approximation of parameters of approximation horizon by the orthogonal polynomials 237
8.4.8 Numerical examples of application of approximation algorithms 238
8.5 Algorithm of Approximation of Formation Velocity with the Use of Areal Observations with Borehole Data 241

9. Elements of applied functional analysis for problem of estimation of the parameters of geophysical objects 247

9.1 Elements of Applied Functional Analysis 247
9.2 Ill-Posed Problems 264
9.3 Statistical Estimation in the Terms of the Functional Analysis . . 272
9.4 Elements of the Mathematical Design of Experiment 289

10. Construction and interpretation of tomographic functionals 293

10.1 Construction of the Model of Measurements 293

10.2 Tomographic Functional 297
10.3 Examples of Construction and Interpretation of Tomographic Functionals 298
10.3.1 Scalar wave equation 298
10.3.2 The Lame equation in an isotropic infinite medium 299
10.3.3 The transport equation of the stationary sounding signal 306
10.3.4 The diffusion equation 308
10.4 Ray Tomographic Functional in the Dynamic and Kinematic Interpretation of the Remote Sounding Data 309
10.5 Construction of Incident and Reversed Wave Fields in Layered Reference Medium 312

11. Tomography methods of recovering the image of medium 315

11.1 Elements of Linear Tomography 315
11.1.1 Change of variables 316
11.1.2 Differentiation of generalized function 317
11.2 Connection of Radon Inversion with Diffraction Tomography . . . 322
11.3 Construction of Algorithms of Reconstruction Tomography 327
11.4 Errors of Recovery, Resolving Length and Backus and Gilbert Method 334
11.5 Back Projection in Diffraction Tomography 340
11.6 Regularization Problems in 3-D Ray Tomography 347
11.7 Construction of Green Function for Some Type of Sounding Signals 351
11.7.1 Green function for the wave equation 351
11.7.2 Green function for "Poisson equation" 353
11.7.3 Green function for Lame equation in uniform isotropic infinite medium 354
11.7.4 Green function for diffusion equation 358
11.7.5 Green function for operator equation of the second genus 359
11.8 Examples of the Recovery of the Local Inhomogeneity Parameters by the Diffraction Tomography Method 360
11.8.1 An estimation of the resolution 360
11.8.2 An estimation of the recovery accuracy of inhomogeneities parameters 361

12. Methods of transforms and analysis of the geophysical data 371

12.1 Fourier Transform 371
12.1.1 Fourier series 371
12.1.2 Fourier integral 372
12.2 Laplace Transform 374
12.3 Z-Transform 375

12.4 Radon Transform for Seismogram Processing 376
12.5 Gilbert Transform and Analytical Signal 378
12.6 Cepstral Transform 380
12.7 Bispectral Analysis 381
12.8 Kalman Filtering 382
12.9 Multifactor Analysis and Processing of Time Series 384
12.10 Wiener Filter 389
12.11 Predictive-Error Filter and Maximum Entropy Principle 390

Appendix A Computer exercises 397

A.1 Statistical Methods 397
A.1.1 Examples of numerical simulation of random values . . . 397
A.1.2 Construction of histogram 398
A.1.3 Description of a random variable 399
A.1.4 Computer simulation of random values 404
A.1.5 Confidence intervals 405
A.1.6 Time series 405
A.2 Transforms 406
A.2.1 Fourier transform 406
A.2.2 Signals and their spectral characteristics 406
A.2.3 Multifactor analysis 407
A.2.4 Cepstral transform 411
A.3 Direct and Inverse Problem Solution 413
A.3.1 Computer simulation of gravitational attraction 413
A.3.2 Computer simulation of magnetic field 414
A.3.3 Computer simulation of the seismic field 414
A.3.4 Deconvolution by the Wiener filter 415
A.3.5 Quantitative interpretation 417
A.3.6 Qualitative interpretation 420
A.3.7 Diffraction tomography 421

Appendix B Tables 425

Bibliography 427

Index 433

Chapter 1

Basic concepts of the probability theory

Subject *of the probability theory* is the calculus of *random events*, which have a statistical stability (statistical stability of frequencies) under given requirements of the idealized experiment.

The probability theory is deductive one and based on the system of postulates. The probability theory is a basis for methods of *mathematical statistics*, which use the inductive method of decision-making about properties of the objects or about hypothesis concerning a nature of an investigated phenomenon with the use of the data obtained as a result of conducting of experiment.

The geophysical observation are the results of random experiments and are random events. Because of the random nature of the observations the question arises as to the probability that these random events occur.

1.1 The Definition of Probability

A central concept in the theory of probability is the *random event.* These random events are the results from measurements or experiments, which are uncertain. It is important to know *the probability* with which random events occur.

1.1.1 *Set of elementary events*

Let some experiment has a finite number of outcomes $\omega_1, \omega_2, \ldots, \omega_n$. The outcomes $\omega_1, \ldots, \omega_n$ are called the *elementary events*, and their set

$$\Omega = \{\omega_1, \omega_2, \ldots, \omega_n\}$$

is called the *space of elementary events* or the *space of outcomes.*

Example 1.1. At single tossing of a coin the space of elementary events $\Omega = \{\mathrm{H}, \mathrm{T}\}$ consists of two points: where H is a head and T is a tail.

Example 1.2. At n-multiple tossing of the coin the space of elementary events Ω consists of combinations of outcomes of the single experiment

$$\Omega = \{\omega : \ \omega = (a_1, a_2, \ldots, a_n) \ \ a_i = \mathrm{H} \text{ or } \mathrm{T}\}$$

with total number of outcomes $N(\omega) = 2^n$.

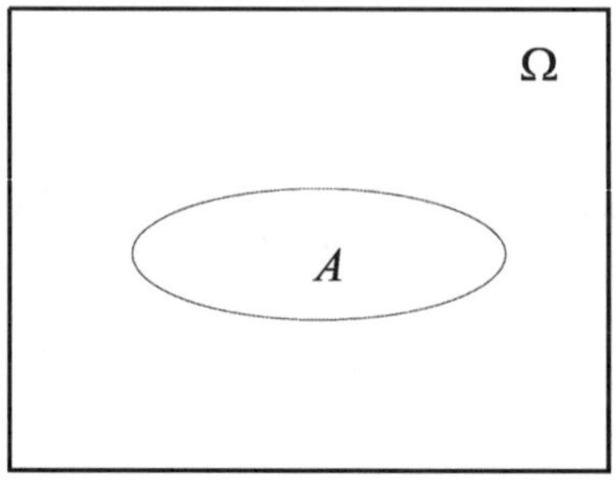

Fig. 1.1 The set A is the subset of Ω.

The *random events* (or *event*) are all of subsets $A \subseteq \Omega$ (Fig. 1.1) for which after experiment it is possible to say: outcome $\omega \in A$ or outcome $\omega \notin A$. Given any set, which one consist of events, it is possible to form the new events, using logical sum, logical product and logical negation.

Definition 1.3 (Union). *The union A_1 and A_2 (Fig. 1.2), written $A_1 \cup A_2$, is the set of points that belong to either A_1 or A_2 or both:*

$$A_1 \cup A_2 = \{\omega \in \Omega : \ \omega \in A_1 \ or \ \omega \in A_2\}.$$

In the terms of the probability theory, the event $A_1 \cup A_2$ consists of the occurrence

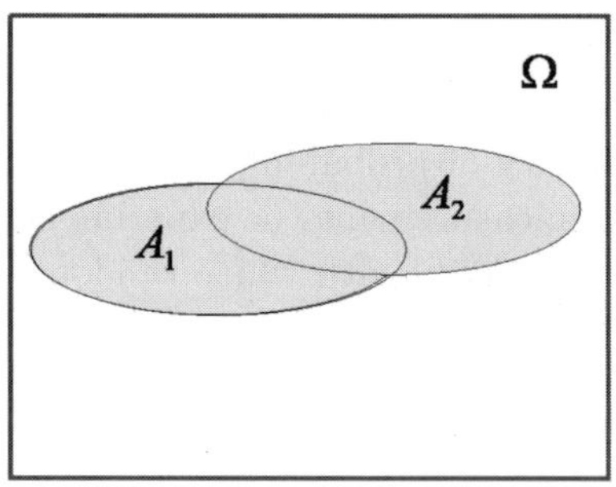

Fig. 1.2 The union of A_1 and A_2.

of the event A_1 or A_2.

Definition 1.4 (Intersection). *The intersection of A_1 and A_2 (Fig. 1.3), written $A_1 \cap A_2$, is the set of points that belong to both A_1 and A_2:*

$$A_1 \cap A_2 \equiv AB = \{\omega \in \Omega : \ \omega \in A_1 \ and \ \omega \in A_2\}.$$

The event $A_1 \cap A_2$ consists of a simultaneous occurrence of the events A_1 and A_2.

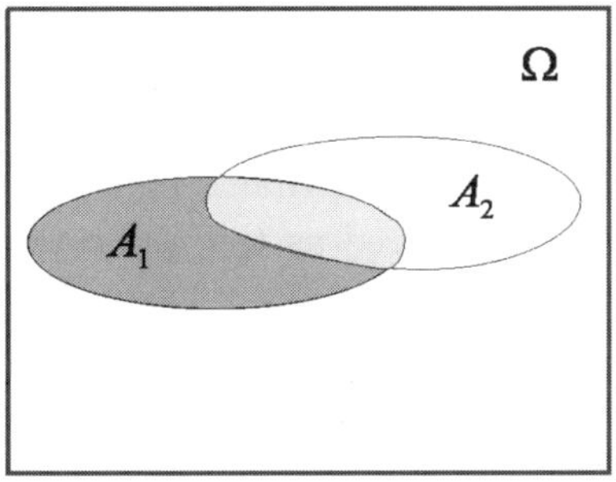

Fig. 1.3 Intersection of two subsets.

Definition 1.5 (Complement). *The complement $\bar{A}$ for a set of A (Fig. 1.4) is the set containing all elements of Ω that do not belong A. The difference $B \setminus A$ is a set containing the elements of B that are not in A. Then $\bar{A} = \Omega \setminus A$ is event, that A does not occur.*

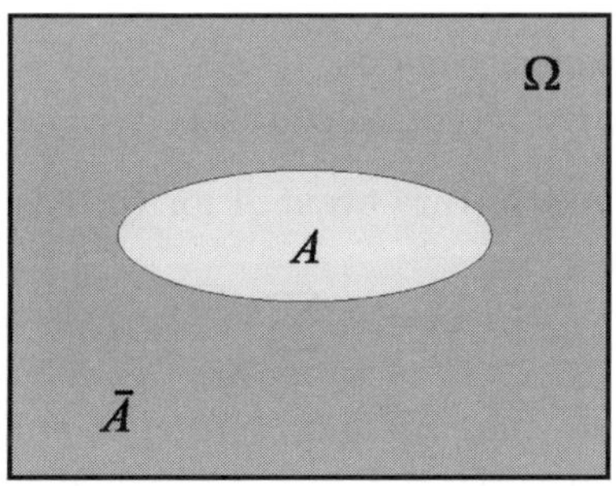

Fig. 1.4 The set $\bar{A}$ is the complement of the set A.

Example 1.6. If the set $A = \{\mathrm{HH}, \mathrm{HT}, \mathrm{TH}\}$, then the event $\bar{A} = \{\mathrm{TT}\}$ consists of a sequential appearance of two "tails". The sets $\bar{A}$ and A do not have the common elements, hence the intersection $A \cap \bar{A}$ is an empty set, i.e. $A \cap \bar{A} = \emptyset$.

In the probability theory the set $\emptyset$ is called the *impossible event* and the set Ω is called the *certain event.*

A class of subsets $\mathcal{A}$ is called *algebra of events*, if it satisfies the following properties

(1) $\Omega \in \mathcal{A}$ (the certain event is contained in $\mathcal{A}$);
(2) $\emptyset \in \mathcal{A}$ (the impossible event is contained in $\mathcal{A}$);
(3) if $A \in \mathcal{A}$, then $\bar{A} \in \mathcal{A}$ ($\bar{A} = \Omega \setminus A$) (is closed under complement);
(4) if $A_1 \in \mathcal{A}$ and $A_2 \in \mathcal{A}$, then $A_1 \cup A_2 \in \mathcal{A}$ and $A_1 \cap A_2 \in \mathcal{A}$ (is closed under the union and intersection).

If the set Ω is finite one, then $\mathcal{A}$ coincides with the class of all subsets of Ω.

Consider the random experiment of measurement of a value ζ, for example, the magnetizing force of the Earth. Let the elementary event is $(\zeta = x)$, where x specified quantity. The set of elementary events can be identified with the point set in the real axis. If it is known a priori, that ζ can take up the values from some set of X, this set should be considered as the set of the elementary events. It is natural to guess a possibility of observation of event $\{a \leq \zeta < b\}$, where $a < b$ are the arbitrary numbers. The finite sums of such half-intervals can be considered as the algebra of events, connected with the experiment.

1.1.2 *Probability model with a finite number of outcomes*

Probability model with a finite number of outcomes includes:

(1) space of elementary events Ω;
(2) subset system $\mathcal{A}$, which is constituted the algebra of events;
(3) probability $p(\omega_i)$ of outcomes ω_i, $\omega_i \in \Omega$ $(i = 1, \dots, n)$ ($p(\omega_i)$ is "weight" of elementary event) with the properties:
    (1) $0 \leq p(\omega_i) \leq 1$ (nonnegativeness);
    (2) $p(\omega_1) + \dots + p(\omega_n) = 1$ (normalization).

The probability $P(A)$ of an arbitrary event A for the case of $A \in \mathcal{A}$ is equal to the expression

$$P(A) = \sum_{\{i:\ \omega_i \in A\}} p(\omega_i).$$

A triad $(\Omega, \mathcal{A}, P)$ determines the *probability model* or *probability space* of an experiment with the outcome space $\Omega = \{\omega_1, \dots, \omega_n\}$ and with the algebra of events $\mathcal{A}$.

From above axioms we conclude that:

(1) the probability of impossible event is equal zero $P(\emptyset) = 0$;
(2) the probability of certain event is equal 1 $P(\Omega) = 1$;
(3) $P(A_1 \cup A_2) = P(A_1) + P(A_2) - P(A_1 \cap A_2)$;
    if the events A_1 and A_2 are mutually exclusive events, i. e. $A_1 \cap A_2 = \emptyset$, then $P(A_1 + A_2) = P(A_1) + P(A_2)$;
(4) for $A_1 \subset A_2$ is valid an inequality $P(A_1) \leq P(A_2)$;
(5) $P(\bar{A}) = 1 - P(A)$;
(6) if $A_1, A_2, \dots, A_n$ are the pairwise disjoint events, then

$$P(\sum_{i=1}^{n} A_i) = \sum_{i=1}^{n} P(A_i);$$

(7) for arbitrary $A_1, A_2, \dots, A_n$ an inequality

$$P(\sum_{i=1}^{n} A_i) \leq \sum_{i=1}^{n} P(A_i)$$

is valid;

(8) consider the events $A_1, A_2, \ldots, A_n$ and introduce denotation $A_{i_1 i_2 \ldots i_k} = A_{i_1} A_{i_2} \ldots A_{i_k}$. The following formula is valid

$$P(\sum_{i=1}^{n} A_i) = \sum_{i=1}^{n} P(A_i) - \sum_{i<j} P(A_i A_j) + \cdots + (-1)^{k-1} \sum_{i_1 < \cdots < i_k} P(A_{i_1 \ldots i_k}) + \cdots + (-1)^{n-1} P(A_{1 \ldots n}).$$

1.1.3 *Relative-frequency definition of probability*

Consider of a sequence of n random experiments. Let's designate their outcomes by points $\omega_1, \ldots, \omega_n$ in the space of elementary events Ω. Let's $\mathcal{A}$ is the algebra of events $A \in \mathcal{A}$, which are observed during experiment. Let's designate a number of occurrence of event A after n experiments as $K_n(A)$. If $\omega_i \in A$, then A occurs after i-th experiment. A value

$$\nu_n(A) = K_n(A)/n$$

is named a *frequency of occurrence of event* A after n experiments. The frequency of event sometimes term as *statistical probability.* At magnification of number of experiments the frequency $\nu_n(A)$ tends to the probability of event A.

1.1.4 *Classical definition of probability*

Consider an experiment with a complete group of the elementary events $E_1, E_2, \ldots, E_n$, which are the components of the set Ω. Then each event from $\mathcal{A}$ looks like

$$A = \sum_{k=1}^{m} E_{i_k},$$

where $(i_1, i_2, \ldots, i_m)$ is a subset of Ω. By using the properties of probability pairwise of not intersected events, we get the formula

$$P(A) = \sum_{k=1}^{m} P(E_{i_k}).$$

Hence, in case of the finite experiment the probability of any event is determined by probabilities of elementary events. For example, taking into account a symmetry of experiment, it is possible to establish a priori, that the elementary events have an equal probability (the probability of the elementary event is equal $1/n$, where n is a number of equally possible outcomes), and the probability of event A is calculated as the relation of number of the favorable outcomes m to number of equally possible outcomes n:

$$P(A) = \frac{m}{n}.$$

Usually calculation of equally possible and favorable outcomes is carried out by the combinatorial methods

Such definition is possible to subject to criticism on the ground that the notion *equally possible* actually means *equality probability*, and reasoning, thus, contains a vicious circle.

1.1.5 *Geometrical definition of probability*

In case of experiments with infinite number of equally possible outcomes, when the effect(result) of the experiment can be connected with a point belonging to R^m, the probabilities of some events can be defined geometrically as the relation of Euclidean volume (area, length) for the part of a figure to volume (area, length) of complete figure. An illustration of geometrical definition of probability is given on Fig. 1.5.

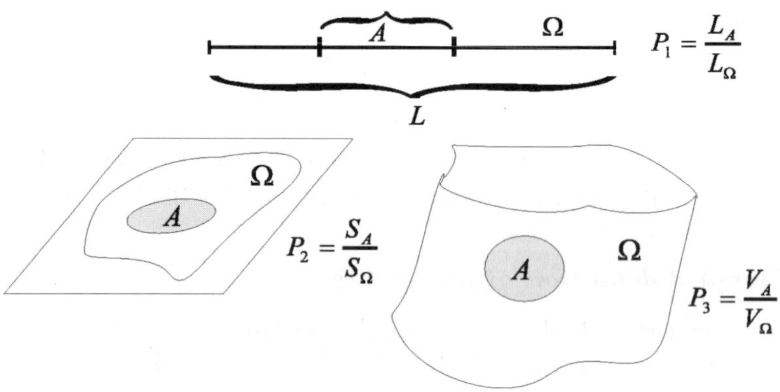

Fig. 1.5 An illustration of geometrical definition of probability.

1.1.6 *Exercises*

(1) Two bones are thrown. To find probability that the total number of pips on the happen to be facets is even, and on a facet even by one of bones will appear six pips.
(2) In the container there are 10 identical devices marked with the numbers $1, 2, \ldots, 10$. The six devices are taken arbitrary. To find probability that among the extracted devices will appear: a) the device with the number 1; b) devices with the numbers 1 and 2.
(3) In a batch from N geophones n geophones are standard ones. The m receivers is selected randomly. To find probability that among select receivers k receivers are standard ones.
(4) The parallel lines are drown on the plane. The distance between the lines is $2a$. A needle of length $2l$ $(l < a)$ is thrown on the plane randomly. To find the

probability that the needle will intercross any line.

(5) On a line segment OA with the length L of a numerical axis $0x$ are arbitrary marked two points $B(x_1)$ and $C(x_2)$. To find probability that it is possible to construct a triangle using three obtained segments.

(6) In the sound detector the signals from two sources are received. An inflow of each of signals is equally possible at any moment of time interval by duration T. There is snap into action, if the difference between the arrival time of signals is equal t ($t < T$). To find probability that the sound detector will work into time T, if each of sources will send one signal.

(7) From a course of the statistical physics it is known, that the indiscernible ultimate particles (electrons, protons, neutrons) satisfy the Pauli's exclusion principle and statistics of Fermi - Dirac. To find a number of possible outcomes of allocation n on M states. (Is possible to use analogy of allocation of n bodies on M without returning.)

(8) Photon and π-meson satisfy to the Bose-Einstein statistics and these particles are considered as indiscernible and do not submit to the Pauli's exclusion principle. To find a number of possible outcomes of allocation n of particles on M states. (It is possible to use analogy of allocation of n indiscernible bodies on M boxes.)

1.2 Basic Properties of Probability

1.2.1 *Addition of probabilities*

If A_1 and A_2 are two mutually exclusive sets, the probability of their sum is equal to the sum of probabilities:

$$P(A_1 + A_2) = P(A_1) + P(A_2), \tag{1.1}$$

or for $A_1, A_2, \ldots, A_n$

$$P(\sum_{i=1}^{n} A_i) = \sum_{i=1}^{n} P(A_i). \tag{1.2}$$

If A_1 and A_2 have common elements, i.e. the elementary events ω_i can belong both A_1 and A_2,

$$P(A_1 + A_2) = P(A_1) + P(A_2) - P(A_1 A_2), \tag{1.3}$$

and for an arbitrary number of sets $A_1, A_2, \ldots, A_n$

$$\begin{aligned} P(\sum_{i=1}^{n} A_i) &= \sum_{i=1}^{n} P(A_i) - \sum_{i}\sum_{j} P(A_i A_j) \\ &+ \sum_{i}\sum_{j}\sum_{k} P(A_i A_j A_k) + \cdots + (-1)^{n-1} P(A_1 A_2 \ldots A_n). \end{aligned} \tag{1.4}$$

The formula (1.4) gives probability of the sum of any number of events through probabilities of products of these events.

Using the formula (1.3) it is possible to express the probability of the product of two events through probabilities of individual events and the probability of the sum of events

$$P(A_1A_2) = P(A_1) + P(A_2) - P(A_1 + A_2). \tag{1.5}$$

The general formula for an arbitrary number of events n looks like

$$\begin{aligned} P(A_1A_2 \dots A_n) &= \sum_i P(A_i) - \sum_i \sum_j P(A_i + A_j) \\ &\quad + \sum_i \sum_j \sum_k P(A_i + A_j + A_k) \\ &\quad + \dots + (-1)^{n-1} P(A_1 + \dots + A_n). \end{aligned} \tag{1.6}$$

The formulae (1.4) and (1.6) can be used at transformation of various expressions containing the sums and products of events.

Example 1.7. The electronic device contains two doubling elements A_1 and A_2 and one element C, which is not doubled. The event B, consisting in failure of the electronic device, corresponds to a situation, when the elements A_1 and A_2 or C are under failure:

$$B = A_1A_2 + C,$$

where A_1 is failure of the element A_1, A_2 is failure of the element A_2 and C is failure of the element C. It is necessary to express probability of the event C through the probabilities of events containing only the sums, instead of product of the elementary events A_1, A_2 and C. Using the formula (1.3) we have

$$P(B) = P(A_1A_2) + P(C) - P(A_1A_2C). \tag{1.7}$$

From the formula (1.5) follows

$$P(A_1A_2) = P(A_1) + P(A_2) - P(A_1 + A_2). \tag{1.8}$$

Further, using the formula (1.6) for three events, we have

$$\begin{aligned} P(A_1A_2C) &= P(A_1) + P(A_2) + P(C) - P(A_1 + A_2) \\ &\quad - P(A_1 + C) - P(A_2 + C) + P(A_1 + A_2 + C). \end{aligned} \tag{1.9}$$

Substituting formulae (1.8) and (1.9) in equality (1.7), we obtain

$$P(B) = P(A_1 + C) + P(A_2 + C) - P(A_1 + A_2 + C).$$

1.2.2 *Nonindependent and independent events*

An event A_1 is called *independent* from an event A_2, if the probability of the event A_1 is independent of occurrence of the event A_2.

Example 1.8. Two coins are flipped. Let consider two events:

- A_1 is a head falling for the 1-st coin,
- A_2 is a tail falling for the 2-nd coin.

The probability of event A_1 does not depend on that, the event A_2 occur or not. The event A_1 is independent of event A_2.

The event A_1 is called *nonindependent* from event A_2, if the probability of the event A_1 varies in depending on that, the event A_2 occur or not.

Example 1.9. The 5 balls, two white and three black, are placed into a urn. The experimenters take out from the urn one by one ball. The events are considered: A_1 is occurrence of a white ball in 1-st experimenter, A_2 is occurrence of a white ball in 2 experimenters. Before something is known about event A_2, the probability of the event A_1 is $P(A_1) = 2/5$. If it is known, that the event A_2 occurs, the the probability of A_1 is $1/4$. Whence follows, that the event A_1 depends on event A_2.

The probability of event A_1, calculated provided that has occurred other event A_2, is called *conditional probability* events A_1 and is denoted as $P(A_1/A_2)$. In the considered example we have: $P(A_1) = 2/5$, $P(A_1/A_2) = 1/4$.

The condition of the independence of the event A_1 from the event A_2 is possible to note as

$$P(A_1/A_2) = P(A_1).$$

The probability of the product of two events is equal to the product of the probability of one of them on conditional probability another provided that the first event is occurred:

$$P(A_1A_2) = P(A_1)P(A_2/A_1), \tag{1.10}$$

or

$$P(A_1A_2) = P(A_2)P(A_1/A_2).$$

If the event A_1 does not depend on event A_2, then the event A_2 does not depend on event A_1.

$$P(A_1/A_2) = P(A_1),$$
$$P(A_2/A_1) = P(A_2).$$

The probability of the product of two independent events is equal to the product of the probabilities of these events

$$P(A_1A_2) = P(A_1)P(A_2). \tag{1.11}$$

The probability of the product of some events is equal to the product of probabilities of these events provided that the previous events are occurred,

$$P(A_1A_2\dots A_n) = P(A_1)P(A_2/A_1)P(A_3/A_2A_1)\dots \\ \dots P(A_n/A_1A_2\dots A_{n-1}). \quad (1.12)$$

The probability of product of independent events is equal to product of probabilities of these events:

$$P(A_1A_2\dots A_n) = P(A_1)P(A_2)\dots P(A_n). \quad (1.13)$$

Example 1.10. The amplifier of seismic station has three basic elements, which ensure a reliable operation during time t. Probabilities of non-failure operation of devices: $p_1 = 0.9$, $p_2 = 0.8$, $p_3 = 0.8$. The amplifier fails at a failure of any of devices. Let's consider the following events:

- B is on-failure operation of the amplifier,
- A_1 is on-failure operation of the 1-st element,
- A_2 is on-failure operation of the 2-nd element,
- A_3 is on-failure operation of the 3-d element,

$$B = A_1A_2A_3.$$

Taking into account that the events are independent, we can write

$$P(B) = P(A_1)P(A_2)P(A_3) = 0.9 \cdot 0.8 \cdot 0.8 = 0.576.$$

1.2.3 *The Bayes formula and complete probability*

Let's find probability of event A, which can occur with one from a group of incompatible events $B_1, B_2, \dots, B_n$, usually termed as hypotheses:

$$P(A) = \sum_{i=1}^{n} P(B_i)P(A/B_i). \quad (1.14)$$

Expression (1.14) is termed as *the formula of complete probability.*

Let's determine the conditional probability of a hypothesis B_i belonging the complete group of incompatible events B_1, B_2, …, B_n, provided that there was an event A. Let's consider known the probability of hypotheses $P(B_1)$, $P(B_2)$, …, $P(B_n)$, which usually are termed as a priori, i.e. given before realization of an experiment.

Using the theorem of multiplication, we obtain the formula

$$P(AB_i) = P(A)P(B_i/A) = P(B_i)P(A/B_i),$$

or

$$P(B_i/A) = \frac{P(B_i)P(A/B_i)}{P(A)}, \quad i = 1, 2, \dots, n.$$

Applying the formula of the complete probability (1.14), we write *the Bayes formula*:

$$P(B_i/A) = \frac{P(B_i)P(A/B_i)}{\sum\limits_{i=1}^{n} P(B_i)P(A/B_i)}. \tag{1.15}$$

The probability of $P(B_i/A)$ is termed as a posterior probability of a hypothesis B_i provided that the event A occurs.

1.2.4 *Exercises*

(1) To prove, that if the event A entails event B, then $P(B) \geq P(A)$.

(2) In a box there are 15 geophones, inclusive of 5 highly sensitive. Three geophones undertake occasionally. To find the probability that even if one of taken three geophones will appear highly sensitive.

(3) The probability of occurrence of a signal reflected from horizon A, is equal P_1, and from horizon B is equal P_2. To find probability of occurrence even if one of these signals, if reflected signals are independent.

(4) To find probability $P(A_1\bar{A}_2)$ by using known probabilities:

$$P(A_1) = P_1, \quad P(A_2) = P_2, \quad P(A_1 + A_2) = P_3.$$

(5) Two of three independent channels of seismic station have failure. To find the probability that the first and second channels have failure, if the probabilities of a failure of the first, second and third channels are accordingly equal 0,2; 0,4 and 0,3.

(6) The occurrence of a reflex signal is equally possible at any moment of time interval $t_2 - t_1 = T$. The probability of occurrence of a signal (for this time interval) is equal P. It is known, that at time $t < T$ the signal will not appear. To find probability of occurrence of a signal in the residuary time interval.

(7) At a seismogram in a given time window the signals reflected from horizons A and B are observed. The statistical properties of noise are those, that the signal from horizon A is distorted on the average with probability $2/5$, and from horizon B with probability $1/5$. The analysis of the seismograms of the neighboring region has shown, that appearance the signal from horizon A is in the relation 3:7 to the signal from horizon B. To find the probabilities
   (a) the locked-on signal is generated by the horizon A,
   (b) the locked-on signal is generated by the horizon B.

1.3 Distribution Functions

1.3.1 *Random variables*

Let $(\Omega, \mathcal{A}, P)$ are a probability model of some experiment with a finite number of outcomes $n(\Omega) < \infty$ and with algebra $\mathcal{A}$ of all subsets Ω. Let's introduce a concept *of a random variable*, subject to measuring in the random experiments. Any numerical function $\xi = \xi(\omega)$, determined on a finite space of elementary events, is termed *as a discrete random variable.*

Example 1.11. Let's determine the random variable for model with double tossing of a coin and with space of outcomes

$$\omega_1 = \text{HH}, \quad \omega_2 = \text{HT}, \quad \omega_3 = \text{TH}, \quad \omega_4 = \text{TT}.$$

To each outcome we shall associate with the numerical characteristic ξ, which determines a number of falling head H:

$$\xi(\omega_1) = 2, \quad \xi(\omega_2) = 1, \quad \xi(\omega_3) = 1, \quad \xi(\omega_4) = 0.$$

Other example of a random variable can be the indicator of some set: $A \in \mathcal{A}$

$$\xi = I_A(\omega), \quad I_A(\omega) = \begin{cases} 1, \, \omega \in A, \\ 0, \, \omega \notin A. \end{cases}$$

Let's introduce a distribution of the probability on a range of a random variable. Since in a considered case Ω consists of a finite number of points, a range of the random variable X is also finite. Let $X = \{x_1, x_2, \dots, x_n\}$, where the various numbers are all values ξ, then

$$P_\xi(x_i) = P\{\omega : \, \xi(\omega) = x_i\}, \quad x_i \in X.$$

The totality of numbers $\{P_\xi(x_1), \dots, P_\xi(x_n)\}$ is termed as *a distribution of probabilities* of random variable ξ. The example of the graphic representation of the distribution of probabilities is given at Fig. 1.6. The probability structure of a discrete

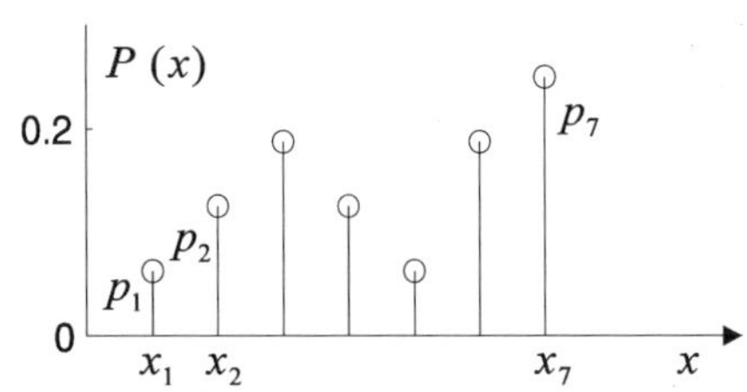

Fig. 1.6 Graphic representation of the distribution of probabilities.

random variable is completely described by a distribution function.

1.3.2 *Distribution function*

If $x \in R^1$, the random variable ξ is defined at real axis $(-\infty, \infty)$. For the description of continuous random variable ξ the *distribution function* is introduced

$$F_\xi(x) = P\{\omega : \, \xi(\omega) \le x\}.$$

Sometimes the distribution function is called as *a cumulative distribution function.*

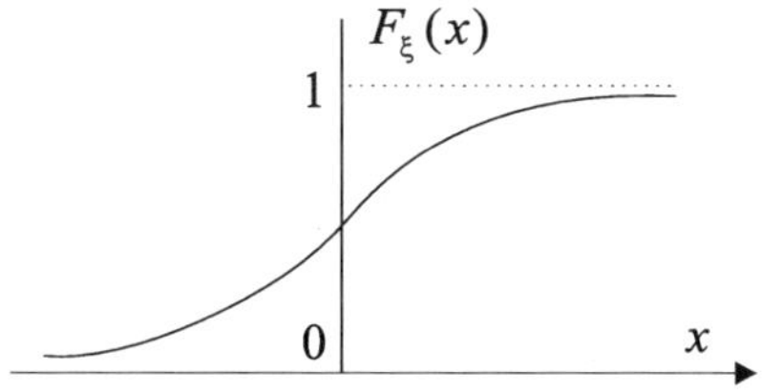

Fig. 1.7 The graphic representation of the distribution function for a continuous random variable.

The graphic representation of the distribution function is given on Fig. 1.7.

The distribution function is the universal characteristic of the random variable, it exists both for continuous, and for discrete random variables:

$$F_\xi(x) = \sum_{\{i:\ x_i \le x\}} P_\xi(x_i),$$

where

$$P_\xi(x_i) = F_\xi(x_i) - F_\xi(x_i^-), \quad F_\xi(x_i^-) = \lim_{y \to x_i} F_\xi(y).$$

If $x_1 < x_2 < \cdots < x_n$ is carried out and is assumed $F_\xi(x_0) = 0$, then

$$P_\xi(x_i) = F_\xi(x_i) - F_\xi(x_{i-1}), \quad i = 1, 2, \ldots, n.$$

An example of graphic representation of the distribution function of the discrete random variable is given in Fig. 1.8.

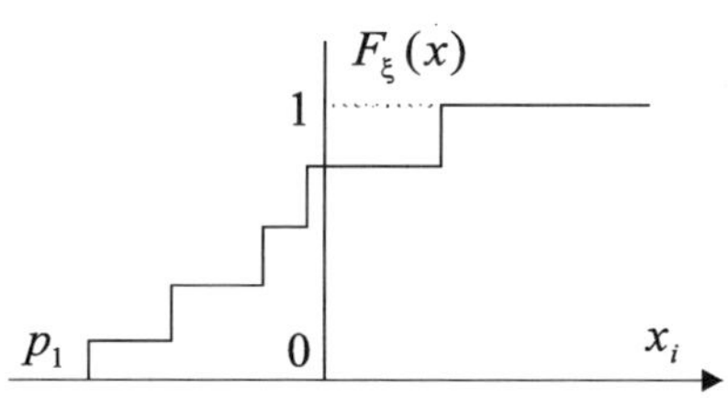

Fig. 1.8 Graphic representation of the distribution function of the discrete random variable.

The distribution function has the following properties:

(1) is non-increasing function of its argument x: $F(x_2) \ge F(x_1)$, $x_2 > x_1$;
(2) is equal to zero for $x = -\infty$: $F_\xi(-\infty) = 0$;
(3) is equal to one for $x = +\infty$: $F_\xi(+\infty) = 1$.

On the basis of the definition of the distribution function it is possible to conclude, that the probability of occurrence of the random variable ξ in the given interval (α, β) is equal an increment of the distribution function on this interval

$$P_\xi(\alpha < \xi \leq \beta) = F_\xi(\beta) - F_\xi(\alpha).$$

1.3.3 *The density function*

The random variable ξ is called *a continuous random variable*, if its distribution function $F_\xi(x)$ is continuous on x. The random variable is called *an absolutely continuous random variable*, if there is a non-negative function $f_\xi(x)$, named the *density function*, such, that

$$F_\xi(x) = \int_{-\infty}^{x} f_\xi(y)dy, \quad x \in R^1.$$

The density function is the derivative of the distribution function

$$f_\xi(x) = F'_\xi(x).$$

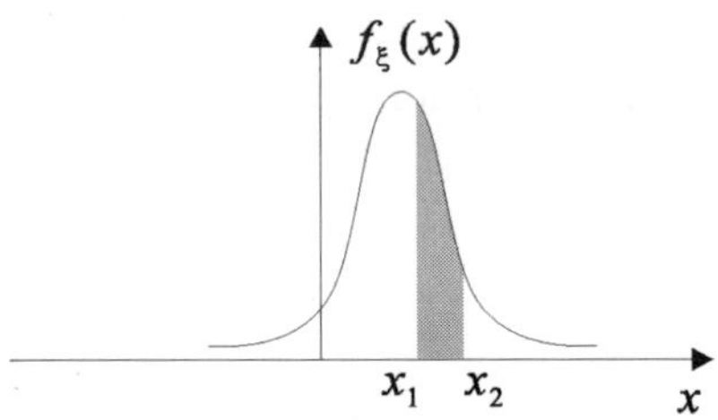

Fig. 1.9 The definite integral from the density function makes sense of probability.

The probability of event $\alpha < \xi \leq \beta$ is given by the definite integral (Fig. 1.9)

$$P(\alpha < \xi \leq \beta) = \int_{\alpha}^{\beta} f_\xi(x)dx.$$

Properties of the density function:

(1) $f_\xi(x) \geq 0$;

(2) $\int_{-\infty}^{\infty} f_\xi(x)dx = 1$.

1.3.4 *The distribution and density of function of one random argument*

Let there is a continuous random variable ξ with the density function $f_\xi(x)$, other random variable η is connected to it by the functional dependence

$$\eta = \varphi(\xi).$$

We shall assume, that the function φ is continuous, differentiable and monotonous on ξ. For a finding of the density function of $f_\eta(y)$ it is necessary to find function ψ, which is reciprocal function of φ, and also its derivative ψ', then

$$f_\eta(y) = f_\xi(\psi(y))|\psi'(y)|.$$

If the function φ is a nonmonotonic one at the interval (a, b) it can treat as follows (Fig. 1.10). Let us draw a straight line l, parallel to an axis x, and choose a segment

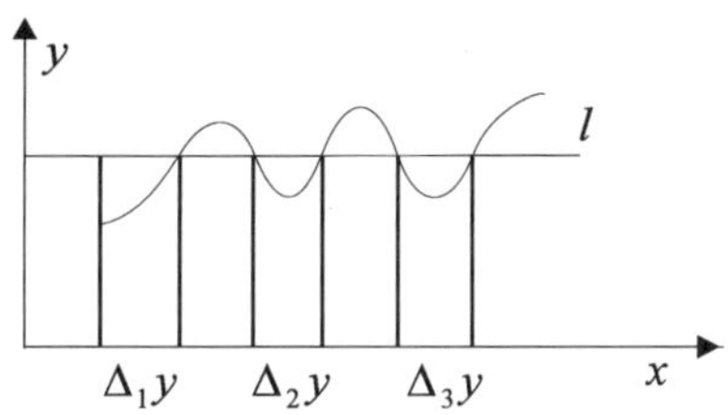

Fig. 1.10 Construction of the density function in case of nonmonotomic dependence $\nu = \varphi(\xi)$.

of a curve $y = \varphi(x)$, on which a condition $\eta < y$ is satisfied. Let us designate these segments on the abscissa axis as the sites $\Delta_1(y), \Delta_2(y), \ldots, \Delta_n(y)$. The event $\eta \leq y$ is equivalent to a hitting of random variable η into one of the sites $\Delta_1(y), \Delta_2(y), \ldots$:

$$\begin{aligned} F_\eta(y) &= P\{\omega : \ \eta(\omega) \leq y\} \\ &= P\{\omega : \ (\xi(\omega) \in \Delta_1(y) + \xi(\omega) \in \Delta_2(y) + \ldots)\} \\ &= \sum_i P\{\omega : \ \xi(\omega) \in \Delta_i(y)\}. \end{aligned}$$

Example 1.12. To find the law of the distribution of a linear function from a random argument, which has the normal law:

$$\xi \in N(m_\xi, \sigma_\xi^2), \quad f_\xi(x) = \frac{\exp\left(-\frac{(x-m_\xi)^2}{2\sigma_\xi^2}\right)}{\sqrt{2\pi}\sigma_\xi},$$

where $\eta = a_1 + a_2\xi$ and a_1, a_2 are nonrandom values. By definition we have $f_\eta(y) = f_\xi(\psi(y))|\psi'(y)|$, whence follows, that $\psi(y) = (y - a_1)/a_2$, $|\psi'(y)| = 1/|a_2|$, and it is the normal law of distribution with the parameters $m_\eta = a_2 m_\xi + a_1$, $\sigma_\eta = |a_2|\sigma_\xi$:

$$f_\eta(y) = \frac{1}{|a_2|\sigma_\xi\sqrt{2\pi}} \exp\left(-\frac{[y - (a_2 m_\xi + a_1)]^2}{2|a_2|^2\sigma_\xi^2}\right).$$

The linear function from a random variable, which distributed under the normal law is distributed also under the normal law.

1.3.5 *Random vectors*

In practice there is a necessity along with the random variable to consider the random vector $\boldsymbol{\xi} = (\xi_1, \ldots, \xi_m)$, which components are random variables. A set of probabilities

$$P_{\xi}(x_1, \ldots, x_m) = P\{\omega : \ \xi_1(\omega) = x_1, \ldots, \xi_m(\omega) = x_m\},$$

where $x_i \in X_i$ are admitted regions of ξ_i, is called a *probability distribution of the random vector* $\boldsymbol{\xi}$. The function

$$F_{\xi}(x_1, \ldots, x_m) = P\{\omega : \ \xi_1(\omega) \leq x_1, \ldots, \xi_m(\omega) \leq x_m\}$$

is called *a distribution function of the random vector* $\boldsymbol{\xi}$. In the applied literature on the theory of probabilities the components of a random vector is named as a *system of random variables* or geometrical interpretation of a random point with coordinates appropriate to components of a vector is used. In the case, when the vector $\boldsymbol{\xi}$ has two components (ξ, η), the distribution function looks like

$$F_{\xi\eta}(x, y) = P\{\omega : \ \xi(\omega) \leq x, \ \eta(\omega) \leq y\}.$$

Let us consider the properties of the distribution function for two random variables.

(1) $F_{\xi\eta}(x, y)$ is non-increasing function of its arguments, i. e.

   (1) if $x_2 > x_1$, $F_{\xi\eta}(x_2, y) \geq F_{\xi\eta}(x_1, y)$,
   (2) if $y_2 > y_1$, $F_{\xi\eta}(x, y_2) \geq F_{\xi\eta}(x, y_1)$.

(2) At equality of one or both arguments to a minus of infinity, the distribution function is equal to zero:

$$F_{\xi\eta}(x, -\infty) = F_{\xi\eta}(-\infty, y) = F_{\xi\eta}(-\infty, -\infty) = 0.$$

(3) At equality of one of arguments to a plus of infinity the distribution function of a system of random variables turns to the distribution function of a random variable appropriate to one argument, i. e.

$$F_{\xi\eta}(x, +\infty) = F_{\xi}(x), \quad F_{\xi\eta}(+\infty, y) = F_{\eta}(y).$$

(4) At equality of both arguments to a plus of infinity, the distribution function is equal 1:

$$F_{\xi\eta}(+\infty, +\infty) = 1.$$

Density function of a random vector, containing two components, we shall name the density function of two variables

$$f_{\xi\eta}(x, y) = \frac{\partial^2 F_{\xi\eta}(x, y)}{\partial x \partial y}.$$

It is possible to write down a probability of an occurrence into a rectangular $\mathcal{D}$, limited by abscissas α and β and ordinates γ and δ, as an integral of the density function

$$P\{\omega:\ (\xi(\omega),\eta(\omega))\in\mathcal{D}\}=\int\limits_{\alpha}^{\beta}\int\limits_{\gamma}^{\delta}f_{\xi\eta}(x,y)dxdy.$$

The distribution function is expressed through the density function as follows:

$$F_{\xi\eta}(x,y)=\int\limits_{-\infty}^{x}\int\limits_{-\infty}^{y}f_{\xi\eta}(x,y)dxdy.$$

1.3.6 *Marginal and conditional distributions*

Having the law of the distribution of two random variables or the random vector with two components, it is possible to receive the law of the distribution of one of random variables

$$F_{\xi}(x)=F_{\xi\eta}(x,\infty),\quad F_{\eta}(y)=F_{\xi\eta}(\infty,y).$$

Let us represent them through the density function:

$$F_{\xi}(x)=\int\limits_{-\infty}^{x}\int\limits_{-\infty}^{\infty}f(x,y)dxdy, \tag{1.16}$$

$$F_{\eta}(y)=\int\limits_{-\infty}^{\infty}\int\limits_{-\infty}^{y}f(x,y)dxdy. \tag{1.17}$$

If we differentiate (1.16) on x and (1.17) on y, we can write

$$f_{\xi}(x)=F'_{\xi}(x)=\int\limits_{-\infty}^{\infty}f(x,y)dy,\quad f_{\eta}(y)=F'_{\eta}(y)=\int\limits_{-\infty}^{\infty}f(x,y)dx.$$

The density functions $f_{\xi}(x)$ and $f_{\eta}(y)$, received by integration in infinite limits of two-dimensional density function $f_{\xi\eta}(x,y)$ on variables y and x correspondingly, are called as *a marginal density function* or *marginal densities. A conditional density function* or *a conditional density* is determined as

$$f_{\xi/\eta}(x/y)=\frac{f_{\xi\eta}(x,y)}{f_{\eta}(y)}=\frac{f_{\xi\eta}(x,y)}{\int\limits_{-\infty}^{\infty}f_{\xi\eta}(x,y)dx},$$

$$f_{\eta/\xi}(y/x)=\frac{f_{\xi\eta}(x,y)}{f_{\xi}(x)}=\frac{f_{\xi\eta}(x,y)}{\int\limits_{-\infty}^{\infty}f_{\xi\eta}(x,y)dy}.$$

A joint density function or *a joint density* can be written down using conditional and marginal densities:

$$f_{\xi\eta}(x,y) = f_\eta(y) f_{\xi/\eta}(x/y), \tag{1.18}$$

$$f_{\xi\eta}(x,y) = f_\xi(x) f_{\eta/\xi}(y/x). \tag{1.19}$$

The conditional density $f_{\xi\eta}(x/y_0)$ can be represented as a section of two-dimensional density $f_{\xi\eta}(x,y)$ by a vertical plane, which is orthogonal to the axis y and crosses a point $y = y_0$ (Fig. 1.11). The components ξ and η of a random vector are called

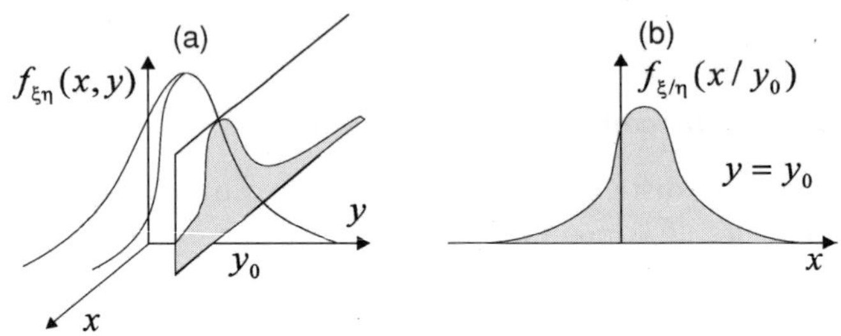

Fig. 1.11 Graphic representation of joint (a) and conditional density functions (b).

independent random variables, if

$$f_{\xi/\eta}(x/y) = f_\xi(x) \quad \text{and} \quad f_{\eta/\xi}(y/x) = f_\eta(y),$$

i. e. conditional densities are equal to the marginal ones. For independent random variables the next equality is valid

$$f_{\xi\eta}(x,y) = f_\xi(x) f_\eta(y).$$

Using the formulae (1.18) and (1.19), it is possible to get *an analog of the Bayes formula for continuous random variable*

$$f_{\xi/\eta}(x/y) = \frac{f_\xi(x) f_{\eta/\xi}(y/x)}{\int\limits_{-\infty}^{\infty} f_{\xi\eta}(x,y)dx}. \tag{1.20}$$

This formula will be used further for a finding the estimations of parameters by the Bayes criterion.

1.3.7 *The distributive law of two random variables*

Let there is a system of two continuous random variables (ξ, η) with the density function $f_{\xi\eta}(x,y)$. The random variable ζ is connected with ξ and η by the functional dependence

$$\zeta = \varphi(\xi, \eta).$$

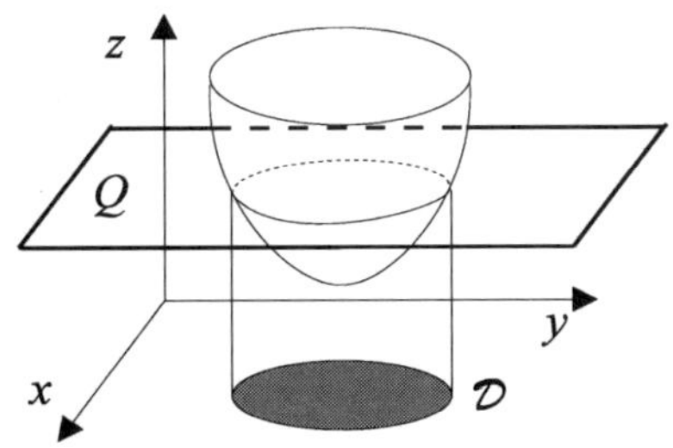

Fig. 1.12 Geometrical illustration to a development of the distributive law of the function of two random arguments.

Let us find the distributive law of ζ. We use geometrical interpretation for an obtaining the distributive law of ζ

$$F_\zeta(z) = P\{\omega : \zeta(\omega) < z\} = P\{\omega : \varphi(\xi(\omega), \eta(\omega)) < z\},$$

which is represented at Fig. 1.12. Let us to pass a plane Q in parallels to a plane $x0y$ on distance z from it. Let us express as $\mathcal{D}$ an area on a plane $x0y$, which is satisfied condition $\varphi(\xi, \eta) < z$. Then

$$F_\zeta(z) = P\{\omega : (\xi, \eta) \in \mathcal{D}\} = \iint\limits_{\mathcal{D}(z)} f_{\xi\eta}(x, y)dxdy.$$

As an example we shall find the distributive law of the sum of two random variables $\zeta = \xi + \eta$. On the plane $x0y$ we pass a line given by the equation $z = x + y$. This straight line divides a plane into two parts: more to the right and above $\xi + \eta > z$, more to the left and below $\xi + \eta < z$ — (Fig. 1.13). Domain $\mathcal{D}$ in our

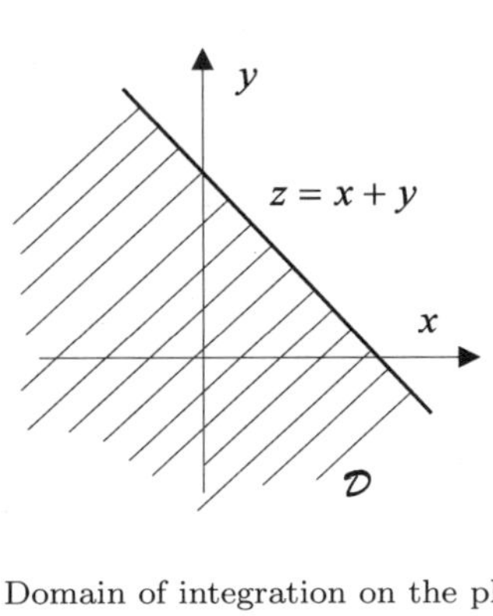

Fig. 1.13 Domain of integration on the plane (x, y).

case is shaded the left and bottom part of a plane. Let us represent the distribution function $F_\zeta(z)$ as integral of density function $f_{\xi\eta}(x,y)$:

$$\begin{aligned} F_\zeta(z) &= \iint\limits_{\mathcal{D}(z)} f_{\xi\eta}(x,y)dxdy = \int\limits_{-\infty}^{\infty}\int\limits_{-\infty}^{z-x} f_{\xi\eta}(x,y)dxdy \\ &= \int\limits_{-\infty}^{\infty}\left\{\int\limits_{-\infty}^{z-x} f_{\xi\eta}(x,y)dy\right\}dx. \end{aligned}$$

Differentiating this expression on z (upper limit of integration of inner integral), we shall get an expression for the density function

$$f_\zeta(z) = \int\limits_{-\infty}^{\infty} f_{\xi\eta}(x, z-x)dx, \tag{1.21}$$

or, starting from a symmetry of the task,

$$f_\zeta(z) = \int\limits_{-\infty}^{\infty} f_{\xi\eta}(z-y, y)dy. \tag{1.22}$$

If the random variables ξ and η are independent, the distributive law of the sum of these variables is represented by a *composition* of distributive laws:

$$f_{\xi\eta}(x,y) = f_\xi(x)f_\eta(y),$$

and formulae (1.21), (1.22) can be written in the following form:

$$f_\zeta(z) = \int\limits_{-\infty}^{\infty} f_\xi(x)f_\eta(z-x)dx,$$

$$f_\zeta(z) = \int\limits_{-\infty}^{\infty} f_\xi(z-y)f_\eta(y)dy.$$

Using denotation $*$ (convolution), we get $f_\zeta(z) = f_\xi * f_\eta$.

Example 1.13. Consider composition of the normal laws:

$$\xi \in N(m_\xi, \sigma_\xi^2), \quad f_\xi(x) = \frac{1}{\sigma_\xi\sqrt{2\pi}}\exp\left(-\frac{(x-m_\xi)^2}{2\sigma_\xi^2}\right),$$

$$\eta \in N(m_\eta, \sigma_\eta^2), \quad f_\eta(y) = \frac{1}{\sigma_\eta\sqrt{2\pi}}\exp\left(-\frac{(y-m_\eta)^2}{2\sigma_\eta^2}\right)$$

and to find the distributive law of

$$\zeta = \xi + \eta.$$

Let us apply the convolution formula for a composition:

$$g(z) = \int_{-\infty}^{\infty} f_\xi(x) f_\eta(z-x)dx$$

$$= \frac{1}{2\pi\sigma_\xi\sigma_\eta}\int_{-\infty}^{\infty} \exp\left(-\frac{(x-m_\xi)^2}{2\sigma_\xi^2} - \frac{(z-x-m_\eta)^2}{2\sigma_\eta^2}\right)dx.$$

Uncovering brackets in an exponent of an exponential curve, we get

$$g(z) = \frac{1}{2\pi\sigma_\xi\sigma_\eta}\int_{(-\infty)}^{\infty} \exp(-a_2x^2 + 2a_1x - a_0)dx,$$

$$a_2 = \frac{1}{2}\frac{\sigma_\xi^2+\sigma_\eta^2}{\sigma_\xi\sigma_\eta}, \quad a_1 = \frac{m_\xi}{2\sigma_\xi^2} + \frac{z-m_\eta}{2\sigma_\eta^2},$$

$$a_0 = \frac{m_\xi^2}{2\sigma_\xi^2} + \frac{(z-m_\eta)^2}{2\sigma_\eta^2}.$$

Using a standard integral

$$\int_{-\infty}^{\infty} \exp(-a_2x^2 + 2a_1x - a_0)dx = \sqrt{\frac{\pi}{a_2}}\exp\left(-\frac{a_2a_0 - a_1^2}{a_2}\right),$$

we obtain after transformations

$$g(z) = \frac{1}{\sqrt{2\pi}\sqrt{\sigma_\xi^2+\sigma_\eta^2}}\exp\left(-\frac{[z-(m_\xi+m_\eta)]^2}{2(\sigma_\xi^2+\sigma_\eta^2)}\right).$$

It is the normal law with the mathematical expectation $m_\zeta = = m_\xi + m_\eta$ and with the variance $\sigma_\zeta^2 = \sigma_\xi^2 + \sigma_\eta^2$.

1.3.8 *Exercises*

(1) The random variable ξ obeys the Cauchy distribution with the density function

$$f_\xi(x) = \frac{1}{\pi(1+x^2)}.$$

The variable η is interlinked with ξ by the relation

$$\eta = 1 - \xi^3.$$

To find a density function of the random variable η.

(2) The random variable ξ obeys the uniform law with the density function on a line segment $(-\pi/2, \pi/2)$:

$$f_\xi = \begin{cases} 1/\pi, \text{ if } |x| < \pi/2, \\ 0, \quad \text{if } |x| > \pi/2. \end{cases}$$

To find a distributive law of $\eta = \cos\xi$.

(3) A random variable ξ obeys the normal law

$$\xi \in N(m_\xi, \sigma_\xi^2), \quad f_\xi(x) = \frac{1}{\sqrt{2\pi}\sigma_\xi} \exp\left\{\frac{(x - m_\xi)^2}{2\sigma_\xi^2}\right\}$$

and a random variable η has the uniform density function

$$f_\eta(y) = \frac{1}{\beta - \alpha} \quad \text{if} \quad \alpha < y < \beta.$$

To find a density function of $\zeta = \eta + \xi$.

(4) The random variable ξ has the uniform law on a line segment

$$f_\xi(x) = \begin{cases} 1/\pi, & \text{if } x \in (-\pi/2, \pi/2), \\ 0, & \text{if } x \notin (-\pi/2, \pi/2). \end{cases}$$

To find a density function of $\eta = \sin\xi$.

(5) The random variable ξ obeys the Rayleigh law with the density function

$$f_\xi(x) = \frac{x}{\sigma^2} \exp\left\{-\frac{x^2}{2\sigma^2}\right\} \quad \text{if} \quad x > 0.$$

To find the density function of $\eta = e^{-\xi^2}$.

(6) At interpretation of the geophysical data frequently there is a necessity of the definition of a distributive law of a random variable η provided that the random variable $\xi = \ln\eta$ has the normal distribution with the parameters m_ξ and σ_ξ^2. To find the density function of a random variable η.

(7) The system of the random variables (ξ, η) obeys the distributive law with the density function $f_{\xi\eta}(x, y)$. To find a density function $f_\zeta(z)$ for the random variable $\zeta = \xi\eta$.

I n s t r u c t i o n . We fix some value z, then on a plane $x0y$ we build a curve with an equation $z = xy$. It is a hyperbola. Distribution function is given by the formula

$$F_\zeta(z) = P((\xi, \eta) \in D) = P(\xi\eta < z), \quad f_\zeta(z) = G'_\zeta(z).$$

(8) The system of random variables (ξ, η) has a joint density function $f_{\xi\eta}(x, y)$. To find the density function $f_\zeta(z)$ of their ratio $\zeta = \eta/\xi$.

(9) To find a distributive law $\zeta = \eta/\xi$ of two independent normally distributed variables ξ, η with parameters $m_\xi = m_\eta = 0$, σ_ξ, σ_η.

(10) To find a distributive law of a random variable $\zeta = \xi + \eta$, if ξ and η are independent random variables, which obey the exponential law with density functions

$$f_\xi(x) = \lambda e^{-\lambda x} \quad (x > 0), \quad f_\eta(y) = \mu e^{-\mu y} \quad (y > 0).$$

1.4 The Numerical Characteristics of Probability Distributions

Let $(\Omega, \mathcal{A}, P)$ a finite probabilistic space and $\xi = \xi(\omega)$ is a random variable with the values belonging to a set $X = \{x_1, \ldots, x_n\}$. If $A_i = \{\omega : \xi = x_i\}$ $(i = 1, \ldots, n)$, then ξ is possible to represent as

$$\xi(\omega) = \sum_{i=1}^{n} x_i I(A_i), \quad I(A_i) = \begin{cases} 1 & \omega \in A_i, \\ 0 & \omega \notin A_i, \end{cases}$$

where $A_1, \ldots, A_n$ is a dissection of the space Ω, i.e. pairwise are not intercrossed, and their sum is equal Ω.

1.4.1 *Mathematical expectation*

Mathematical expectation, or (*expectation, expected value, mean of distribution*) of a random variable $\xi = \sum_{i=1}^{n} x_i I(A_i)$, is named a value

$$M\xi = \sum_{i=1}^{n} x_i P(A_i).$$

To take into account $A_i = \{\omega : \xi(\omega) = x_i\}$ and $P_\xi(x_i) = P(A_i)$, we can obtain

$$M\xi = \sum_{i=1}^{n} x_i P_\xi(x_i). \tag{1.23}$$

If the random variable $\xi \in R^1$ is continuous one, its mathematical expectation is defined as

$$M\xi = \int_{-\infty}^{\infty} x f_\xi(x) dx. \tag{1.24}$$

The mathematical expectation is a measure of location of a random variable ξ and characterizes "center" of its distribution. The examples of the density functions with various mathematical expectations is given in a Fig. 1.14.

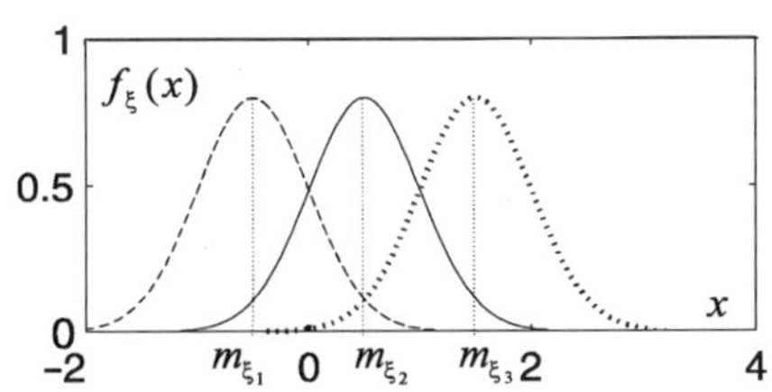

Fig. 1.14 Normal density functions with various mathematical expectation. Mathematical expectation: $m_{\xi_1} = -0.5$, $m_{\xi_2} = 0.5$, $m_{\xi_3} = 1.5$. Standard deviation: $\sigma_{\xi_1} = \sigma_{\xi_2} = \sigma_{\xi_3} = 0.5$.

The mathematical expectation of a random variable has the following properties:

(1) if $\xi \geq 0$, then $M\xi \geq 0$;

(2) $M(a\xi + b\eta) = aM\xi + bM\eta$, a and b are constants;
(3) if $\xi \geq \eta$, then $M\xi \geq M\eta$;
(4) $|M\xi| \leq M|\xi|$;
(5) if ξ and η are independent variables, then $M(\xi \cdot \eta) = M\xi \cdot M\eta$;
(6) $(M(\xi \cdot \eta))^2 \leq M\xi^2 \cdot M\eta^2$ is the Cauchy–Bunyakovskii inequality.

Along with the entered above name of the mathematical expectation, the following names are used: $\langle \xi \rangle$, m_ξ.

For a function of the random variable $\varphi(\xi)$ the mathematical expectation is defined as

$$M\varphi(\xi) = \sum_i \varphi(x_i) P_\xi(x_i) \tag{1.25}$$

(for a discrete case) and

$$M\varphi(\xi) = \int_{-\infty}^{\infty} \varphi(x) f_\xi(x) dx \tag{1.26}$$

(for a continuous case).

1.4.2 *Variance and correlation coefficients*

The variance of a random variable $\xi(\omega)$ is defined as

$$D\xi = M(\xi - M\xi)^2. \tag{1.27}$$

The value $\sigma_\xi = \sqrt{D\xi}$ is named the *standard deviation* or *mean-square deviation.* The examples of uncorrelated Gaussian time series with various standard deviations are given in Fig. 1.15.

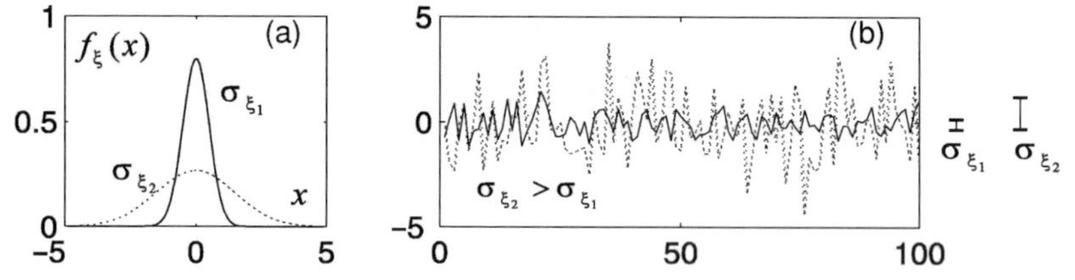

Fig. 1.15 Normal density functions (a) and uncorrelated time series, answering them, (b). Mathematical expectation: $m_{\xi_1} = m_{\xi_2} = 0$. Standard deviation: $\sigma_{\xi_1} = 0.5$, $\sigma_{\xi_2} = 1.5$.

The variance of a random variable has the following properties:

(1) $D\xi = M\xi^2 - (M\xi)^2$;
(2) $D(\xi) \geq 0$;
(3) $D(a_1 + a_2\xi) = a_2^2 D\xi$, a_1 and a_2 are constants. $Da_1 = 0$, $D(a_2\xi) = a_2^2 D\xi$;

(4) $D(\xi+\eta) = M((\xi - M\xi) + (\eta - M\eta))^2$
$= D\xi + D\eta + 2M(\xi - M\xi)(\eta - M\eta)$.

Let us define $\text{cov}(\xi, \eta) = R_{\xi\eta} = M(\xi - M\xi)(\eta - M\eta)$. This quantity is named as a *covariance of random variables* ξ and η (frequently for the covariance use a notation $R_{\xi\eta}$). As $D\xi > 0$, $D\eta > 0$, it is possible to introduce a normalized covariance

$$r(\xi, \eta) = \frac{\text{cov}(\xi, \eta)}{\sqrt{D\xi}\sqrt{D\eta}}, \tag{1.28}$$

which is named a *correlation coefficient* of the random variables ξ and η. The correlation coefficient varies inside an interval

$$-1 \le r(\xi, \eta) \le 1,$$

And the sign of an equality is reached under condition of, if ξ and η are connected by a linear dependence

$$\eta = a_1\xi + a_2.$$

If $a_1 > 0$, then $r(\xi, \eta) = 1$. If $a_1 < 0$, then $r(\xi, \eta) = -1$. If ξ and η are independent, then

$$\text{cov}(\xi, \eta) = M(\xi - M\xi) \cdot M(\eta - M\eta) = 0.$$

For independent ξ and η the variance of the sum is equal to the sum of variances:

$$D(\xi + \eta) = D\xi + D\eta.$$

The samples of a normal bivariate distribution at various values of the correlation coefficient are represented at Fig. 1.16. Taking into account the definition of a

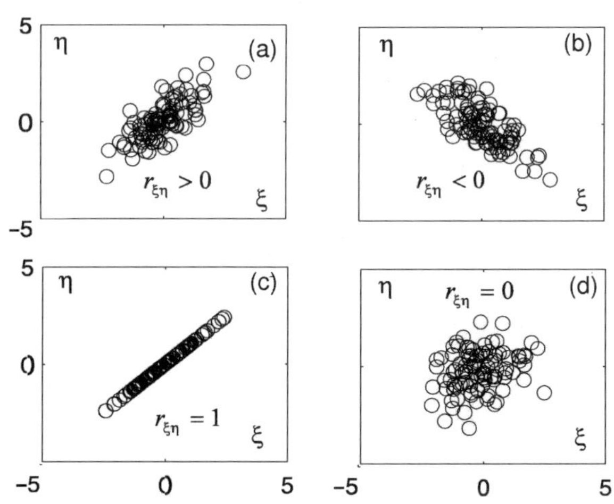

Fig. 1.16 Graphic representation of the samples from two-dimensional Gaussian distribution ξ and η. $M\xi = M\eta = 0$; $r_{\xi\eta} = 0.75$ (*a*); $r_{\xi\eta} = -0.75$ (*b*); $r_{\xi\eta} = 1$ (*c*); $r_{\xi\eta} = 0$ (*d*). Sample size is equal 100.

covariance, the mathematical expectation of the product of random variables ξ and η is possible to write as

$$M(\xi \cdot \eta) = M\xi \cdot M\eta + \mathrm{cov}(\xi, \eta).$$

In the case of uncorrelated random variables is valid $M(\xi \cdot \eta) = M\xi \cdot M\eta$. The variance of a linear function of random argument ξ_i $(i = 1, \ldots, n)$ becomes

$$D\left(\sum_{i=1}^{n} a_i\xi_i + a_0\right) = \sum_{i=1}^{n} a_i^2 D\xi_i + 2\sum_{i<j} a_i a_j \mathrm{cov}(\xi_i, \xi_j).$$

If ξ_i and ξ_j uncorrelated, then

$$D\left(\sum_{i=1}^{n} a_i\xi_i + a_0\right) = \sum_{i=1}^{n} a_i^2 D\xi_i.$$

If ξ and η are independent random variables, then

$$D(\xi \cdot \eta) = D\xi \cdot D\eta + M\xi \cdot D\eta + M\eta \cdot D\xi.$$

1.4.3 *Quantiles*

The quantile of order p of one-dimensional distribution is such a value x_p of the random variable ξ, for which

$$P(\xi < x_p) = F_\xi(x_p) = p, \quad 0 < p < 1,$$

(Fig. 1.17).

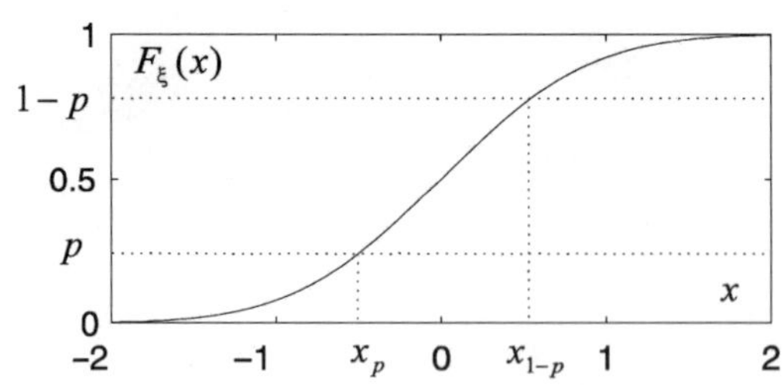

Fig. 1.17 Distribution function and quantiles of the orders p and $1 - p$. Normal distribution: $m_\xi = 0$, $\sigma_\xi = 1$.

A value $x_{1/2}$ is named as a median of a distribution (Fig. 1.18). The geometrical sense of the median is an abscissa of a point, in which area under the density curve is bisected. *Quartile* $x_{1/4}$, $x_{1/2}$, $x_{3/4}$, *decile* $x_{0.1}$, $x_{0.2}$, ..., $x_{0.9}$ and *percentile* $x_{0.01}$, $x_{0.02}$, ..., $x_{0.99}$ divide a domain of variation of x accordingly into 4, 10 and 100 intervals, the hits in which have the equal probabilities (Fig. 1.19).

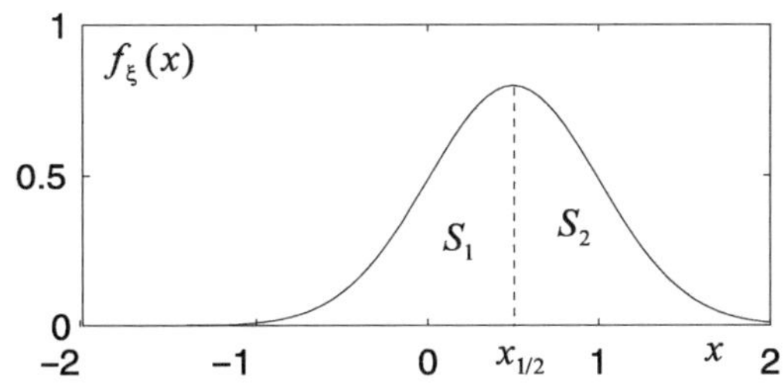

Fig. 1.18 The density function and the median. Normal distribution: $m_\xi = 0.5$, $\sigma_\xi = 0.5$.

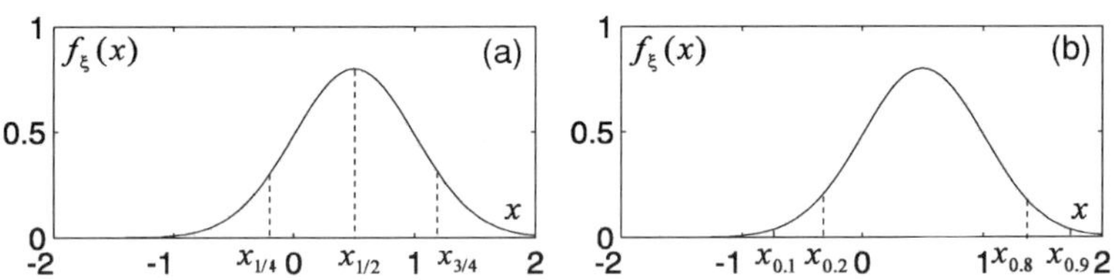

Fig. 1.19 The density function and its characteristics: quartiles (a) and deciles (b). Normal distribution: $m_\xi = 0.5$, $\sigma_\xi = 0.5$.

1.4.4 *Characteristics of a density function*

The characteristic of location of a distribution of a random variable ξ in addition to the mathematical expectation $M\xi$ and the median $x_{1/2}$ is the *mode* of a continuous distribution, which is a point of a maximum of the density function. The distributions having one, two and more modes, is named one-modal, two-modal and multimodal (Fig. 1.20). In addition to the entered above variance $D\xi$ and a

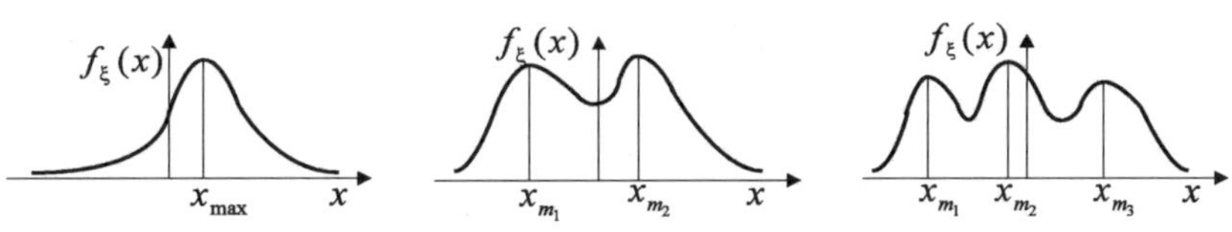

Fig. 1.20 The density functions with a various number of the modes.

standard deviation σ_ξ, as the characteristics of a dispersion of a random variable ξ are used the following quantities:

- *coefficient of variation* $\sigma_\xi / M\xi$;
- *mean absolute deviation* $M|\xi - M\xi|$;
- *interquartile distance* $x_{3/4} - x_{1/4}$ (Fig.1.21);
- *10 — 20 percent distance* $x_{0.9} - x_{0.1}$.

For a discrete random variable is determined the *range* $|x_{\max} - x_{\min}|$ of the probability series $P_\xi(x_i)$ (Fig. 1.22).

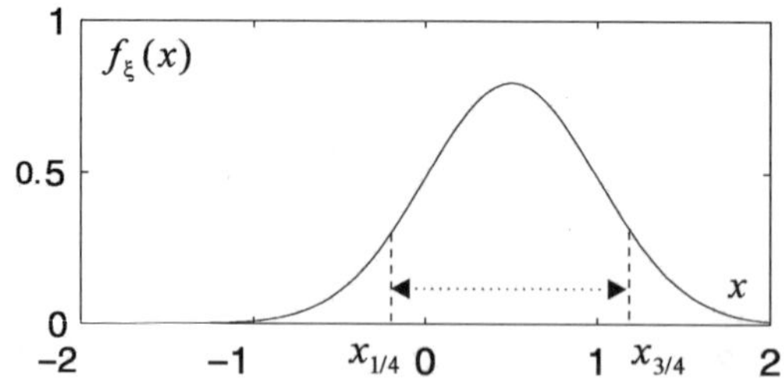

Fig. 1.21 Density function and interquartile distance. The normal distribution: $m_\xi = 0.5$, $\sigma_\xi = 0.5$.

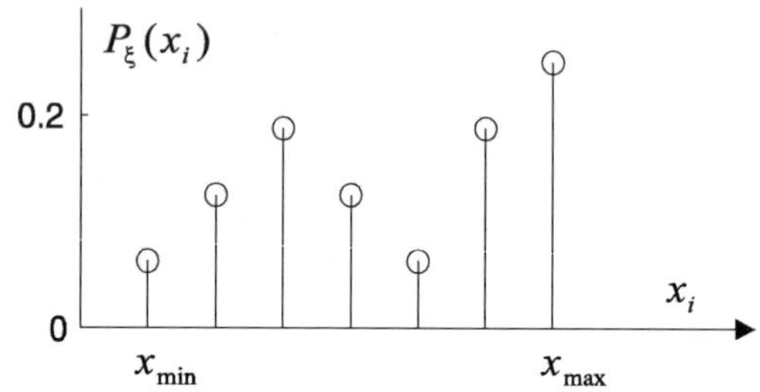

Fig. 1.22 Probability distribution of a discrete random variable and the range.

The *initial moment* of the random variable ξ of s range is called the mathematical expectation of s power of the random variable:

$$\alpha_s = M\xi^s.$$

The *centered random variable* is called the random variable $\xi - M\xi$. The *Central moment* of a range s is called the mathematical expectation of s power for the centered random variable

$$\mu_s(\xi) = M(\xi - M\xi)^s,$$

at $s = 2$ we get a value of a variance $D\xi$. There is a connection between the initial and central moments

$$\begin{aligned} \mu_1 &= 0, \\ \mu_2 &= \alpha_2 - m_\xi^2, \\ \mu_3 &= \alpha_3 - 3m_\xi\alpha_2 + 2m_\xi^2, \\ &\ldots\ldots\ldots\ldots\ldots \end{aligned}$$

The third central moment μ_3 characterizes an asymmetry of a distribution. If the distribution is symmetrical with respect to m_ξ, all moments of the odd order are equal to zero. Therefore as a measure of the asymmetry is taken the third moment. A *nondimensional asymmetry coefficient* is introduced: $\gamma_1 = \mu_3/\sigma^3$, Fig. 1.23. The fourth central moment use as a characteristic of "steepness" or peakedness of the

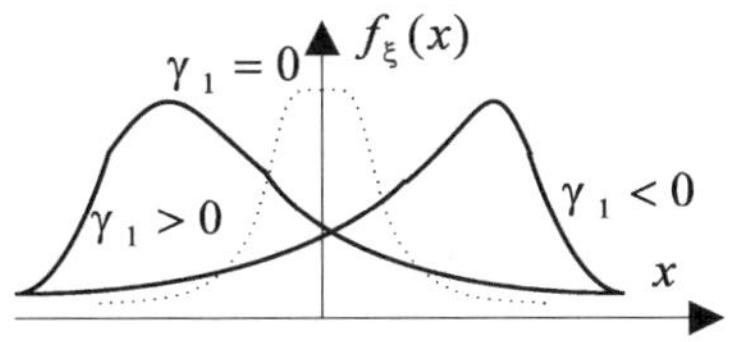

Fig. 1.23 Density functions with different asymmetry coefficients.

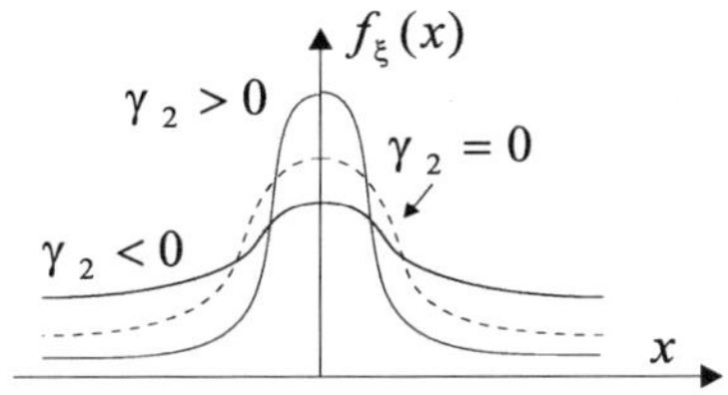

Fig. 1.24 A density function and excess.

density function. An *excess* of a random variable is introduced: $\gamma_2 = (\mu_4/\sigma^4) - 3$ (Fig. 1.24). Let us note, that for the normal distribution the equality $\mu_4/\sigma_4 = 3$ is valid. So the excess is a measure of a diversion from a normal distribution, for which it is equal to zero. If the density curve is more peaked one, than in case of the normal distribution, then $\gamma_2 > 0$, if the density curve is more flat-topped one, then $\gamma_2 < 0$.

1.4.5 *Exercises*

(1) The random variable ξ obeys the binomial distribution

$$P(\xi = m) = C_n^m p^m (1-p)^{n-m}, \quad m = 0, 1, \ldots, n.$$

To find an mathematical expectation and variance of a random variable $\eta = \exp(a\xi)$.

(2) The J-scope of a navigation system, which works on a geophysical vessel, is a circle of radius a. Owing to noises the stain with center in any point of this circle can appear. To find the mathematical expectation and variance of a distance of the center of a stain from center of a circle.

(3) The random variable ξ has a density function $f_\xi(x) = = 0.5 \cdot \sin x$ in an interval $(0, \pi)$. Outside of this interval $f_\xi = 0$ is valid. To find the mathematical expectation of $\eta = \xi^2$.

(4) The random variable ξ has the density function $f_\xi(x) = 2\cos 2x$ in an interval $(0, \pi/4)$. Outside of this interval $f_\xi(x) = 0$ is valid. To find the mode and median of ξ.

(5) To prove, that the mathematical expectation of a continuous random variable is enclosed between its least and greatest values.

(6) The random variable ξ has a density function

$$f_\xi(x) = \begin{cases} x^n \exp(-x)/n!, & x \geq 0, \\ 0, & x < 0. \end{cases}$$

To find the mathematical expectation and variance.

K e y . It is expedient to take the expression for the gamma-function

$$\Gamma(n) = \int_0^\infty e^{-x} dx.$$

(7) To prove, that if the random variables η and ξ are interlinked by a linear relation $\eta = a_1\xi + a_0$, their correlation coefficients are equal (± 1) depending on a sign a_1.
(8) To prove, that for any random variables $|r_{\xi\eta}| \leq 1$ is valid.
(9) The random variable ξ is in accord with the normal distribution with the mathematical expectation $m_\xi = 0$. The interval (α, β), not including the coordinate origin is given. At what value of the standard deviation σ the probability of hit of a random quantity ξ into an interval (α, β) will reach a maximum.
(10) The random variable ξ is in accord with the normal distribution with the mathematical expectation m_ξ and variance σ_ξ^2. It is required to exchange the normal distribution by the uniform distribution in an interval (α, β) approximately. The cell boundary α and β should be selected so that to remain a constant the mathematical expectation and variance of random variable ξ.

1.5 Characteristic and Generating Functions

The characteristic function of a random variable ξ is called the following function

$$g(t) = M(\exp(it\xi)), \tag{1.29}$$

where i is the imaginary unit. If ξ is a discrete random variable, its characteristic function is written as

$$g(t) = \sum_{k=1}^{n} \exp(itx_k) P_\xi(x_k). \tag{1.30}$$

If ξ is a continuous random variable with the density function f_ξ, its characteristic function looks like

$$g(t) = \int_{-\infty}^{\infty} \exp(itx) f_\xi(x) dx. \tag{1.31}$$

The expression (1.31) is possible to consider (accurate to a sign of an exponent of an exponential curve) as the Fourier transform of a density function. Then the calculation of a density function $f_\xi(x)$ using the known characteristic function is reduced to the inverse Fourier transform:

$$f_\xi(x) = \int_{-\infty}^{\infty} \exp(-itx)g(t)dt. \quad (1.32)$$

Properties of the characteristic functions:

(1) If ξ and η are interlinked by the relation $\eta = a\xi$ (a is a constant), then $g_\eta(t) = g_\xi(at)$,

$$g_\eta(t) = M\exp(it\eta) = M\exp(ita\xi) = M\exp(i(at)\xi) = g_\xi(at).$$

(2) The characteristic function of the sum of independent random variables is equal to product of characteristic functions of addends. Let $\xi_1, \xi_2, \ldots, \xi_n$ are independent random variables with characteristic functions $g_{\xi_1}(t)$, $g_{\xi_2}(t)$, $\ldots$, $g_{\xi_k}(t)$, then the characteristic function of their sum $\eta = \sum\limits_{k=1}^{n} \xi_k$ looks like

$$g_\eta(t) = M(\exp(it\eta)) = M\left(\exp\left(it\sum_k \xi_k\right)\right) = M\left[\prod_k \exp(it\xi_k)\right] = \prod_k g_{\xi_k}(t).$$

(3) Let ξ and η are independent random variables with the distributions $f_\xi(x)$ and $f_\eta(y)$ accordingly. Let us discover a density function of a random variable $\zeta = \xi + \eta$. In subsection 1.3.7 is shown, that density function $f_\zeta(z)$ is a result of the convolution of the density functions $f_\xi(x)$ and $f_\eta(y)$, that answers a product of the appropriate characteristic functions, i. e.

$$g_\zeta(t) = g_\xi(t) \cdot g_\eta(t).$$

The density function $f_\zeta(z)$ can be obtained, by taking the inverse Fourier transform of the function $g_\zeta(t)$

$$f_\zeta(z) = \int_{-\infty}^{\infty} g_\zeta(t)\exp(-itz)dt,$$

being product of characteristic functions $g_\xi(t)$ and $g_\eta(t)$.

1.5.1 *Moment generating function*

The set of the moments $\{\alpha\}$ characterizes the distribution of a random variable. It is expedient to enter a function, which would depend on all moments and would give a simple technique of a finding of the moment of any order.

The *generating function of the initial moments* of the random variable ξ is called the function

$$\alpha_\xi(t) = M(\exp(\xi t)) = \begin{cases} \int\limits_{-\infty}^{\infty} \exp(xt)f_\xi(x)dx, & \text{continuous r.v.,} \\ \sum\limits_{k=1}^{n} \exp(x_k t)P_\xi(x_k), & \text{discrete r.v.} \end{cases}$$

After decomposition an exponential curve in a series, we shall obtain

$$\alpha_\xi(t) = M\left[1 + \xi t + \frac{(\xi t)^2}{2!} + \frac{(\xi t)^3}{3!} + \dots\right]$$
$$= 1 + \alpha_1 t + \alpha_2 \frac{t^2}{2!} + \alpha_3 \frac{t^3}{3!} + \dots \quad (1.33)$$

So, the initial moment about s is a coefficient at $t^s/s!$ in expression (1.33). The second technique of a finding of the initial moment about s using a generating function of the moments is reduced to the differentiation $\alpha_\xi(t)$ on t s times:

$$\frac{\partial^s \alpha_\xi(t)}{\partial t^s} = \int_{-\infty}^{\infty} x^s \exp(xt) f_\xi(x) dx,$$

from which

$$\alpha_s = \left.\frac{\partial^s \alpha_\xi(t)}{\partial t^s}\right|_{t=0} = \int_{-\infty}^{\infty} x^s f_\xi(x) dx. \quad (1.34)$$

The *generating function of the central moments* of random variable ξ is similarly defined:

$$\mu_\xi(t) = M\left[\exp\{(\xi - m_\xi)t\}\right].$$

After decomposition an exponential curve in a series, we shall obtain

$$\mu_\xi(t) = M\left[1 + (\xi - m_\xi)t + (\xi - m_\xi)^2 \frac{t^2}{2!} + \dots\right]$$
$$= 1 + \mu_2 \frac{t^2}{2!} + \mu_3 \frac{t^3}{3!} + \dots$$

The central moment μ_s is the coefficient at $t^s/s!$.

1.5.2 *Probability generator*

Probability generator is called the function

$$G_\xi(t) = M(t^\xi) = \begin{cases} \int_{-\infty}^{\infty} t^x f_\xi(x) dx, & \text{continuous r.v.,} \\ \sum_{k=1}^{n} t^k P_\xi(x_k), & \text{discrete r.v.,} \end{cases}$$

where $G_\xi(1) = 1$ by the normality condition. It is easy to establish a connection of the probability generator with the moments. Having designated $\partial G/\partial t$ through G', we shall obtain

$$\alpha_1 = G'_\xi(1), \quad \alpha_2 = G''_\xi(1) + G'_\xi(1),$$
$$\alpha_3 = G'''_\xi(1) + 3G''_\xi(1) + G'_\xi(1) \dots,$$

and for the variance

$$\mu_2 = D\xi = G''_\xi(1) + G'_\xi(1) - [G'_\xi(1)]^2.$$

1.5.3 *Semi-invariants or cumulants*

If for one-dimensional distribution the moments of s-th order exist, then *semi-invariants* (*cumulants*) exist also, which are defined as coefficients $\varkappa_1, \varkappa_2, \ldots, \varkappa_s$ of expansion of the logarithm of the moment generating function in Taylor series

$$\ln \alpha_\xi(t) = \varkappa_1 t + \varkappa_2 \frac{t^2}{2!} + \varkappa_3 \frac{t^3}{3!} + \ldots$$

It follows that

$$\varkappa_s = \left[\frac{\partial^s}{\partial t^s} \ln \alpha_\xi(t)\right]_{t=0},$$

where $\varkappa_1 = m_\xi$, $\varkappa_2 = D\xi$, $\varkappa_3 = \mu_3$, $\varkappa_4 = \mu_4 - 3\mu_2^2$ and so on.

1.5.4 *Exercises*

(1) To find the characteristic function $g_\xi(t)$ of the random variable ξ with a probability density

$$f_\xi(x) = \frac{1}{2} e^{-|x|}.$$

(2) The random variable ξ has the characteristic function

$$g_\xi(t) = \frac{1}{1+t^2}.$$

To find the density function of the random variable.

(3) The random variable ξ has the normal density function

$$f_\xi(x) = \frac{1}{\sqrt{2\pi\sigma_\xi^2}} \exp\left\{-\frac{(x-m_\xi)^2}{2\sigma_\xi^2}\right\}.$$

To find its characteristic function.

(4) Using the characteristic function to find a composition of two normal distributions $\xi \in N(0, \sigma_\xi)$, $\eta \in N(0, \sigma_\eta)$. To find $g_\zeta(t)$ for $\zeta = \xi + \eta$.

(5) The discrete random variable ξ satisfies to the Poisson's distribution.

$$P(\xi = m) = \frac{a^m}{m!} e^{-a}.$$

To find

(1) the characteristic function;
(2) $M\xi$ and $D\xi$, with the use of $g(z)$.

(6) To find the characteristic function and the initial moments of a random variable with the density function

$$f_\xi(x) = \begin{cases} \exp(-x) & \text{for } x \geq 0, \\ 0 & \text{for } x < 0. \end{cases}$$

(7) To find the characteristic function and complete set of the initial moments of the random variable which satisfies the uniform distribution at the interval (a, b).

(8) The random variable ξ has the density function

$$f_\xi(x) = 2h\tfrac{x}{2}e^{-h^2x^2} \quad (x \geq 0).$$

To find its characteristic function.

(9) To find the characteristic function of the random variable with the density function

$$f_\xi(x) = \frac{1}{\pi\sqrt{a^2 - x^2}} \quad (|x| < a).$$

(10) The random variable ξ satisfies to the Cauchy distribution

$$f_\xi(x) = \frac{a}{\pi}\frac{1}{(x-c)^2 + a^2}.$$

To find its characteristic function.

1.6 The Limit Theorems

The analysis of mass phenomena, which are randomicity ones, shows, that the stability of the central tendency is observed, i.e. at major number of chance phenomena their central tendency ceases to be random and can be predicted with a major degree of the definiteness.

This fact determines the physical content of the *law of averages*. The opportunities of predictions in the field of mass casual appearances are expanded, if it is possible to find the limiting laws. It allows to make by the group of the theorems integrated under a title *central limit theorems*.

1.6.1 *Convergence in probability*

The random variable ξ_n converges in probability to a quantity a, if at magnification n the probability that ξ_n and a will be arbitrary close, unrestrictedly close to 1. It means, that at sufficiently great n

$$P(|\xi_n - a| < \varepsilon) > 1 - \delta,$$

where ε and δ are arbitrary small values.

1.6.2 *Chebyshev inequality*

Let there is a random variable ξ with the mathematical expectation m_ξ and variance $D\xi$. For arbitrary $\alpha > 0$ the follow inequalities are true

$$P(|\xi - m_\xi| \geq \alpha) \leq \frac{D\xi}{\alpha^2} \tag{1.35}$$

(in the case of a discrete random variable ξ) and

$$P(|\xi - m_\xi| > \alpha) \leq \frac{D\xi}{\alpha^2} \tag{1.36}$$

(in the case of a continuous random variable ξ). Let us prove an inequality (1.36) for the case of a continuous random variable

$$P(|\xi - m_\xi| > \alpha) = \int\limits_{|\xi - m_\xi|>\alpha} f_\xi(x)dx,$$

$$D\xi = \int\limits_{-\infty}^{\infty} (x - m_\xi)^2 f_\xi(x)dx = \int\limits_{-\infty}^{\infty} |x - m_\xi|^2 f_\xi(x)dx$$

$$\geq \int\limits_{|x - m_\xi|>\alpha} |x - m_\xi|^2 f_\xi(x)dx.$$

Exchanging $|x - m_\xi|$ under the integral on α:

$$D\xi \geq \alpha^2 \int\limits_{|x - m_\xi|>\alpha} f_\xi(x)dx = \alpha^2 P(|\xi - m_\xi| > \alpha),$$

we obtain the inequality (1.36).

1.6.3 *The law of averages (Chebyshev's theorem)*

At sufficiently large number of independent trials the arithmetic average of observed random variables converges in probability to its mathematical expectation.

Let ξ is a random variable with the mathematical expectation m_ξ and variance $D\xi$. Let us consider n of such independent random variables $\xi_1, \ldots, \xi_n$ as a model of the experiment with n independent trials of an initial random variable. Let us write the arithmetic average n of these random variables

$$\eta = \frac{\sum\limits_{i=1}^{n} \xi_i}{n}.$$

The mathematical expectation m_η and variance $D\eta$ look like

$$m_\eta = M\eta = \frac{1}{n} \sum_{i=1}^{n} M\xi_i = \frac{1}{n} n m_\xi = m_\xi,$$

$$D\eta = \frac{1}{n^2} \sum_{i=1}^{n} D\xi_i = \frac{D\xi}{n}.$$

Let us note the Chebyshev's theorem in the following form

$$P\left(\left| \frac{1}{n} \sum_{i=1}^{n} \xi_i - m_\xi \right| < \varepsilon \right) > 1 - \delta. \tag{1.37}$$

Proof. Let us consider the random variable η together with the Chebyshev inequality (1.35), by setting $\alpha = \varepsilon$, we shall obtain

$$P(|\eta - m_\eta| \geq \varepsilon) \leq \frac{D\eta}{\varepsilon^2} = \frac{D\xi}{n\varepsilon^2}.$$

Whatever small ε may be, it is possible to take n by such great, that the inequality is fulfilled

$$\frac{D\xi}{n\varepsilon^2} < \delta,$$

where δ is any small quantity. Then

$$P\left(\left|\frac{1}{n}\sum_{i=1}^{n}\xi_i - m_\xi\right| \geq \varepsilon\right) < \delta,$$

or passing to complement event, we shall obtain

$$P\left(\left|\frac{1}{n}\sum_{i=1}^{n}\xi_i - m_\xi\right| < \varepsilon\right) > 1 - \delta,$$

which was to be proved. □

1.6.4 *Generalised Chebyshev's theorem*

If $\xi_1, \xi_2, \ldots, \xi_n$ are independent random variables with the mathematical expectations

$$m_{\xi_1}, m_{\xi_2}, \ldots, m_{\xi_n}$$

and the variances

$$D_{\xi_1}, D_{\xi_2}, \ldots, D_{\xi_n},$$

and if all variances are restricted by the same quantity C

$$D\xi_i < C \quad (i = 1, 2, \ldots, n),$$

That at increase of n the arithmetic average of the observed variables $\xi_1, \xi_2, \ldots, \xi_n$ converges in probability to the arithmetic average of their mathematical expectations

$$P\left(\left|\frac{1}{n}\sum_{i=1}^{n}\xi_i - \frac{1}{n}\sum_{i=1}^{n}m_{\xi_i}\right| < \varepsilon\right) > 1 - \delta.$$

The proof is similar to the proof of the Chebyshev's theorem for one random variable.

1.6.5 *Markov's theorem*

If $\xi_1, \xi_2, \ldots, \xi_n$ are independent random variables, and if $D(\sum\limits_{i=1}^{n}\xi_i)/n^2 \to 0$ under $n \to \infty$, the arithmetic average of observed random variables $\xi_1, \xi_2, \ldots, \xi_n$ converges in probability to the arithmetic average of their mathematical expectations.

Proof. Let us consider the random variable η having the variance $D\eta$:

$$\eta = \frac{\sum_{i=1}^{n} \xi_i}{n}, \quad D\eta = \frac{D\left[\sum_{i=1}^{n} \xi_i\right]}{n^2}.$$

Consider the random variable η together with the Chebyshev inequality

$$P(|\eta - m_\eta| \geq \varepsilon) \leq \frac{D\eta}{\varepsilon^2}.$$

As $D\eta \to 0$ under $n \to \infty$, for sufficiently great n, then

$$P(|\eta - m_\eta| \geq \varepsilon) < \delta,$$

or passing to complement event, we shall obtain

$$P(|\eta - m_\eta| < \varepsilon) = P\left(\left|\frac{1}{n}\sum_{i=1}^{n} \xi_i - \frac{1}{n}\sum_{i=1}^{n} m_{\xi_i}\right| < \varepsilon\right) > 1 - \delta.$$

□

1.6.6 *Bernoulli theorem*

Let n of independent trials are carry out. The probability of realization of the event A in the each trial is equal p. The Bernoulli theorem states, that at unbounded increasing of number of the trials n the frequency of occurrence $\widetilde{p}$ the events A converges in probability to the probability of its realization p,

$$P(|\widetilde{p} - p| < \varepsilon) > 1 - \delta,$$

where ε and δ are arbitrary small positive numbers.

1.6.7 *Poisson theorem*

Let n of independent trials are carry out. The probability of occurrence of the event A in i trial is equal p_i. Under increasing of n the frequency of occurrence of A converges in probability to the arithmetic average of probabilities p_i:

$$P\left(\left|\widetilde{p} - \frac{1}{n}\sum_{i=1}^{n} p_i\right| < \varepsilon\right) > 1 - \delta.$$

The Poisson theorem is important for the practical application of the probability theory, as it takes into account various experimental conditions.

1.6.8 *The central limit theorem*

If $\xi_1, \xi_2, \ldots, \xi_n$ are independent random variables having the same distribution law with the mathematical expectation m_ξ and the variance σ_ξ^2, then at unrestricted increase of n the distribution law of the sum $\eta_n = \sum_{i=1}^{n} \xi_i$ comes arbitrary close to the normal distribution.

Proof. According to the property of the characteristic function

$$g_\eta(t) = [g_\xi(t)]^n.$$

The first three terms of expansion of $g_\xi(t)$ in a Maclaurin's series in a vicinity of $t = 0$ read as

$$g_\xi(t) \approx g_\xi(0) + g'_\xi(0)t + [g''_\xi(0)/2 + \alpha(t)]t^2,$$

where $\alpha(t) \to 0$ under $t \to 0$, $g_\xi(0) = 1$, $g'_\xi(0) = im_\xi$, let $m_\xi = 0$, then $g'_\xi(0) = 0$, $g''_\xi(0) = -\sigma^2$. To rewrite the expression of $g_\xi(t)$ with the use of the formulae of $g_\xi(0)$, $g'_\xi(0)$ and $g''_\xi(0)$:

$$g_\xi(t) = 1 - \left[\frac{\sigma^2}{2} - \alpha(t)\right] t^2.$$

Let us introduce a normalized random variable $\zeta_n = \eta_n/(\sigma\sqrt{n})$. Let us show, that the characteristic function η_n under increasing of n tends to the characteristic function of the normal law

$$g_\zeta(t) = g_\eta\left(\frac{t}{\sigma\sqrt{n}}\right) = \left[g_\xi\left(\frac{t}{\sigma\sqrt{n}}\right)\right]^n,$$

or, using expansion of $g_\xi(t)$, we obtain

$$g_\zeta(t) = \left\{1 - \left[\frac{\sigma^2}{2} - \alpha\left(\frac{t}{\sigma\sqrt{n}}\right)\right]\frac{t^2}{\sigma^2 n}\right\}^n.$$

After taking the logarithm of $g_\eta(t)$ and introducing a notation

$$\beta = \left[\frac{\sigma^2}{2} - \alpha\left(\frac{t}{\sigma\sqrt{n}}\right)\right]\frac{t^2}{\sigma^2 n},$$

we can obtain

$$\ln g_\zeta(t) = n\ln(1 - \beta).$$

At increasing of n β will tend to zero, then it is possible to be restricted to the first term of the expansion

$$\ln(1 - \beta) \approx -\beta.$$

Passing to a limit

$$\lim_{n\to\infty} \ln g_\zeta(t) = \lim_{n\to\infty} n(-\beta) = \lim\left\{-\frac{t^2}{2} + \alpha\left(\frac{t}{\sigma\sqrt{n}}\right)\frac{t^2}{\sigma^2}\right\},$$

$$\lim_{n\to\infty} \alpha\left(\frac{t}{\sigma\sqrt{n}}\right) = 0, \quad \text{and} \quad \lim_{n\to\infty} \ln g_\zeta(t) = \frac{-t^2}{2},$$

we obtain

$$\lim_{n\to\infty} g_\zeta(t) = \exp\left(\frac{-t^2}{2}\right),$$

but this expression is the characteristic function of the normal law with parameters $m_\zeta = 0$, $\sigma_\zeta = 1$. □

1.6.9 *Exercises*

(1) Using the Chebyshev inequality to find the upper estimate of the probability that the random quantity ξ, having an ensemble average m_ξ and variance σ_ξ^2, will deviate from m_ξ on value less than $3\sigma_\xi$.

(2) The large number n of independent trials is yielded, in each of trial we have the realization of the random variable ξ, which has a uniform distribution at an interval (1, 2). We shall consider the arithmetic average $\eta = \sum_{i=1}^{n} \xi_i/n$ of observed quantities of a random variable ξ. On the basis of the law of averages to find out, to what number a the value η will converge in probability under $n \to \infty$. To estimate a maximum (practically possible) error of the equality $\eta \approx a$.

(3) The sequence n of random variables $\xi_1, \xi_2, \ldots, \xi_n$, which have a uniform distribution in intervals $(0,1)$, $(0,2)$, $\ldots$, $(0,n)$ is considered. What will happen to their arithmetic average $\eta = \sum_{i=1}^{n} \xi_i/n$ under increasing of n?

(4) The random variables $\xi_1, \xi_2, \ldots, \xi_n$ are distributed uniformly on the intervals $(-1,1)$, $(-2,2)$, $\ldots$, $(-n,n)$. Whether will be arithmetic average $\eta = \sum_{i=1}^{n} \xi_i/n$ of random variables $\xi_1, \xi_2, \ldots, \xi_n$ to converge in probability to zero under increasing of n?

(5) At the spaceship the geiger for the definition of a number of hitting of the cosmic particles with the spaceship for some interval of time T is installed. The stream of cosmic particles is the Poisson flow (exponential arrivals) with the intensity λ, each particle is registered by the geiger with probability p. The geiger is switched on for the random time T which value is distributed under the exponential law with parameter μ. A random quantity ξ is a number of the registered particles. To find a distribution law of a random variable ξ.

1.7 Discrete Distribution Functions

The distribution law (distribution series) of discrete random variable is called a set of its possible values and the probabilities, which correspond to them. The distribution law of a discrete random variable can be given as the table (Table 1.1) where $\sum_i p_i = 1$, or in analytic form

$$P(\xi = x_i) = \varphi(x_i).$$

The distribution law can be represented by graph (Fig. 1.25).

Table 1.1 Representation of the distribution law of a discrete random variable.

ξ	x_1	x_2	$\dots$	x_n
P	p_1	p_2	$\dots$	p_n

,

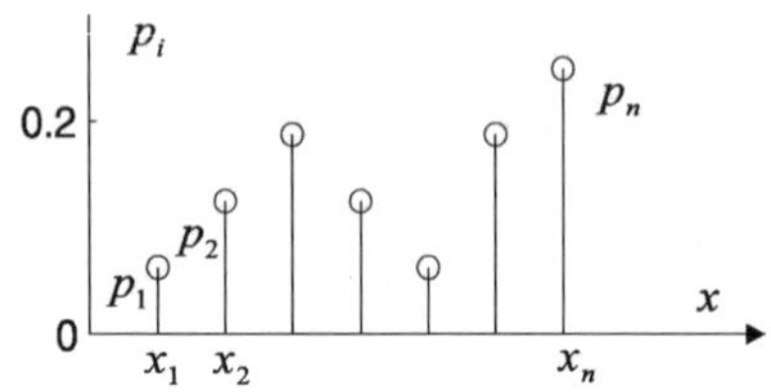

Fig. 1.25 Graphic presentation of the probability distribution of the discrete random variable.

1.7.1 *Binomial distribution*

The random variable ξ has the binomial distribution, if its possible quantities $0, 1, \dots, m, \dots, n$ correspond to the quantities of probability

$$P_m = P(\xi = m) = C_n^m p^m q^{n-m}, \tag{1.38}$$

where $0 < p < 1$, $q = 1 - p$, $m = 0, 1, \dots, n$ (Fig. 1.26). The binomial distribution

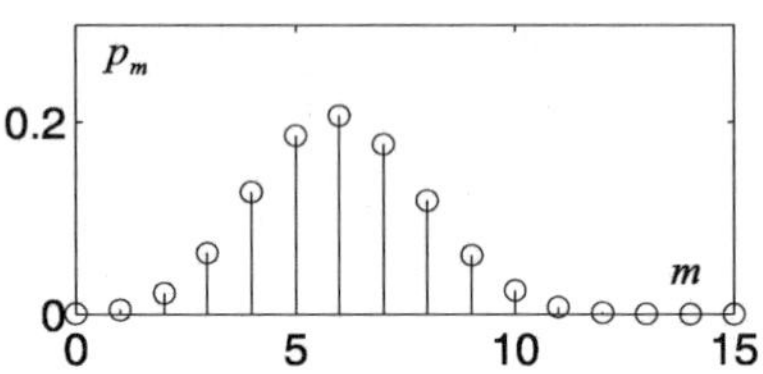

Fig. 1.26 Binomial distribution ($p = 0.4$, $n = 15$).

depends on two parameters n and p.

- Mathematical expectation $M\xi = np$.
- Variance $D\xi = np(1-p)$.
- Asymmetry coefficient $\gamma_1 = (1-2p)/[np(1-p)]^{1/2}$.
- Excess $\gamma_2 = [1 - 6p(1-p)]/[np(1-p)]$.

The binomial distribution gives the probability of m successful trials at total number of trials n, when the probability of a success in one trial is equal p.

Example 1.14. Probability to register a stream of ultrarays in a given energy range (event A) using one satellite observation is equal p. To find a number of observations such, that the probability of the occurrence even one event A is equal α. So, it is necessary to find n, for which

$$P(\xi \geq 1) \geq \alpha,$$

$$1 - P(\xi = 0) \geq \alpha, \quad P(\xi = 0) \leq 1 - \alpha.$$

Using formula (1.38) under $m = 0$, we obtain

$$(1 - p)^n \leq 1 - \alpha,$$

whence

$$n \geq \ln(1 - \alpha)/\ln(1 - p).$$

1.7.2 *Poisson distribution*

The discrete random variable ξ has the Poisson distribution, if the probability of $\xi = m$ $(m = 0, 1, \ldots, m, \ldots)$ can be given by the formula

$$p_m = P(\xi = m) = \frac{a^m}{m!} e^{-a}, \quad m = 0, 1, 2, \ldots, \tag{1.39}$$

where $a > 0$ (Fig. 1.27).

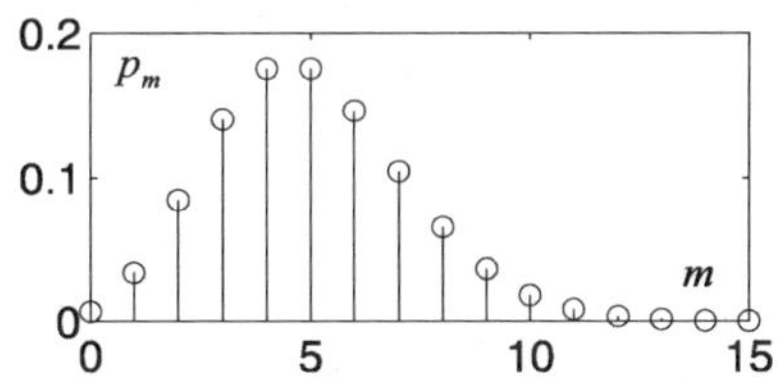

Fig. 1.27 Poisson distribution ($a = 5$).

The Poisson distribution depends on one parameter a.

- Mathematical expectation $M\xi = a$.
- Variance $D\xi = a$.
- Asymmetry coefficient $\gamma_1 = 1/\sqrt{a}$.
- Excess $\gamma_2 = 1/a$.

The Poisson distribution gives the probability of observation of m events in a given time interval, on condition that the events are independent. This distribution is an extreme case for a binomial distribution at $p \to 0$, $n \to \infty$, if $np = a =$ const. The Poisson distribution can use approximately in cases, when the major number of independent trials are carried out and the stated event happens to small probability. The Poisson distribution is used for the description of the flow of events. *The flow of events* is called a sequence of homogeneous events occurring one behind another in random time. The average of events λ per unit time, is called *a rate of occurrence.* The value λ can be both stationary value, and variable.

The flow of events is called *the flow without a consequence*, if the probability of hit of this or that number of events on any time interval does not depend on that, how many events have hitted on any other interval, not traversed with it.

The flow of events is called *ordinary*, if the probability of the occurrence on a elementary interval Δt of two or more events is a negligible quantity in comparison with the probability of occurrence of one event.

The ordinary flow of events without consequences is called *the Poisson flow*. If the events produce the Poisson flow, then the number ξ of events hitting on any time slice $(t_0, t_0 + \tau)$ has the Poisson distribution

$$p_m = \frac{a^m}{m!}e^{-a} \quad (m = 0, 1, 2, \ldots),$$

where a is the mathematical expectation of a number of hitting into time slice:

$$a = \int_{t_0}^{t_0+\tau} \lambda(t)dt.$$

If $\lambda = \text{const}$, the Poisson flow is called *a stationary flow*, in this case of a number of events hitting on a time slice τ, has the Poisson distribution with the parameter $a = \lambda/\tau$.

The Poisson flow can be used for the description of an arrival time of seismic waves on a seismogram. It is possible to show, that the number of wave events for great number practically of interesting cases is well described by the Poisson distribution.

Example 1.15. At a seismogram there is a reflected wave train produced by a layered unit, which is a stationary Poisson flow with the intensity λ is filed. To find the probability that during τ there are following events:

(1) A (nothing arrivals);
(2) B (no less than three arrivals);
(3) C (three arrivals).

Mathematical expectation of number of waves is $a = \lambda\tau$, then

$$P(A) = P_0 = e^{-\lambda\tau}, \quad P(C) = ((\lambda\tau)^3/3!)e^{-\lambda\tau},$$

$$P(B) = 1 - (P_0 + P_1 + P_2) = 1 - e^{-\lambda\tau}[1 - \lambda\tau - 0.5(\lambda\tau)^2].$$

1.7.3 *Geometrical distribution*

The random variable ξ has the geometrical distribution, if its possible quantities are equal 0, 1, 2, ..., m, ..., and the probabilities of these quantities are given by the formula

$$p_m = q^{m-1}p \quad (m = 0, 1, 2, \ldots), \tag{1.40}$$

where $0 < p < 1$, $q = 1 - p$ (Fig. 1.28). The geometrical distribution gives probability to get m unsuccessful trials before the first successful trial provided that probability of success in one trial is equal p. The probability p_m for a series of sequential quantities m forms an indefinitely decreasing geometrical progression with a denominator q.

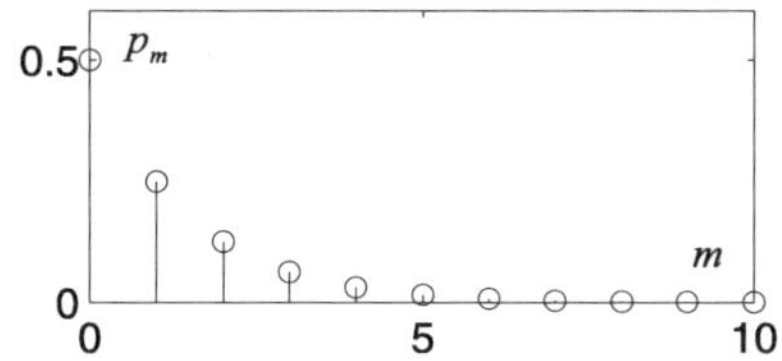

Fig. 1.28 Geometrical distribution ($p = 0.5$).

- Mathematical expectation $M\xi = 1/p$.
- Variance $D\xi = (1-p)/p^2$.
- Asymmetry coefficient $\gamma_1 = (2-p)/(1-p)^{1/2}$.
- Excess $\gamma_2 = (p^2 - 6p + 6)/(1-p)$.

Example 1.16. During a transmission of an information with the use of the radiochannel there are the noises, which make difficulties for decoding the information. With the probability p the information fails to be decoded. The information is transmitted so long as it will not be decoded. The duration of transmission of the information is equal 2 minutes. To find the mathematical expectation of time, which need for the transmission of the information.

Random variable ξ is a number of "attempts" of the information transmission, which has a geometrical distribution with $T = 2\xi$ of minutes. The distribution T can be written as follows:

$$\begin{matrix} 2 & 4 & 6 & \dots & 2m & \dots, \\ q & pq & p^2q & \dots & p^{m-1}q & \dots \end{matrix}$$

Since $m_\xi = 1/p$, then $M[T] = 2/p$.

1.7.4 *Exercises*

(1) At seismic stations A and B the seismograms are registered. Let x_1 is the event consisting in a signal extraction with the probability p_1 at a given time interval on a seismogram A, and x_2 is the event of a signal extraction with the probability p_2 at the station B. It is required to find a probability distribution of a random variable $z = x_1 + x_2$, i.e. probability of a signal extraction or at the station A, or at the station B.

(2) A subunit of a seismic station operates trouble-free during a random time T, the subunit renews after a failure. A flow of failures is a stationary one with the intensity μ. To find the probability of events $A = \{$in time τ a subunit it is not necessary to renews $\}$, $B = \{$the device should be renewed three times $\}$.

(3) The block of a seismic station consists of three subblocks. The first subblock consist of $n_1$elements, the second one consists of n_2 elements, the third one consist of n_3 elements. The first subblock is unconditionally necessary for the operation, second and third duplicate each other. The failure flow is a stationary

one; for the elements which are included in the first subblock, the intensity of the failure flow is equal λ_1, in the second or third subblocks the intensity of the failure flow is equal λ_2. The first subblock fails, if not less than two elements fail. The second (third) subblock fails if even of one element fails. The block of the seismic station fails if the first subblock or second and third subblocks fail together. To find the probability that during a time τ the block of the seismic station will leave out of the operation.

(4) The artificial satellite revolving during n of day, may collide randomly with meteorites. The meteorites are traversing an orbit and colliding with the satellite, form a stationary Poisson flow with intensity $\varkappa$ (meteorites per day). The meteorite which has hitted the satellite, punches its envelope with the probability p_0. The meteorite, which has punched an envelope, puts out of the action the devices of the satellite with probability p_1. To find the probability of the following events:

(A) — {the envelop is punched during flight time};
(B) — {the devices put out of action during flight time};
(C) — {the envelop is punched but the devices do not put out of the action during the flight time}.

(5) The hunters are drawn up in a chain by a random fashion so, that they form on an axis $0x$ a stationary flow of points with the intensity λ (λ is a number of the hunters on unity of length). The wolf runs perpendicularly to the chain. Any hunter shoots at the wolf only in the event that the wolf is distant from it no greater than R_0, and then to hit the mark with the probability p. To find the probability that the wolf will cross the chain, if it does not know, where the hunters are located and if the chain has a sufficient length.

(6) The random events ξ and η are considered which possess the values $(0,1)$ (presence and absence of a signal on two recording areas of a seismogram) with a probability distribution accordingly

$$\xi : \begin{array}{|c|c|} 0 & 1 \\ \hline q_\xi & p_\xi \end{array} \quad \text{and} \quad \eta : \begin{array}{|c|c|} 0 & 1 \\ \hline q_\eta & p_\eta \end{array} .$$

To draw the probability distributions:

(1) for the sum $\zeta = \xi + \eta$;
(2) for the difference $\zeta = \xi - \eta$;
(3) for the product $\zeta = \xi \cdot \eta$.

(7) In a memory cell of a computer n-bit binary number is recorded. Each bit position of this number, irrespective of the rest of bit positions, has the value of 0 or 1 with the equal probability. A random variable ξ is a number of bit positions, which is equal "1". To find the probabilities of the events $\{\xi = m\}$, $\{\xi \geq m\}$, $\{\xi < m\}$.

(8) On a communication bus k messages are transmitted, which containing n_1, n_2, …, n_k digits ("0" or "1"). The digits with probability 0.5 are equal to the

values 0 or 1 independent from each other. Each digit is distorted with the probability p. At coding the messages the error-correcting code in one or two digits is applied. This code operates practically with a complete reliability. The presence of an error even in one digit (after correction) leads to the erroneous message. To find probability that even one of k messages will be erroneous.

1.8 Continuous Distributions

If the distribution function $F_\xi(x)$ of random variable ξ for arbitrary x is continuous and has derivative $F'_\xi(x)$, the random variable ξ is called as continuous.

The probability of each value of a continuous random variable is equal to zero. Probability of the hit ξ in an interval (α, β) is equal to

$$P(\alpha < \xi \leq \beta) = F_\xi(\beta) - F_\xi(\alpha).$$

The quantity $f_\xi(x)dx$ in the case of continuous random variable is called *an elementary probability* ($f_\xi(x) = F'_\xi(x)$ is the density function).

1.8.1 *Univariate normal distribution*

The random variable ξ has the *normal distribution*, if its density function reads as

$$f_\xi(x) = N(m_\xi, \sigma_\xi^2) = \frac{1}{\sqrt{2\pi}\sigma_\xi} \exp\left\{-\frac{(x - m_\xi)^2}{2\sigma_\xi^2}\right\}, \tag{1.41}$$

where m_ξ is the mathematical expectation σ_ξ^2 is the variance of the random variable ξ. The distribution function of the normal distribution looks as

$$F_\xi(x) = \Phi\left(\frac{x - m_\xi}{\sigma_\xi}\right),$$

where

$$\Phi(z) = \frac{1}{\sqrt{2\pi}} \int\limits_{-\infty}^{z} \exp\left\{-\frac{1}{2}y^2\right\} dy \tag{1.42}$$

is the Laplace function (Fig. 1.29), which has the next properties:

(1) $\Phi(-\infty) = 0$;
(2) $\Phi(-x) = 1 - \Phi(x)$;
(3) $\Phi(\infty) = 1$.

The asymmetry coefficient and excess of the normal distribution are accordingly equal $\gamma_1 = 0$ and $\gamma_2 = 0$, and the characteristic function is given by the expression

$$g(t) = \exp\left\{itm_\xi - \frac{1}{2}t^2\sigma_\xi^2\right\}.$$

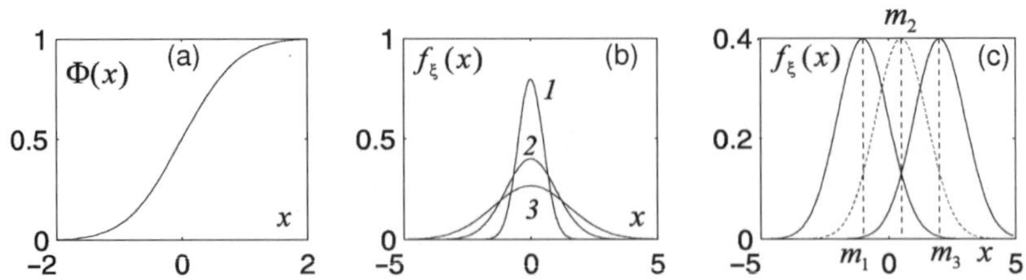

Fig. 1.29 The Laplace function and the normal density function. The Laplace function (a): $m = 0$, $\sigma = 0.5$; The normal density function (b): $m_1 = m_2 = m_3 = 0$; $\sigma_1 = 0.5$, (*1*); $\sigma_2 = 1.0$, (*2*); $\sigma_3 = 1.5$, (*3*); The normal density function (c): $m_1 = -1$; $m_2 = 0.5$; $m_3 = 2$; $\sigma_1 = \sigma_2 = \sigma_3 = 1$.

The moments of the normal distribution are written as follows:

$$\mu_{2s} = \frac{(2s)!}{2^s s!}\sigma^{2s}, \quad \mu_{2s+1} = 0 \quad (s \geq 1).$$

The density function of the normal or *Gaussian* distribution $N(m, \sigma^2)$ is the most important theoretical distribution in the mathematical statistics. The normal distribution arises, when the variable ξ is formed as a result of a sum of a great number of independent (or weakly dependent) random addends, which have a comparable contribution to the sum.

The probability of the hit into a symmetric interval $(-l, l)$ around m_ξ is equal

$$P(|\xi - m_\xi| < l) = 2\Phi\left(\frac{l}{\sigma_\xi}\right) - 1.$$

Probabilities of the hit into the intervals

$$P(-1.64 < \frac{\xi - m_\xi}{\sigma_\xi} < 1.64) = 0.90,$$

$$P(-1.96 < \frac{\xi - m_\xi}{\sigma_\xi} < 1.96) = 0.95,$$

$$P(-2.58 < \frac{\xi - m_\xi}{\sigma_\xi} < 2.58) = 0.99$$

are used most frequently. Function $N(0, 1)$ is called the *standard normal density function*, and

$$\Phi(x) = \frac{1}{\sqrt{2\pi}} \int_{-\infty}^{x} \exp\left\{-\frac{1}{2}t^2\right\} dt$$

is called the *standard normal distribution function*. The functions $N(0, 1)$ and $\Phi(x)$ are tabulated.

If the random variables $\xi_1, \xi_2, \ldots, \xi_n$ are independent and have the normal distribution, their linear combinations have the following properties.

(1) An arbitrary linear combination of the random variables ξ_i has a normal distribution. Let us assume, that the mathematical expectations and the variances of

random variables ξ_1, ξ_2 are denoted accordingly as m_1, m_2 and σ_1^2, σ_2^2, then $a_1\xi_1$ and $a_2\xi_2$ are independent normal variables with the characteristic functions

$$g_{a_1\xi_1}(t) = \exp(ita_1m_1 - \frac{1}{2}t^2a_1^2\sigma_1^2),$$
$$g_{a_2\xi_2}(t) = \exp(ita_2m_2 - \frac{1}{2}t^2a_2^2\sigma_2^2).$$

The characteristic function $\eta = a_1\xi_1 + a_2\xi_2$ is equal

$$g_\eta(t) = g_{a_1\xi_1} \cdot g_{a_2\xi_2} = \exp[it(a_1m_1 + a_2m_2) - \frac{1}{2}t^2(a_1^2\sigma_1^2 + a_2^2\sigma_2^2)],$$

i.e. η has the normal distribution with the mathematical expectation $m_\eta = a_1m_1 + a_2m_2$ and the variance $\sigma_\eta^2 = a_1^2\sigma_1^2 + a_2^2\sigma_2^2$. It is possible to show, that $\eta = \sum\limits_{i=1}^{n} a_i\xi_i$ have the normal distribution with the mathematical expectation $m_\eta = \sum\limits_{i=1}^{n} a_im_i$ and variance $\sigma_\eta^2 = \sum\limits_{i=1}^{n} a_i^2\sigma_i^2$.

(2) The sample mean

$$\bar{\xi} = \frac{1}{n}\sum_{i=1}^{n}\xi_i$$

and sample variance

$$s^2 = \frac{1}{n}\sum_{i=1}^{n}(\xi_i - \bar{\xi})^2$$

are independent random variables, if ξ_i have an equal normal distributions (with equal m_ξ and σ_ξ). This is a property of the normal distribution only.

1.8.2 *Multivariate normal distribution*

System of random variables $\xi_1, \xi_2, \ldots, \xi_n$ or n-dimensional vector $\boldsymbol{\xi} = (\xi_1, \xi_2, \ldots, \xi_n)$ has the normal distribution, if its density function $f_{\boldsymbol{\xi}}(\boldsymbol{x})$ reads as

$$f_{\boldsymbol{\xi}}(\boldsymbol{x}) = (2\pi)^{-n/2}|D_\xi|^{-1/2}\exp\left\{-(1/2)(\boldsymbol{x} - \boldsymbol{m}_\xi)^T D_\xi^{-1}(\boldsymbol{x} - \boldsymbol{m}_\xi)\right\},$$

where the mathematical expectation $M\boldsymbol{\xi} = \boldsymbol{m}_\xi$ and covariance matrix with the elements $[D_\xi]_{ij} = M[(\xi_i - m_{\xi i})(\xi_j - m_{\xi j})]$ determine the distribution completely. The characteristic function is written as follows:

$$g_{\boldsymbol{\xi}}(\boldsymbol{t}) = \exp\left\{i\boldsymbol{t}^T\boldsymbol{m}_\xi - (1/2)\boldsymbol{t}^T D_\xi \boldsymbol{t}\right\}.$$

In case of a two-component random vector the density function looks like

$$f_{\xi_1\xi_2}(x_1, x_2) = \frac{1}{2\pi\sigma_1\sigma_2\sqrt{1-r^2}}\exp\Big\{-\frac{1}{2(1-r^2)}\Big[\frac{(x_1-m_1)^2}{\sigma_1^2} - 2r\frac{(x_1-m_1)(x_2-m_2)}{\sigma_1\sigma_2} + \frac{(x_2-m_2)^2}{\sigma_2^2}\Big]\Big\}, \quad (1.43)$$

$$D = \begin{Vmatrix} \sigma_1^2 & r\sigma_1\sigma_2 \\ r\sigma_1\sigma_2 & \sigma_2^2 \end{Vmatrix}, \quad r = \frac{M[(\xi_1 - m_1)(\xi_2 - m_2)]}{\sigma_1\sigma_2},$$

where r is the correlation coefficient of the random variables ξ_1 and ξ_2 (Fig. 1.30). To find the marginal density function $f_{\xi_1}(x_1)$:

$$f_{\xi_1}(x_1) = \int_{-\infty}^{\infty} f_{\xi_1\xi_2}(x_1, x_2)dx_2.$$

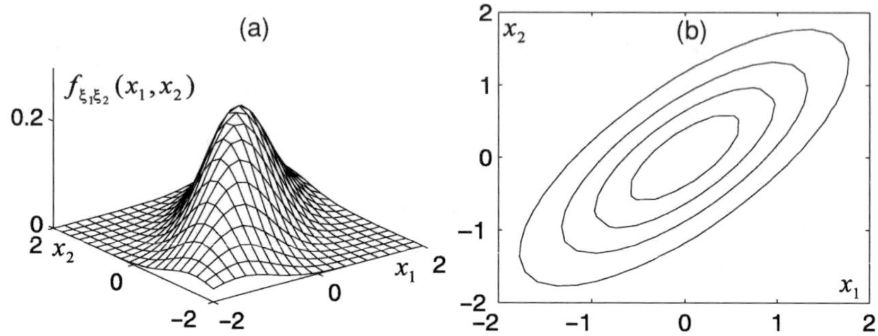

Fig. 1.30 Two-dimensional normal density function for the random variables ξ_1 and ξ_2: $m_1 = m_2 = 0$; $\sigma_1 = \sigma_2 = 1$; $r = 0.75$: (a) is a three-dimensional representation; (b) is a representation at the (x_1, x_2) plane.

To perform an interaction of the density function (1.43), we obtain

$$f_{\xi_1}(x_2) = \frac{1}{\sigma_1\sqrt{2\pi}} \exp\left\{-\frac{(x_1 - m_1)^2}{2\sigma_1^2}\right\},$$

i.e. the variable ξ_1 has the normal distribution with the mathematical expectation m_1 and the variance σ_1^2. It is similarly possible to show, that

$$f_{\xi_2}(x_2) = \frac{1}{\sigma_2\sqrt{2\pi}} \exp\left\{-\frac{(x_2 - m_2)^2}{2\sigma_2^2}\right\}.$$

To find conditional density function

$$\begin{aligned} f_{\xi_2/\xi_1}(x_2/x_1) &= \frac{f_{\xi_1\xi_2}(x_1, x_2)}{f_{\xi_1}(x_1)} \\ &= \frac{1}{\sqrt{2\pi}\sigma_2\sqrt{1 - r^2}} \exp\left\{\frac{-1}{2(1 - r^2)}\left[\frac{(x_2 - m_2)}{\sigma_2} - r\frac{(x_1 - m_1)}{\sigma_1}\right]^2\right\} \\ &= \frac{1}{\sqrt{2\pi}\sigma_2\sqrt{1 - r^2}} \exp\left\{\frac{-1}{2(1 - r^2)\sigma_2^2}\left[(x_2 - m_2) - r\frac{\sigma_2}{\sigma_1}(x_1 - m_1)\right]^2\right\} \end{aligned}$$

and, accordingly,

$$\begin{aligned} f_{\xi_1/\xi_2}(x_1/x_2) &= \frac{1}{\sqrt{2\pi}\sigma_1\sqrt{1 - r^2}} \\ &\times \exp\left\{-\frac{1}{2(1 - r^2)\sigma_1^2}\left[(x_1 - m_1) - r\frac{\sigma_1}{\sigma_2}(x_2 - m_2)\right]^2\right\}. \end{aligned}$$

After an analysis of $f_{\xi_2/\xi_1}(x_2/x_1)$ we can conclude, that this is a normal distribution with the mathematical expectation

$$m_{\xi_2/\xi_1} = m_2 + r(\sigma_2/\sigma_1)(x_1 - m_1)$$

and the standard deviation

$$\sigma_{\xi_2/\xi_1} = \sigma_2\sqrt{1 - r^2},$$

Thus m_{ξ_2/ξ_1} is called the *conditional mathematical expectation*, and $\sigma^2_{\xi_2/\xi_1}$ is called the *conditional variance*. The dependence m_{ξ_2/ξ_1} is possible to represent on a plane as a line, which is called the line of a regression ξ_2 on ξ_1 (Fig. 1.31). In the case

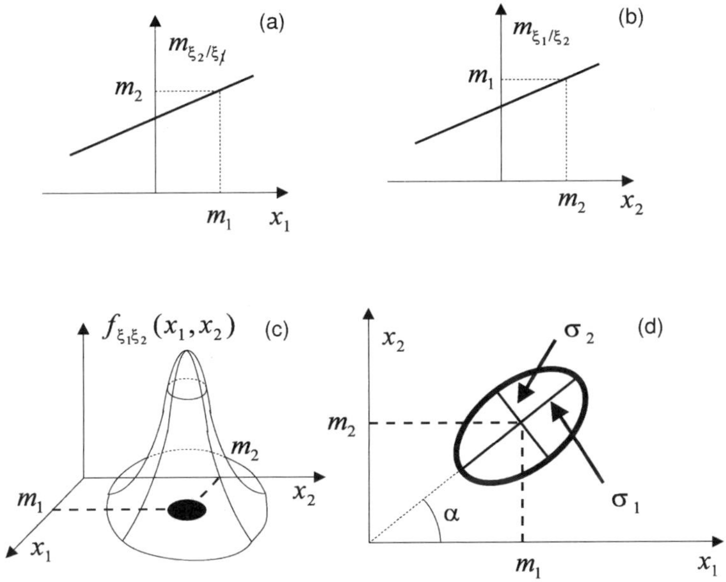

Fig. 1.31 Two-dimensional normal distribution: (a) and (b) are conditional mathematical expectations ξ_1 and ξ_2 accordingly; (c) is a representation of the density function; (d) is a contour of the density function at a plane (x_1, x_2).

of independent random variables ξ_1 and ξ_2 the lines of a regression are parallel to the coordinate axises. The section of a surface of a bivariate density function $f_{\xi_1\,\xi_2}(x_1, x_2)$ by the plane, which is parallel to the axis of values $f_{\xi_1\,\xi_2}(x_1, x_2)$, gives curve, similar Gaussian curve. The section of the density function by plane, which is parallel to the plane $x_1 0 x_2$, gives an ellipse

$$\frac{(x_1 - m_1)^2}{\sigma_1^2} - 2r\frac{(x_1 - m_1)(x_2 - m_2)}{\sigma_1\sigma_2} + \frac{(x_2 - m_2)^2}{\sigma_2^2} = \lambda^2.$$

The center of the ellipse is in a point with coordinates (m_1, m_2). A direction of symmetry axes make with an axis $0x_1$ a corner α:

$$\tan\,2\alpha = \frac{2r\sigma_1\sigma_2}{\sigma_1^2 - \sigma_2^2}.$$

Such ellipse are called *the equiprobability ellipse* or *the ellipse of dispersion* (see Fig. 1.31(d)).

Yielding transfer of an origin of coordinates to a point (m_1, m_2) and rotational displacement with a corner α, we get a canonical form of an ellipse.

The set of variables, when each of them is a linear function of variables with the normal distribution, also has the multivariate normal distribution.

Example 1.17. The random point (ξ_1, ξ_2) has the normal distribution on a plane with parameters $m_1 = 1$, $m_2 = -2$, $\sigma_1 = 1$, $\sigma_1 = 4$, $r = 0$. To find a probability that the random point will hit inside of the area $\mathcal{D}$, which restricted by the ellipse

$$(x_1 - 1)^2 + \frac{(x_2 + 2)^2}{16} = 1.$$

To write the density function as

$$f_{\xi_1\,\xi_2}(x_1, x_2) = \frac{1}{2\pi\sigma_1\sigma_2} \exp\left\{-\frac{1}{2}\left[\frac{(x_1-1)^2}{\sigma_1^2} + \frac{(x_2+1)^2}{\sigma_2^2}\right]\right\}.$$

The ellipse of dispersion $\mathcal{D}_k$ is determined by the formula

$$\frac{(x_1-1)^2}{\sigma_1^2} + \frac{(x_2+1)^2}{\sigma_2^2} = k^2.$$

Let us make a change of variables $(x_1 - 1)/\sigma_1 = u_1$, $(x_2 + 1)/\sigma_2 = u_2$ and we transform an ellipse $\mathcal{D}_k$ to a circle c_k of the radius k. Then the probability of the hit can be written as follows:

$$P((\xi_1, \xi_2) \in \mathcal{D}_k) = \frac{1}{2\pi} \iint\limits_{c_k} \exp\left\{-\left(\frac{u_1^2}{2} + \frac{u_1^2}{2}\right)\right\} du_1 du_2.$$

Let us transfer to the polar frame

$$u_1 = \rho\cos\theta, \quad u_2 = \rho\sin\theta.$$

The Jacobian of this transformation is equal ρ. After the integration, we obtain

$$\begin{aligned} P((\xi_1, \xi_2) \in \mathcal{D}_k) &= \frac{1}{2\pi} \int\limits_{-\pi}^{\pi} \int\limits_{0}^{k} \rho \exp\left\{-\frac{\rho^2}{2}\right\} d\rho d\theta \\ &= \int\limits_{0}^{k} \rho \exp\left\{-\frac{\rho^2}{2}\right\} d\rho = 1 - \exp\left\{-\frac{k^2}{2}\right\}, \end{aligned}$$

and for the case $k = 1$ we have

$$P((\xi_1, \xi_2) \in \mathcal{D}_1) = 1 - \exp\left\{-\frac{1}{2}\right\} \approx 0.393.$$

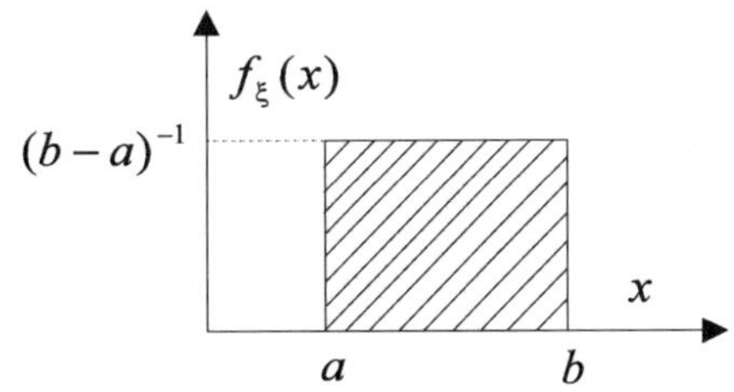

Fig. 1.32 Uniform density function.

1.8.3 *Uniform distribution*

The random variable ξ has the uniform distribution on the interval from a to b, if its density function

$$f_\xi(x) = \begin{cases} (b-a)^{-1}, & \text{if } x \in (a,b), \\ 0, & \text{if } x \notin (a,b) \end{cases} \tag{1.44}$$

on this interval is constant (Fig. 1.32).

- Mathematical expectation: $M\xi = (a+b)/2$.
- Variance: $D\xi = (b-a)^2/12$.
- Asymmetry coefficient: $\gamma_1 = 0$.
- Excess $\gamma_2 = -1.2$.

The typical requirements of origin of the uniform distribution consist in the following: the point M is tossed randomly on an axis $0x$, which is divided on equal intervals of length l. The lengths of random segments ξ_1 and ξ_2, on which the point M divides an interval (on which it has hitted), has the uniform distribution on the interval $(0, l)$.

1.8.4 χ^2*–distribution*

The random variable $\eta = \sum\limits_{i=1}^{n} \xi_i^2$, which is a sum of random variables having the standard normal distribution $\xi_i \in N(0,1)$, have the χ^2–distribution with n degrees of freedom (Fig. 1.33):

$$f_\eta(x) = \begin{cases} \left[(x/2)^{(n/2)-1}\exp(-x/2)\right]/(2\Gamma(n/2)), & x > 0, \\ 0, & x < 0. \end{cases} \tag{1.45}$$

The density function $f_\eta(x)$ depends on one parameter n, which is called the number of degrees of freedom.

- Mathematical expectation: $M\xi = n$.
- Variance: $D\xi = 2n$.
- Asymmetry coefficient: $\gamma_1 = 2\sqrt{2/n}$.

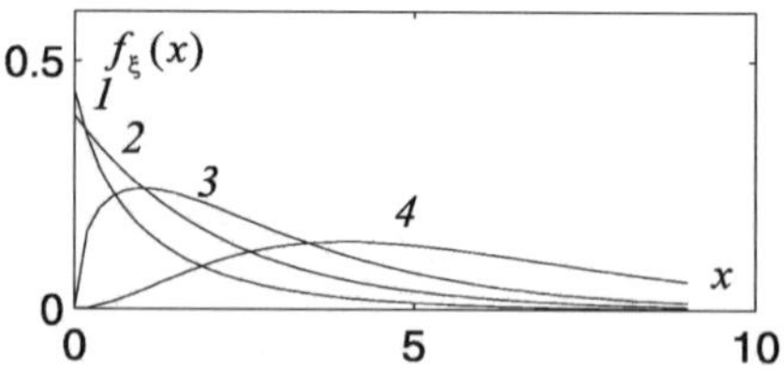

Fig. 1.33 χ^2–distribution: *1* — $n = 1$; *2* — $n = 2$; *3* — $n = 3$; *4* — $n = 6$.

- Excess $\gamma_2 = 12/n$.

The characteristic function of χ^2–distribution reads as

$$g(t) = (1 - 2it)^{-n/2}.$$

At the limit of $n \to \infty$ the χ_n^2–distribution tends to a normal distribution, and at $n > 30$ with an enough good accuracy it is possible to consider this distribution as the normal one. The χ^2–distribution has found wide application in problems of a test of hypothesis.

If $\xi_1, \xi_2, \ldots, \xi_n$ is a sample belongs to the normal distribution $N(m, \sigma^2)$ with the sample expectation

$$\bar{\xi} = \frac{1}{n} \sum_{i=1}^{n} \xi_i$$

and the sample variance

$$s^2 = \frac{1}{n-1} \sum_{i=1}^{n} (\xi_i - \bar{\xi})^2,$$

then the random variable

$$\eta = \frac{(n-1)s^2}{\sigma^2} = \frac{1}{\sigma^2} \sum_{i=1}^{n} (\xi_i - \bar{\xi})^2 \in \chi_{n-1}^2$$

has the χ^2-distribution with $n - 1$ degrees of freedom.

1.8.5 *Student's distribution (t-distribution)*

If $\xi_1, \xi_2, \ldots, \xi_n$ are the random variables and each of them has the normal distribution $N(m, \sigma^2)$, then the random variable

$$\eta = \frac{\sqrt{n}(\bar{\xi} - m_\xi)}{s}, \tag{1.46}$$

where

$$\bar{\xi} = \frac{1}{n} \sum_{i=1}^{n} \xi_i, \quad s^2 = \frac{1}{n-1} \sum_{i=1}^{n} (\xi_i - m_\xi)^2$$

are the sample expectation and sample variance accordingly, have the Student's distribution with $n-1$ degrees of freedom. The density function for the Student's distribution is given by the formula (Fig. 1.34):

$$f_\eta(t) = \frac{\Gamma((n+1)/2)}{\sqrt{n\pi}\Gamma(n/2)}\left(1+\frac{t^2}{n}\right)^{-(n+1)/2}. \tag{1.47}$$

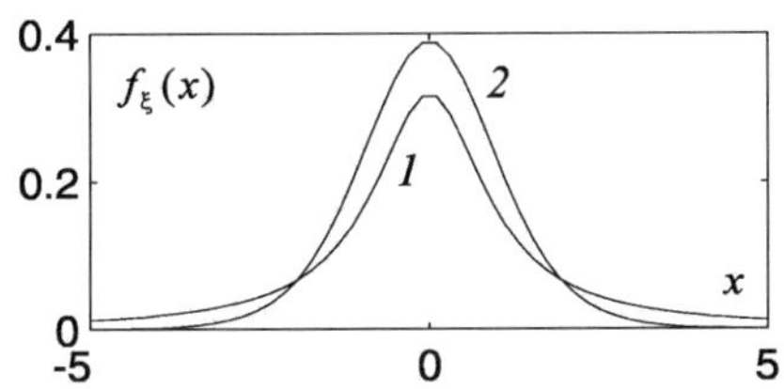

Fig. 1.34 Student's distribution: *1* — $n=1$; *2* — $n=10$.

- Mathematical expectation: $M\eta = 0$.
- Variance: $D\eta = n/(n-2)$, $n>2$.
- Asymmetry coefficient: $\gamma_1 = 0$.
- Excess: $\gamma_2 = 6/(n-4)$, $n>4$.

At the numerator of η (from (1.46)) is placed the random variable

$$\eta_1 = \sqrt{n}(\bar{\xi} - m_\xi),$$

which has the normal distribution $N(0,\sigma^2)$. A random variable η_2, which is the sample variance, is placed at the denominator of (1.46) and, as mentioned above, η_2 is independent from random variable η_1. The random variable $(n-1)s^2/\sigma^2$ has the χ^2-distribution with $(n-1)$ degrees of freedom.

The Student's distribution is symmetric relatively of $t=0$. If $n=1$ then the Student's distribution transforms to the Cauchy distribution and at $n\to\infty$ it tends to the standard normal distribution $N(0,1)$. Using the Student's distribution function is possible to draw the confidence interval for the mathematical expectation m_ξ.

1.8.6 *Fisher distribution and Z-distribution*

Let us $\xi_1, \xi_2, \dots, \xi_{n_1}$ and $\eta_1, \eta_2, \dots, \eta_{n_2}$ are the random variables which belong to normal distribution $\xi_i \in N(m_1, \sigma_1^2)$ and $\eta_i \in N(m_2, \sigma_2^2)$ accordingly. Their sample variances

$$s_1^2 = \frac{1}{n_1-1}\sum_{i=1}^{n_1}(\xi_i - \bar{\xi})^2 \quad \text{and} \quad s_2^2 = \frac{1}{n_2-1}\sum_{i=1}^{n_2}(\eta_i - \bar{\eta})^2$$

are independent and the random values $n_1 s_1^2/\sigma_1^2$ and $n_2 s_2^2/\sigma_2^2$ have the distributions $\chi^2_{n_1}$ and $\chi^2_{n_2}$ accordingly. It is said, that random variable

$$\zeta = \frac{s_1^2}{\sigma_1^2}\frac{\sigma_2^2}{s_2^2} \tag{1.48}$$

has the Fisher distribution or F-distribution (or distribution of ratio of variances) with (n_1, n_2) degrees of freedom (Fig. 1.35):

$$f_\zeta(F) = \frac{n_1^{n_2/2} n_2^{n_1/2} F(\frac{n_1+n_2}{2})}{\Gamma(\frac{n_1}{2})\Gamma(\frac{n_2}{2})}\frac{F^{(n_1/2)-1}}{(n_1+n_2F)^{(n_1+n_2)/2}},$$

where degrees of freedom n_1 and n_2 are the parameters of distribution.

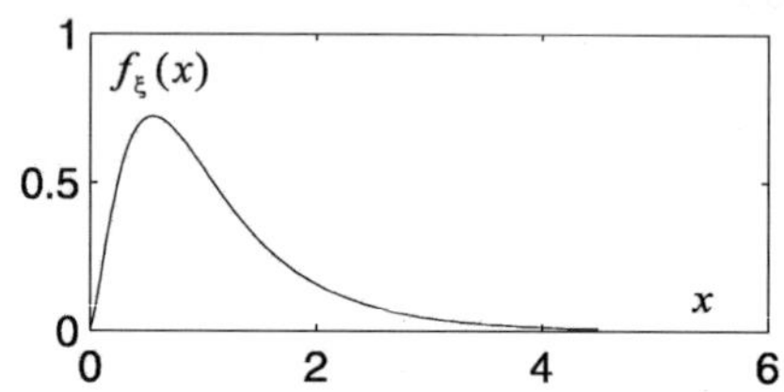

Fig. 1.35 Fisher distribution ($n_1 = 5$, $n_2 = 50$).

- Mathematical expectation: $M\zeta = n_2/n_2 - 2$, $n_2 > 2$.
- Variance:

$$D\zeta = \frac{2n_2^2(n_1+n_2-2)}{n_1(n_2-2)^2(n_2-4)}, \quad n_2 > 4.$$

The Fisher distribution is positive and asymmetric one and tends to the normal distribution for $n_1, n_2 \to \infty$, but rather slowly. If $(n_1, n_2 > 50)$ then the Fisher distribution is close to the normal distribution. The Fisher distribution is used in tasks of a test of hypothesis, for example, at the definition of a degree of an polynomial approximation.

The random variable $Z = (1/2)\ln F$ (Z is a real number) has a distribution which is considerably close to the normal one and the distribution is called Z-distribution. The parameters of Z-distribution are the degrees of freedom n_1 and n_2. The density function reads as

$$f(Z) = \frac{2n_1^{n_1/2} n_2^{n_2/2}}{B(n_1/2, n_2/2)}\frac{e^{n_1 z}}{[n_1 e^{2z} + n_2]^{(n_1+n_2)/2}},$$

The mathematical expectation and variance are given accordingly by the formulae

$$M\zeta = \frac{1}{2}[n_2^{-1} - n_1^{-1}] - \frac{1}{6}[n_1^{-2} - n_2^{-2}]$$

and

$$D\zeta = \frac{1}{2}[n_1^{-1} + n_2^{-1}] + \frac{1}{2}[n_1^{-2} + n_2^{-2}] + \frac{1}{3}[n_1^{-3} + n_2^{-3}].$$

Z-distribution is used for an estimation of the confidence interval for the correlation coefficient.

1.8.7 *Triangular distribution*

The random variable ξ has the triangular distribution, if its distribution function describes by formula

$$f_\xi(x) = \begin{cases} r^{-1} + |x - m|/r^2, & \text{if } m - r < x < m + r, \\ 0, & \text{for all others } x. \end{cases}$$

The parameters of distribution: m, $r > 0$.

- Mathematical expectation: $m_\xi = m$.
- Variance: $D\xi = r^2/6$.
- Asymmetry coefficient: $\gamma_1 - 0$.
- Excess $\gamma_2 = -0.6$.

As an example of appearance of the triangular distribution it is possible to consider a sum of two independent random variables ξ_1 and ξ_2, which have the uniform distribution.

1.8.8 *Beta distribution*

The random variable ξ has the beta distribution, if its density function can be represented as

$$f_\xi(x) = \begin{cases} \frac{\Gamma(n+m)}{\Gamma(n)\Gamma(m)} x^{m-1}(1-x)^{n-1}, & \text{if } 0 < x < 1, \\ 0, & \text{for all others } x, \end{cases}$$

where n and m are positive integers.

- Mathematical expectation: $M\xi = m/(m + n)$.
- Variance:

$$D\xi = \frac{mn}{(m+n)^2(m+n+1)}.$$

The beta distribution is used in the mathematical statistics, when the random variables with double limiting, for example $0 < x < 1$, are considered.

1.8.9 *Exponential distribution*

The random variable ξ has the exponential distribution, if its density function can be represented as

$$f_\xi(x) = \begin{cases} \lambda \exp(-\lambda x), & \text{if } x \geq 0, \\ 0, & \text{if } x < 0, \end{cases} \tag{1.49}$$

where $\lambda > 0$ is a positive number (the parameter of distribution) (Fig. 1.36).

- Mathematical expectation: $M\xi = \lambda^{-1}$.
- Variance: $D\xi = \lambda^{-2}$.
- Asymmetry coefficient: $\gamma_1 = 2$.

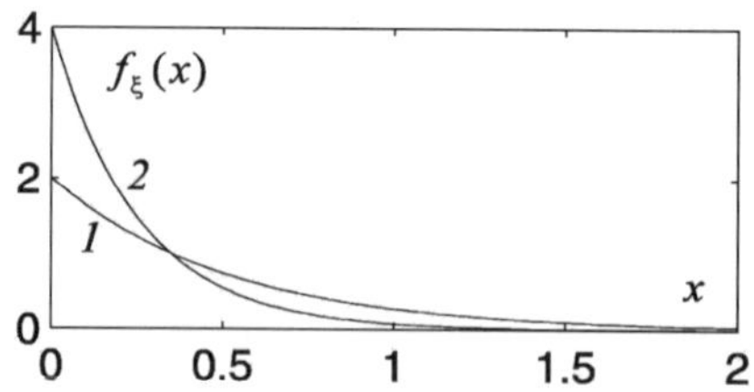

Fig. 1.36 Exponential distribution: *1* — $\lambda = 2$, *2* — $\lambda = 4$.

- Excess: $\gamma_2 = 6$.

The characteristic function is given by the formula

$$g(t) = (1 - it/\lambda)^{-1}.$$

The exponential distribution has significance in the theory of Markovian processes. If on the time axis there is a stationary Poisson flow with the intensity λ, the interval of time between two adjacent events has an exponential distribution with the parameter λ.

1.8.10 *Laplace distribution*

The random variable ξ has the Laplace distribution, if its density function can be written as

$$f_\xi(x) = (\lambda/2)\exp(-\lambda|x|), \tag{1.50}$$

where λ is a parameter of the distribution (Fig. 1.37).

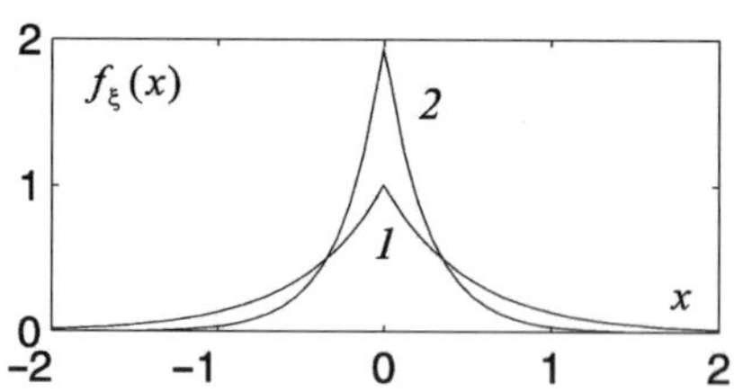

Fig. 1.37 Laplace distribution: *1* — $\lambda = 2$, *2* — $\lambda = 4$.

- Mathematical expectation: $M\xi = 0$.
- Variance: $D\xi = 2/\lambda^2$.

The Laplace distribution plays a vital part at the robust estimation, when the models with a great spread of data are considered.

1.8.11 *Cauchy distribution*

The random variable ζ with the density function

$$f_\zeta(x) = (1/\pi)(1/(1+x^2)) \tag{1.51}$$

has the Cauchy distribution (Fig. 1.38). This distribution can be obtained as a

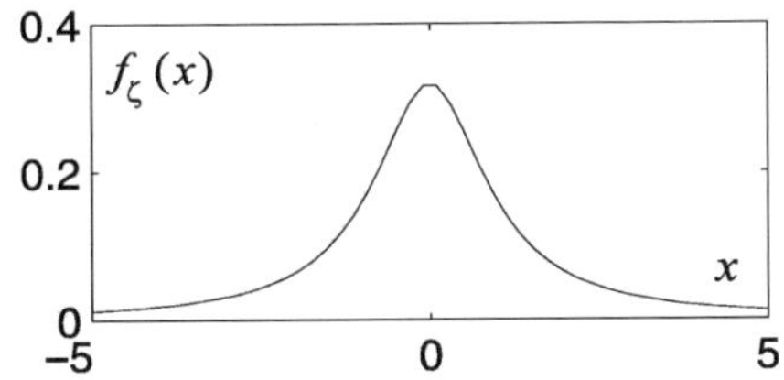

Fig. 1.38 Cauchy distribution.

distribution of a ratio $\zeta = \xi/\eta$ of two random variables with standard normal distributions: $\xi \in N(0,1)$, $\eta \in N(0,1)$.

The mathematical expectation $M\zeta$ is undefined and variance, asymmetry coefficient, excess are represented by the divergent integrals. The characteristic function is written as

$$g(t) = \exp(-|t|).$$

The Cauchy distribution has the special place in the probability theory and applications – its mathematical expectation is undefined and all other moments are divergent.

1.8.12 *Logarithmic normal distribution*

If the random variable $\eta = \ln \xi$ has the normal distribution, then the density function $f_\xi(x)$ is called logarithmic normal and is written as (Fig. 1.39):

$$f_\xi(x) = \frac{1}{\sqrt{2\pi}\sigma x} \exp\left\{ \frac{-1}{2\sigma^2} (\ln x - m)^2 \right\}, \tag{1.52}$$

where m, σ are the distribution parameters.

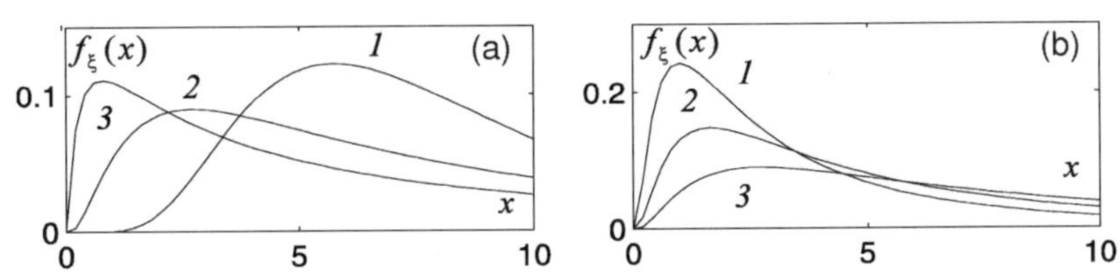

Fig. 1.39 Logarithmic normal distribution. (a) — $m = 2.0$, $\sigma = 0.5$ (*1*); 1.0 (*2*); 1.5 (*3*); (b) — $\sigma = 1.0$, $m = 1.0$ (*1*); 1.5 (*2*); 2.0 (*3*).

- Mathematical expectation: $M\xi = \exp(m - \sigma^2/2)$.
- Variance: $D\xi = \exp(2m + \sigma^2)(\exp(\sigma^2) - 1)$.
- Asymmetry coefficient: $\gamma_1 = (\exp(\sigma^2) - 1)^{1/2}(\exp(\sigma^2) + 2)$.
- Excess:

$$\gamma_2 = (\exp(\sigma^2) - 1)(\exp(3\sigma^2) + 3\exp(2\sigma^2) + 6\exp(\sigma^2) + 6).$$

The logarithmic normal distribution is used as a model for the description of errors of a random process, which include a great number of multiplicative errors.

1.8.13 *Significance of the normal distribution*

The mentioned above distributions, which using more frequently for the description of the geophysical models, are connected with the normal distribution. The most part of these distributions either asymptotically tend to the normal distribution or appearance of them is induced by the normal distribution. The asymptotic properties of the distributions and their connection with the normal distribution are represented at Fig. 1.40.

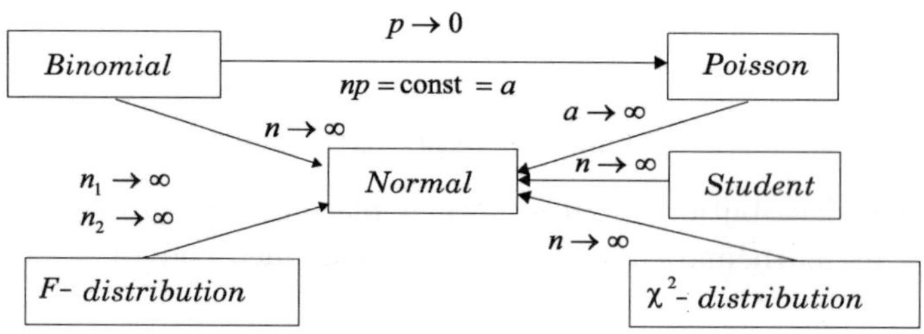

Fig. 1.40 The asymptotic properties of the distributions and their connection with the normal distribution.

The probability distributions of random variables being the functions of random variables with the normal distribution, are represented in Fig. 1.41.

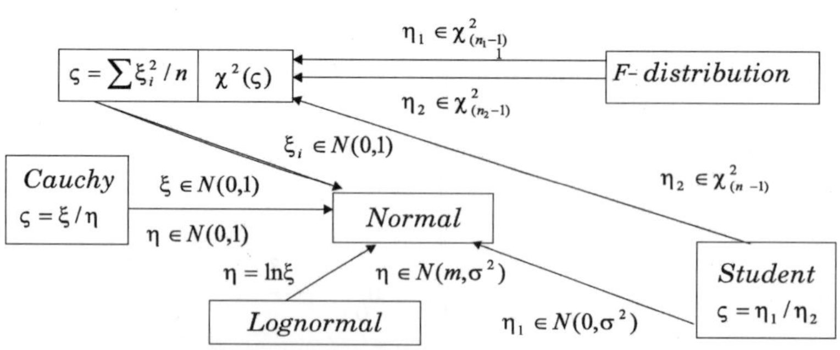

Fig. 1.41 Probability distributions, defined on the basis of the normal distribution.

1.8.14 *Confidence intervals*

Owing to a randomness of outcomes of the random experiment it is impossible to establish narrow limits for possible deviations of an estimate of the parameter from its true value. Therefore there is a problem of a definition by the results of experiments of such limits, which the error of the estimation would not go out with a given probability.

The random interval, which covers the unknown parameter ρ with the given probability β

$$P(|\tilde{\rho} - \rho| < \varepsilon) = \beta, \quad P(\tilde{\rho} - \varepsilon < \rho < \tilde{\rho} + \varepsilon) = \beta,$$

is called the *confidence interval* for the given parameters, and the probability β is called the *confidence probability*. The quantity $1 - \beta$ is called the *significance level*. The endpoints of the confidence interval $I_\beta = (\tilde{\rho} - \varepsilon, \tilde{\rho} + \varepsilon)$ are determined by the *confidence bounds*.

Let us consider the creation of the confidence interval for the mathematical expectation and variance. It should be noted, that the problem of determination of such confidence intervals for an arbitrary number of the experiments n is solved for the case of the normal distribution of the random variable ξ. At creation of the confidence interval we shall use a sample estimate for the mathematical expectation $\tilde{\rho}$:

$$\tilde{\rho} = \bar{\xi} = \frac{1}{n} \sum_{i=1}^{n} \xi_i,$$

which is the arithmetic mean of the quantities ξ_i, which obtained from n independent experiments.

If the variance σ_ξ^2 of the random value ξ is known, the sample mean $\bar{\xi}$ has the normal distribution with the mathematical expectation m_ξ and the standard deviation $\sigma_m = \sigma_\xi / \sqrt{n}$, therefore

$$P\{-l \leq (\bar{\xi} - m_\xi)\sqrt{n}/\sigma_\xi \leq l\} = 2\Phi(l) - 1,$$
$$P\{\bar{\xi} - l\sigma_\xi/\sqrt{n} \leq m_\xi \leq \bar{\xi} + l\sigma_\xi/\sqrt{n}\} = 2\Phi(l) - 1.$$

If the variance σ_ξ^2 is unknown, let us consider the random value

$$\eta = \sqrt{n}(\bar{\xi} - m_\xi)/s,$$

where

$$s^2 = \frac{1}{n-1} \sum_{i=1}^{n} (\xi_i - \bar{\xi})^2 \tag{1.53}$$

is the sample variance of random variable ξ. It is possible to prove, that the random variable η has the Student's distribution and it leads to the determination of the confidence interval, $P\{|\eta| < t_\beta\} = \beta$:

$$P\{\bar{\xi} - t_\beta s/\sqrt{n} \leq m_\xi \leq \bar{\xi} + t_\beta s/\sqrt{n}\} = \beta.$$

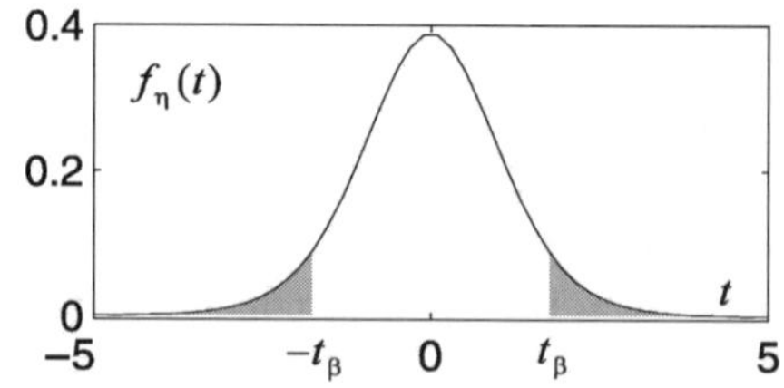

Fig. 1.42 Student's distribution (n=10) and the confidence interval ($\beta = 0.95$).

At Fig. 1.42 the confidence interval for the Student's distribution is represented by the confidence probability $\beta = 0.95$.

To find the confidence interval for the variance σ_ξ^2 of random variable ξ ($\xi_i \in N(m_\xi, \sigma_\xi^2)$). The quantities $\xi_1, \xi_2, \ldots, \xi_n$ are obtained as the results of n independent experiments. It is possible to prove, that the random value

$$\frac{(n-1)s^2}{\sigma_\xi^2}$$

has χ^2–distribution with $n-1$ degrees of freedom. To take into account the follow equalities

$$1 - P(\chi^2 > \chi_1^2) = P(\chi^2 < \chi_1^2) = \alpha/2, \quad P(\chi^2 > \chi_2^2) = \alpha/2,$$
$$P(\chi_1^2 < (n-1)s^2/\sigma^2 < \chi_2^2) = 1 - P(\chi^2 < \chi_1^2) - P(\chi^2 > \chi_2^2) = 1 - \alpha,$$

let to write an expression, which connects the length of the confidence interval with the significance level α:

$$P\left(\frac{(n-1)s^2}{\chi_2^2} < \sigma^2 < \frac{(n-1)s^2}{\chi_1^2}\right) = 1 - \alpha.$$

At Fig. 1.43 the areas with the significance level of α=0.13 are marked.

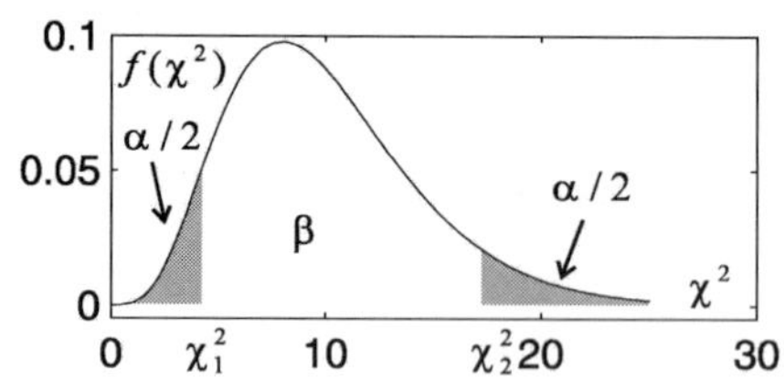

Fig. 1.43 χ^2–distribution: α=0.13.

1.8.15 *Exercises*

(1) To find the distribution function $F_\xi(x)$ for the random variable ξ, which has the uniform density function on the interval (a, b).

(2) The random variable has the Simpson's distribution (triangular distribution) on the interval $(-a, a)$. To draw the distribution function graph. To calculate $M\xi$, $D\xi$, γ_1. To find the probability of hitting of the random variable ξ into the interval $(a/2, a)$.

(3) The quality control of detonators for an explosive source on seismic survey is implemented in the following way. If a resistance R of the detonator is satisfied the inequalities $R_1 < R < R_2$, then the detonator can be used, but if the inequalities are not valid, then the detonator rejects as defective. It is known, that the resistance of the detonator has the normal distribution with the mathematical expectation $MR = (R_1 + R_2)/2$ and standard deviation $\sigma_R = (R_2 - R_1)/4$. To find the probability of the rejecting of the detonator.

(4) For the case of the above considered task, to find the standard deviation of the resistance of the detonator σ_R, if it is known, that rejection is 10 % of all detonators.

(5) During a device operation at the random moments can be malfunctions. The time T of the device operation up to the first malfunction is distributed under the exponential law $f_\xi(t) = \nu \exp\{-\nu t\}$ with parameter ν $(t > 0)$. The malfunction is immediately discovered and the device acts in repair. The duration of repair is equal t_0, then the device again acts in operation. To find the density function $f_\zeta(t)$ and cumulative distribution function $F_\zeta(t)$ of a time interval ζ between two neighboring defects. To find its mathematical expectation and variance. To find probability that ζ will be greater than $2t_0$.

(6) Consider the Poisson field of points with the constant density λ. To find a distribution law and the numerical characteristics m_r, D_r for a distance apart arbitrary point and a nearest neighbor point.

(7) In the some star set the stars are three-dimensional Poisson field of points with the density λ (the mathematical expectation of a number of stars per unit volume). Let us to fix one (arbitrary) star and to consider the nearest star, the next (second) star with greater distance, third star and so on. To find the distribution low for the distance between the fixed star and the n-the star in this line.

(8) The iron-manganese concretions are placed at the bottom of ocean in the random points and forms the Poisson field of points with a density λ (mathematical expectation of number of concretions per unit area). The arbitrary point O at the bottom is chosen. Consider the random variable R_1, which is a distance between the point O and the nearest concretion. To find the distribution of R_1.

(9) Let us consider an amplitude of seismic signal at the time point t_0 as a random variable ξ which has the normal distribution and the mathematical expectation $m_\xi = 0$. There is an interval (α, β) and origin of coordinates does not belong to this interval. At what value of the standard deviation the probability of hitting the random variable ξ inside the interval (α, β) will be maximum?

1.9 Information and Entropy

A physical matter of the information appears from thermodynamics and statistical physics and is based on a notion of entropy, which is a measure of the state of indeterminacy of the physical system.

1.9.1 *Entropy of the set of discrete states of system*

In a case of a finite set of the possible states of a system x_1, x_2, ..., x_n (with probabilities p_1, p_2,..., p_n,), which forms a complete group of the disjoint events, the entropy is defined by the formula

$$H = -\sum_{i=1}^{n} p_i \log p_i. \tag{1.54}$$

A choice of the base of the logarithm is determined a unit of the entropy If the base of logarithm is 2, then we deal with binary unit of information *bit.*[1], If the bases of logarithm are e or 10 we deal with Napierian or Hartley units correspondingly. If we choose 2 as the base of logarithm, then unit value of entropy corresponds to a simplest system of two equally possible states

x	x_1	x_2
p	$1/2$	$1/2$

,

and

$$H = -\left(\frac{1}{2}\log\frac{1}{2} + \frac{1}{2}\log\frac{1}{2}\right) = 1$$

this is the entropy of the bit site if it can be 0 or 1 with the equal probability.

If the system has n equiprobable states x_i with probabilities $p_i = 1/n$, then the entropy of the system is equal

$$H = -n\frac{1}{n}\log\frac{1}{n} = \log n.$$

If we know the state of the system a priori, then its entropy is equal to $H = 0$. In this case all states have the probabilities $p_i = 0$, except for one state $p_{i_0} = 1$. The entropy of a system with a finite set of states reaches a maximum, when all states are equiprobable.

The entropy can be represented as the mathematical expectation

$$H_\xi = M[-\log p_\xi], \tag{1.55}$$

where $\log p_\xi$ is a logarithm of the probability of the random state of the system, which is considered as the random variable ξ.

[1]A word *bit* is made up the words *binary digit.*

1.9.2 *Entropy of the complex system*

Let us consider a union of two systems ξ and η with the possible states $x_1, x_2, \ldots, x_n$ and $y_1, y_2, \ldots, y_n$ correspondingly. The states (x_i, y_i) of the complex system (ξ, η) are the complex combinations of x_i, y_i and together with appropriate probabilities (Table 1.2).

Table 1.2 Representation of the joint distribution function for a case of two discrete random variables x_i and y_j.

$y_i \backslash x_i$	x_1	x_2	$\ldots$	x_{n-1}	x_n
y_1	p_{11}	p_{12}	$\ldots$	$p_{1\,n-1}$	$p_{1\,n}$
y_2	p_{21}	p_{22}	$\ldots$	$p_{2\,n-1}$	$p_{2\,n}$
$\vdots$	.	.	$\ldots$	.	.
y_m	p_{m1}	p_{m2}	$\ldots$	$p_{m\,n-1}$	$p_{m\,n}$

The entropy of the complex system is written as

$$H_{\xi\eta} = -\sum_{i=1}^{n}\sum_{j=1}^{m} p_{ij}\log p_{ij}, \qquad H_{\xi\eta} = M[-\log p(\xi,\eta)].$$

For a case of independent random variables we have

$$p(\xi,\eta) = p(\xi)p(\eta)$$

and, hence,

$$H_{\xi\eta} = H_\xi + H_\eta, \tag{1.56}$$

or extending on a case of an arbitrary number of independent systems, we shall obtain

$$H_{\xi_1,\xi_2,\ldots,\xi_n} = H_{\xi_1} + H_{\xi_2} + \cdots + H_{\xi_n}.$$

in a case of a union of independent systems their entropies are summed.

In case of dependent systems the notion of the conditional entropy of a system ξ with respect to a state y_j of the system η is introduced:

$$H_{\xi/y_j} = -\sum_{i=1}^{n} p(x_i/y_j)\log p(x_i/y_j), \tag{1.57}$$

or

$$H_{\xi/y_j} = M[-\log p(\xi/y_j)]. \tag{1.58}$$

The mean entropy or the total entropy is defined as

$$H_{\xi/\eta} = \sum_{j=1}^{m} p_j H_{\xi/y_j} = \sum_{i=1}^{n}\sum_{j=1}^{m} p_j p(x_i/y_j)\log p(x_i/y_j). \tag{1.59}$$

Taking into account that $p_j p(x_i/y_j) = p_{ij}$ (joint probability), we obtain

$$H_{\xi/\eta} = -\sum_{i=1}^{n}\sum_{j=1}^{m} p_{ij}\log p(x_i/y_j). \tag{1.60}$$

The entropy of a joint system (ξ, η) is determined by the formula

$$H_{\xi\eta} = H_\xi + H_{\eta/\xi}, \quad \text{or} \quad H_{\xi\eta} = H_\eta + H_{\xi/\eta}, \tag{1.61}$$

that can be proof using the notion of the entropy and formula for multiplication of probabilities

$$p(\xi, \eta) = p(\xi)p(\eta/\xi) = p(\eta)p(\xi/\eta).$$

1.9.3 *Shannon information (discrete case)*

A notion of the amount of information is tightly connected with a notion of the entropy being measure of the indeterminacy. The deriving of the information is accompanied by a diminution of the indeterminacy, therefore the amount of the information can be measured by a diminution of the indeterminacy, i.e. entropy.

Let us ξ is an observed random variable, that describes a physical system. The entropy before observation (a priori) we shall determine H_ξ^{apr}, and entropy after observation (a posteriori) we shall determine H_ξ^{apost}, then the Shannon information is defined as follows:

$$I_\xi^{\text{Sh}} = H_\xi^{\text{apr}} - H_\xi^{\text{apost}}. \tag{1.62}$$

In that special case, when as a result of the observation ξ the state of a system was completely spotted, i.e. $H_\xi^{\text{apost}} = 0$, the information is equal

$$I_\xi^{\text{Sh}} = H_\xi^{\text{apr}} = -\sum_{i=1}^{n} p_i \log p_i$$

or

$$I_\xi^{\text{Sh}} = M[-\log p(\xi)]. \tag{1.63}$$

However a direct observation of the random variable ξ frequently appears impossible and the inference about a state of the system described by ξ, is implemented by observations of another random variable η which is connected with ξ. So for example, in a radiolocation an operator observes a reflected signal η by a screen of the locator, instead of the direct observation of an airplane ξ. In a seismic exploration the inference about an interior structure of the Earth crust ξ is yielded on the basis of the observation of a seismic field η on original ground. The amount of the information about the system ξ at the observation of the system η is defined as a decreasing of the entropy of the system ξ at observation of the system η:

$$I_{\eta\to\xi}^{\text{Sh}} = H_\xi - H_{\xi/\eta}, \tag{1.64}$$

$I_{\xi\to\eta}^{\text{Sh}}$ is called the complete (or mean)information, which contains in the system η about the system ξ. It is easy to show, that

$$I_{\xi\to\eta}^{\text{Sh}} = H_\eta - H_{\eta/\xi}. \tag{1.65}$$

Taking into account the equalities

$$H_{\xi\eta} = H_\xi + H_{\eta/\xi} = H_\eta + H_{\xi/\eta},$$

we obtain

$$I^{\mathrm{Sh}}_{\xi\to\eta} = I^{\mathrm{Sh}}_{\eta\to\xi} = I^{\mathrm{Sh}}_{\eta\leftrightarrow\xi},$$

where $I^{\mathrm{Sh}}_{\eta\leftrightarrow\xi}$ is called *the complete mutual information.* If the random variables ξ and η (that describe the system) are independent variables, then $I_{\eta\leftrightarrow\xi} = 0$, since $H_{\eta/\xi} = H_\eta$, $H_{\xi/\eta} = H_\xi$. It is possible to represent the information $I_{\eta\leftrightarrow\xi}$ using the entropy of the joint system

$$I_{\eta\leftrightarrow\xi} = H_\xi + H_\eta - H_{\xi\eta} \tag{1.66}$$

or, with the help of expectation operator, we obtain

$$I_{\eta\leftrightarrow\xi} = M[-\log p(\xi) - \log p(\eta) + \log p(\xi,\eta)],$$

$$I_{\eta\leftrightarrow\xi} = M\left[\log\frac{p(\xi,\eta)}{p(\xi)p(\eta)}\right], \tag{1.67}$$

$$I_{\eta\leftrightarrow\xi} = \sum_{i=1}^{n}\sum_{j=1}^{m} p_{ij}\log\frac{p_{ij}}{p_{\xi i}p_{\eta j}}. \tag{1.68}$$

In a case of the practical problems solution there can be a necessity for an estimation *of the partial information* $I_{y_j\to x}$ about the system ξ, contained in the individual message, connected with a specific state y_j, where the complete information $I_{\eta\leftrightarrow\xi}$ is the mathematical expectation of all messages y_j $j = 1, 2, \ldots, m$, over partial information

$$I_{\eta\leftrightarrow\xi} = \sum_{j=1}^{m} p_j I_{y_j\to\xi}, \tag{1.69}$$

where

$$I_{y_j\to\xi} = \sum_{i=1}^{n} p(x_i/y_j)\log\frac{p(x_i/y_j)}{p_{\xi i}}. \tag{1.70}$$

The expression (1.70), which defines the partial information, it is possible to rewrite as:

$$I_{y_j\to\xi} = M_{y_j}\left[\log\frac{p(\xi/y_j)}{p(\xi)}\right], \tag{1.71}$$

where M_{y_j} is a conditional expectation operation. It could show, that partial information $I_{y_j\to\xi}$, as the complete information $I_{\eta\to\xi}$, can not be negative one. The partial information about an event x_i, which is obtained with the help of a message about the other event y_j, can be represented as follows

$$I_{y_j\to x_i} = \log\frac{p(x_i/y_j)}{p_{\xi_i}}. \tag{1.72}$$

It is necessary to note, that the such partial information from an event to an other event, as against entered above the complete $I_{\eta\leftrightarrow\xi}$ and the partial $I_{y_j\to\xi}$ information, can be both positive and negative.

1.9.4 *Entropy and information for systems with a continuous set of states*

The measure of indeterminacy of a system, which is described by a continuous random variable ξ with the given density function, is the *differential entropy* H_ξ, defined by the formula

$$H_\xi = -\int\limits_{-\infty}^{\infty} f_\xi(x)\log f_\xi(x)dx,$$

or

$$H_\xi = M[-\log f_\xi(x)]. \tag{1.73}$$

In a case of two continuous random variables (or two systems) ξ and η, their density function $f_{\xi\eta}(x,y)$, marginal densities $f_\xi(x)$, $f_\eta(y)$ and conditional densities $f_{\eta/\xi}(y/x)$, $f_{\xi/\eta}(x/y)$, it is possible to introduce the partial conditional entropy $H_{\eta/x}$, i.e. the entropy of the system η on condition that the system ξ is in a definite condition

$$H(\eta/x) = -\int\limits_{-\infty}^{\infty} f_{\eta\xi}(y/x)\log f_{\eta\xi}(y/x)dy.$$

The total or *mean conditional entropy* is defined as

$$H(\eta/\xi) = -\iint\limits_{-\infty}^{\infty} f_\xi(x)f_{\eta/\xi}(y/x)\log f_{\eta/\xi}(y/x)dxdy,$$

or, taking into account, that $f_{\xi\eta}(x,y) = f_\xi(x)f_{\eta/\xi}(y/x)$, we obtain

$$H(\eta/\xi) = -\iint\limits_{-\infty}^{\infty} f_{\xi\eta}(x,y)\log f_{\eta/\xi}(y/x)dxdy.$$

Similarly, how it was made for a discrete case, it is possible to show, that the entropy of the system (ξ,η) is equal

$$H(\xi,\eta) = H(\xi) + H(\eta/\xi), \tag{1.74}$$

and in a case of independent random variables (or independent states of the system) ξ and η:

$$H(\xi,\eta) = H(\xi) + H(\eta). \tag{1.75}$$

Example 1.18. In the class of continuous distributions $f_\xi(x)$ which belong to the random variable ξ, to find such, that for the given mathematical expectation m_ξ and the variance σ_ξ^2 provides a maximum of the entropy H_ξ $(\max H_\xi)$.

It is given

$$\int\limits_{-\infty}^{\infty} f_\xi(x)dx = 1, \quad \int\limits_{-\infty}^{\infty} xf_\xi(x)dx = m_\xi,$$

$$\int_{-\infty}^{\infty} (x - m_\xi)^2 f_\xi(x)dx = \sigma_\xi^2.$$

To find

$$\hat{f}_\xi(x) \Rightarrow \max\left[-\int_{-\infty}^{\infty} f_\xi(x)\log f_\xi(x)dx\right].$$

Finding $\hat{f}_\xi(x)$ is reduced to the solution of a variational problem

$$\hat{f}_\xi(x) \Rightarrow \max\left[\int_{-\infty}^{\infty} \Phi(f_\xi(x), x)dx\right]$$

with the side conditions

$$\int_{-\infty}^{\infty} \varphi_k(f_\xi(x), x)dx = c_k \quad (k = 1, 2, 3),$$

where

$$\begin{aligned}
&\Phi(f_\xi(x), x) = -f_\xi(x)\log f_\xi(x),\\
&\varphi_1 = f_\xi(x), \quad c_1 = 1,\\
&\varphi_2 = x f_\xi(x), \quad c_2 = m_\xi,\\
&\varphi_3 = (x - m_\xi)^2 f_\xi(x), \quad c_3 = \sigma_\xi^2.
\end{aligned}$$

To write down the Euler equations:

$$\begin{aligned}
&\partial\Phi_1/\partial f = 0,\\
&\Phi_1(f, x) = -f\log f + \lambda_1 f + \lambda_2 x f + \lambda_3(x - m_\xi)^2 f.
\end{aligned}$$

After the differentiation we obtain the equation

$$-1 - \log f + \lambda_1 + \lambda_2 x + \lambda_3(x - m_\xi)^2 = 0,$$

with a solution relatively to f:

$$f_\xi(x) = \exp(\lambda_1 + \lambda_2 x + \lambda_3(x - m_\xi)^2 - 1).$$

Using additional conditions

$$\int_{-\infty}^{\infty} \exp(\lambda_1 + \lambda_2 x + \lambda_3(x - m_\xi)^2 - 1)dx = 1,$$

$$\int_{-\infty}^{\infty} x\exp(\lambda_1 + \lambda_2 x + \lambda_3(x - m_\xi)^2 - 1)dx = m_\xi,$$

$$\int_{-\infty}^{\infty} (x - m_\xi)^2\exp(\lambda_1 + \lambda_2 x + \lambda_3(x - m_\xi)^2 - 1)dx = \sigma_\xi^2,$$

we obtain

$$\lambda_1 = 1 - \ln(\sqrt{2\pi}\sigma_\xi), \quad \lambda_2 = 0, \quad \lambda_3 = -(2\sigma_\xi^2)^{-1}$$

and

$$\hat{f}_\xi(x) = \frac{1}{\sigma_\xi\sqrt{2\pi}} \exp\left\{-\frac{(x-m_\xi)^2}{2\sigma_\xi^2}\right\}.$$

Thus, the normal distribution corresponds to the peak entropy in a class of distribution with the predetermined first two moments.

The expression for the complete mutual information contained in two continuous systems ξ and η, is similar to expression the (1.68) for a discrete case, but with the replacement of the probabilities of discrete events by the distribution functions, can be written as

$$I_{\eta\leftrightarrow\xi} = \iint\limits_{-\infty}^{\infty} f_{\xi\eta}(x,y) \log \frac{f_{\xi\eta}(x,y)}{f_\xi(x) f_\eta(y)} dxdy,$$

or, using the expectation operator,

$$I_{\eta\leftrightarrow\xi} = M\left[\log(f_{\xi\eta}/f_\xi f_\eta)\right], \tag{1.76}$$

where $I_{\eta\leftrightarrow\xi}$ is the nonnegative value, which is equal to zero only for independent random variables ξ and η.

Example 1.19. Let us consider a simple model

$$\eta = \xi + \zeta,$$

where η is a recorded signal, ξ is a transmitted signal with the normal distribution $N(m_\xi = 0, \sigma_\xi^2)$, ζ is an additive noise with the normal distribution $N(m_\zeta, \sigma_\zeta^2)$, where ξ and ζ are independent random variables.
To find an amount of the information $I_{\eta\leftrightarrow\xi}$ about the signal ξ, which is contained in the recorded signal η.

By definition we have

$$I_{\eta\leftrightarrow\xi} = M\left[\log\frac{f_{\xi\eta}}{f_\xi f_\eta}\right] = M\left[\log\frac{f_{\eta/\xi}}{f_\eta}\right] = M\kappa, \quad \kappa = \log\frac{f_{\eta/\xi}}{f_\eta}.$$

By problem situation $\sigma_\eta^2 = \sigma_\xi^2 + \sigma_\zeta^2$,

$$f_\eta(y) = \frac{1}{\sqrt{2\pi}\sqrt{\sigma_\xi^2 + \sigma_\zeta^2}} \exp\left\{-\frac{1}{2}\frac{y^2}{(\sigma_\xi^2 + \sigma_\zeta^2)}\right\},$$

$$f_{\eta/\xi}(y/x) = \frac{1}{\sqrt{2\pi}\sigma_\eta\sqrt{1-r^2}} \exp\left\{-\frac{1}{2(1-r^2)}\left[\frac{y}{\sigma_\eta} - r\frac{x}{\sigma_\xi}\right]^2\right\}.$$

To find the correlation coefficient of η and ξ taking into account that $\eta = \xi + \zeta$:

$$r = \frac{1}{\sigma_\eta\sigma_\xi} M(\eta\cdot\xi) = \frac{1}{\sigma_\eta\sigma_\xi}\left[M(\xi^2) + M(\xi\zeta)\right] = \frac{\sigma_\xi}{\sigma_\eta}.$$

By substituting r into $f_{\eta/\xi}(y/x)$, we obtain

$$f_{\eta/\xi}(y/x) = \frac{1}{\sqrt{2\pi}\sigma_\zeta} \exp\left\{-\frac{(y-x)^2}{2\sigma_\zeta^2}\right\} = \frac{1}{\sqrt{2\pi}\sigma_\zeta} \exp\left\{-\frac{z^2}{2\sigma_\zeta^2}\right\}.$$

We use the obtained relations for finding the quantity κ:

$$\kappa = \log\frac{\sqrt{\sigma_\xi^2+\sigma_\zeta^2}}{\sigma_\zeta} + \frac{1}{\ln 2}\left[\frac{\zeta^2}{2\sigma_\zeta^2} - \frac{\eta^2}{2(\sigma_\xi^2+\sigma_\zeta^2)}\right].$$

Now it is necessary to calculate the mathematical expectation

$$M\kappa = I_{\eta\leftrightarrow\xi} = \log\frac{\sqrt{\sigma_\xi^2+\sigma_\zeta^2}}{\sigma_\zeta} + \frac{1}{\ln 2}\left[\frac{M\zeta^2}{2\sigma_\zeta^2} - \frac{M\eta^2}{2(\sigma_\xi^2+\sigma_\zeta^2)}\right].$$

Taking into account that $M\zeta^2 = \sigma_\zeta^2$, $M\eta^2 = \sigma_\xi^2 + \sigma_\zeta^2$, we obtain

$$I_{\eta\leftrightarrow\xi} = \log\frac{\sqrt{\sigma_\xi^2+\sigma_\zeta^2}}{\sigma_\zeta}.$$

So, the information on a signal ξ, contained in the recorded signal η, is equal to the logarithm of the standard deviation of the signal η (σ_η) to the standard deviation of the noise ζ (σ_ζ).

1.9.5 *Fisher information*

Let us a random variable ξ has the density function $f_\xi(x,\theta)$, where θ is a parameter. Let us assume that $f_\xi(x,\theta)$ is differentiable on θ, then the *Fisher information* about the unknown parameter θ, which is contained in a random variable ξ or in its density function $f_\xi(x,\theta)$, is treated as the diminution of indeterminacy about the unknown value θ after an observation of the random variable ξ. The amount of the Fisher information about θ, contained in a random variable ξ, is defined as

$$I_\xi^{(F)}(\theta) = M[(\partial \log f_\xi(x,\theta)/\partial\theta)^2]. \tag{1.77}$$

In a case of a random vector $\boldsymbol{\xi} = (\xi_1, \xi_2, \ldots, \xi_n)$ and a parametric vector $\boldsymbol{\theta} = (\theta_1, \theta_2, \ldots, \theta_s)$ the *Fisher matrix* with the dimension $[S \times S]$ and the elements

$$I_{ss'}^{(F)}(\boldsymbol{\theta}) = M\left[\frac{\partial \log f_{\boldsymbol{\xi}}(x_1,x_2,\ldots,x_n,\boldsymbol{\theta})}{\partial\theta_s}\frac{\partial \log f_{\boldsymbol{\xi}}(x_1,x_2,\ldots,x_n,\boldsymbol{\theta})}{\partial\theta_{s'}}\right] \tag{1.78}$$

is introduced.

The properties of the Fisher information.

(1) Let ξ_1 and ξ_2 are independent random variables, $I_{\xi_1}^{(F)}(\theta)$, $I_{\xi_2}^{(F)}(\theta)$ are the quantities of information about θ, which are contained in the pair (ξ_1, ξ_2). Then the following equality is valid

$$I_{\xi_1\xi_2}^{(F)}(\theta) = I_{\xi_1}^{(F)}(\theta) + I_{\xi_2}^{(F)}(\theta).$$

Proof.

$$I^{(F)}_{\xi_1\xi_2}(\theta) = M\left\{\frac{\partial}{\partial\theta}\left[\log f_{\xi_1}(x,\theta)\cdot f_{\xi_2}(x,\theta)\right]\right\}^2$$
$$= M\left[\left(\frac{\partial \log f_{\xi_1}(x,\theta)}{\partial\theta}\right)^2\right] + M\left[\left(\frac{\partial \log f_{\xi_2}(x,\theta)}{\partial\theta}\right)^2\right]$$
$$+2M\left[\frac{\partial \log f_{\xi_1}(x,\theta)}{\partial\theta}\cdot\frac{\partial \log f_{\xi_2}(x,\theta)}{\partial\theta}\right]$$
$$= I^{(F)}_{\xi_1}(\theta) + I^{(F)}_{\xi_2}(\theta),$$

since

$$M\left[\frac{\partial \log f_{\xi_1}(x,\theta)}{\partial\theta}\cdot\frac{\partial \log f_{\xi_2}(x,\theta)}{\partial\theta}\right]$$
$$= M\left[\frac{\partial \log f_{\xi_1}(x,\theta)}{\partial\theta}\right]\cdot M\left[\frac{\partial \log f_{\xi_2}(x,\theta)}{\partial\theta}\right] = 0.$$

(2) Let $\xi_1, \xi_2, \dots, \xi_n$ are independent, identically distributed random variables and $I^{(F)}_{\xi}(\theta)$ the amount of information about θ, which is contained in an each random variable. Then the amount of information in $\xi_1, \xi_2, \dots, \xi_n$ is equal to $nI^{(F)}_{\xi}(\theta)$. Analogously to the scalar case it is possible to show:

(1) If $I^{(F)}_{\xi_1}(\boldsymbol{\theta})$ and $I^{(F)}_{\xi_2}(\boldsymbol{\theta})$ are information matrices, which are connected with the random variables ξ_1 and ξ_2, then the information matrix for the pair of random variables (ξ_1, ξ_2) is equal to

$$I^{(F)}_{\xi_1\xi_2}(\boldsymbol{\theta}) = I^{(F)}_{\xi_1}(\boldsymbol{\theta}) + I^{(F)}_{\xi_2}(\boldsymbol{\theta}).$$

(2) If $\xi_1, \xi_2, \dots, \xi_n$ are independent, random variables with equal distributive law and $I^{(F)}_{\xi}(\boldsymbol{\theta})$ are amount of information about $\boldsymbol{\theta}$, which are contained in an each random variable ξ_i, then the amount of information contained in $\xi_1, \xi_2, \dots, \xi_n$ is equal to $nI^{(F)}_{\xi}(\boldsymbol{\theta})$.

(3) Let us a function $T = T(\xi)$ of a random variable ξ is given, then

$$I^{(F)}_{\xi}(\boldsymbol{\theta}) - I^{(F)}_{T}(\boldsymbol{\theta}) \geq 0,$$

i.e. the matrix is nonnegative defined and in a case of a single parameter θ

$$I^{(F)}_{\xi} \geq I^{(F)}_{T},$$

i.e. the amount of information about θ which is contained in a random variable, ever more or equal than the amount of information, which is contained in the function of the random variable $T = T(\xi)$.

When it is possible by using an observed value of a random variable with the probability 1 to restore precisely a value of parameter θ, in this case we tell, that the random variable contains the greatest possible information about parameter. But if the distribution is a constant at all values of parameter, it is impossible to judge about the value of θ by the results of observations of the random variable. The sensitivity of a random variable with respect to a parameter can be measured by the quantity of the change of the distribution of this random variable at change of the value of the parameter.

In the case of a vector parameter $\boldsymbol{\theta}$ a distance between distributions at transition from $(\theta_1, \theta_2, \ldots, \theta_s)$ to $(\theta_1 + \delta\theta_1,\ \theta_2 + \delta\theta_2, \ldots, \theta_s + \delta\theta_s)$ describes by a quadratic form

$$\sum_s \sum_{s'} I_{ss'}^{(F)}(\boldsymbol{\theta})\delta\theta_s\delta\theta_{s'}.$$

The formula (1.77) for the Fisher information can be rewritten as

$$I_\xi^{(F)}(\theta) = M\left[-\frac{\partial^2 \log f_\xi(x,\theta)}{\partial\theta^2}\right], \tag{1.79}$$

since

$$\frac{\partial^2 \log f}{\partial\theta^2} = -\frac{1}{f^2}\left(\frac{\partial f}{\partial\theta}\right)^2 + \frac{1}{f}\frac{\partial^2 f}{\partial\theta^2}; \quad \left(\frac{1}{f}\frac{\partial f}{\partial\theta}\right)^2 = \left(\frac{\partial \log f}{\partial\theta}\right)^2,$$

$$\begin{aligned} M\left(\frac{1}{f}\frac{\partial^2 f}{\partial\theta^2}\right) &= \int_{-\infty}^{\infty} f_\xi(x,\theta)\frac{1}{f_\xi(x,\theta)}\frac{\partial^2 f_\xi(x,\theta)}{\partial\theta^2}dx \\ &= \frac{\partial^2}{\partial\theta^2}\int_{-\infty}^{\infty} f_\xi(x,\theta)dx = 0. \end{aligned}$$

In the case of the vector parameter $(\boldsymbol{\theta})$, using analogous relations, we shall obtain

$$I_{ss'}^{(F)}(\boldsymbol{\theta}) = M\left[-\frac{\partial^2 \log f(x,\boldsymbol{\theta})}{\partial\theta_s\partial\theta_{s'}}\right]. \tag{1.80}$$

Example 1.20. Let us the random variable ξ belongs to the normal distribution with the known variance σ_ξ^2 and the unknown mathematical expectation m_ξ. To find the amount of information about m_ξ, which is contained in a single observation of ξ. The Fisher information about m_ξ, which is contained in a single observation is equal

$$\begin{aligned} I_\xi^{(F)}(m_\xi) &= -\int_{-\infty}^{\infty} \frac{\partial^2}{\partial m_\xi^2}\left[-\frac{(x-m_\xi)^2}{2\sigma_\xi^2} - \ln\sigma_\xi\sqrt{2\pi}\right] \\ &\times \frac{1}{\sqrt{2\pi}\sigma_\xi}\exp\left[-\frac{(x-m_\xi)^2}{2\sigma_\xi^2}\right]dx \\ &= \int_{-\infty}^{\infty} \frac{1}{\sigma_\xi^2\sqrt{2\pi}\sigma_\xi}\exp\left[-\frac{(x-m_\xi)^2}{2\sigma_\xi^2}\right]dx = \frac{1}{\sigma_\xi^2}. \end{aligned}$$

For the case of n observations $(x_1, \ldots, x_n)$, we have

$$I^{(F)}_{\xi_1,\ldots,\xi_n}(m) = -\int\limits_{-\infty}^{\infty} \cdots \int\limits_{-\infty}^{\infty} \frac{\partial^2}{\partial m_\xi^2}\left[\sum_{i=1}^{n}\left(-\frac{(x_i - m_\xi)^2}{\sigma_\xi^2} l \ln(\sigma_\xi\sqrt{2\pi})\right)\right]$$
$$\times \prod_{i=1}^{n} \frac{1}{\sqrt{2\pi}\sigma_\xi} \exp\left[-\frac{1}{2}\frac{(x_i - m_\xi)^2}{\sigma_\xi^2}\right] dx_1 \ldots dx_n = \frac{n}{\sigma_\xi^2}.$$

1.9.6 *Exercises*

(1) To find the entropy of the random variable ξ, which corresponds to a phase of a seismic signal, on condition that ξ belongs to the uniform distribution at the interval $(-\pi/2, \pi/2)$.
(2) To find the entropy of the random variable ξ, which corresponds to an amplitude of a seismic signal, on condition that ξ belongs to the normal distribution $N(0, \sigma^2)$.
(3) Let us specify a state of a seismic signal by the three values: an amplitude ξ_A, belonging to the normal distribution $N(0, \sigma_A^2)$; a phase ξ_φ, belonging to the uniform distribution at an interval $(0, \pi)$; an arrival time, belonging to the uniform distribution $(0, T)$. To find the entropy of the system, which corresponds to the seismic signal.
(4) At an urn is contained five white and six black spheres. Five spheres have been extracted from the urn, they are three black and two white ones. To find the information contained at the observation of an event A_1 with respect to an event A_2: the next extracted sphere will have a black colour.
(5) At one square of chess-board a chess piece is situated. A priori any arrangement of the chess piece is supposed. To find the information, which contains in a statement about the location of the chess piece.

1.10 Random Functions and its Properties

The function $\xi(t)$ is called the random function of the argument t, which value at any value t is a random variable. The argument t is considered as a nonrandom value.

In many geophysical problems the argument t is the time, for example, at a record of seismic waves (seismic trace) or record of an electromagnetic field in a given spatial point. Therefore in the further the argument t we shall call as time.

Usually two types of argument of a random function are considered

(1) The argument of a random function t can possess any values in a given interval, which can be finite or infinite.
(2) The argument of a random function t can possess only definite discrete values.

In the first case $\xi(t)$ is called as a random process, in the second case it is called as a random sequence. An example of the random process can be a graphic presentation of a seismic trace and an example of a random sequence can be digital records of a seismic trace. Let's consider a random function $\xi(t)$ and we shall assume, that for its study is yielded n independent trials, for example, the seismic traces from n identical explosions is filed. In each trial the *realization* of the random function $x_1(t), x_2(t), \ldots, x_n(t)$ is obtained. Let's fix an instant $t = t_1$. If we were interested with properties of $\xi(t_1)$ only in an instant t_1, then for a continuous random variable $\xi_1 = \xi(t_1)$ a complete description is reduced to the density function $f(x_1/t_1)$. An example of two realizations of random process is given in Fig. 1.44. For more detailed description of a random function we shall choose two points t_1

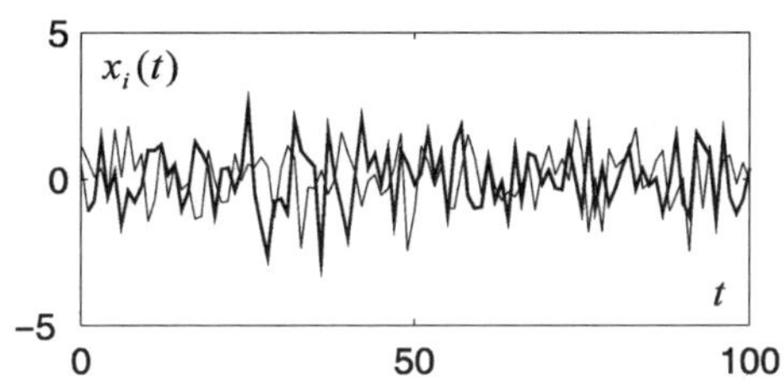

Fig. 1.44 An example of two realizations of the random process.

and t_2. The ordinates, relevant to them, $\xi(t_1)$ and $\xi(t_2)$ will be random variables ξ_1 and ξ_2, which are completely characterised by a two-dimensional density function $f(x_1, x_2/t_1, t_2)$. The random function is specified, if the multivariate density function $f(x_1, x_2, \ldots, x_n/t_1, t_2, \ldots, t_n)$ is given. Though such description of a random function is complete, in practice, usually, only the first two moments are considered. *The first moment*

$$m_{\xi_1} = M[\xi(t_1)] = \langle \xi(t_1) \rangle$$

is the mathematical expectation of the ordinate of the random function at an arbitrary time t_1. Omitting an index 1 at t, we shall write

$$m_\xi(t) = M[\xi(t)] = \langle \xi(t) \rangle.$$

The function $m_\xi(t)$ is not random and is completely determined by a distribution law $f(x/t)$:

$$m_\xi(t) = \int_{-\infty}^{\infty} x f(x/t) dx.$$

The central moments of the second order.

(1) *Variance* of a random function $\xi(t)$ at fixed timing t:

$$D[\xi(t)] = M[(\xi(t) - m_\xi(t))^2] = \langle (\xi(t) - m_\xi(t))^2 \rangle.$$

(2) *Covariance moment* of the random functions $\xi(t_1)$ and $\xi(t_2)$:

$$R(t_1,t_2) = \mathrm{cov}(\xi(t_1),\xi(t_2)) = M[(\xi(t_1) - m_\xi(t_1)) \times (\xi(t_2) - m_\xi(t_2))] = \langle(\xi(t_1) - m_\xi(t_1))(\xi(t_2) - m_\xi(t_2))\rangle.$$

The function $R(t_1,t_2)$ in the theory of random functions is called *a correlation function* or *an autocorrelation function.* Writing down the explicit expression for the mathematical expectation, we shall obtain

$$D[\xi(t)] = \int_{-\infty}^{\infty} (x(t) - m_\xi(t))^2 f(x/t)dx,$$

$$R(t_1,t_2) = \iint_{-\infty}^{\infty} (x(t_1) - m_\xi(t_1)) \times (x(t_2) - m_\xi(t_2)) f(x_1,x_2/t_1,t_2)dx_1dx_2.$$

A branch of the theory of random functions operating only the moments of the first two orders is called as the correlation theory of random functions. For any random process $\xi(t)$ the time average (on parameter t) is determined by the formula

$$\langle\xi(t)\rangle_T = \lim_{T\to\infty} (1/2T) \int_{-T}^{T} \xi(t)dt,$$

if this limit exists. As a result $\langle\xi(t)\rangle_T$ is a random quantity.

1.10.1 *Properties of random functions*

For the case of *the stationary random function* all multivariate distribution laws depend only on a cross location of instants $t_1, t_2, \ldots, t_n$, i.e. for the stationarity of a random function at arbitrary n should be fulfilled the next relation

$$f(x_1,\ldots,x_n/t_1,\ldots,t_n) = f(x_1,\ldots,x_n/t_1 + t_0,\ldots,t_n + t_0),$$

were t_0 is an arbitrary number. In the specific case at $n = 1$ and $n = 2$, assuming $t_0 = -t_1$, for stationary functions we shall have accordingly

$$f(x_1/t_1) = f(x_1/0) = f(x_1)$$

and

$$f(x_1,x_2/t_1,t_2) = f(x_1,x_2/0,\tau),$$

where $\tau = t_2 - t_1$, i.e. the one-dimensional distribution law of an ordinate of a random function does not depend on an instant of time, for which this ordinate is chosen, and the two-dimensional law depends only on a difference of instants. Using expressions for distribution laws of stationary random functions, we shall obtain

$$m_\xi(t) = \int_{-\infty}^{\infty} x f(x) dx = \text{const},$$

$$D[\xi(t)] = \int_{-\infty}^{\infty} (x(t) - m_\xi(t))^2 f(x) dx = \text{const},$$

$$R(t_1, t_2) = \int_{-\infty}^{\infty} (x(t_1) - m_\xi(t_1))(x(t_2) - m_\xi(t_2)) \times f(x_1, x_2/t_1, t_2) dx_1 dx_2 = R(t_2 - t_1) = R(\tau).$$

The conditions $m_\xi = \text{const}$, $D[\xi(t)] = \text{const}$ and $R(t_1, t_2) = R(t_2 - t_1) = R(\tau)$ are the necessary conditions of the stationarity, but not the sufficient conditions, for they can be carried out, but, beginning at some n, the distribution law will cease to satisfy to a condition $f(x_1, \ldots, x_n/t_1, \ldots, t_n) = f(x_1, \ldots, x_n/t_1 + t_0, \ldots, t_n + t_0)$, so the random function $\xi(t)$ will be non-stationary. However for the solution of practical problems frequently it is restricted to the application of the correlation theory, therefore A.Ya. Khinchin has entered concept *the stationarity in wide sense.* The random function is called stationary in the wide sense, if its mathematical expectation and variance are constant, and the correlation function depends only on a difference of instants, for which the ordinates of a random function are taken. The examples non-stationary in the wide sense of random functions are submitted on Fig. 1.45.

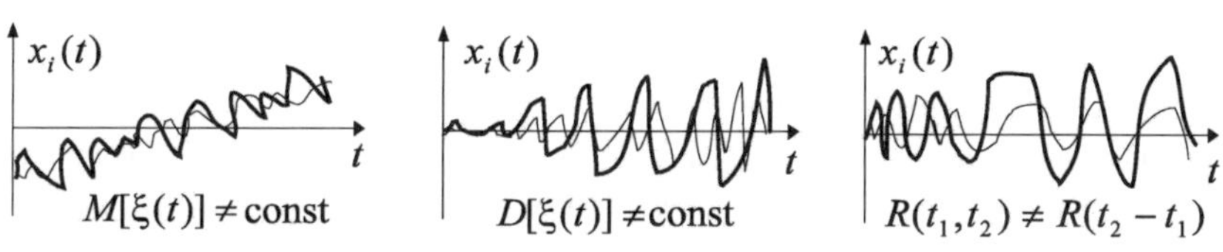

Fig. 1.45 Examples of non-stationary processes.

The normal distribution law is used most frequently at the study of the random functions. The total characteristic of the normal random process are the mathematical expectation and correlation function. The values of ordinates of a normal random function in arbitrary instants $t_1, t_2, \ldots, t_n$ are completely described by the mathematical expectation and covariance matrix

$$R_{ij} = R(t_i - t_j).$$

If the probability properties of a random function in the subsequent time interval are completely determined by the value of ordinate of this function in a given instant and do not depend on its values at the previous moments, such random function is called *Markovian*, otherwise the random function is called non-Markovian.

The stationary random process $\xi(t)$ satisfies to the *ergodic hypothesis*, if with the probability 1 the time average is equal to the assembly average (on a set of observations)

$$P\left[\langle\xi(t)\rangle_T = M\left[\xi(t)\right]\right] = 1,$$

always supposing that these averages exist. The application of an ergodic hypothesis enables to spot the statistical properties of the process on one member function, that is very important for the practical applications.

1.10.2 *Properties of the correlation function*

(1) Autocorrelation function is an even function:

$$R(\tau) = R(-\tau).$$

(2) Ordinate of a real autocorrelation function does not exceed a variance of a random function:

$$R(0) \geq R(\tau).$$

(3) Multiplication of $[\xi(t) - m_\xi(t)]$ on real function $z(t)$ gives a following inequality

$$\int_a^b \int_a^b z(t_1)z(t_2)R(t_2 - t_1)dt_1dt_2 \geq 0.$$

At sufficiently great value of an interval $\tau = t_2 - t_1$ the diversion of an ordinate of a random function from its mathematical expectation in an instant t_2 becomes practically independent from the value of this deviation in an instant t_1. In this case the function $R(\tau)$, giving value of connection between $\xi(t_2)$ and $\xi(t_1)$, at $\tau \to \infty$ will tend to zero. Most frequently used approximations of autocorrelation functions are submitted in a Fig. 1.46.

(4) The adding to a random function of a nonrandom quantity or a nonrandom function does not change a value of the correlation function. Let us

$$\eta(t) = \xi(t) + \varphi(t),$$

where $\varphi(t)$ is a nonrandom function. Taking into account

$$m_\eta(t) = m_\xi(t) + \varphi(t),$$

we obtain

$$\begin{aligned} R_\eta(t_1, t_2) &= M[(\eta(t_1) - m_\eta(t_1)) \times (\eta(t_2) - m_\eta(t_2))] \\ &= M[(\xi(t_1) - m_\xi(t_1))(\xi(t_2) - m_\xi(t_2))] = R_\xi(t_1, t_2). \end{aligned}$$

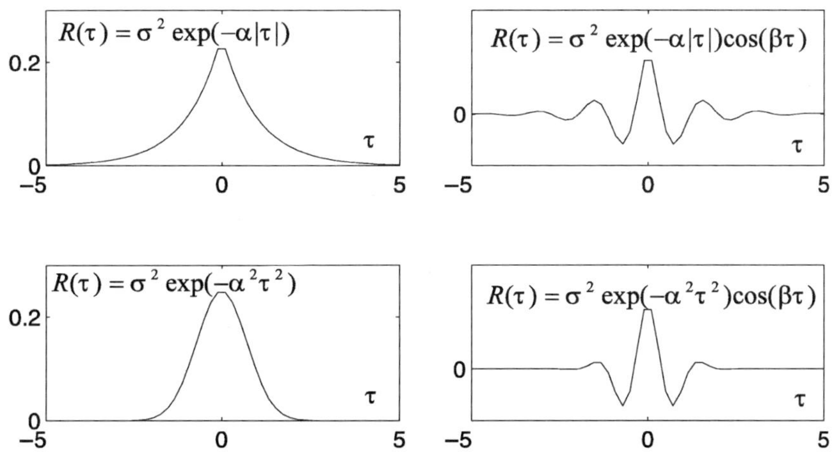

Fig. 1.46 Examples of approximating functions for the autocorrelation function ($\sigma = 0.5$, $\alpha = 1.0$, $\beta = 4.0$).

1.10.3 *Action of the linear operator on a random function*

An linear homogeneous operator L is called the operator which preserves linear operations

(1) $L[c\xi(t)] = cL[\xi(t)]$, where c is a constant.
(2) $L[\xi_1(t) + \xi_2(t)] = L\xi_1(t) + L\xi_2(t)$.

Let us $\xi(t)$ is a random function with the mathematical expectation $m_\xi(t)$ and the correlation function $R_\xi(t_1, t_2)$. Let us $\eta(t)$ is one more correlation function

$$\eta(t) = L[\xi(t)].$$

To find the mathematical expectation

$$m_\eta(t) = M[L[\xi(t)]] = L[M[\xi(t)]] = Lm_\xi(t)$$

and the correlation function

$$\begin{aligned} R_\eta(t_1, t_2) &= M[(\eta(t_1) - m_\eta(t_1))(\eta(t_2) - m_\eta(t_2))] \\ &= M\{(L[\xi(t_1)] - L[m_\xi(t_1)])(L[\xi(t_2)] - L[m_\xi(t_2)])\} \\ &= L_{t_1} L_{t_2} M[(\xi(t_1) - m_\xi(t_1))(\xi(t_2) - m_\xi(t_2))] \\ &= L_{t_1} L_{t_2} R_\xi(t_1, t_2). \end{aligned}$$

1.10.4 *Cross correlation function*

Let us consider a system of two random functions $\xi(t)$, $\eta(t)$ and to define a cross correlation function as following

$$R_{\xi\eta}(t_1, t_2) = M[(\xi(t_1) - m_\xi(t_1))(\eta(t_2) - m_\eta(t_2))].$$

In a case of the real functions we obtain

$$R_{\xi\eta}(t_1,t_2)= R_{\eta\xi}(t_2,t_1),$$

and for the stationary random functions:

$$R_{\xi\eta}(\tau)= R_{\eta\xi}(-\tau).$$

According to the definition of the correlation function we can write

$$R_{\xi\eta}(t_1,t_2) = \iint\limits_{-\infty}^{\infty} (x(t_1) - m_\xi(t_1))(y(t_2) - m_\eta(t_2)) \times f_{\xi\eta}(x,y/t_1,t_2)dxdy.$$

Using the Cauchy-Bunyakovskii inequality we can write

$$\left[\iint\limits_{-\infty}^{\infty} (x(t_1) - m_\xi(t_1))(y(t_2) - m_\eta(t_2))f(x,y/t_1,t_2)dxdy\right]^2$$
$$\leq \iint\limits_{-\infty}^{\infty} (x(t_1) - m_\xi(t_1))^2 f(x,y/t_1,t_2)dxdy \times \iint\limits_{-\infty}^{\infty} (y(t_2) - m_\eta(t_2))^2 f(x,y/t_1,t_2)dxdy,$$

whence follows

$$[R_{\xi\eta}(t_1,t_2)] \leq \sqrt{R_\xi(t_1,t_1)R_\eta(t_2,t_2)},$$

and

$$R_{\xi\eta}(\tau) \leq \sqrt{D_\xi D_\eta}$$

for the stationary case. Thus, for non-dimensional cross correlation function

$$r_{\xi\eta}(t_1,t_2) = \frac{R_{\xi\eta}(t_1,t_2)}{\sqrt{R_\xi(t_1,t_1)}\sqrt{R_\eta(t_2,t_2)}}$$

the following inequalities are carried out

$$-1 \leq r_{\xi\eta}(t_1,t_2) \leq 1.$$

1.10.5 *Wiener–Khinchin theorem and power spectrum*

In the most of applications the function $\xi(t)$, describing the stationary random process, is a real function with the following properties:

(1) There are the finite values of quantities

$$\langle|\xi|\rangle_T = \lim_{T\to\infty}(1/2T)\int_{-T}^{T}|x(t)|dt$$

and

$$\langle|\xi|^2\rangle_T = \lim_{T\to\infty}(1/2T)\int_{-T}^{T}|x(t)|^2 dt.$$

A quantity $\langle|\xi|^2\rangle_T$ is called the average power.

(2) In each finite interval the function $\xi(t)$ is a function with a constrained variation.

The correlation function (with time dependence) for a real random function $\xi(t)$ is defined by the formula

$$R_\xi(\tau) = \lim_{T\to\infty}\frac{1}{2T}\int_{-T}^{T}\xi(t)\xi(t+\tau)dt.$$

This limit exists always, when exists $\langle|\xi|^2\rangle_T$. The function $R_\xi(\tau)$ is real at real $\xi(t)$. For an ergodic random processes the function $R_\xi(\tau)$ with probability 1 coincides with the correlation function $R_\xi(\tau)$ on a set of observations.

The cross correlation function (with time dependence) for two real functions $\xi(t)$, $\eta(t)$ is defined by the formula

$$R_{\xi\eta}(\tau) = \lim_{T\to\infty}\frac{1}{2T}\int_{-T}^{T}\xi(t)\eta(t+\tau)dt.$$

This limit exists always, when exists $\langle|\xi|^2\rangle_T$, $\langle|\eta|^2\rangle_T$. The function $R_{\xi\eta}(\tau)$ is real at real $\xi(t)$, $\eta(t)$. A spectral density $R_\xi(\omega)$ for a function $\xi(t)$ and a cross spectral density $R_{\xi\eta}(\omega)$ for a pair of functions $\xi(t)$ and $\eta(t)$ are determined with the help of Wiener–Khinchin relations (*the Wiener–Khinchin theorem*):

$$R_\xi(\omega) = \int_{-\infty}^{\infty}R_\xi(\tau)\exp\{-i\omega\tau\}d\tau = 2\int_{0}^{\infty}R_\xi(\tau)\cos\omega\tau d\tau,$$

$$R_\xi(\tau) = \frac{1}{2\pi}\int_{-\infty}^{\infty}R_\xi(\omega)\exp\{i\omega\tau\}d\omega = \frac{1}{\pi}\int_{0}^{\infty}R_\xi(\omega)\cos\omega\tau d\omega,$$

$$R_{\xi\eta}(\omega) = \int_{-\infty}^{\infty}R_{\xi\eta}(\tau)\exp\{-i\omega\tau\}d\tau,$$

$$R_{\xi\eta}(\tau) = \frac{1}{2\pi}\int_{-\infty}^{\infty}R_{\xi\eta}(\omega)\exp\{i\omega\tau\}d\omega,$$

where both integrals with cosine in the right hand sides of two first relations are applicable only in case of real $R_\xi(\tau)$. For the real process $\xi(t)$ the spectral density $R_\xi(\omega)$ is an even function

$$R_\xi(\omega) = R_\xi(-\omega).$$

Assuming $\tau = 0$, it is possible to get an expression for a variance of the process through the spectral density

$$D_\xi = R_\xi(0) = \frac{1}{2\pi}\int_{-\infty}^{\infty} R_\xi(\omega)d\omega.$$

Let's consider three examples of an evaluation of the spectral density.

Example 1.21. Let the correlation function is approximated by the expression (see Fig. 1.47)

$$R_\xi(\tau) = \sigma_\xi^2 \exp\{-\alpha|\tau|\}, \tag{1.81}$$

Then its spectrum is given by the expression

$$\begin{aligned} R_\xi(\omega) &= \sigma_\xi^2 \int_{-\infty}^{\infty} \exp\{-i\omega\tau - \alpha|\tau|\}d\tau \\ &= \sigma_\xi^2\left[\int_0^{\infty} \exp\{-(i\omega+\alpha)\tau\}d\tau + \int_{-\infty}^{0} \exp\{(\alpha - i\omega)\tau\}d\tau\right] \\ &= \sigma_\xi^2\left[\frac{1}{\alpha+i\omega} + \frac{1}{\alpha - i\omega}\right] = 2\sigma_\xi^2\frac{\alpha}{\alpha^2+\omega^2}. \end{aligned} \tag{1.82}$$

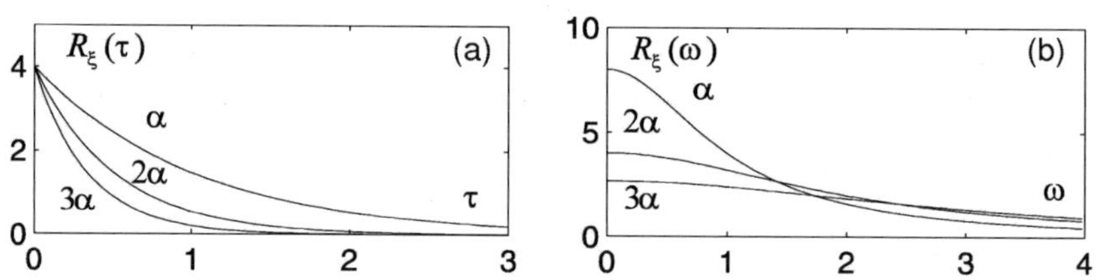

Fig. 1.47 Graphic representation: (a) for correlation function (1.81), (b) for spectrum of the correlation function (1.82), ($\sigma = 2$, $\alpha = 1$).

Example 1.22. The correlation function reads as

$$R_\xi(\tau) = \sigma_\xi^2 \exp\{-\alpha|\tau|\}\cos\omega_0\tau.$$

Taking into account that

$$\cos\omega_0\tau = \frac{1}{2}(\exp\{i\omega_0\tau\} + \exp\{-i\omega_0\tau\}),$$

let evaluate its spectral density:

$$R_\xi(\omega) = \frac{\sigma_\xi^2}{2}\left[\int_{-\infty}^{\infty} \exp\{-i\omega\tau - \alpha|\tau| + i\omega_0\tau\}d\tau\right.$$
$$\left. + \int_{-\infty}^{\infty} \exp\{-i\omega\tau - \alpha|\tau| - i\omega_0\tau\}d\tau\right]$$
$$= \sigma_\xi^2\left[\frac{\alpha}{(\omega-\omega_0)^2+\alpha^2} + \frac{\alpha}{(\omega+\omega_0)^2+\alpha^2}\right].$$

Example 1.23. The correlation function is given by the formula

$$R_\xi(\tau) = \sigma_\xi^2 \exp\{-\alpha^2\tau^2\}\cos\omega_0\tau.$$

Expressing $\cos\omega_0\tau$ through the exponentials, we shall have for the spectral density

$$R_\xi(\omega) = \frac{\sigma_\xi^2}{2}\left[\int_{-\infty}^{\infty} \exp\{-\alpha^2\tau^2 + i\omega_0\tau - i\omega\tau\}d\tau\right.$$
$$\left. + \int_{-\infty}^{\infty} \exp\{-\alpha^2\tau^2 - i\omega_0\tau - i\omega\tau\}d\tau\right].$$

Each of integrals can be reduced to the Poisson integral, if an exponent in the integrand results in the form

$$\left(\alpha\tau + i\frac{\omega}{2\alpha} \mp i\frac{\omega_0}{2\alpha}\right)^2 + \frac{1}{4\alpha^2}(\omega\pm\omega_0)^2.$$

Evaluating these integrals we shall obtain

$$R_\xi(\omega) = \frac{\sigma_\xi^2\sqrt{\pi}}{2\alpha}\left[\exp\left\{\frac{-(\omega+\omega_0)^2}{4\alpha^2}\right\} + \exp\left\{\frac{-(\omega-\omega_0)^2}{4\alpha^2}\right\}\right].$$

1.10.6 *Definition of estimations of the characteristics of random variables*

Let a realization of a random process is registered in the digital form with a step of discreteness Δt — $x(t_j)$, $j = 1,\dots,m$. The estimate of the mathematical expectation is determined by the formula

$$\hat{m}_\xi(t_j) = \frac{1}{n}\sum_{i=1}^{n} x_i(t_j).$$

An estimate for the correlation function is given by the formula

$$\hat{R}_\xi(t_j,t_k) = \frac{1}{n-1}\sum_{i=1}^{n}[(x_i(t_j) - \hat{m}_\xi(t_j))(x_i(t_k) - \hat{m}_\xi(t_k))],$$

or

$$\hat{R}_\xi(t_j, t_k) = \frac{n}{n-1}\sum_{i=1}^{n} x_i(t_j)x_i(t_k) - \frac{n}{n-1}m_\xi(t_j)m_\xi(t_k).$$

In a case of a stationary ergodic random processes an estimate of the mathematical expectation is written as:

$$\hat{m}_\xi = \frac{1}{m}\sum_{j=1}^{m} x(t_j),$$

or, change over integral, we shall obtain

$$\hat{m}_\xi = \frac{1}{T}\int_0^T x(t)dt.$$

For the autocorrelation function the estimates in the discrete and integral forms, taking into account the accepted assumptions, can be written accordingly as follows:

$$\hat{R}_\xi(\tau) = \frac{1}{m-l-1}\sum_{j=1}^{m-l}(x(t_j) - \hat{m}_\xi)(x(t_j+\tau) - \hat{m}_\xi),$$

where $\tau = \Delta tl = t_l$, and

$$\hat{R}_\xi(\tau) = \frac{1}{T-\tau}\int_0^{T-\tau} [x(t) - \hat{m}_\xi][x(t+\tau) - \hat{m}_\xi]dt.$$

Chapter 2

Elements of mathematical statistics

The mathematical statistics studies both statistical description of experimental observations, and checkout of competing mathematical models describing natural physical process. The mathematical statistics is the basic tool of processing of results of physical, including geophysical, experiments with the minimum losses of the information about investigated object.

The basic notion of the mathematical statistics is the *sample* or block of observations $\boldsymbol{\xi} = (x_1, x_2, \ldots, x_n)$ of any quantitative variable. If this variable is a random one, the sample $\boldsymbol{\xi} = (x_1, x_2, \ldots, x_n)$ is a random vector. The number n is called as the *sample size*. The sample is called *repeated*, if the components of the vector $\boldsymbol{\xi}$ are independent and identically distributed. Any function $g(x_1, x_2, \ldots, x_n)$ of observations $(x_1, x_2, \ldots, x_n)$ is called the *statistics*.

The examples of the statistics:

- sample mean $\bar{x} = (1/n) \sum\limits_{i=1}^{n} x_i$;
- sample variance $s^2 = (1/n) \sum\limits_{i=1}^{n} (x_i - \bar{x})^2$;
- sample variance coefficient $v = s/\bar{x}$;
- greatest sample element $x_{\max}$;
- least sample element $x_{\min}$.

The modern level of a geophysical experiment is characterized by the great data flow. So, at realization of seismic operations by the sea for one day 10^{10} bit of the information is filed. Thus the visualization of the data, even without interpretation, is impossible without application of statistical methods of processing of the seismic data.

2.1 The Basic Concepts of the Decision Theory

The results (measurement data) of a real geophysical experiment are always random because of an inevitable presence of errors of measuring. These data represent values of a random vector $\boldsymbol{\xi}$ with the distribution $f(\boldsymbol{x}, \boldsymbol{\theta})$ which is considered partially known. It is supposed, that parameter θ is an element of a set Ω (space of parameters). The observations $\boldsymbol{\xi}$ usually are used for deriving the information about the distribution of $\boldsymbol{\xi}$ or about a value of the parameter θ (vector of parameters $\boldsymbol{\theta}$), on which it depends. The implementation of the statistical analysis is necessary in connection with the indeterminacy of an interpretation of results of observations, arising because of the ignorance of the distribution of $\boldsymbol{\xi}$ and consequently the ignorance of basic elements of the mathematical model of a phenomenon. Therefore there is an indeterminacy in a choice of best "activities" according to results of observations. The problem is to find a rule erecting correspondence between results of observations and reached decision. This correspondence is given by the *decision function* $\delta(x_1, \ldots, x_n)$ (decision rule), which to each sample of experimental data $x_1, \ldots, x_n$ assigns a particular expedient of an activity or a decision $d = \delta(x_1, \ldots, x_n)$, and the domain of the definition of this function is the set X, and a range of values is the set of the possible decisions $\mathcal{D}$. In the decision theory for a choice of the function δ the *loss function* $\varphi(\theta, d)$ is introduced, which describes the losses, bound with a decision making d provided that the distribution of ξ is equal $f(\boldsymbol{x}, \theta)$ and θ is considered as a true parameter.

It is possible to emphasize the following basic elements, which define a solution of problems of the mathematical statistician.

(1) Distribution class $\mathcal{F} = \{f(\boldsymbol{x}, \boldsymbol{\theta}), \boldsymbol{\theta} \in \Omega\}$, in which distribution of $\boldsymbol{\xi}$ is included.
(2) Structure of the space $\mathcal{D}$ of possible decisions d, $d \in D$.
(3) The shapes of a loss function $\varphi(\theta, d)$.

In this connection three various spaces are used. *The observation space* X contains all possible results of observations. *The parameter space* Ω contains all possible values of the parameter θ or vector of parameters $\boldsymbol{\theta} = (\theta_1, \ldots, \theta_n)$. *The decision space* $\mathcal{D}$ contains all possible decisions. The connections between spaces X, D and Ω are shown at Fig. 2.1.

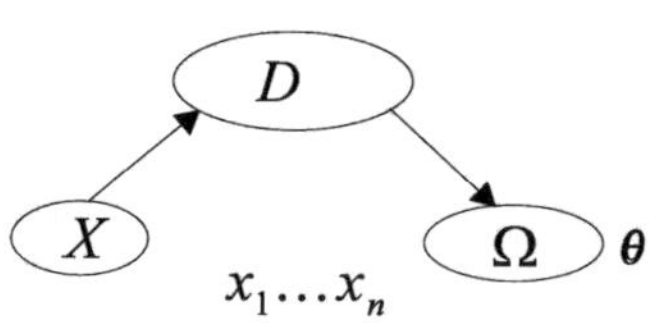

Fig. 2.1 The connection between spaces X, D and Ω.

The loss function $\varphi(\theta, d)$ is a random value, because $d = \delta(\boldsymbol{\xi})$ depends on a random vector $\boldsymbol{\xi}$. It is possible to define an average loss, which arises from the application of the decision rule $d = \delta(\boldsymbol{\xi})$ at major number of repeated observations in an experiment, approximately as the mathematical expectation of the loss function:

$$R_\delta(\theta) = M\left[\varphi(\theta, \delta(\boldsymbol{\xi}))\right] = \int \cdots \int \varphi(\theta, \delta(\boldsymbol{\xi})) f_\xi(\boldsymbol{x}, \theta) dx_1 \ldots dx_n.$$

The function $R_\delta(\theta)$ is called the *risk function* for the decision rule δ.

If we consider θ as a random variable and there is a priori density function $f(\theta)$, then the notion of *Bayes risk* is introduced:

$$r_{f(\theta)}(\delta) = \int R_\delta(\theta) f(\theta) d\theta,$$

or

$$r_{f(\theta)}(\delta) = M_\theta \left[M_\xi \left[\varphi(\theta, \delta(\boldsymbol{\xi})) \right] \right].$$

Let us consider the main elements, which define the statistical problem definition.

2.1.1 *Distribution class of the statistical problem*

Definition of the distribution class $f(x, \theta)$ is based on a priori information about properties of the random observations and an experiment environment.

Example 2.1. If in a fixed time or spatial interval the probability of an occurrence more than one event (seismic wave) in a small time interval is quantity of higher order of smallness, than probability of the occurrence of one event (seismic wave), and if such events are nonoverlapping intervals statistically independent, at these guesses the number of occurrence of events (seismic waves) is described by the Poisson distribution:

$$P(\xi = m) = \frac{a^m}{m!} e^{-a} \quad (m = 1, 2 \ldots).$$

Example 2.2. The study of a microseism shows, that they are a sum of a great number of the independent factors, and the influence of each of them it is very small in comparison with all total, hence, at the realization of conditions of the central limit theorem it is possible to make the guess about a normal distribution of microseism $\boldsymbol{\xi} \in N(\boldsymbol{m}_\xi, D_\xi)$.

Usually type of distribution is chosen from the preliminary analysis of the experiment statement and the physical nature of investigated fields and errors of measurements. The vector of parameters $\boldsymbol{\theta}$ is determined during the solution of a problem.

2.1.2 *The structure of the decision space and the loss function*

The structure of a decision space substantially determines a type of a current statistical problem, i.e. it underlies an expedient of the classification of statistical problems.

Example 2.3. Let us $(x_1, x_2, \ldots, x_n)$ is a sample belonging to a sample population with the distribution $f(\boldsymbol{x}, \theta)$, where θ is an unknown parameter of the distribution, as, for example, in the cases of the considered earlier Poisson distribution $(\theta = a)$

and the Gaussian distribution $\boldsymbol{\theta} = (m_\xi, D_\xi)$, and let us $\gamma = \gamma(\theta)$ is a real function of the parameter.

(A) If it is necessary to determine, exceeds $\gamma(\theta)$ some threshold value γ_0 whether or not, then pick one from the solutions: $d_0 : \ \gamma > \gamma_0$ or $d_1 : \ \gamma \leq \gamma_0$. The loss function $\varphi(\gamma, d)$ is picked usually so that it is equal to zero, if the right decision is chosen, and it is a steadily increasing function $|\gamma - \gamma_0|$ at the incorrect decision. Such statement agrees with a task of a single signal extraction (on a noise background). In this case γ is a relation) signal/noise, γ_0 is the threshold relation signal/noise, d_0 is a presence of the signal, d_1 is an absence of the signal.

(B) If it is necessary to find numerical value of $\gamma(\theta)$, then in this case, the decision d is equal to an estimate of $\hat{\gamma}(\theta)$. The loss function is usually represented as $\varphi(\gamma, d) = w(\gamma)\psi(|d - \gamma|)$, where $\psi(|d - \gamma|)$ is the steadily increasing error function $|\hat{\gamma} - \gamma|$, relevant to a deviation of an estimate from a true value. More often in the estimation theory the square-law loss function is used. So, for $\gamma(\theta) = \theta$, $d = \hat{\theta}$ we shall obtain

$$\varphi(\boldsymbol{\theta} - \hat{\boldsymbol{\theta}}) = (\boldsymbol{\theta} - \hat{\boldsymbol{\theta}})^T W (\boldsymbol{\theta} - \hat{\boldsymbol{\theta}})$$

for the vector parameter and

$$\varphi(\theta - \hat{\theta}) = w(\theta - \hat{\theta})^2$$

for the scalar parameter θ.

(C) The parametrisation of a signal which is received on a noise background can be considered as an example of the point estimation problem: $u_i = A\varphi(t_i - \tau) + \varepsilon_i$ with the vector parameter $\boldsymbol{\theta} = (A, \tau)$.

(D) If at decision-making it is necessary to estimate an interval (α, β), in which with a given probability the estimate $\hat{\gamma}(\theta)$ lays, then in this case we come to a problem of an interval estimation. As a posteriori density $f(\theta/\boldsymbol{x})$ is the most complete characteristic of the parameter θ at a given sample of observations $(x_1, \ldots, x_n)$, the problem of an interval estimation can be formulated as a problem of a choice of an interval (α, β) of values θ, which in the best way describes $f(\theta/\boldsymbol{x})$. In this case the loss function $\varphi(\theta; \alpha, \beta)$ is a function of the parameter θ and the interval. For example, the loss function can look like

$$\varphi(\theta; \alpha, \beta) = \begin{cases} w_1 \ (\beta - \alpha)^2, \text{ if } \theta \in (\alpha, \beta), \\ w_2 \ (\theta - \alpha)^2, \text{ if } \theta < (\alpha, \beta), \\ w_3 \ (\theta - \beta)^2, \text{ if } \theta > (\alpha, \beta), \end{cases}$$

and a posteriori loss is equal

$$M_{\theta/\xi}[\varphi(\theta;\alpha,\beta)] = w_1 \int_{\alpha}^{\beta} (\beta-\alpha)^2 f(\theta/x_1,\ldots,x_n)d\theta + w_2 \int_{\alpha}^{\beta} (\theta-\alpha)^2 f(\theta/x_1,\ldots,x_n)d\theta + w_3 \int_{\alpha}^{\beta} (\theta-\beta)^2 f(\theta/x_1,\ldots,x_n)d\theta.$$

The decision-making is connected with a value of an interval minimizing a posteriori loss.

(E) From the practical point of view, the problem is important, in which it is necessary to implement decision-making about a state of the object and to yield an estimate of desired parameters, i.e. to combine the problems such as A and B in one problem. As an example it is possible to give a problem of the signal extraction with a simultaneous estimate of its parameters (an amplitude and a time of the arrival).

The statistical classification of foregoing problems consists in the following:

The problems such as A with two decisions are stated usually in the terms of hypothesis checking, which should be accepted or is rejected. Such problems belong to a wide class of problems, which is called *the test of hypothesis.*

The problems such as B, in which it is necessary to find numerical value of parameter, belongs to a class of problems of the *point statistical estimation.* Problems such as C, in which the interval covering an estimate of the parameter with the given probability is determined, belongs to a class of problems of the *interval statistical estimation.*

In the statistical theory of the interpretation of geophysical observations the following classification of foregoing problems is used.

(1) Problems such as A, in which it is necessary to determine a quantitative state of geophysical object (there is a seismic signal on a given interval of a seismogram or there is only seismic noise; whether there is a magnetized object of the given shape or it is realization of a noise registered, for example, on a vertical component of a magnetic field etc.), have received a title of problems of *the qualitative interpretation.*

(2) Problems such as B, in which the point parameter estimation of geophysical objects (the determination of an amplitude and a time of the arrival of a seismic signal; a finding of a magnetization and an occurrence depth of the upper edge of a magnetic body; the determination of an electron concentration at given height etc.) is called as problems of *the quantitative interpretation* or *point-quantitative interpretation.*

(3) Problems such as C, in which the interval parameter estimation of geophysical objects (the determination of an interval of the occurrence depth of a reflecting seismic horizon; the finding of an interval of an occurrence depth of the upper edge of the magnetized body; the determination of an interval for quantity of an electron concentration at a given height etc.), is called as problems of *the interval-quantitative interpretation.*
(4) It is necessary to estimate problems such as D, in which both a qualitative state of the object, and parameters, describing it (a signal extraction together with the determination of its parameters – an amplitude and a time of the arrival; the detection of the magnetized body and the determination of its parameters – magnetization, the occurrence depth etc.), is called as problems of *the quantitative-qualitative interpretation* or *by the combined interpretation.*

2.1.3 *Decision rule*

The classical approach to a choice of a decision rule $\delta(\boldsymbol{\xi})$ is based on a risk function $R_\delta(\theta)$. The best decision rule minimize the risk. Let δ_1 and δ_2 are two various decision rules,

$$R_{\delta_1}(\theta) < R_{\delta_2}(\theta) \quad \text{for all } \theta,$$

then δ_1 is the best decision rule in the comparison with δ_2.

Let's consider an example of the analysis of three risk functions, which are connected with three decision rules.

In a given range of values θ the decision rule δ_1 is more preferable than δ_2. While the decision rule δ_3 at some values of the parameter has the least value of a risk function in the comparison with δ_1 and δ_2, and at other values of the parameter it is greater, it is impossible to find a rule, which is the best at all θ, however at fixed θ it is possible to find a sole best rule. The basic indeterminacy in a such choice is arises from an unknown value of θ.

The decision rule δ $(\delta \in \mathcal{D})$ is called *admissible*, if the decision rule δ_1 $(\delta_1 \in \mathcal{D})$ with the inequality

$$R_\delta(\theta) \geq R_{\delta_1}(\theta),$$

which is valid for all θ, does not exist. Thus, if the solving rule is admissible, then in a class of the decision rules $\mathcal{D}$ the decision rule δ_1 does not exist, which is not worse than δ for all θ. Usually admissible estimates meet much, therefore it is necessary to offer a criterion of a choice of the best rule among admissible rules.

Bayes strategy for a choice of a decision rule δ for a priori density $f(\theta)$ is based on a function of a posteriori risk $r_{f(\theta)}(\delta)$. The best, from the point of view of the Bayes strategy, is a decision rule relevant to the minimum of a posteriori loss

$$r^{(B)}_{f(\theta)}(\hat{\delta}) \leq r_{f(\theta)}(\delta)$$

for arbitrary δ, or

$$\hat{\delta} = \arg\min_{\delta \in \mathcal{D}} \int_{\theta} R_{\delta}(\theta) f(\theta) d\theta.$$

The minimum risk $r^{(B)}_{f(\theta)}(\hat{\delta})$ is called the Bayes risk for a priori density $f(\theta)$. To use this criterion, it is necessary to assume, that the parameter θ is a random one and its a priori distribution is known. The knowledge of a such distribution is not always possible in practice. Frequently $f(\theta)$ is interpreted as a weight function, which is determined a degree of a significance of various values of the parameter.

If a priori information about the parameter θ misses, it is expedient to view a maximum of a risk function as its most important characteristic. Among two functions is more preferable one which has the minimum value of a maximum of risk. The decision strategy ensuring a minimum of the maximum risk are termed *the minimax procedure*:

$$\hat{\delta}^{\,\mathrm{minmax}} = \arg_{\delta \in \mathcal{D}} \min \max_{\theta \in \Omega} R_{\delta}(\theta).$$

As the maximum designates most major (on the average) losses, which can arise at the use of the given procedure, the minimax decision in the comparison with other decisions gives the greatest protection against great losses.

Other criterion of a choice of the decision rule is based on the *likelihood function*. Let us consider a sample of n independent observations $x_1, x_2, \ldots, x_n$ of the components of a random vector $\boldsymbol{\xi}$. As $x_1, x_2, \ldots, x_n$ are independent, then a joint density function is equal

$$L(\boldsymbol{x}, \theta) = \prod_{i=1}^{n} f(x_i, \theta).$$

The function $L(\boldsymbol{x}, \theta)$ is called the likelihood function for θ, if we consider it as a function of the parameter θ or the vector parameter $\boldsymbol{\theta}$ at stipulated $x_1, x_2, \ldots, x_n$, which are registered at an experiment.

If the decision is reduced to a problem of the determination of a value of the parameter θ, then the natural estimate of θ can be picked on the basis of the condition of the greatest probability at the given sample of experimental observations $x_1, x_2, \ldots, x_n$, i.e. the estimate maximizes the likelihood function

$$\hat{\Theta}_{\mathrm{MLM}} = \arg\max_{\theta \in \Omega} L(\boldsymbol{x}, \theta).$$

The method of an estimation, when an estimate of the parameter brings the greatest value of the likelihood function is called as the *maximum likelihood method*. Using the likelihood function, it is possible to formulate the procedure of a decision not in the terms of a loss strategy, and in the terms of a winning strategy. The winning is equal to zero ($\alpha(\theta) = 0$), in a case of the improper decision, and it is positive $\alpha(\theta) > 0$, for the proper decision, where θ is a true quantity of the parameter. Then it is advisable to supply the likelihood function with a weight function connected with winnings, and to determine the value of θ, maximizing a product $\alpha(\theta)L(\boldsymbol{x},\theta)$.

If we consider θ as random variable and designate as $\alpha(\theta)$ a priori distribution of the parameter, then a posteriori distribution of parameter the θ is written

$$p(\theta/\boldsymbol{x}) = \frac{1}{p(\boldsymbol{x})}\alpha(\theta)L(\boldsymbol{x},\theta).$$

The maximization of a posteriori probability can be considered as a strategy of the estimation

$$\hat{\Theta}_{\text{MAP}} = \arg\max_{\theta\in\Omega} p(\theta/\boldsymbol{x}) = \arg\max_{\theta\in\Omega} \alpha(\theta)L(\boldsymbol{x},\theta)$$

(the multiplier $p(\boldsymbol{x})$ does not influence a value of the argument at the maximum point of the likelihood function, therefore we have rejected it). The estimate $\hat{\Theta}_{\text{MAP}}$ is called *the maximum a posteriori probability estimate.*

Consider a two-alternative problem. Let ω_0 and ω_1 designate a range θ, for which d_0 and d_1 accordingly are the proper decisions and $\alpha(\theta) = \alpha_0$ at $\theta \in \omega_0$ and $\alpha(\theta) = \alpha_1$ at $\theta \in \omega_1$ are valid. We make a decision d_0, if

$$\alpha_1 \max_{\theta\in\omega_1} L(\boldsymbol{x},\theta) < \alpha_0 \max_{\theta\in\omega_0} L(\boldsymbol{x},\theta),$$

$$\frac{\max_{\theta\in\omega_1} L(\boldsymbol{x},\theta)}{\max_{\theta\in\omega_0} L(\boldsymbol{x},\theta)} < \frac{\alpha_0}{\alpha_1},$$

and the decision d_1, if

$$\alpha_1 \max_{\theta\in\omega_1} L(\boldsymbol{x},\theta) > \alpha_0 \max_{\theta\in\omega_0} L(\boldsymbol{x},\theta),$$

$$\frac{\max_{\theta\in\omega_1} L(\boldsymbol{x},\theta)}{\max_{\theta\in\omega_0} L(\boldsymbol{x},\theta)} > \frac{\alpha_0}{\alpha_1}.$$

A such procedure is called the *likelihood ratio method.* If θ it is possible to consider as a random parameter, then α_0 and α_1 it is possible to interpret as a priori probabilities of belonging of the parameter to the field of ω_0 and ω_1 accordingly.

2.1.4 *Sufficient statistic*

A high profile in the practical sense has such a transformation of observations $x_1,\ldots,x_n$, which gives in cutting number of experimental data, but thus the information about parameter θ (with distribution $f(\boldsymbol{x},\theta)$)or vector of the parameters $\boldsymbol{\theta}$ (with the distribution $f(\boldsymbol{x},\boldsymbol{\theta})$) is completely maintained. Such type of a transformation is connected with the concept of the notion of *the sufficient statistic* $T(x_1,\ldots,x_n)$, which contains the complete information on the parameter θ. The statistics $T(\boldsymbol{x})$ is called sufficient for θ, if the conditional density function $\boldsymbol{x}$ at given T is independent from θ, and T and θ can be multivariate and have various dimensionalities. The necessary and sufficient condition, that $T(\boldsymbol{x})$ is the sufficient statistic for θ, leads to the next writing down of the likelihood function

$$L(\boldsymbol{x},\theta) = g(T,\theta)h(\boldsymbol{x}),$$

where $h(\boldsymbol{x})$ does not depend from θ and $g(T,\theta)$ and it is proportional to $f(T/\theta)$, which is the conditional density of T at the given parameter θ.

Example 2.4. Let $\boldsymbol{\xi}$ is in the accord with to the Gaussian distribution $N(m_\xi, \sigma_\xi^2)$, then the likelihood function

$$L(\boldsymbol{x}, m_\xi, \sigma_\xi^2) = (2\pi\sigma_\xi^2)^{-n/2} \exp\left[-\frac{1}{2\sigma_\xi^2}\sum_{i=1}^{n}(x_i - m_\xi)^2\right]$$
$$= (2\pi\sigma_\xi^2)^{-n/2} \exp\left[-\frac{n}{2\sigma_\xi^2}(\bar{x} - m_\xi)^2\right] \exp\left[-\frac{1}{2\sigma_\xi^2}\sum_{i=1}^{n}(x_i - \bar{x})^2\right],$$

where $\bar{x} = (1/n)\sum_{i=1}^{n} x_i$. If σ_ξ^2 is known, then $\bar{x}$ is in the accord with the Gaussian distribution $N(m_\xi, \sigma_\xi^2/n)$ and the likelihood function can be represented as

$$L(\boldsymbol{x}, m_\xi) = g(\bar{x}, m_\xi) h(\boldsymbol{x}),$$

where

$$g(\bar{x}, m_\xi) = f(\bar{x}, m_\xi) = \left(\frac{n}{2\sigma_\xi^2}\right)^{1/2} \exp\left[-\frac{n}{2\sigma_\xi^2}(\bar{x} - m_\xi)^2\right],$$
$$h(\boldsymbol{x}) = n^{-1/2}(2\pi\sigma_\xi^2)^{-(n-1)/2} \exp\left[-\frac{1}{2\sigma_\xi^2}\sum_{i=1}^{n}(x_i - \bar{x})^2\right].$$

Consequently, $\bar{x}$ is a sufficient statistic for m_ξ. Using the sufficient statistics, we reduce a volume of data, which is necessary for the parameter estimation. It is especially essential in a case of a great information flow.

Example 2.5. Let $x_1, \ldots, x_n$ are an independent components of vector $\boldsymbol{\xi}$ with Gaussian distribution. The likelihood function in this case reads as

$$L(\boldsymbol{x}, m_\xi, \sigma_\xi^2) = (2\pi\sigma_\xi^2)^{-n/2} \exp\left(-\frac{1}{2\sigma_\xi^2}\sum_{i=1}^{n} x_i^2 + \frac{m_\xi}{\sigma_\xi^2}\sum_{i=1}^{n} x_i - \frac{n}{2\sigma_\xi^2} m_\xi^2\right).$$

The factorization criterion shows that the statistics $\boldsymbol{T}(x) = (\sum_i x_i, \sum_i x_i^2)$ is sufficient for σ_ξ^2, m_ξ, i.e. for the estimation of the parameters of the Gaussian distribution is sufficient to have a sum of observations and a sum of squared observations. Thus, it is necessary to store in a computer memory only two numbers — $\sum_i x_i$ and $\sum_i x_i^2$, that can appear essential at a great sample size n.

2.2 Estimate Properties

The wide class properties of important problems of processing of the geophysical information brings to finding of estimates of desired parameters (an amplitude and an arrival time of a seismic signal, a magnetization and an occurrence depth of the magnetized body, an electron concentration at the given height etc.). Because the estimate is a statistics, then to it the concept of the sufficiency surveyed in the Sec. 2.1.4 is applicable. The finding of the estimate is reduced to a choice of a function according to the accepted criterion. The quality of a criterion is determined by the following basic properties: a consistency, an unbiasedness, an efficiency and a robustness.

2.2.1 *Consistency*

The estimate $\hat{\theta}_n$ for the parameter θ is termed *a consistent*, if $\hat{\theta}_n$ converges to a true value of the parameter θ_0 with increasing of number of observations. It is possible to define different types of a competence, using various definition of convergence. In the mathematical statistics at the definition of a consistency a probability convergence is used more often. The estimate $\hat{\theta}_n$ is consistent one, if if there is such N for arbitrary small $\varepsilon > 0$, $\eta > 0$, that

$$P(|\hat{\theta}_n - \theta_0| < \varepsilon) > 1 - \eta$$

for all $n > N$. In this case it is said, that $\hat{\theta}_n$ converges on the probability to θ_0 at the increase n (Fig. 2.2).

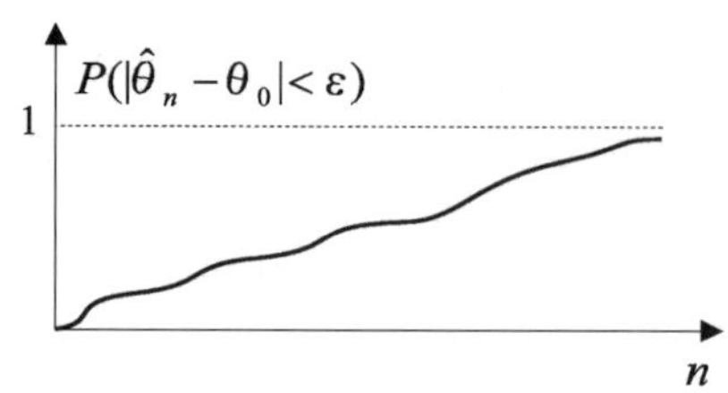

Fig. 2.2 The illustration of the convergence in probability.

2.2.2 *Bias*

A bias $b_n(\hat{\theta})$ of the estimate $\hat{\theta}$, obtained by n observations, means a divergence of the mathematical expectation of the estimate of a true value of the parameter:

$$b_n(\hat{\theta}) = M[\hat{\theta}_n] - \theta_0.$$

An estimate $\hat{\theta}$ is called *unbiased*, if for all n

$$b_n(\hat{\theta}) = 0 \quad \text{or} \quad M[\hat{\theta}] = \theta_0.$$

At Fig. 2.3 a changing of the distribution function at the increasing of a number of observations n (marked by arrow) under conditions of the consistency and the unbias and without a such condition is demonstrated. Among the estimates of the

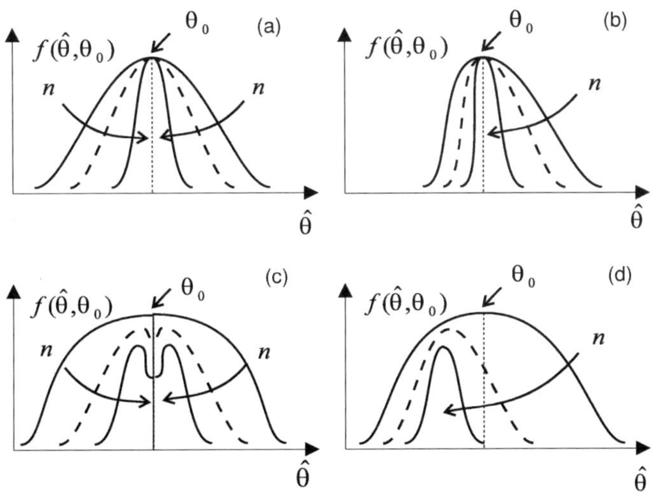

Fig. 2.3 Consistency and unbias. An estimate of a parameter $\hat{\theta}$: unbias and consistent (a); consistent, but bias (b); unbias, but inconsistent (c); inconsistent and unbias (d).

parameter θ, which can be bias, there is an *asymptotically bias* estimate, which satisfies the condition $M[\hat{\theta}_n] \overset{n\to\infty}{\longrightarrow} \theta_0$. The mathematical expectation of such estimates is tend to a right quantity of the parameter.

Example 2.6. Let for a repeated sample $(x_1, x_2, \ldots, x_n)$ of normal observations $\xi_i \in N(m_\xi, \sigma_\xi^2)$ to find an estimate for σ_ξ^2.

Using the relation for a sample variance

$$s^2 = \frac{1}{n}\sum_{i=1}^{n}(x_i - \bar{x})^2 = \frac{1}{n}\sum_{i=1}^{n} x_i^2 - \bar{x}^2,$$

we obtain

$$M(s^2) = \frac{1}{n}\sum_{i=1}^{n} M[x_i^2] - M[\bar{x}^2] = \left(1 - \frac{1}{n}\right)\sigma_\xi^2.$$

So, the estimate of σ_ξ^2 is biased in the direction of smaller values. Obviously, that the unbiased estimate of σ_ξ^2 can be written as

$$S^2 = \frac{n}{n-1}s^2 = \frac{1}{n-1}\sum_{i=1}^{n}(x_i - \bar{x})^2,$$

because

$$M[S^2] = \frac{n}{n-1}M[s^2] = \sigma_\xi^2.$$

The estimate s^2 is an asymptotically unbiased estimate of σ_ξ^2, because $M[s^2] - \sigma_\xi^2 = -\sigma_\xi^2/n \to 0$ at $n \to \infty$.

2.2.3 *Rao–Cramer inequality. Efficiency*

An accuracy of the parameters estimation at a given number of observations and an existence of the lower edge for the variance of the parameter was solved by S. Rao and G. Cramer independently. They have proved the existence of the lower edge for the variance of the estimate (Rao, 1972; Cramer, 1946).

Let us consider a case of a single parameter. Let the likelihood function $L(x_1, x_2, \ldots, x_n, \theta)$ for a sample $x_1, x_2, \ldots, x_n$ depends on the parameter θ, and $\hat{\theta}$ is an unbias estimate $\hat{\theta} = \hat{\theta}(x_1, x_2, \ldots, x_n)$ of θ. The likelihood function satisfies the normality condition

$$\int \cdots \int L(x_1, x_2, \ldots, x_n, \theta) dx_1, \ldots, dx_n = 1. \tag{2.1}$$

Assuming permissible parameter differentiation of integrand, we obtain

$$\int \cdots \int \frac{\partial L}{\partial \theta} dx_1, \ldots, dx_n = 0. \tag{2.2}$$

As the estimate $\hat{\theta}$ ($M\hat{\theta} = \theta$) is unbias, we have

$$\int \cdots \int \hat{\theta} L dx_1, \ldots, dx_n = \theta, \tag{2.3}$$

from which

$$\int \cdots \int \hat{\theta} \frac{\partial L}{\partial \theta} dx_1, \ldots, dx_n = 1. \tag{2.4}$$

Taking into account the relations (2.2) and (2.4), we can write

$$\int \cdots \int [\hat{\theta} - \theta] \frac{\partial L(x_1, \ldots, x_n, \theta)}{\partial \theta} dx_1, \ldots, dx_n = 1, \tag{2.5}$$

or

$$\int \cdots \int [\hat{\theta} - \theta] \left[\frac{1}{L} \frac{\partial L}{\partial \theta} \right] L dx_1, \ldots, dx_n = 1. \tag{2.6}$$

To rewrite the equality (2.6) as

$$M \left[(\hat{\theta} - \theta) \left(\frac{1}{L} \frac{\partial L}{\partial \theta} \right) \right] = 1. \tag{2.7}$$

To denote $\hat{\theta} - \theta = V$ and $(\frac{1}{L})(\partial L / \partial \theta) = W$, and using an analog of the Cauchy–Bunyakovckii inequality

$$M[V^2] \cdot M[W^2] \geq (M[V \cdot W])^2,$$

we obtain

$$M[V^2] \cdot M[W^2] \geq 1 \quad \text{and} \quad M\left[(\hat{\theta} - \theta)^2 \right] \cdot M \left[\frac{1}{L} \frac{\partial L}{\partial \theta} \right]^2 \geq 1. \tag{2.8}$$

To implement a transformation of the expression $M\left[(1/L)(\partial L/\partial\theta)\right]$. Let $L \neq 0$, then

$$\frac{\partial \ln L}{\partial \theta} = \frac{1}{L}\frac{\partial L}{\partial \theta},$$
$$\frac{\partial^2 \ln L}{\partial \theta^2} = -\frac{1}{L^2}\left[\frac{\partial L}{\partial \theta}\right]^2 + \frac{1}{L}\frac{\partial^2 L}{\partial \theta^2}. \tag{2.9}$$

We multiply the expression (2.9) on L and, taking into account

$$M\left[\frac{1}{L}\frac{\partial^2 L}{\partial \theta^2}\right] = \int \cdots \int \frac{\partial^2 L}{\partial \theta^2} dx_1, \ldots, dx_n = 0,$$

we obtain the equality for the mathematical expectations

$$M\left[\frac{1}{L}\frac{\partial L}{\partial \theta}\right]^2 = -M\left[\frac{\partial^2 \ln L}{\partial \theta^2}\right]. \tag{2.10}$$

Taking into account the equality (2.10), the inequality (2.8) can be rewritten as

$$D(\hat{\theta}) = M(\hat{\theta} - \theta)^2 \geq \frac{1}{M[-\partial^2 \ln L/\partial \theta^2]}. \tag{2.11}$$

The inequality (2.11), which determine the lower edge for the variance of the parameter estimation, is called *the Rao–Cramer inequality.* A quantity

$$M\left[-\frac{\partial^2 \ln L}{\partial \theta^2}\right] = M\left[\frac{\partial \ln L}{\partial \theta}\right]^2 = I^{(F)}(\theta) \tag{2.12}$$

is the quantity of Fisher information, which has been introduced in Sec. 1.9.5. So, the quantity of the Fisher information describes the lower edge for the variance of the parameter:

$$D[\hat{\theta}] \geq (I^{(F)}(\theta))^{-1}. \tag{2.13}$$

As mentioned above, this quantity does not depend on the way of the estimate $\hat{\theta}$ and it is the lower edge for the accuracy of an arbitrary estimate So, at given a sample size the accuracy of the estimate has the bottom limit. In a case of the bias estimate with bias $b(\theta)$ the Rao–Cramer inequality reads as

$$D[\hat{\theta}] \geq \frac{1 + db/d\theta}{M\left[\partial \ln L/\partial \theta\right]}, \tag{2.14}$$

where $b(\theta) = M[\hat{\theta} - \theta]$.

In the case of repeated sample is valid

$$L(x_1, \ldots, x_n, \theta) = f_\xi(x_1, \theta) f_\xi(x_2, \theta) \ldots f_\xi(x_n, \theta)$$

$f_\xi(x_i, \theta) \neq 0$, then

$$\frac{\partial^2 \ln L}{\partial \theta^2} = \sum_{i=1}^{n} \frac{\partial^2 \ln f(x_i, \theta)}{\partial \theta^2}$$

and

$$M\left[\frac{\partial^2 \ln L}{\partial\theta^2}\right] = \sum_{i=1}^{n} M\left[\frac{\partial^2 \ln f(x_i,\theta)}{\partial\theta^2}\right] = nM\left[\frac{\partial^2 \ln f(x_i,\theta)}{\partial\theta^2}\right].$$

Since all quantities $M\left[\partial^2 \ln f(x_i,\theta)/\partial\theta^2\right]$ are equal for all i, we obtain

$$D[\hat{\boldsymbol{\theta}}] \geq \frac{1}{nM\left[-\partial^2 \ln f(x,\theta)/\partial\theta^2\right]} = \frac{1}{nI^{(F)}(\theta)}. \tag{2.15}$$

In a case of a few parameters $\boldsymbol{\theta} = (\theta_1, \ldots, \theta_S)$ the Rao–Cramer inequality reads as

$$D[\hat{\boldsymbol{\theta}}] \geq (I^{(F)}(\boldsymbol{\theta}))^{-1}, \tag{2.16}$$

where

$$D[\hat{\boldsymbol{\theta}}] = \|D_{ss'}\|_{s,s'=1}^{S} \quad \text{and} \quad I^{(F)}(\boldsymbol{\theta}) = \|I_{ss'}^{(F)}\|_{s,s'=1}^{S}$$

accordingly the covariance matrix for the estimates of parameters

$$D_{ss'} = M\left[(\hat{\theta}_s - \theta_s)(\hat{\theta}_{s'} - \theta_{s'})\right]$$

and the Fisher information matrix

$$I_{ss'}^{(F)} = M\left[-\frac{\partial^2 \ln L(\boldsymbol{x},\boldsymbol{\theta})}{\partial\theta_s\partial\theta_{s'}}\right] = M\left[\frac{\partial \ln L(\boldsymbol{x},\boldsymbol{\theta})}{\partial\theta_s}\frac{\partial \ln L(\boldsymbol{x},\boldsymbol{\theta})}{\partial\theta_{s'}}\right].$$

An estimate of the parameter $\hat{\theta}$ is called *an efficient estimate*, if its variance reaches own lower edge, i. e. the Rao–Cramer inequality transforms to the equality

$$D[\hat{\theta}] = (I^{(F)}(\theta))^{-1}. \tag{2.17}$$

As the efficiency measure a function

$$e(\hat{\theta}) = (I^{(F)}(\theta)D(\hat{\theta}))^{-1}, \tag{2.18}$$

is used. This function has values from 0 up to 1, moreover for the efficient estimate $e(\hat{\theta}) = 1$. In the case of repeated observations the efficiency can be written as

$$e(\hat{\theta}) = nI^{(F)}(\theta)D[\hat{\theta}]. \tag{2.19}$$

The estimate is called as *an asymptotic efficient estimate*, if $e(\hat{\theta}) \to 1$ at $n \to \infty$.

Consider a case of the vector parameter $\boldsymbol{\theta}$. Let for the vector $\boldsymbol{V} = \|V_s\|_{s=1}^{S}$ an equality $\sum_{s=1}^{S} V_s^2 = 1$ is valid. Let project on it a random vector of estimates of parameters $\hat{\boldsymbol{\theta}} = (\hat{\theta}_1, \ldots, \hat{\theta}_S)$. We get a random variable with a variance $\boldsymbol{V}^T D(\hat{\boldsymbol{\theta}})\boldsymbol{V}$ and in accordance with the Rao–Cramer inequality we can write

$$\boldsymbol{V}^T D(\hat{\boldsymbol{\theta}})\boldsymbol{V} \geq \boldsymbol{V}^T[I^{(F)}(\boldsymbol{\theta})]^{-1}\boldsymbol{V}. \tag{2.20}$$

It means, that the correlation ellipsoid $\boldsymbol{V}^T D^{-1}(\hat{\boldsymbol{\theta}})\boldsymbol{V} = 1$ of the random vector $\hat{\boldsymbol{\theta}}$ envelops a fixed ellipsoid $\boldsymbol{V}^T I^{(F)}(\boldsymbol{\theta})\boldsymbol{V} = 1$. If the correlation ellipsoid of the system of unbias estimates $(\hat{\theta}_1, \ldots, \hat{\theta}_S)$ can be written as $\boldsymbol{V}^T D^{-1}(\hat{\boldsymbol{\theta}})\boldsymbol{V} = 1$ and coincides with an ellipsoid $\boldsymbol{V}^T I^{(F)}(\boldsymbol{\theta})\boldsymbol{V} = 1$, then $(\hat{\theta}_1, \ldots, \hat{\theta}_S)$ is called the system of *jointly efficient* estimates. The system of *jointly asymptotic efficient estimates* is defined as the consequence of the system of estimates $(\hat{\theta}_{1n}, \ldots, \hat{\theta}_{Sn})$ for the parameters $(\hat{\theta}_1, \ldots, \hat{\theta}_S)$, which depend on a sample size n, asymptotically unbias and their correlation ellipsoids $\boldsymbol{V}^T D^{-1}(\hat{\boldsymbol{\theta}})\boldsymbol{V} = 1$, at $n \to \infty$ asymptotically tend to the ellipsoid $\boldsymbol{V}^T I^{(F)}(\boldsymbol{\theta})\boldsymbol{V} = 1$, which gives by the Fisher information matrix.

2.2.4 *Sufficiency*

An estimate $\hat{\theta}(x_1, \ldots, x_n)$ of the parameter θ is called *a sufficient estimate*, if the likelihood function for a sample $\boldsymbol{x}$ can be represented as

$$L(\boldsymbol{x}, \theta) = g(\hat{\theta}, \theta)h(\boldsymbol{x}), \tag{2.21}$$

where $h(\boldsymbol{x})$ does not depend on θ. In this case the conditional distribution of a vector $\boldsymbol{x}$ at fixed $\hat{\theta}$ does not depend on θ, so the sufficient estimate $\hat{\theta}$ of the parameter θ accumulates the complete information on θ, contained in the sample. The effective estimate necessarily is sufficient. The set of joint sufficient estimates $(\hat{\theta}_1, \ldots, \hat{\theta}_S)$ leads to the representation of the likelihood function as a product of functions

$$L(\boldsymbol{x}, \boldsymbol{\theta}) = g(\hat{\boldsymbol{\theta}}, \boldsymbol{\theta})h(\boldsymbol{x}), \tag{2.22}$$

where, as before, $h(\boldsymbol{x})$ does not depend on θ.

2.2.5 *Asymptotic normality*

Let us present n measurements $(x_1, \ldots, x_n)$ of random variables ξ with a density function $f_\xi(x, \theta)$, where θ is an unknown parameter. Let's consider asymptotic properties of the distribution of an estimate, which we shall consider as consistent. Let's take the advantage of the central limit theorem. From this theorem follows, that

$$\frac{1}{n}\sum_{i=1}^{n} g(x_i) \tag{2.23}$$

at an asymptotic bound $(n \to \infty)$ has the normal distribution with the mathematical expectation $M[g(\boldsymbol{\xi})]$ and the variance $D[g(\boldsymbol{\xi})]/n$, always supposing, that $D[g(\boldsymbol{\xi})]$ is a finite value. From the law of large numbers follows, that

$$\frac{1}{n}\sum_{i=1}^{n} g(x_i) \overset{n\to\infty}{\longrightarrow} M[g(\xi)] = \int g(x)f(x, \theta)dx. \tag{2.24}$$

Supposing there is a such function $g(x)$, that the right hand side of the equation (2.24) is a known function $q(\theta)$:

$$M[g(\xi)] = \int g(x)f(x, \theta)dx = q(\theta). \tag{2.25}$$

If the equality (2.25) is valid for the true value θ_0, there is an inverse function q^{-1} for q, determined by the expression

$$q^{-1}[q(\theta_0)] \equiv \theta_0. \tag{2.26}$$

Using the equality (2.24), we obtain

$$\theta_0 = q^{-1}\{M[g(\xi)]\}. \tag{2.27}$$

For a finding of an estimate $\hat{\theta}$ we shall take the advantage of the expression

$$\hat{\theta} = q^{-1}\left\{\frac{1}{n}\sum_{i=1}^{n} g(x_i)\right\}. \tag{2.28}$$

The expression (2.28) it is possible to take Taylor about a point $M[g(\xi)]$:

$$\hat{\theta} = q^{-1}\left\{\frac{1}{n}\sum_{i=1}^{n} g(x_i)\right\} = q^{-1}\{M[g(\xi)]\} + \frac{\partial q^{-1}}{\partial g}\{M[g(\xi)]\}$$
$$\times\left\{\frac{1}{n}\sum_{i=1}^{n} g(x_i) - M[g(\xi)]\right\} + O(1/n).$$

At such order of the accuracy the estimate $\hat{\theta}$ asymptotically distributed under the normal law $N(\theta_0, D(\hat{\theta}))$ with an asymptotic variance

$$D[\hat{\theta}] = M\left[\hat{\theta} - \theta_0)^2\right] = \frac{1}{n}\left[\frac{\partial q^{-1}}{\partial g}\right]^2 D[g],$$

where $D[g]$ is a variance $g(\xi)$.

In the case of a vector of parameters $\boldsymbol{\theta} = (\theta_1, \ldots, \theta_S)$, the functions g and q^{-1} become the vector functions $\boldsymbol{g}$ and $\boldsymbol{q}^{-1}$ accordingly. In this case matrix of the second moments of an asymptotic distribution $\hat{\boldsymbol{\theta}}$ reads as

$$D[\hat{\boldsymbol{\theta}}] = M\left[(\hat{\boldsymbol{\theta}} - \boldsymbol{\theta}_0)(\hat{\boldsymbol{\theta}} - \boldsymbol{\theta}_0)^T\right] = \frac{1}{n}\left[\frac{\partial \boldsymbol{q}^{-1}}{\partial \boldsymbol{g}}\right] D[\boldsymbol{g}]\left[\frac{\partial \boldsymbol{q}^{-1}}{\partial \boldsymbol{g}}\right]^T,$$

where $D[\boldsymbol{g}]$ is a matrix of second moments $\boldsymbol{g}$, $\partial\boldsymbol{q}^{-1}/\partial\boldsymbol{g}$ is a matrix with the elements $\partial\boldsymbol{q}_i^{-1}/\partial g_j$ and $q_i^{-1}(\boldsymbol{g}) = \theta_{0i}$.

2.2.6 *Robustness*

Robust estimation is a such kind of estimation, which does not depend on a type of the distribution or insensitive to the deviation from the guessed distribution. As a rule, at processing the experimental geophysical data the normality of an errors of measurements is supposed. The great outliers lead to the deviation from the normal distribution. The robust procedure allows to exclude the influence of great errors (examples of robust procedures see in the Sec. 6.14).

In addition to the mentioned above desirable properties of estimates of sought for parameters (consistence, unbiasedness, efficiency, normality, robustness), at the practical realization of a method of an estimation the essential factors are computational complexity of a method at the processing of the bulk of the geophysical information and the computation time.

Chapter 3

Models of measurement data

At the design of the geophysical experiment and the interpretation of its data the concept of a model is a basic concept. The success of the prognosis of a geological structure of an explored parcel of the Earth crust substantially depends on the chosen model. In the geophysical literature there are various aspects of the use of the concept of the model. At the first, the notion of the model is connected with the concrete a priori assumptions concerning a structure of one or another geophysical object. For example, the layered model of the medium is a basis of many methods of the processing of the seismic data: build-up of a pseudo-acoustic impedance section with the purpose of the prediction of a geological section; the migration of a seismic time section in the depth section etc. In more complicated geological conditions is possible to use models of an anisotropic layered medium, in which the velocity of the propagation of the seismic waves depends on a direction of the propagation. In a case of a magnetometry and a gravimetry an example of such models can be an object with a priori assigned shape, which creates an anomalous magnetic or a gravitational field. Secondly, the concept of the model includes a priori assumptions about the connection between the observed experimental field and parameters, interesting for the investigator, and conditions of a parcels of the Earth's crust, thus the random character of experimental data should be taken into account. The explored geological object is a complicated multiparameter system, and for its study it is necessary to take all available geological and geophysical information. At the analysis of real objects the principle of a "black box", which guess a complete absence of a priori information is completely unacceptable.

3.1 Additive Model

Let's suppose that the observation of a geophysical field u will be carried out in a Cartesian frame (x, y, z) in discrete points of the space (x_k, y_l, z_m) in instants t_i, where $k = 1, \dots, K$, $l = 1, \dots, L$, $m = 1, \dots, M$, $i = 1, \dots, n$. If a step at measuring is uniform, then $x_k = k\Delta x$, $y_l = l\Delta y$, $z_m = m\Delta z$, $t_i = i\Delta t$, where Δx, Δy, Δz, and Δt are the quantization steps on the space and time accordingly. The important special case of a system of observations are the measuring along a linear profile, when, combining an axis $0x$ with a direction of the profile, we shall obtain for a seismic field $u(x_k, t_i)$ and for a magnetic or a gravitational field we shall obtain $u(x_k)$.

On the basis of the available a priori information an investigator creates the model of the medium, which allows with the use of the physical laws to establish a functional connection between desired parameters of the medium $\boldsymbol{\theta} = (\theta_1, \dots, \theta_S)$, where S is a number of desired parameters, and a model field on a observation plane $f(\boldsymbol{\theta}, x_k, y_l, z_m, t_i)$. Thus, a random component of the model $\varepsilon(x_k, y_l, z_m, t_i)$ is functionally connected with the the model field and the observed field:

$$u(x_k, y_l, z_m, t_i) = \Phi(f(x_k, y_l, z_m, t_i, \boldsymbol{\theta}), \varepsilon(x_k, y_l, z_m, t_i)). \qquad (3.1)$$

The additive model has a greatest prevalence in a practice of the processing of the geophysical data. In a case of this model a functional connection Φ corresponds to a simple superposition of a model field f and a random component of the model ε:

$$u(x_k, y_l, z_m, t_i) = f(x_k, y_l, z_m, t_i, \boldsymbol{\theta}) + \varepsilon(x_k, y_l, z_m, t_i). \qquad (3.2)$$

Such representation points to the independence of sources of a model field and noise. Using the formula (3.2) for a seismic trace, we guess, that f includes the wave field, computed according to the accepted model of the medium, and ε is the contribution of a microseism, a noise of the received chain, geological inhomogeneities unaccounted in a model field. For problems of the magnetometric prospecting and gravimetric prospecting the possible example of a such model can be written as a superposition of a model field f of the object of the known sample shape (for example, layer, bench etc.) and a random field ε, which is caused by errors of observations, the influence of non interpretive sources of a field, and also errors, which is connected with a replacement of the real magnetic object by its physical analog.

The model (3.2) can be obtained from a general model (3.1) by a linearization at a small enough ε. The parameters $\boldsymbol{\theta}$, included in model, depending on a physical statement of a problem can be either unknowns values, or random values, thus the various algorithms of estimations of parameters are used. At the formalization of a model, together with the physical basis, it is necessary to take into account a computing complexity and a practical realizability of the chosen approach. Further the models as with known (but not random), and with random parameters will be used. Let's mark, that the representation about a random character of an experimental material underlies the statistical theory of the interpretation and determines the concrete algorithm and the efficiency of a processing algorithm. It is necessary to underline, that the information substance of the interpretation becomes up to the extremely clear only within the framework of the statistical theory, in which the gained information is determined by a difference of entropies of a priori probability distributions (before interpretation) and a posteriori probability distributions (after interpretation) about a state of the object. So a random character of inferences thus admits which is a direct consequence of a random character of measurements.

3.2 Models of the Quantitative Interpretation

The model (3.2) belongs to the set of models of the quantitative interpretation. The function $f(x_k, y_l, z_m, t_i, \boldsymbol{\theta})$ is supposed to be known and is determined by the physical problem statement. The problem consists in a finding of an estimate of parameters $\hat{\boldsymbol{\theta}}$ on a given experimental material $u(x_k, y_l, z_m, t_i)$, that corresponds to procedure of a point estimation of the parameters in the terms of a mathematical statistics.

As an example of a such model we shall consider a model of a seismic trace:

$$u(t_i) = \sum_{\mu=1}^{m} A_\mu \varphi(t_i - \tau_\mu) + \varepsilon(t_i), \tag{3.3}$$

with the vector of sought for parameters $\boldsymbol{\theta} = \|A_\mu, \tau_\mu\|_{\mu=1}^{M}$. At that time-continuous analog of the model (3.3) can be written as

$$u(t) = \sum_{\mu=1}^{m} A_\mu \varphi(t - \tau_\mu) + \varepsilon(t). \tag{3.4}$$

An example of the seismic trace is represented at Fig. 3.1.

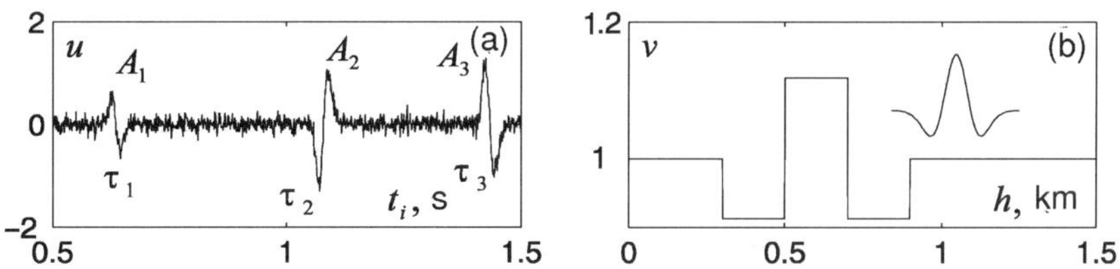

Fig. 3.1 An example of the seismic trace: (a) it is seismic trace with the Gaussian noise ($N(0, 0.1)$); (b) depth dependence of velocity of the wave propagation and the time dependence in a seismic source. Observation point is located near the free surface.

In some problems of the data processing it is expedient to carry out the analysis in the frequency domain. Using the Fourier transform, the model (3.4) is possible to rewrite as

$$u(\omega) = \sum_{\mu=1}^{m} A_\mu \Phi(\omega) \exp\{-i\omega\tau_\mu\} + E(\omega), \tag{3.5}$$

where

$$u(\omega) = \int_{-\infty}^{\infty} u(t) \exp\{-i\omega t\} dt,$$

$$\Phi(\omega) = \int_{-\infty}^{\infty} \varphi(t) \exp\{-i\omega t\} dt,$$

$$E(\omega) = \int_{-\infty}^{\infty} \varepsilon(t) \exp\{-i\omega t\} dt.$$

In the case of the model (3.3) we use the discrete Fourier transform or Z-transform $(Z = e^{-i\omega\Delta t})$:

$$u(z) = \sum_{\mu=1}^{m} A_\mu \Phi(z) z^{\tau_\mu/\Delta t} + E(z),$$

$$u(z) = \sum_{i=0}^{n} u_i z^i, \quad \Phi(z) = \sum_{i=1}^{n} \varphi_i z^i, \quad E(z) = \sum_{i=0}^{n} \varepsilon_i z^i. \tag{3.6}$$

Let's consider an example from a magnetometry. Let a vertical component of a magnetic field $u(x_k)$ is registered along a linear lateral profile in points x_k is caused by a vertical magnetic dipole, then a linear model of the experimental field can be written as

$$u(x_k) = \frac{M(2h^2 - x_k^2)}{(x_k^2 + h^2)^{5/2}} + \varepsilon(x_k), \tag{3.7}$$

where M is a magnetic moment of the dipole, h is an occurrence depth (Fig. 3.2).

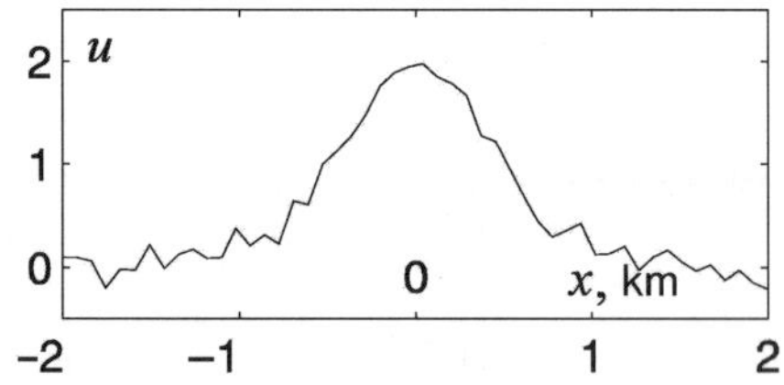

Fig. 3.2 Magnetic field with the Gaussian $(N(0, \sigma^2))$ noise (3.7) along the linear lateral profile $(h = 1,\ M = 1,\ \sigma = 0.1)$.

3.3 Regression Model

At the processing of the geophysical information the interpreter frequently confronts with a problem of the smoothing of the initial or the intermediate data by functions of a given degree of a smoothness. So, for example, at the calculation of the normal-moveout spectrum of velocities with the purpose of the definition of the velocity V_{CDP} the hodograph of a wave is approximated by a parabola of the second order the similar approximation of a hodograph is used also at a static correction. For extraction of the local magnetic or gravity anomalies it is necessary to remove the influence of a regional background, which can be circumscribed by a smooth surface. The similar problem arises at the analysis of structural maps obtained on the data of a seismic exploration, where it is necessary to extract an anticline with a small amplitude on a background of a regional raising. These and many other problems are solved by the regression analysis.

The classical model of a linear regression can be obtained as a particular case of (3.1), with a linear dependence of the model on the desired vector of the parameters:

$$u(x_k, y_l, z_m, t_i) = \sum_{j=0}^{S} \theta_j \psi_j(x_k, y_l, z_m, t_i) + \varepsilon(x_k, y_l, z_m, t_i). \tag{3.8}$$

The formula (3.8) can be considered as a basis of four dimensional space regression, which is applicable at a numerical realization of the ray method. At the processing of the geophysical data at present time three dimensional spatial model and the three dimensional space-time model are widespread:

$$u(x_k, y_k, z_m) = \sum_{j=0}^{S} \theta_j \psi_j(x_k, y_k, z_m) + \varepsilon(x_k, y_k, z_m), \tag{3.9}$$

$$u(x_k, y_k, t_i) = \sum_{j=0}^{S} \theta_j \psi_j(x_k, y_k, t_i) + \varepsilon(x_k, y_k, t_i). \tag{3.10}$$

The formula (3.9) is used for the description of a formation velocity, values of magnetic and gravitational fields at an areal spread on a ground surface. The expression (3.10) is applied as a model of a seismic trace, thus function $\psi(x_k, t_i)$ describes the shape of a signal with unknown moments of the arrival θ_j.

In the geophysical practice one-dimensional models are widespread

$$u(x_k) = \sum_{j=0}^{S} \theta_j \psi_j(x_k) + \varepsilon(x_k), \tag{3.11}$$

$$u(t_i) = \sum_{j=0}^{S} \theta_j \psi_j(t_i) + \varepsilon(t_i). \tag{3.12}$$

The formula (3.11) is used, for example, for the velocity smoothing in CDP or seismic horizon smoothing along a profile of observations. The representation (3.12) is an initial model of a seismic trace for a great number of processing algorithms, such as the predictive deconvolution and adaptive filtration. In the matrix-vector form the formulas (3.11), (3.12) look like

$$\boldsymbol{u} = \psi \boldsymbol{\theta} + \boldsymbol{\varepsilon}. \tag{3.13}$$

For example, in the case of the model (3.11) the vectors $\boldsymbol{u}$ and $\boldsymbol{\varepsilon}$ have the components

$$\boldsymbol{u} = [u_1, \dots u_K], \quad \boldsymbol{\varepsilon} = [\varepsilon_1, \dots, \varepsilon_K],$$

$$\psi = \begin{pmatrix} \psi_{11} & \dots & \psi_{1S} \\ \dots & \dots & \dots \\ \psi_{K1} & \dots & \psi_{KS} \end{pmatrix}.$$

The matrix ψ is called *a design matrix.*

3.4 The Models of Qualitative Interpretation

If as a result of the interpretation a geophysicist can obtain the information only about a qualitative state of a target object, the appropriate model of experimental data is called as a model of qualitative interpretation:

$$\begin{array}{l} u(x_k, y_l, z_m, t_i) \\ \nu = 1, 2, \dots, N \end{array} = \begin{cases} \text{either } f_1(x_k, y_l, z_m, t_i, \boldsymbol{\theta}) + \varepsilon(x_k, y_l, z_m, t_i), \\ \quad\text{or } \quad f_2(x_k, y_l, z_m, t_i, \boldsymbol{\theta}) + \varepsilon(x_k, y_l, z_m, t_i), \\ \quad\dots \qquad\qquad \dots \\ \quad\text{or } \quad f_N(x_k, y_l, z_m, t_i, \boldsymbol{\theta}) + \varepsilon(x_k, y_l, z_m, t_i), \end{cases} \tag{3.14}$$

where $f_\nu(x_k, y_l, z_m, t_i, \boldsymbol{\theta})$ is a field of the idealized object of the investigation with a state with a number $\nu = 1, 2, \dots, N$. As an example we can consider a problem of the detection of a seismic signal with the noise background using an area of a seismic trace

$$u(t_i) = \begin{cases} \text{either } A\varphi(t_i - \tau) + \varepsilon(t_i), \\ \text{or} \qquad \varepsilon(t_i), \end{cases} \tag{3.15}$$

where A, τ are known signal parameters (Fig. 3.3).

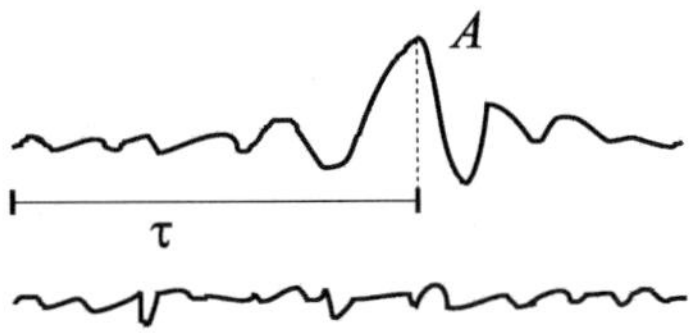

Fig. 3.3 Qualitative interpretation (either "signal and noise" or "noise").

In the case of a choice between two signals we have (see Fig. 3.4)

$$u(t_i) = \begin{cases} \text{either } A_1\varphi(t_i - \tau_1) + \varepsilon(t_i), \\ \text{or} \qquad A_2\varphi(t_i - \tau_2) + \varepsilon(t_i). \end{cases} \tag{3.16}$$

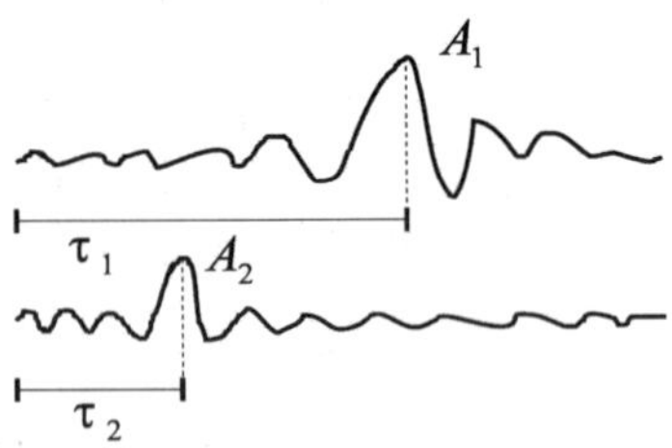

Fig. 3.4 Qualitative interpretation ("signal (A_1) and nose" or "signal (A_2) and noise").

The important class of problems of the resolution of seismic signals is founded on the following model of a seismogram (Fig. 3.5)

$$u(t_i) = \begin{cases} \text{either } A_0\varphi_0(t_i - \tau_0) + \varepsilon(t_i), \\ \text{or} \qquad A_1\varphi_1(t_i - \tau_1) + A_2\varphi_2(t_i - \tau_2) + \varepsilon(t_i). \end{cases} \tag{3.17}$$

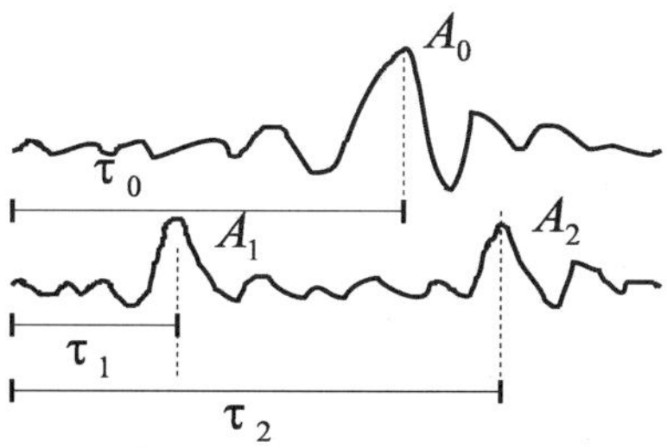

Fig. 3.5 Qualitative interpretation (either "signal (A_1) and noise" or "two signals (A_1 and A_2) and noise").

The statistical hypothesis test will allow to establish (with a given significance level) consistency of hypothesis that on the given interval of a seismogram there is only one wave or, this hypothesis is rejected.

The statistical hypothesis test can be applied also to the spectral representation of a seismogram. In this case the problem consists in the examination of the resolution at the separation of two spectral lines:

$$u(\omega) = \begin{cases} \text{either } A_0\Phi_0(\omega - \omega_0) + E(\omega), \\ \text{or} \quad A_1\Phi_1(\omega - \omega_1) + A_2\Phi_2(\omega - \omega_2) + E(\omega), \end{cases} \tag{3.18}$$

where ω_0, ω_1, ω_2 is a frequency delays, which determine a location of the signals Φ_0, Φ_1, Φ_2 (Fig. 3.6).

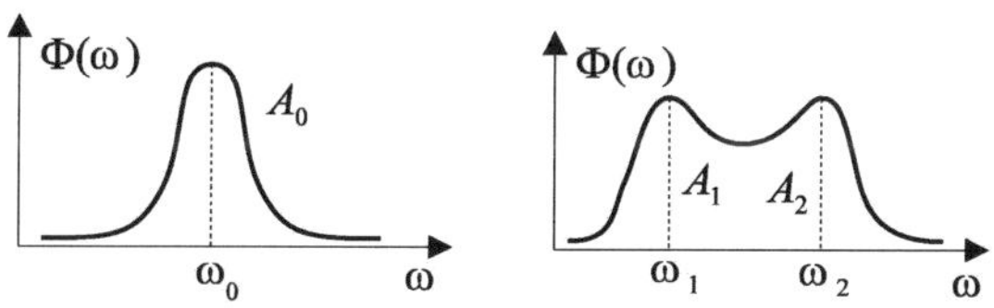

Fig. 3.6 Qualitative interpretation in the frequency domain (either "signal (A_0) and noise" or "two signals (A_1 and A_2) and noise").

3.5 The Models of Qualitative-Quantitative Interpretation

At the solution of composite interpretation problems of a modern geophysics the investigator deals with models of a type

$$\begin{array}{l} u(x_k, y_l, z_m, t_i) \\ \nu = 1, 2, \ldots, N \end{array} = \begin{cases} \text{either} f_1(x_k, y_l, z_m, t_i, \boldsymbol{\theta}_1) + \varepsilon(x_k, y_l, z_m, t_i), \\ \text{or} \quad f_2(x_k, y_l, z_m, t_i, \boldsymbol{\theta}_2) + \varepsilon(x_k, y_l, z_m, t_i), \\ \ldots \qquad\qquad \ldots \\ \text{or} \quad f_N(x_k, y_l, z_m, t_i, \boldsymbol{\theta}_N) + \varepsilon(x_k, y_l, z_m, t_i), \end{cases} \tag{3.19}$$

where the state of an interesting object is characterized by N various deterministic components f_ν ($\nu = 1, \ldots, N$), each of which depends on the vector $\boldsymbol{\theta}_\nu$ of unknown parameters with components $\theta_{\nu s}$ $(s = 1, \ldots, S)$. The problem consists in finding an optimum procedure of a choice among probable states and an estimation of unknown parameters on a given experimental material. The competing states of the object, for which the models of an experimental material are under the construction, should form a complete system of alternative (incompatible) events.

Let's note, that at the formalization of the model of an experimental material it is necessary to determine properties of a probability distribution of a random component of the model, that can be implemented on the basis of available a priori data.

3.6 Random Components of Model and its Properties

The properties of a random component ε in surveyed above models substantially determine the structure of algorithms of a finding of an estimate of required parameters. Therefore at processing of a real geophysical material it is necessary carefully to explore properties of a random component of the model. Let's consider the variants of the representation of a random component having the greatest practical interest at the solution of approximation problems of the geophysical data.

The central place at the determination of properties of a random component is a choice of its distribution law. The geophysical noise field has a composite character and it formed by the great number of the independent factors. So, for example, at the description of a seismic field the random component of the model can conventionally divide on the following basic components.

- Noise, which is connected with the passage of a signal through the seismic channel.
- Microseism is a weak oscillation of the ground determined by a state of a surrounding environment. The level of the microseism is determined by meteorological conditions, the motion of a transport, and also series of the geological factors; for example, the water-saturated sedimentary rock produces 100 – 1000 times greater noise than the solid basement rock.
- The assembly of a partial regular and not enough extended waves formed owing to an inhomogeneity of the medium, an inclusion, a roughness of seismic interfaces, and also the great number of multiple and partial multiple waves with a small energy create the complex interference and irregular field of a seismogram. This field it is permissible to consider as a sampling of some random function. Taking into account a physical nature of the noise, it is possible, basing on the central limit theorem, which proves an asymptotic normality of the sum of independent random quantities to make the guess about the normality of a random component.

It is necessary to note, that at an increasing of a number of observations the majority of distributions tend to the normal distribution (Student's distribution, binomial etc.). If it is known the first two moments — (mathematical expectation ε_0 and variance σ^2) only, the maximum entropy principle, which leads to the solution of a variation problem with constraints, gives a normal distribution.

We write an explicit form for a random vector, which belongs to the normal distribution $\boldsymbol{\varepsilon} = [\varepsilon_1, \ldots, \varepsilon_n]$:

$$p(\boldsymbol{\varepsilon}) = (2\pi)^{-n/2}|R|^{-1/2} \exp\left\{-\frac{1}{2}(\boldsymbol{\varepsilon} - \boldsymbol{\varepsilon}_0)^T R^{-1}(\boldsymbol{\varepsilon} - \boldsymbol{\varepsilon}_0)\right\}, \tag{3.20}$$

where T is a sign of the transposition, $\boldsymbol{\varepsilon}_0$ is a vector of the mathematical expectations R is a covariance matrix with elements $R_{ii'}$, $|R|$ is a determinant of the covariance matrix, R^{-1} is an inverse covariance matrix. The main diagonal of the covariance matrix contains the variances σ_i^2 of the random values ε_i, non diagonal elements $R_{ii'}$ describe a linear relations between values ε_i and $\varepsilon_{i'}$.

Along with the covariance matrix R a correlation matrix with the elements

$$r_{ii'} = \frac{R_{ii'}}{\sqrt{R_{ii}}\sqrt{R_{i'i'}}}.$$

is used often enough. The main diagonal of the correlation matrix contains units and nondiagonal elements $r_{ii'}$ are the correlation coefficients of the values ε_i and $\varepsilon_{i'}$.

Let's consider special cases of the covariance matrix R, which are very important from the practical point of view.

1. The vector $\boldsymbol{\varepsilon}$ has a property of a stationarity in the wide sense, i.e. the values of a vector of the mathematical expectation and the variance are constant, and the correlation matrix depends only on a difference of indexes i and i', thus on diagonals, parallel to the main diagonal, the identical values are placed

$$r = \begin{bmatrix} 1 & r_1 & r_2 & \ldots & r_n \\ \ldots & \ldots & \ldots & \ldots & \ldots \\ r_n & r_{n-1} & r_{n-2} & \ldots & 1 \end{bmatrix}.$$

2. In a case of a noncorrelated random component the covariance matrix can be written as

$$R_{ii'} = \sigma_i^2 \delta_{ii'},$$

where $\delta_{ii'}$ is the Kronecker delta:

$$\delta_{ii'} = \begin{cases} 1, & \text{if } i = i', \\ 0, & \text{if } i \neq i'. \end{cases}$$

At equal accuracy measurements

$$R = \sigma^2 I,$$

where I is an identity matrix. In this case it is easy to calculate the inverse matrix

$$R^{-1} = \frac{1}{\sigma^2} I.$$

Now the density of probabilities is possible to write as

$$p(\boldsymbol{\varepsilon}) = (2\pi\sigma^2)^{-n/2} \exp\left\{-\frac{1}{2\sigma^2}(\boldsymbol{\varepsilon} - \boldsymbol{\varepsilon}_0)^2\right\}.$$

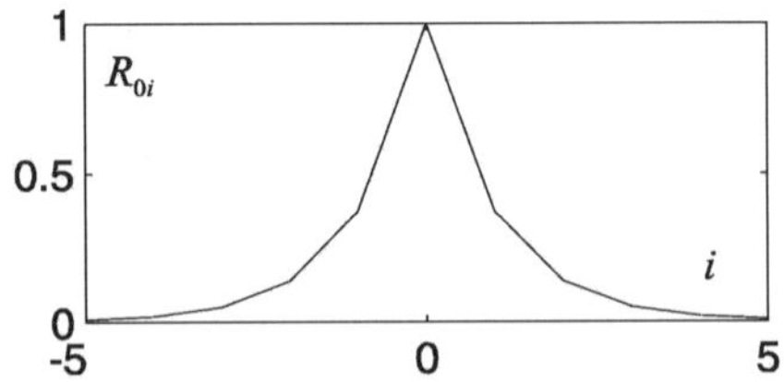

Fig. 3.7 The elements of the covariance matrix in the case the Markovian process (3.21) ($\sigma = 1$, $c = 1$).

3. To consider the *Markovian type of the correlation*. In this case correlation of random quantities ε_i and $\varepsilon_{i'}$ is described by the following function

$$R_{ii'} = \sigma^2 \exp\{-c|i - i'|\} \tag{3.21}$$

(Fig. 3.7). Introducing the designation $\gamma = \exp(-c)$ we obtain the next representation for the matrix R:

$$R = \sigma^2 \begin{bmatrix} 1 & \gamma & \gamma^2 & \dots & \gamma^n \\ \gamma & 1 & \gamma & \dots & \gamma^{n-1} \\ \dots & \dots & \dots & \dots & \dots \\ \gamma^n & \gamma^{n-1} & \gamma^{n-2} & \dots & 1 \end{bmatrix}. \tag{3.22}$$

the determinant of R can be written as

$$|R| = \sigma^{2n}(1 - \gamma^2)^{n-1}. \tag{3.23}$$

An essential advantage of a matrix R is that it can be inverted irrespectively of the quantity n, thus the inverse matrix looks like

$$R^{-1} = \frac{1}{\sigma^2(1 - \gamma^2)} \begin{bmatrix} 1 & -\gamma & 0 & \dots & 0 \\ -\gamma & 1 + \gamma^2 & -\gamma & \dots & 0 \\ \dots & \dots & \dots & \dots & \dots \\ 0 & 0 & 0 & \dots & 1 \end{bmatrix}. \tag{3.24}$$

The formula (3.20) for the density function of ε taking into account (3.23) and (3.24) can be rewritten as

$$p(\boldsymbol{\varepsilon}) = (2\pi\sigma^2)^{-n/2}(1 - \gamma^2)^{(n-1)/2} \exp\{-\frac{1}{2\sigma^2(1 - \gamma^2)}$$
$$\times [(1 + \gamma^2)\sum_i \varepsilon_i^2 - 2\gamma \sum_i \varepsilon_i \varepsilon_{i+1} - \gamma^2(\varepsilon_1^2 + \varepsilon_n^2)]\}.$$

The simplicity of the matrix inversion (3.22) allows to use the Markovian type of a correlation for the approximate description of more complex correlations.

The practical interest, for example, for the model (3.10), represents the following variant of the assignment of the random component ε_i: the mathematical expectation is equal to zero and finite interval of the time correlation

$$\langle \varepsilon_i \rangle, \quad \langle \varepsilon_i \varepsilon_{i-j} \rangle, \quad j \geq n_1,$$

where $\langle\cdot\rangle$ is an averaging operator. In this case for $\boldsymbol{\varepsilon}$ the next representation is valid

$$\boldsymbol{\varepsilon}_i = \boldsymbol{\varepsilon}_i + \sum_{j=1}^{n_1} D_j \boldsymbol{\varepsilon}_{i-j},$$

where $\langle \boldsymbol{\varepsilon}_i \rangle = 0$, $\langle \boldsymbol{\varepsilon}_i \boldsymbol{\varepsilon}_{i-j} \rangle = \sigma^2 \delta_{ij}$ (δ_{ij} is the Kronecker delta function). The matrix D_j can be diagonal as well as nondiagonal. Such description of the random component can be helpful for the sesimogram processing.

Along with stationary or quasi-stationary components of a random variable we shall consider the non-stationary random component using an example of the one-dimensional model

$$u_i = f_i + \sigma_i(f_i)\varepsilon_i,$$

where $f_i = \sum_j \psi_{ij}\theta_j$, ε_i are independent and identically distributed random variables with the mathematical expectation 0 and an unknown symmetric density function. According to a priori receipt about the character of a stationarity choose a view of a functional dependence of $\sigma_i(f_i)$. For example,

$$\sigma_i = \sigma(1 + |f_i|)^\lambda, \quad \sigma_i = \sigma \exp(\lambda f_i), \quad \sigma_i = \sigma |f_i|^\lambda,$$
$$\sigma_i = \sigma(1 + \lambda f_i^2), \quad \sigma_i = \sigma(1 + \lambda f_i^2)^{1/2},$$

where λ, σ are unknown parameters, which estimate together with $\boldsymbol{\theta}$.

Let's note, that the made above assumption about a normality of estimates in practice is not always observed. The presence of sharp outliers in a noise realization can give an essential deviation from the normal distribution, thus statistical procedures basing on the moments of the first and second orders, can lose the optimum properties. Under this conditions it is expedient to make the assumption, that the variance of noise is known only on average. Applying the maximum entropy principle, we shall discover a distribution, which at the accepted assumption ensures the greatest indeterminacy of the system, i.e. it is necessary to find density $\hat{p}(\varepsilon)$, which gives a maximum of an entropy,

$$H_\varepsilon = -\int_{-\infty}^{\infty} \hat{p}(\varepsilon) \ln \hat{p}(\varepsilon) d\varepsilon, \quad \hat{p}(\varepsilon) = \arg\max H_\varepsilon.$$

Let's we have a normal distribution for a random component, but the standard deviation is a random variable with the mathematical expectation σ_0. In this case a desired density function $\hat{p}(\varepsilon)$ can be written as follows

$$\hat{p}(\varepsilon) = \frac{1}{\sqrt{2\pi}} \int_0^\infty \frac{1}{\sigma} \exp\left\{-\frac{(\varepsilon - \varepsilon_0)^2}{2\sigma^2}\right\} p(\sigma) d\sigma = \int_0^\infty l(\varepsilon, \sigma) p(\sigma) d\sigma,$$

where ε_0 is the mathematical expectation of the random variable ε. The density function of σ satisfies the conditions

$$\int_0^\infty \sigma p(\sigma) d\sigma = \sigma_0, \tag{3.25}$$

$$\int_0^\infty p(\sigma)d\sigma = 1. \tag{3.26}$$

For finding the density function $\hat{p}(\varepsilon)$ it is necessary at first to determine $\hat{p}(\sigma)$. The maximum entropy method, taking into consideration the constrains (3.25), (3.26) is reduced to finding of a conditional extremum. For its finding, we shall take an advantage of the Lagrange method. The functional subject to a maximization, looks like

$$L(p(\sigma)) = -\int_{-\infty}^{\infty}\left\{\int_0^\infty l(\varepsilon,\sigma)p(\sigma)d\sigma\left[\ln\int_0^\infty l(\varepsilon,\sigma)p(\sigma)d\sigma\right]\right\}d\varepsilon$$
$$+\lambda_1\left[\int_0^\infty p(\sigma)d\sigma - 1\right] + \lambda_2\left[\int_0^\infty \sigma p(\sigma)d\sigma - \sigma_0\right].$$

The necessary condition leads to the solution of the equation

$$\frac{\partial L}{\partial p} = 0,$$

$$\int_{-\infty}^{\infty} l(\varepsilon,\sigma)\left[\ln\int_0^\infty l(\varepsilon,\sigma)p(\sigma)d\sigma\right]d\varepsilon + 1 - \lambda_1 - \lambda_2\sigma = 0. \tag{3.27}$$

It is easy to test, that the solution of the equation (3.27) with conditions (3.25), (3.26) is a density function

$$p(\sigma) = (2/\lambda^2)\sigma\exp\{-\sigma^2/\lambda^2\}, \tag{3.28}$$

where $\lambda = 2\sigma_0/\sqrt{\pi}$. Then

$$\hat{p}(\varepsilon) = \int_0^\infty l(\varepsilon,\sigma)p(\sigma)d\sigma = \frac{1}{2\alpha}\exp\left\{\frac{-|\varepsilon-\varepsilon_0|}{\alpha}\right\},$$

the density function of the standard deviation σ is equal

$$p(\sigma) = \frac{2\sigma}{\lambda^2}\exp\{-\sigma^2/\lambda^2\}.$$

The obtained function $\hat{p}(\varepsilon)$ is the Laplace density function. "The Tails" of this density function decrease more slowly, than "tails" of the normal density function, i.e. the values of errors ε with the appreciable deviation from the mathematical expectation ε_0 can appear with higher the probability. Therefore the assumption about an error of the distribution of observations ε under the Laplace law is more cautious in comparison with the assumption of the normal law, if the dispersion of the noise is not accurately known or unstable. In a Fig. 3.8 the density functions of the normal and the Laplace distributions are represented.

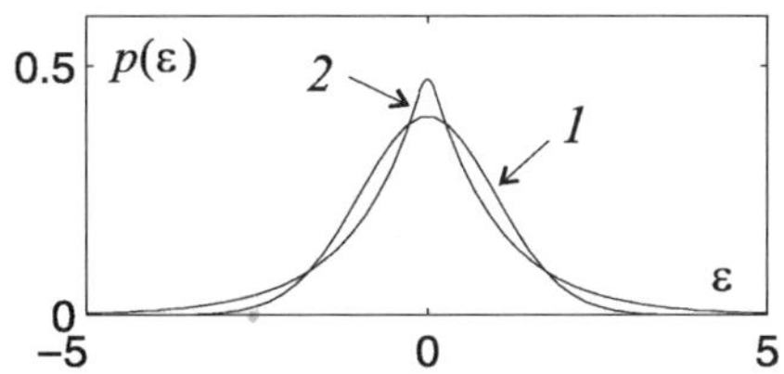

Fig. 3.8 Density functions: *1* — normal ($\varepsilon_0 = 0$, $\sigma = 1$); *2* — Laplace ($\varepsilon_0 = 0$, $\alpha = 1$).

3.7 Model with Random Parameters

Along with the model (3.1), in which the vector of parameters is considered as unknown, but nonrandom, the model with random parameters of signals can be considered

$$u(x_k, t_i) = \sum_{\mu=1}^{M} A_\mu \varphi(t_i - \tau_{\mu k}) + \varepsilon(x_k, t_i), \tag{3.29}$$

where A_μ, $\tau_{\mu k}$ are the random amplitude and moment of the arrival of the wave with a number μ. Let's consider the distribution of these arguments as normal with the mathematical expectation 0 and variance σ_A^2, σ_τ^2. The shape of the signal $\varphi(t_i)$ is supposed the known determined function. A random component ε_{ki} has also a normal distribution with the mathematical expectation 0 and covariance matrix R.

3.8 A Priori Information

Usually at conducting geophysical operations the interpreter has a priori information, given by past geophysical explorations in this or adjoined regions, and also given by the well logging on available deep wells. The engaging of this information at creating algorithms of the interpretation and for making the automatized system of the processing of geophysical data requires its formalization in the pure mathematical shape.

In general case a priori information can be introduced by a system of algebraic equations or inequalities:

$$\boldsymbol{\Phi}(\boldsymbol{\theta}) = \boldsymbol{c}, \tag{3.30}$$

$$\boldsymbol{c}_1 \le \boldsymbol{\Phi}(\boldsymbol{\theta}) \le \boldsymbol{c}_2, \tag{3.31}$$

where $\boldsymbol{\theta}$ is a vector of unknown parameters, $\boldsymbol{\Phi}(\boldsymbol{\theta})$ is a given vector function.

For many practical applications it is possible to be confined to a special case of the formula (3.30), by the system of the linear equations:

$$A\boldsymbol{\theta} = \boldsymbol{V}, \tag{3.32}$$

where A and $\boldsymbol{V}$ are the given matrix and vector of absolute terms. Special case of the formula (3.31) is the linear system of inequalities

$$\boldsymbol{V}_1 \leq A\boldsymbol{\theta} \leq \boldsymbol{V}_2, \tag{3.33}$$

where $\boldsymbol{V}'$ and $\boldsymbol{V}''$ are low and upper bounds of inequalities. As the problem of an estimation, as a rule, consist in finding an extremum in the space of required arguments, the account of a priori information leads to the problem of finding of a conditional extremum, for finding of which one the method of Lagrange multipliers is usually used.

If the model parameters are considered as random, the natural shape of the representation of a priori information is the assignment of a priori probability distribution of required parameters. As such distributions the normal or uniform distributions are used more often. So, if for each of arguments, which one are assumed independent, it is possible to specify a priori intervals of their variation, and the arguments with the equal probability accept any values in these intervals, a priori distribution of arguments will be uniform:

$$p(\boldsymbol{\theta}) = \begin{cases} \prod_{s=1}^{S} 1/(\theta_s'' - \theta_s'), & \theta_s' \leq \theta_s \leq \theta_s'', \\ 0, & \theta_s, \end{cases} \tag{3.34}$$

where θ_s'' and θ_s' are the low bound and upper bound of a priori variation of parameter θ_s correspondingly.

At presence of great number of independent random factors affecting arguments, as a priori distribution the normal distribution is usually chosen, issuing from the central limit theorem. If we shall assume, that the arguments are independent, then a priori distribution will be written as

$$p(\boldsymbol{\theta}) = \prod_{s=1}^{S} (2\pi\sigma_{\theta_s}^2)^{-1/2} \exp\left\{-\frac{1}{2}\sum_{s=1}^{S} \frac{(\theta_s - \langle\theta_s\rangle)^2}{\sigma_{\theta_s}^2}\right\}. \tag{3.35}$$

Let's assume, that the wave amplitude has the normal distribution with the mathematical expectation $\langle A\rangle$ and the variance σ_A^2. While the moment of the arrival time t_0 and the apparent slowness γ have an uniform distribution with intervals $[t_0', t_0'']$ and $[\gamma', \gamma'']$. In this case the joint distribution function will be written as

$$p(A, t_0, \gamma) = \begin{cases} [\sqrt{2\pi}\sigma_A(t_0'' - t_0')(\gamma'' - \gamma')]^{-1} \times & \text{if } t_0' \leq t_0 \leq t_0'' \\ \times \exp\{-(A - \langle A\rangle)^2/2\sigma_A^2\}, & \text{and } \gamma' \leq \gamma \leq \gamma'', \\ 0, & \text{for other } t_0,\ \gamma. \end{cases}$$

It is possible other alternatives of the assignment of a priori information, however described here distributions are used most frequently in practice at constructing algorithms of the interpretation of the seismic data.

The considered models of an experimental material do not encompass all variety of practical situations. But they full enough illustrate conditions and opportunities, connected with the statistical approach. At the solution of a concrete geophysical problem it is necessary with the maximum attention and carefulness to choose the model of an experimental material. The development of an optimum model should be carried out together by geophysicists and mathematicians, it should be included "art", intuition and rich a priori information of the interpreter formalized in the rigorous mathematical form.

Chapter 4

The functional relationships of sounding signal fields and parameters of the medium

Basic part of remote sounding tasks are the mathematical models of the propagation of the sounding signals of the different physical nature in the media under the investigation. In the present chapter the laws of elastic, acoustic, electromagnetic wave field propagation are formulated and the ray energy or homoenergetic neutrons transfer equation are considered.

4.1 Seismology and Seismic Prospecting

As a sounding signal in the seismology and in seismic prospecting the elastic waves, excited by earthquakes (natural sources of the elastic waves), the shots, vibroplates and other artificial sources are used (Aki and Richards, 2002; Jensen *et al.*, 2000; Karus *et al.*, 1986). The elastic wave filed is registered by the seismograph, carries the information on parameters of the medium, which determine the propagation of elastic waves. So, in a Fig. 4.1 the examples of the tomography experiment by elastic waves are represented ((a) — the scheme of the vertical seismic profiling; and the sources are located on a free surface E; (b) — a section of the vertical seismic profiling in a plane $x0z$; (c) – the observed data and the sources on a free surface).

The main aim of the seismology and the seismic prospecting consists in the restoration the Earth internal structure in accordance with signals recorded in the borehole or at the Earth surface. The most effective method of the seismic prospecting is the reflection method, which is prevailing for the exploration activity and the mineral exploration both in the land and the sea. The method is based on the registration of the seismic waves reflected by some interface, which correspond to some variation of the wave impedance of the geologic media. Those interfaces are usually connected with lithological and geologic borders. On carrying out seismic prospecting in the new area, where the seismic prospecting has never been previously applied to, it is primary necessary to build the macromodel of the medium, i.e. to consider the main reflected horizons and to determine the average values of elastic parameters (possibly with their gradients) inside of each geological horizon. When working out the deposits the macromodel is usually known and goal of seis-

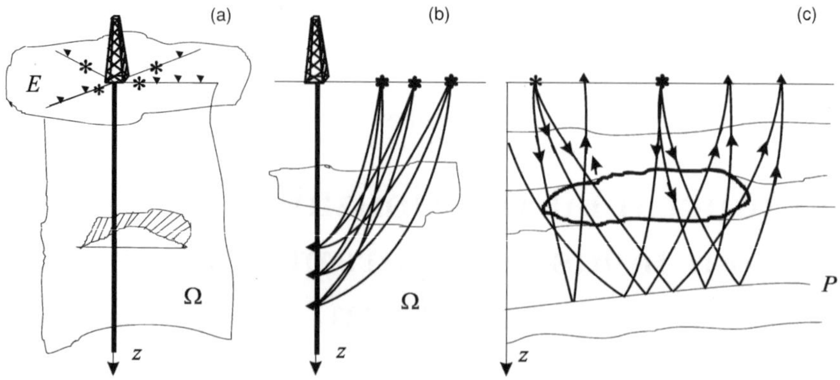

Fig. 4.1 Examples of the tomography sounding by elastic waves.

mic prospecting is to obtain the detailed and reliable information about the elastic properties of the medium.

The solution of problems of the seismology and seismic prospecting depends on the connection between the media properties and the corresponding transformation of the sounding signal, i.e. *the propagation equation.*

The foundation of the seismic waves propagation problem usually is the model of *the ideal elastic body,* built on the linear connection between the *strain* (change of a shape or volume of an elastic body) and *the backmoving elastic force.* This model contains implicitly the statement about so called 'short-range acting' between pieces of solid, i.e. it is assumed that the interacting of two different parts of solid accomplishes only through their interface. It is possible to consider the deformation of the rock far enough from the center of an earthquake or a shotpoint being small. Hence, it is possible to use the model of the ideal elastic body.

Let us write the basic relations of *the mathematical theory of the elasticity.* The displacement could be induced by a volume force.

$$d\mathbf{f} = \mathbf{s}dV,$$

where $\mathbf{s} = \mathbf{s}(x) \triangleq d\mathbf{f}/dV$ is a density of the volume force $\mathbf{f}$ (or the surface force $d\mathbf{f} = \mathbf{t}d\sigma$). Here $\mathbf{t}$ is the density of the surface force, acting to the piece of the surface $d\sigma$ of the considered volume dV, which is a called *stress*:

$$\mathbf{t} = \mathbf{t}(x) \triangleq \frac{d\mathbf{f}}{d\sigma}.$$

Let us denote the value of the displacement of the solid volume dV (with the mass density ρ) from the equilibrium state as $\boldsymbol{\varphi} = \boldsymbol{\varphi}(x)$ and write the equation of the motion

$$\rho\frac{\partial^2\boldsymbol{\varphi}}{\partial t^2} = \mathbf{s} + \mathbf{s}', \tag{4.1}$$

where $\mathbf{s}' = K\boldsymbol{\varphi}$ is a density of the backmoving elastic force, induced by the displacement $\boldsymbol{\varphi}$, which is a result of the integration of the stress vector acting to the

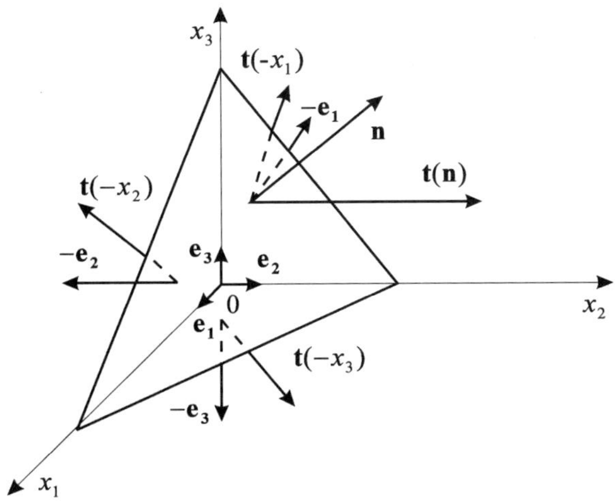

Fig. 4.2 Distribution of the stress on elementary tetrahedron.

chosen volume

$$\iiint \mathbf{s}' dV = \iint \mathbf{t} d\sigma. \tag{4.2}$$

Taking into account that $\mathbf{t} = \mathbf{t}(\mathbf{n})$, where $\mathbf{n}$ is an external (with respect to the considered volume) normal vector:

$$\mathbf{t}(\mathbf{n}) = -\mathbf{t}(-\mathbf{n}).$$

Taking into account the condition

$$\frac{|\iint \mathbf{t}(\mathbf{n})d\sigma|}{\iint d\sigma} \xrightarrow{\Delta V \to 0} 0$$

for the distribution of the stress on an elementary tetrahedron (Fig. 4.2) we obtain:

$$\frac{\mathbf{t}(\mathbf{n})\Delta\sigma_n + \mathbf{t}(-\mathbf{e}_1)\Delta\sigma_1 + \mathbf{t}(-\mathbf{e}_2)\Delta\sigma_2 + \mathbf{t}(\mathbf{e}_3)\Delta\sigma_3}{\Delta\sigma_n + \Delta\sigma_1 + \Delta\sigma_2 + \Delta\sigma_3} \xrightarrow{\Delta V \to 0} 0.$$

We can then derive

$$\mathbf{t}(\mathbf{n}) = \sum_j \mathbf{t}(\mathbf{e}_j)(\mathbf{e}_j \cdot \mathbf{n}) = \sum_j t(\mathbf{e}_j) n_j, \tag{4.3}$$

The relation (4.3) determines the connection between the density of the surface force (stress) $\mathbf{t}$, works into the $\mathbf{n}$ direction, and the density of the surface force works into the unit vector $\mathbf{e}$. The relation (4.3) gives us an opportunity to introduce *a stress tensor*:

$$\mathbf{t} = \hat{\tau}\mathbf{n}, \tag{4.4}$$

i.e. $t_i = \sum_j \tau_{ij} n_j$, where $\tau_{ij} = t_i(\mathbf{e}_j)$ i-th component of the stress vector, induced by the external force of solid pieces laying outside of the outlined volume an acting into plane, the normal to the unit vector $\mathbf{e}_j$, Not that $\hat{\tau} = \hat{\tau}(x)$, i.e. deformed solid is described by means of the stress tensor field. The symmetrical tensor $\hat{\tau}$ it is possible to transform to the diagonal form: $\tau_{ij} = \tau_i^0 \delta_{ij}$, where τ_i^0 are principal stress axises, $\delta_{ij} = 1$, $i = j$, $\delta_{ij} = 0$, $i \neq j$ (δ_{ij} is Kronecker delta). Using the expression (4.4) and applying the Gauss theorem, we obtain

$$\iint \mathbf{t}(\mathbf{n}) d\sigma = \iint \hat{\tau}\mathbf{n} d\sigma = \iiint (\boldsymbol{\nabla} \cdot \hat{\tau}) dV,$$

whence for the density of the volume backmoving elastic force (4.2) is found

$$\mathbf{s}' = (\boldsymbol{\nabla} \cdot \hat{\tau}). \tag{4.5}$$

In the linear elasticity the connection between the stress and the strain (displacement $\boldsymbol{\varphi}$) assumed to be linear (the Hooke law), i.e.

$$\hat{\tau} = \tilde{K}\hat{\varepsilon}, \tag{4.6}$$

where $\hat{\varepsilon}$ is the stress tensor, defined as

$$d\boldsymbol{\varphi} \triangleq \hat{\xi} d\mathbf{x} \triangleq \hat{\varepsilon} d\mathbf{x} - \hat{\eta} d\mathbf{x}. \tag{4.7}$$

In this relation the general form of the tensor $\hat{\xi}$ is represented as a decomposition into two parts: the symmetric tensor

$$\hat{\varepsilon} = \hat{\varepsilon}^+, \qquad \varepsilon_{ij} = \frac{1}{2}\left(\frac{\partial \varphi_i}{\partial x_j} + \frac{\partial \varphi_j}{\partial x_i}\right),$$

and the antisymmetric tensor

$$\hat{\eta} = -\hat{\eta}, \qquad \eta_{ij} = \frac{1}{2}\left(\frac{\partial \varphi_i}{\partial x_j} - \frac{\partial \varphi_j}{\partial x_i}\right).$$

The antisymmetric tensor $\hat{\eta} = (\boldsymbol{\nabla} \times \boldsymbol{\varphi} x)/2$ describes the rotation under the condition $|\partial \varphi_i / \partial x_j| \ll 1$ the tensor K is a tensor of 4-th rank of the elastic modulus

$$\tau_{ij} = \sum_{kl} K_{ijkl} \varepsilon_{kl} \quad (i, j, k, l = 1 \div 3), \tag{4.8}$$

determines the elastic properties of the medium and has got the next properties of the symmetry due to the symmetry of the stress and the strain tensors:

$$K_{ijkl} = K_{jikl} = K_{ijlk} = K_{jilk}, \tag{4.9}$$

i.e. the tensor K has 21 independent components.

The model of the isotropic elastic medium with a local invariance relatively to the rotation is usually used for the solution of the seismic problems. Let us consider the model with coinciding the main axises for the strain and stress tensors. Let, for example, the tension stress is coincided with the direction of the homogeneous rod, which becomes longer longitudinal direction and becomes clenched in transversal direction. If the first Cartesian ort is directed along the rod, then the relation (4.6) can be written as following

$$\varepsilon_{11} = \frac{1}{E}\tau_{11}, \tag{4.10}$$

$$\varepsilon_{22} = \varepsilon_{33} = -\frac{\sigma}{E}\tau_{11}. \qquad (4.11)$$

The coefficient E is called *the Young modulus*, σ is *the Poisson's ratio*, thus E and σ are constants of a material of a rod describing its elastic properties. All reductants of a stress tensor, except for τ_{11}, and three components of the strain tensor ε_{23}, ε_{12}, ε_{13} are equal to zero. If to the stress tensor τ_{11} "to add" the tensors τ_{22} and τ_{33}, then the strain ε_{11} from the expression (4.10) will vary according to the accepted linear model and the requirement (4.11) as follows:

$$\varepsilon_{11} = \frac{1}{E}\tau_{11} - \frac{\sigma}{E}\tau_{22} - \frac{\sigma}{E}\tau_{33}, \qquad (4.12)$$

with $\tau_{23} = 0$ and $\varepsilon_{23} = 0$. We can rewrite the expression (4.12) in the form:

$$\mathrm{E}\varepsilon_{11} = (1+\sigma)\tau_{11} - \sigma(\tau_{11} + \tau_{22} + \tau_{33}),$$

$$\text{or} \quad \mathrm{E}\hat{\varepsilon} = (1+\sigma)\hat{\tau} - \sigma \mathrm{sp}\hat{\tau}\hat{I} \qquad (4.13)$$

($\hat{I}$ is the identity matrix). As it is visible, the connection of tensors of the stress and the strain in case of an isotropic elastic medium is determined by two parameters, for example, by the Young modulus and the Poisson ratio. By convolving tensors in the left-hand and right-hand sides of the formula (4.13) we obtain

$$\mathrm{Esp}\hat{\varepsilon} = (1+\sigma)\mathrm{sp}\hat{\tau} - 3\sigma\mathrm{sp}\hat{\tau} = (1-2\sigma)\mathrm{sp}\hat{\tau},$$

therefore formula (4.13) can be reduced to

$$(1+\sigma)\hat{\tau} = \mathrm{E}\hat{\varepsilon} + \frac{\sigma E}{1-2\sigma}\,\mathrm{sp}\hat{\varepsilon}\hat{I},$$

and

$$\hat{\tau} = \frac{E}{1+\sigma}\hat{\varepsilon} + \frac{\sigma E}{(1+\sigma)(1-2\sigma)}\mathrm{sp}\hat{\varepsilon}\hat{I} \triangleq 2\mu\hat{\varepsilon} + \lambda\mathrm{sp}\hat{\varepsilon}\hat{I}. \qquad (4.14)$$

The parameters $\mu = E/[2(1+\sigma)]$ and $\lambda = \sigma E/[(1+\sigma)(1-2\sigma)]$ are known as *Lame parameters*. The tensor K of the elasticity modulus from the relation (4.6) for the isotropic medium has a form

$$K_{ijkl} = \lambda\delta_{ij}\delta_{kl} + \mu(\delta_{ik}\delta_{jl} + \delta_{il}\delta_{jk}).$$

Let us consider the physical meaning of $\mathrm{sp}\hat{\varepsilon}$ from (4.7):

$$\mathrm{sp}\hat{\varepsilon} = (\nabla\cdot\boldsymbol{\varphi}) \simeq \frac{\prod_i(1+\partial\varphi_i/\partial x_i)dx_i - \prod_i dx_i}{\prod_i dx_i} = \frac{dV - dV_0}{dV_0},$$

i.e. $\mathrm{sp}\hat{\varepsilon}$ is numerically equal to the relative variation of the volume, named *the dilatation*, and it is clear a sense of the Lame parameter λ (λ is a coefficient of the dilatation).

The free elastic energy $F = K\hat{\varepsilon}/2$ (accurate within terms of the second order) expresses through Lame parameters as follows

$$F = \frac{\lambda}{2}(\mathrm{sp}\hat{\varepsilon})^2 + \mu\sum_{\substack{ik\\ i\neq ik}}\varepsilon_{ik}^2.$$

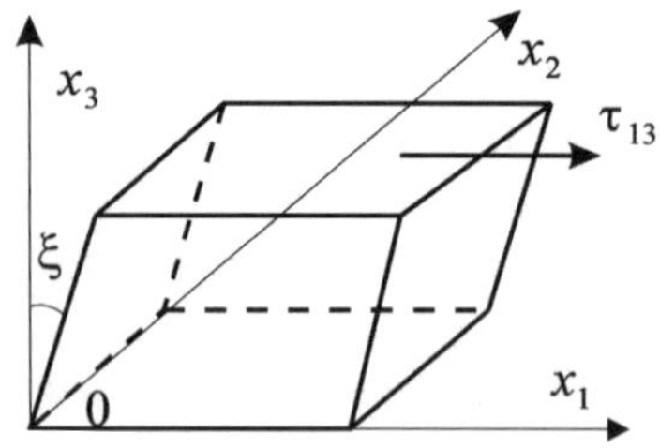

Fig. 4.3 Shear deformation.

Note, that any strain can be introduced as the sum of strains of the simple shear and the uniform pressing ($\hat{\varepsilon} = \text{const}\hat{I}$), we can write

$$\hat{\varepsilon} = \left(\hat{\varepsilon} - \frac{1}{3}\hat{I}\text{sp}\hat{\varepsilon}\right) + \frac{1}{3}\hat{I}\text{sp}\hat{\varepsilon}.$$

The spur of the tensor which was written to round brackets, obviously it is equal to zero, therefore F it is possible to introduce in the form

$$F = \mu \sum_{\substack{ik \\ i \neq ik}} \left(\varepsilon_{ik} - \frac{1}{3}\delta_{ik}\text{sp}\hat{\varepsilon}\right)^2 + \frac{k}{2}(\text{sp}\hat{\varepsilon})^2,$$

where $k = \lambda + 2\mu/3$ is *the uniform pressure modulus.* Let's write the equation of the motion (4.1), using the formula for the backmoving elastic force (4.5) and the Hooke law (see relation (4.6)):

$$\rho\frac{\partial^2 \boldsymbol{\varphi}}{\partial t^2} = \mathbf{s} + \boldsymbol{\nabla} \cdot \tilde{K}\hat{\varepsilon} = \mathbf{s} + \boldsymbol{\nabla} \cdot \tilde{K}\frac{\partial \boldsymbol{\varphi}}{\partial \mathbf{x}}, \tag{4.15}$$

as from formula (4.7) follows

$$\frac{\partial \boldsymbol{\varphi}}{\partial \mathbf{x}} = \hat{\varepsilon} - \hat{\eta} = \hat{\varepsilon} - \left[\frac{1}{2}(\boldsymbol{\nabla} \times \boldsymbol{\varphi})x\right],$$

$$\boldsymbol{\nabla} \cdot [\tilde{K}(\boldsymbol{\nabla} \times \boldsymbol{\varphi})x] \equiv 0.$$

Equation (4.15) is called *the Lame equation*, and the operator

$$L \triangleq \rho\partial_t^2 - \boldsymbol{\nabla} \cdot \tilde{K}_x\hat{\partial}_{\mathbf{x}} \tag{4.16}$$

is called *the Lame operator.* In the operator form (4.15) reads as

$$L\boldsymbol{\varphi} = \mathbf{s}.$$

We shall name an operator Ł including boundary and (linear) initial conditions lain down to $\boldsymbol{\varphi}$ as *a generalized Lame operator*:

$$\hat{L}\boldsymbol{\varphi} = \mathbf{s}, \quad \hat{\Gamma}\boldsymbol{\varphi} = \mathbf{g}.$$

After this we can write

$$\hat{\mathcal{L}} = \left\| \begin{matrix} \hat{L} \\ \hat{\Gamma} \end{matrix} \right\|.$$

The lame equation for the isotropic case:

$$\rho\frac{\partial^2\boldsymbol{\varphi}}{\partial t^2} - (\lambda+\mu)\nabla\nabla\cdot\boldsymbol{\varphi} - \mu\Delta\boldsymbol{\varphi} - \nabla\lambda\nabla\cdot\boldsymbol{\varphi}$$
$$-\nabla\mu\times\nabla\times\boldsymbol{\varphi} - 2(\nabla\mu\cdot\nabla)\boldsymbol{\varphi} = \mathbf{s}. \tag{4.17}$$

In the case *of the uniform isotropic medium* ($\lambda = \text{const}, \mu = \text{const}, \rho = \text{const}$) formula (4.17) can be written down as

$$\rho\frac{\partial^2\boldsymbol{\varphi}}{\partial t^2} - (\lambda+\mu)\nabla\nabla\cdot\boldsymbol{\varphi} - \mu\Delta\boldsymbol{\varphi} = \mathbf{s}. \tag{4.18}$$

Taking into account the equality

$$\nabla\cdot\nabla\boldsymbol{\varphi} = \nabla\nabla\cdot\boldsymbol{\varphi} - \nabla\times\nabla\times\boldsymbol{\varphi},$$

we can write down the equation (4.17) in the form

$$\rho\frac{\partial^2\boldsymbol{\varphi}}{\partial t^2} - (\lambda+2\mu)\nabla\nabla\cdot\boldsymbol{\varphi} - \mu\nabla\times\nabla\times\boldsymbol{\varphi} = \mathbf{s}. \tag{4.19}$$

Using the Helmholtz theorem, the field of the vector $\boldsymbol{\varphi}$ can be introduced as the sum of the potential ($\boldsymbol{\varphi}_p$) and the solenoidal ($\boldsymbol{\varphi}_s$) field of vectors:

$$\boldsymbol{\varphi} \triangleq \boldsymbol{\varphi}_p + \boldsymbol{\varphi}_s = -\nabla\Phi + \nabla\times\mathbf{A}, \tag{4.20}$$

where $\mathbf{A} : \nabla\cdot\mathbf{A} = 0$; Φ is a scalar potential; $\mathbf{A}$ is a vector potential. Substituting $\boldsymbol{\varphi} = \boldsymbol{\varphi}_p + \boldsymbol{\varphi}_s$ and $\mathbf{s} = \mathbf{s}_p + \mathbf{s}_s$ to (4.19), and applying operator $\nabla\times(\cdot)$ and $\nabla\cdot(\cdot)$ to its left hand side and right-hand side, we obtain the system of linear equations for $\boldsymbol{\varphi}_p$:

$$\nabla\times\boldsymbol{\varphi}_p = 0 \Rightarrow \frac{\partial^2}{\partial t^2}\boldsymbol{\varphi}_p - \frac{\lambda+2\mu}{\rho}\nabla\nabla\cdot\boldsymbol{\varphi}_p \triangleq \left(\frac{\partial^2}{\partial t^2} - v_p^2\Delta\right)\boldsymbol{\varphi}_p = \mathbf{s}_p \tag{4.21}$$

and for $\boldsymbol{\varphi}_s$:

$$\nabla\times\boldsymbol{\varphi}_s = 0 \Rightarrow \frac{\partial^2}{\partial t^2}\boldsymbol{\varphi}_s - \frac{\mu}{\rho}\nabla\nabla\cdot\boldsymbol{\varphi}_s \triangleq \left(\frac{\partial^2}{\partial t^2} - v_s^2\Delta\right)\boldsymbol{\varphi}_s = \mathbf{s}_s. \tag{4.22}$$

Substituting the partial solution of the systems (4.21), (4.22) in the form

$$\boldsymbol{\varphi}_{ps} = \mathbf{e}_{ps} f_{ps}(\mathbf{n}\cdot\mathbf{x} - v_{ps}t), \tag{4.23}$$

where $|\mathbf{e}_{p,s}| = 1$, and taking into account that conditions $\nabla\times\boldsymbol{\varphi}_p \equiv 0$ follows $[\mathbf{n}\times\mathbf{e}_p]f' = 0$, and the condition $\nabla\cdot\boldsymbol{\varphi}_s \equiv 0$ leads to $[\mathbf{n}\times\mathbf{e}_s]f' = 0$, we can interpret $\boldsymbol{\varphi}_p$ as *the longitudinal* wave with the velocity of the propagation $v_p = ((\lambda+2\mu)/\rho)^{1/2}$, moreover polarization vector $\mathbf{e}_p$ is codirectional with the unit vector $\mathbf{n}$, which is a normal to the wave front. We can interpret $\boldsymbol{\varphi}_s$ as *a shear wave* with the velocity $v_s = (\mu/\rho)^{1/2}$. The displacement vector $\mathbf{e}_s$ of such wave is orthogonal to direction of its propagation. Because of conditions $\nabla\cdot\boldsymbol{\varphi}_s = \text{sp}\hat{\varepsilon}$ and $\nabla\cdot\boldsymbol{\varphi}_s \equiv 0$ the shear wave $\boldsymbol{\varphi}_s$ is not connected with the change of the volume. At the same time valid $\nabla\times\boldsymbol{\varphi}_p \equiv 0$, the longitudinal wave is a wave of the dilatation. It is necessary to note, that the representation about the longitudinal and shear waves is a consequence of the translation (on time and space) invariance.

Let us consider the case of the uniform isotropic medium (Petrashen, 1978, 1980; Petrashen and Kashtan, 1984). The tensor K of the elastic module is able to be represented in the form of the block symmetric matrix due to symmetry relations (4.8), (4.9):

$$\tilde{\mathbf{K}} \to \mathbf{K}_{\alpha\beta} = \begin{Vmatrix} K_{11} & K_{12} & \dots & K_{16} \\ K_{21} & K_{22} & \dots & K_{26} \\ . & . & \dots & . \\ K_{61} & K_{62} & \dots & K_{66} \end{Vmatrix}, \tag{4.24}$$

$$K_{\alpha\beta} = K_{\beta\alpha},$$

and, if between pairs ij and kl from relations (4.8), (4.9) and α, β respectively the links is introduced, for example: $i, i \leftrightarrow i$; $(2,3) = (3,2) \leftrightarrow 4$; $(3,1) = (1,3) \leftrightarrow 5$; $(1,2) = (2,1) \leftrightarrow 6$. It is seen from (4.24) that tensor $\tilde{K}$ has no more than 21 independent components in general case. By the virtue of translation invariance three degrees of freedom are determined by a choice of the orientation of a coordinate frame, therefore 18 components are independent.

In the notation (4.24) the tensor $\tilde{K}$ of the isotropic medium has a form:

$$\tilde{K}_{\alpha\beta} = \begin{Vmatrix} \lambda + 2\mu & \lambda & \lambda & 0 & 0 & 0 \\ \lambda & \lambda + 2\mu & \lambda & 0 & 0 & 0 \\ \lambda & \lambda & \lambda + 2\mu & 0 & 0 & 0 \\ 0 & 0 & 0 & \mu & 0 & 0 \\ 0 & 0 & 0 & 0 & \mu & 0 \\ 0 & 0 & 0 & 0 & 0 & \mu \end{Vmatrix}.$$

Let us write down the motion equation without the source ($\mathbf{s} = 0$):

$$\rho \frac{\partial^2 \varphi_i}{\partial t^2} = \sum_{k,l,j} K_{ijkl} \frac{\partial}{\partial x_l} \frac{\partial}{\partial x_j} \varphi_k. \tag{4.25}$$

Substituting into (4.25) the partial solution in the form of the plane wave: $\boldsymbol{\varphi} = \mathbf{e} f(t - (\mathbf{p} \cdot x))$, where $\mathbf{p}$ is *a slowness vector* (gradient of eikonal), we obtain the condition:

$$\det \left| \rho \delta_{ik} - \sum_{j,l} K_{ijkl} p_j p_l \right| = 0. \tag{4.26}$$

The general solution of the equation (4.26) for an arbitrary anisotropic medium misses.

In an isotropic medium the equation (4.26) is reduced to three quadratic equations for the slowness components: a longitudinal wave and two shear waves.

4.2 Acoustics of the Ocean

The acoustic method has widely application at learning the water mass of the ocean and the ocean sedimentary cover. During studies were found out, that the severe ocean has a various bottom topography. The main structural elements of the ocean bottom topography are: the abyss plains, middle-ocean mountain ridges, the small shapes (middle-ocean island and shelfs of archipelagoes), large faults. The special attention and detailed learning of these structure factors to take on the special significance in connection with a progress of geodynamics, and also at the solution of applied problems (for example, by looking up of ferro-manganese concretions, oil and gas in a shelf). The presence of a wave guide in the near surface ocean mass, supplying the long-range propagation of acoustic signals, stipulates unique possibilities for the sounding of the vast water areas. The analysis of the underwater propagation of the acoustic signal is founded on the laws of theoretical acoustics (Brekhovskikh, 1960; Brekhovskikh and Godin, 1998; Brekhovskikh and Lysanov, 1991). The propagation of the acoustic wave is described by the wave equation with the appropriate boundary conditions on a surface and at the bottom of the ocean. The problem of the acoustic tomography is reduced to the restoration of a velocity of the propagation of a sound, i.e. to the restoration of the wave operative using values of the acoustic pressure in some space points ("receiver or observation points").

Let's write the equation of the motion of a small liquid piece (dV) in the form of the second Newton's law:

$$\frac{d}{dt}\int_V \rho \mathbf{v} dV = \mathbf{S} + \mathbf{S}', \tag{4.27}$$

where $\mathbf{S} = \int_V \mathbf{s}\rho dV$ is an external body force, by analogy with (4.1); $\mathbf{S}' = \int_V \mathbf{s}' dV$ resultant applied force of the surrounding liquid to the considered volume, i.e

$$\mathbf{S}' = -\oint_{\partial V} p d\boldsymbol{\sigma} = -\int_V \boldsymbol{\nabla} p dV \tag{4.28}$$

(p is a pressure). To write down the total time derivative from the left-hand side of (4.27), let's remember the connection of the substantial derivative (d/dt is a trajectory derivative) and the partial derivative ($\partial/\partial t$):

$$\frac{d}{dt} f(x(t), t) = \frac{\partial}{\partial t} f + \frac{\partial f}{\partial x_1}\frac{dx_1}{dt} + \frac{\partial f}{\partial x_2}\frac{dx_2}{dt} + \frac{\partial f}{\partial x_3}\frac{dx_3}{dt},$$

$$\text{i. e.} \quad \frac{d}{dt} = \frac{\partial}{\partial t} + (\mathbf{v}\cdot\boldsymbol{\nabla}). \tag{4.29}$$

In the next turn, let us make the expression for the substantial time derivative of the integral over a small volume V, moving together with a liquid element i.e. consist of the same particles

$$\frac{d}{dt}\int\limits_V \mathbf{F}dV = \frac{d}{dt}\int\limits_{\tilde{V}} \mathbf{F}Jd\tilde{V} = \int\limits_{\tilde{V}} \frac{d}{dt}(\mathbf{F}J)d\tilde{V}$$
$$= \int\limits_V \frac{1}{J}\frac{d}{dt}(\mathbf{F}J)DV = \int\limits_V \left[\frac{d}{dt}\mathbf{F} + \mathbf{F}\left(\frac{1}{J}\frac{dJ}{dt}\right)\right]dV$$
$$= \int\limits_V \left[\frac{\partial}{\partial t}\mathbf{F} + (\mathbf{v}\cdot\boldsymbol{\nabla})\mathbf{F} + (\boldsymbol{\nabla}\cdot\mathbf{v})\mathbf{F}\right]dV.$$

Here J is an Jacobian and in the last step uses the equality

$$\frac{1}{J}\frac{dJ}{dt} = \boldsymbol{\nabla}\cdot\mathbf{v}.$$

Taking into account (4.29) and (4.28) the left-hand side of (4.27) becomes

$$\int\limits_V \frac{d}{dt}(\rho\mathbf{v})dV = \int\limits_V \left(\mathbf{v}\frac{d}{dt}\rho + \rho\frac{d}{dt}\mathbf{v}\right)dV$$
$$= \int\limits_V \left\{\mathbf{v}\left[\frac{\partial}{\partial t}\rho + (\mathbf{v}\cdot\nabla)\rho + (\nabla\cdot\mathbf{v})\rho\right] + \rho\frac{d}{dt}\mathbf{v}\right\}dV. \tag{4.30}$$

The equation (4.27) contains the functions p, ρ, $\mathbf{v}$. To obtain the close equation it needs to add to the equation (4.27) the relations connecting these functions. Taking into account, that the thermodynamic state equation is generally represented as $p = p(\rho, H)$ (H is an entropy of the system), shall write the complete system of an equation of motion as

$$\frac{d}{dt}(\rho\mathbf{v}) + \boldsymbol{\nabla}p = \mathbf{s}; \tag{4.31}$$

for continuity equation:

$$\frac{\partial}{\partial t}\rho + \boldsymbol{\nabla}\cdot(\rho\mathbf{v}) = 0 \tag{4.32}$$

following the condition

$$\frac{\partial}{\partial t}\int\limits_V \rho dV = -\oint\limits_{\partial V} \rho\mathbf{v}\cdot d\sigma = -\int\limits_V (\boldsymbol{\nabla}\cdot\rho\mathbf{v})dV$$

$$\text{or}\quad \frac{\partial}{\partial t}\rho + \rho(\boldsymbol{\nabla}\cdot\mathbf{v}) + (\mathbf{v}\cdot\boldsymbol{\nabla})\rho = 0);$$

for the state equation:

$$p = p(\rho, H); \tag{4.33}$$

and for the iso-entropy (adiabatic) equation:

$$\frac{\partial}{\partial t}H + \boldsymbol{\nabla}\cdot(H\mathbf{v}) = 0 \tag{4.34}$$

(the liquid has no thermal conductivity).

By virtue of the continuity equation (4.32) the expression in square brackets in the formula (4.30) will be equal to zero, and the equation of motion (4.31) (the Euler equation) becomes

$$\frac{\partial}{\partial t}\mathbf{v} + (\mathbf{v}\cdot\boldsymbol{\nabla})\mathbf{v} + \frac{1}{\rho}\boldsymbol{\nabla}p = \mathbf{s}.$$

The linear equations of motion can be obtained, using a few simplifying assumptions: incompressibilities of liquid ($\boldsymbol{\nabla}\cdot\mathbf{v} = 0$); its barotropies ($p = p(\rho(t))$); small perturbations of the stationary steady-state flow ($p = p_0 + p_1$, $\rho = \rho_0 + \rho_1$, $\mathbf{v} = \mathbf{v}_0 + \mathbf{v}_1$). Let us replace the initial system of equations (4.31) —(4.34) by another one, using above mentioned assumptions

$$\frac{\partial}{\partial t}\mathbf{v} + (\mathbf{v}\cdot\boldsymbol{\nabla})\mathbf{v} + \frac{1}{\rho}\boldsymbol{\nabla}p = \mathbf{s}_0 + \varepsilon\mathbf{s}_1 \tag{4.35}$$

(we shall consider, that the solution of system (4.31) — (4.34): $p_0 = 0, \rho_0 = 0, \mathbf{v}_0 = 0$ corresponds to a hydrostatic state and is aroused by an action of the mass force $\mathbf{s}_0$, for example, by the gravity field $\mathbf{g} \equiv \mathbf{s}_0$, and the presence of perturbations p, ρ, $\mathbf{v}$ is a result of the perturbing force $\varepsilon\mathbf{s}_1$ containing a small parameter). The equation

$$\frac{\partial}{\partial t}\rho + (\mathbf{v}\cdot\boldsymbol{\nabla})\rho = 0 \tag{4.36}$$

corresponds to the equation (4.32) under the condition of incompressibility ($\boldsymbol{\nabla}\cdot\mathbf{v} = 0$). But there would be a system like

$$p = p(\rho). \tag{4.37}$$

Then the system of equations for the perturbed medium becomes (with loss of index 1: $p_1 \Rightarrow p$, $\rho_1 \Rightarrow \rho$, $\mathbf{v}_1 \Rightarrow \mathbf{v}$, $\mathbf{s}_1' \Rightarrow \mathbf{s}$):

$$\frac{\partial}{\partial t}\mathbf{v} - \frac{\rho}{\rho_0^2}\boldsymbol{\nabla}p_0 + \frac{1}{\rho_0}\boldsymbol{\nabla}p = \mathbf{s}; \tag{4.38}$$

$$\frac{\partial}{\partial t}\rho + \rho_0(\boldsymbol{\nabla}\cdot\mathbf{v}) + (\mathbf{v}\cdot\boldsymbol{\nabla})\rho_0 = 0; \tag{4.39}$$

$$p = \frac{dp}{d\rho}\rho \triangleq c^2\rho. \tag{4.40}$$

As a rule, in problems of acoustics of ocean the measured value is the pressure. To derive of the closed equation relative to p we turn down in last system (4.38), (4.39) the terms containing $\boldsymbol{\nabla}p_0$ and $\boldsymbol{\nabla}\rho_0$ (neglect an action of volume forces such as gravity). Differentiate on time the left-hand side of the equation (4.39):

$$\frac{\partial^2}{\partial t^2}\rho + \rho_0(\boldsymbol{\nabla}\cdot\frac{\partial}{\partial t}\mathbf{v}) = 0$$

and the substituting expression from (4.38) instead of $\partial/\partial t\mathbf{v}$, and expression from (4.40) ($\rho = c^{-2}p$) instead of ρ, we obtain

$$\Delta p - \frac{1}{c^2}\frac{\partial^2}{\partial t^2}p = \boldsymbol{\nabla}\cdot\mathbf{s}. \tag{4.41}$$

So, *the wave equation* (4.41) describes a wave propagation aroused by the perturbing force (**s**), thus the velocity of the acoustic wave propagation is $c = (\partial p/\partial \rho)^{1/2}$.

Let's note, that by using the known solution (4.41) $p = p(x,t)$ it is possible by the integration of the equation (4.38) at $\nabla p_0 = 0$ to obtain a field of velocities $\mathbf{v} = \mathbf{v}(x,t)$:

$$\mathbf{v}(x,t) = \mathbf{v}(x,t_0) - \frac{1}{\rho_0}\nabla \int_{t_0}^{t} p(x,\tau)d\tau.$$

The law of the propagation of an acoustic signal (4.41) can be introduced now in the operator form, similarly to the law of elastic waves propagation (4.16): $L\varphi = \mathbf{s}$, here $L \Rightarrow \Delta - [c^2(x)]^{-1}\partial_t^2$, $\varphi \Rightarrow p$, $s \Rightarrow \nabla \cdot \mathbf{s}$.

4.3 Wave Electromagnetic Fields in Geoelectrics and Ionospheric Sounding

The electromagnetic methods of the prospecting are based on the influence of the medium parameters on the propagation of electromagnetic fields. The primary electromagnetic fields can be excited, passing an alternating current through a loop composed of many coils of a wire, or large loop. An occurrence of secondary electromagnetic fields is an echo of the medium. The arising fields can be registered with the help of alternating currents induced by them in a receiving loop under influencing of an electromagnetic induction. The primary electromagnetic field is diffused from a generating loop to the receiver loop both above and under the ground surface. These fields at the propagating in the presence of a homogeneous medium have a small difference. But at the presence of a conductive body of a magnetic component of the electromagnetic field, penetrating on the medium, induces in the body the eddy currents. The eddy currents generate a natural secondary electromagnetic field, coming to a measuring device, which one fixes the signals of primary and secondary fields. These signals differ both on a phase, and on an amplitude from a signal produced by only the primary field. Such distinctions between transmitted and received signals of electromagnetic fields are conditioned by the presence of a conductor and carry the information on its geometrical and electric properties (Zhdanov *et al.*, 1988).

The anomalous areas with a high conductivity produce strong secondary electromagnetic fields. Since 1970 years for electromagnetic sounding of the Earth the heavy-lift impulsive hydromagnetics oscillators are used. One hydromagnetics oscillator, allows geophysical mapping in vast territory and to reach penetration depths about the first tens kilometers in areas with high resistivity.

The exploring of the the ionosphere by means of electromagnetic methods has also a great significance. With the help of these methods the basic patterns of relationships of the structure of ionosphere are found.

The mathematical model of the electromagnetic signal propagation can be derived on the basis of the Maxwell equations. Let us denote the basic physical laws underlie the Maxwell system of equations.

From *the Coulomb law*

$$\mathbf{E} = (q/|\mathbf{r}|^3)\mathbf{r},$$

using the superposition principle ($\mathbf{E}$ is a vector of an electric intensity; q is a quantity of an electric charge) we can obtain:

$$\oint_{\partial V} \mathbf{E} \cdot d\boldsymbol{\sigma} = 4\pi \int_V q dV. \tag{4.42}$$

Applying the Gauss theorem

$$\oint_{\partial V} \mathbf{E} \cdot d\boldsymbol{\sigma} = \int_V \boldsymbol{\nabla} \cdot \mathbf{E} dV,$$

we can write down the relation of (4.42) in the differential form:

$$\boldsymbol{\nabla} \cdot \mathbf{E} = 4\pi q. \tag{4.43}$$

The experimental confirmation of magnetic charges being absent (H is a curl field):

$$\oint \mathbf{H} \cdot d\boldsymbol{\sigma} \equiv 0, \tag{4.44}$$

and its differential representation:

$$\boldsymbol{\nabla} \cdot \mathbf{H} = 0. \tag{4.45}$$

The *generalized Faraday law* (connection of an electromotive force in a loop subtending a surface S, with the variation of a magnetic flux through this surface); in a Fig. 4.4 the differentials of vector quantities in the Stokes formula are shown:

$$\oint \mathbf{E} \cdot d\mathbf{l} = -\frac{1}{c}\frac{\partial}{\partial t} \int \mathbf{H} \cdot d\sigma. \tag{4.46}$$

In accordance with the Stocks theorem $\oint \mathbf{E} \cdot d\mathbf{l} = \int (\boldsymbol{\nabla} \times \mathbf{E}) \cdot d\boldsymbol{\sigma}$, and we can overwrite

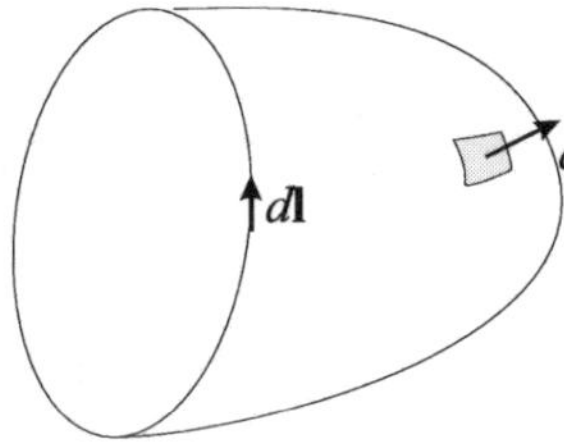

Fig. 4.4 The differentials of vector quantities from the Stokes formula.

(4.46) in the next

$$\int (\nabla \times \mathbf{E}) \cdot d\boldsymbol{\sigma} = -\frac{1}{c}\frac{\partial}{\partial t}\int \mathbf{H} \cdot d\boldsymbol{\sigma}, \tag{4.47}$$

and the differential form of the above expression is:

$$\boldsymbol{\nabla} \times \mathbf{E} = -\frac{1}{c}\frac{\partial}{\partial t}\mathbf{H}. \tag{4.48}$$

The *Maxwell hypothesis* consists in the assumption, that the circulation of a magnetic field is connected with the variation of the flux (by analogy to the formula (4.47)):

$$\boldsymbol{\nabla} \times \mathbf{H} = \frac{1}{c}\frac{\partial}{\partial t}\mathbf{E}. \tag{4.49}$$

The Oersted law (the connection of the charger's motion with the originating of a magnetic field in a loop which enclose a current):

$$\oint \mathbf{H} \cdot d\mathbf{l} = \frac{4\pi}{c}\int \mathbf{j} \cdot d\boldsymbol{\sigma} = \frac{4\pi}{c}J, \tag{4.50}$$

and its differential shape:

$$\boldsymbol{\nabla} \times \mathbf{H} = \frac{4\pi}{c}\mathbf{j}. \tag{4.51}$$

If to accept the Maxwell hypothesis, the analysis of expressions (4.49) and (4.51) demonstrates, that the circulation of a magnetic field is connected as to currents $\mathbf{j}$, and with the variation of $\mathbf{E}$ ("displacement current")

$$\nabla \times \mathbf{H} = \frac{1}{c}\frac{\partial}{\partial t}\mathbf{E} + \frac{4\pi}{c}\mathbf{j}. \tag{4.52}$$

Let's note, that the Maxwell hypothesis (4.49) can be validated with applying of a continuity equation ("electric charges do not arise and do not disappear"), i.e. the variation of a charge inside the chosen volume V is connected only with a charge stream through a covered surface of V:

$$\frac{\partial}{\partial t}\int\limits_V q dV = -\oint\limits_{\partial V}(q\mathbf{v})d\boldsymbol{\sigma} \triangleq -\int\limits_{\partial V}\mathbf{j} \cdot d\boldsymbol{\sigma} = -\int\limits_V \boldsymbol{\nabla} \cdot \mathbf{j} dV, \tag{4.53}$$

and in the differential form:

$$\frac{\partial}{\partial t}q + \boldsymbol{\nabla} \cdot \mathbf{j} = 0. \tag{4.54}$$

If to use for the connection of the circulation of a magnetic field and a current only the Oersted law (4.50), that applying the operator of the divergence to the right and left-hand sides of this expression, we write down:

$$\boldsymbol{\nabla} \cdot \boldsymbol{\nabla} \times \mathbf{H} = \frac{4\pi}{c}\boldsymbol{\nabla} \cdot \mathbf{j}.$$

Here the left-hand side of the equation is identically equal to zero, and the right-hand side is not equal to zero. In accordance with the equation of the continuity

(4.54), we should add the expression $\partial/\partial t\, q$ to the right-hand side of this equation. So we have

$$\boldsymbol{\nabla}\cdot\boldsymbol{\nabla}\times\mathbf{H}=\frac{4\pi}{c}\boldsymbol{\nabla}\cdot\mathbf{j}+\frac{\partial}{\partial t}q=\frac{4\pi}{c}\boldsymbol{\nabla}\cdot\mathbf{j}+\frac{\partial}{\partial t}\frac{1}{4\pi}\boldsymbol{\nabla}\cdot\mathbf{E}.$$

Here we have used the Coulomb law (4.42) for the latest transformation. The most simple dynamic equations of electromagnetic fields can be introduced with the use of the *field potential.* Owing to (4.45) and the identity $\boldsymbol{\nabla}\cdot\boldsymbol{\nabla}\times\mathbf{A}\equiv 0$, $\forall\mathbf{A}$, vector $\mathbf{H}$ can always be represented as:

$$\mathbf{H}=\boldsymbol{\nabla}\times\mathbf{A}, \tag{4.55}$$

where $\mathbf{A}$ is the *vector potential.* When $\mathbf{A}$ is used, (4.48) has a form

$$\boldsymbol{\nabla}\times\mathbf{E}=-\frac{1}{c}\left(\boldsymbol{\nabla}\times\frac{\partial}{\partial t}\mathbf{A}\right), \tag{4.56}$$

i.e.

$$\boldsymbol{\nabla}\times\left(\mathbf{E}+\frac{1}{c}\frac{\partial}{\partial t}\mathbf{A}\right)=0.$$

Taking into account the last equality and the identity $\boldsymbol{\nabla}\times\boldsymbol{\nabla}\varphi=0$, $\forall\varphi$ we can write the next representation

$$\mathbf{E}+\frac{1}{c}\frac{\partial}{\partial t}\mathbf{A}=-\boldsymbol{\nabla}\varphi, \tag{4.57}$$

where φ is the scalar potential.

We use formulas (4.43) and (4.52) for the determination $\mathbf{A}$ and φ. Substituting the expression for $\mathbf{E}$ from (4.57) into (4.43), we obtain

$$\Delta\varphi=-4\pi q-\frac{1}{c}\frac{\partial}{\partial t}\boldsymbol{\nabla}\cdot\mathbf{A}. \tag{4.58}$$

Substituting (4.55) and (4.57) into the formula (4.52), we (by analogy) obtain:

$$\boldsymbol{\nabla}\times\boldsymbol{\nabla}\times\mathbf{A}=-\frac{1}{c^2}\frac{\partial^2}{\partial t^2}\mathbf{A}-\frac{1}{c}\boldsymbol{\nabla}\frac{\partial}{\partial t}\varphi+\frac{4\pi}{c}\mathbf{j}. \tag{4.59}$$

Using the identity $\boldsymbol{\nabla}\times\boldsymbol{\nabla}\times\mathbf{A}=\boldsymbol{\nabla}\cdot\boldsymbol{\nabla}\cdot\mathbf{A}-\Delta\mathbf{A}$ we rewrite the formula (4.59) in the form

$$\Delta\mathbf{A}\frac{1}{c^2}\frac{\partial^2}{\partial t^2}\mathbf{A}=-\frac{4\pi}{c}\mathbf{j}+\boldsymbol{\nabla}\left(\boldsymbol{\nabla}\cdot\mathbf{A}+\frac{1}{c}\frac{\partial\varphi}{\partial t}\right). \tag{4.60}$$

At last, using a freedom in a choice of $\mathbf{A}$ and φ, we shall accept a following requirement:

$$\boldsymbol{\nabla}\cdot\mathbf{A}+\frac{1}{c}\frac{\partial\varphi}{\partial t}=0$$

is the "Lorentz gauge". In this calibration the expression (4.60) becomes a *vector wave equation*:

$$\Delta\mathbf{A}-\frac{1}{c^2}\frac{\partial^2\mathbf{A}}{\partial t^2}=-\frac{4\pi}{c}\mathbf{j},$$

and the expression (4.58) becomes *a scalar wave equation:*

$$\Delta\varphi - \frac{1}{c^2}\frac{\partial^2}{\partial t^2}\varphi = -4\pi q.$$

Let us write the Maxwell equations (4.43) — (4.52) for an isotropic medium:

$$\nabla \cdot \varepsilon \mathbf{E} = 4\pi q, \tag{4.61}$$

$$\nabla \cdot \mu \mathbf{H} = 0, \tag{4.62}$$

$$\nabla \times \mathbf{E} = -\mu \frac{1}{c}\frac{\partial}{\partial t}\mathbf{H}, \tag{4.63}$$

$$\nabla \times \mathbf{H} = \frac{4\pi}{c}\mathbf{j} + \varepsilon \frac{1}{c}\frac{\partial}{\partial t}\mathbf{E}. \tag{4.64}$$

Here ε is a dielectric permittivity and μ is a magnetic permeability. It is a common practice to supply the equations of (4.61)–(4.64) by the Ohm law

$$\mathbf{j} = \sigma \mathbf{E}, \tag{4.65}$$

where σ is an electrical conductivity.

In the uniform isotropic medium (σ = const, ε = const, μ = const). At the absence of extraneous currents ($\mathbf{j}$) and free charges (q) the fields $\mathbf{E}$ and $\mathbf{H}$ satisfy the homogeneous *telegraph equations*

$$\Delta \mathbf{E} - \varepsilon \frac{\partial^2}{\partial t^2}\mathbf{E} - \sigma \frac{\partial \mathbf{E}}{\partial t} = 0, \tag{4.66}$$

$$\Delta \mathbf{H} - \mu \frac{\partial^2}{\partial t^2}\mathbf{H} - \sigma \frac{\partial \mathbf{H}}{\partial t} = 0. \tag{4.67}$$

In connection with the problems of geoelectrics it is possible to represent *quasi-stationary areas (field)* ($\partial^2/\partial t^2 \mathbf{H} \ll \partial/\partial t\, \mathbf{H}$, $\partial^2/\partial t^2 \mathbf{E} \ll \partial/\partial t\, \mathbf{E}$), so the telegraph equations transform to the homogeneous *diffusion equation*

$$\Delta \mathbf{E} - \sigma \frac{\partial}{\partial t}\mathbf{E} = 0, \tag{4.68}$$

$$\Delta \mathbf{H} - \frac{\partial}{\partial t}\mathbf{H} = 0. \tag{4.69}$$

Let's remind, as from the equations (4.49) and (4.52), it is possible to obtain the connection between *the energy* and the vector of the *energy flux* (*Poynting's vector*). Let's implement a scalar multiplication of the equation (4.49) on $\mathbf{H}$ and the equation (4.52) on $\mathbf{E}$. And after the summation of these equations, we shall write

$$\mathbf{E} \cdot \frac{\partial}{\partial t}\mathbf{E} + \mathbf{H} \cdot \frac{\partial}{\partial t}\mathbf{H} = c\mathbf{E} \cdot \nabla \times \mathbf{H} - c\mathbf{H} \cdot \nabla \times \mathbf{H} - \mathbf{E} \cdot 4\pi \mathbf{j}. \tag{4.70}$$

Taking into account that $\mathbf{E} \cdot \nabla \times \mathbf{H} - \mathbf{H} \cdot \nabla \times \mathbf{E} \equiv -c\nabla \cdot (\mathbf{E} \times \mathbf{H})$, we integrate both parts of the expression (4.70) over the volume:

$$\frac{\partial}{\partial t}\int_V \frac{(\mathbf{E} \cdot \mathbf{E}) + (\mathbf{H} \cdot \mathbf{H})}{8\pi} dV = \int_V \nabla \cdot \left(\frac{c}{4\pi}(\mathbf{E} \times \mathbf{H})\right) dV - \int_V \mathbf{j} \cdot E dV. \tag{4.71}$$

The equation (4.71) becomes the continuity equation for the conserved quantity $\mathcal{E}$, which has the physical sense of the energy:

$$\frac{\partial}{\partial t}\int_V d\mathcal{E} = -\int_V \nabla\cdot\mathbf{R}dV - \int_V \mathbf{j}\cdot\mathbf{E}dV,$$

where

$$d\mathcal{E} = \frac{E^2+H^2}{8\pi}dV; \quad \mathbf{R} = \frac{c}{4\pi}(\mathbf{E}\times\mathbf{H})$$

is *an energy flux vector* (owing to $\int_V \boldsymbol{\nabla}\cdot\mathbf{R}dV = -\oint_{\partial V}(\mathbf{R}\cdot d\boldsymbol{\sigma})$). The last expression has a physical sense of losses on the Joule heat (at $\mathbf{j} = q\mathbf{v}$ Coulomb force $\mathbf{F} = q\mathbf{E}$, and the work in a unit time is equal $\mathbf{F}\cdot v = q\mathbf{E}\cdot v = \mathbf{j}\cdot\mathbf{E}$).

Considering (4.65), we get in the similar way from (4.63), (4.64) the expression for an isotropic medium:

$$\frac{\partial}{\partial t}\int_V \frac{\varepsilon E^2+\mu H^2}{8\pi}dV = -\oint_{\partial V}(\mathbf{R}\cdot d\boldsymbol{\sigma}) - \int_V \sigma E^2 dV,$$

i.e. the energy density is described by

$$\frac{\varepsilon E^2+\mu H^2}{8\pi} = \frac{\mathbf{E}\cdot\mathbf{D}+\mathbf{H}\cdot\mathbf{B}}{8\pi}.$$

Here the Poynting vector has the same form, as in the vacuum: $\mathbf{R} = [c/(4\pi)]\ ,\mathbf{E}\times\mathbf{H}$; the loss power on the Joule heat is given by:

$$\mathbf{j}\cdot\mathbf{E} = (\sigma\mathbf{E})\cdot\mathbf{E} = \sigma\mathbf{E}^2.$$

The expressions introduced in this section, form the basis mathematical model of the propagation of the electromagnetic wave fields in the vacuum and the isotropic medium.

4.4 Atmospheric Sounding

The remote sensing is a significant tool for the investigation of a structure and dynamics of the ionosphere. So, in Fig. 4.5 the scheme of the sounding of the atmosphere using a satellite (the geometry of the observation of a limb) is represented. The received information is used for the improvement of the quality of a weather forecast and development of methods of the artificial affecting on it, for the study of atmospheric pollution. The remote sensing of the atmosphere is grounded on the phenomenon *of the carrying* of the solar radiation or the heat radiation, generated by the atmosphere, which has electromagnetic nature. The phenomenological transport theory can be considered as an extreme case of the rigorous statistical theory basing on the stochastic wave equation (Kravtsov and Apresyan, 1996). The transport theory operates with the so-called *photometric* concept (the authors of (Kravtsov and Apresyan, 1996) point out that the photometric quantities with the

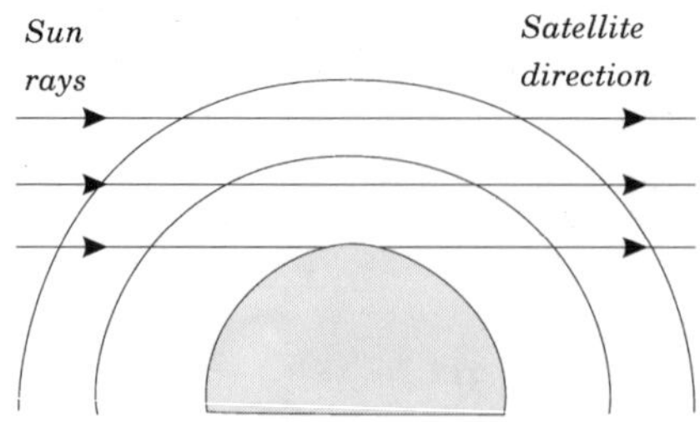

Fig. 4.5 The schema of sounding of an atmosphere using satellite.

equal success can be called *acoustic-metric*) quantities). The applicability of the phenomenological transport theory is ensured with following requirements: 1) the wave field appears as ray one, i.e. the requirements of the applicability of the geometrical optics are carried out; 2) a complete incoherence of the rays (the additivity of the transferred field), i.e. the interference is excluded; 3) the observed value is a time average and a space average process (i.e. the square quantities of wave fields are considered); 4) the radiating is supposed stationary and ergodic.

Let us consider a radiant flux dP at the point X, passing in the bodily cone $d\Omega$ through the area element $d\sigma$ in the direction $\boldsymbol{n}$ (Fig. 4.6):

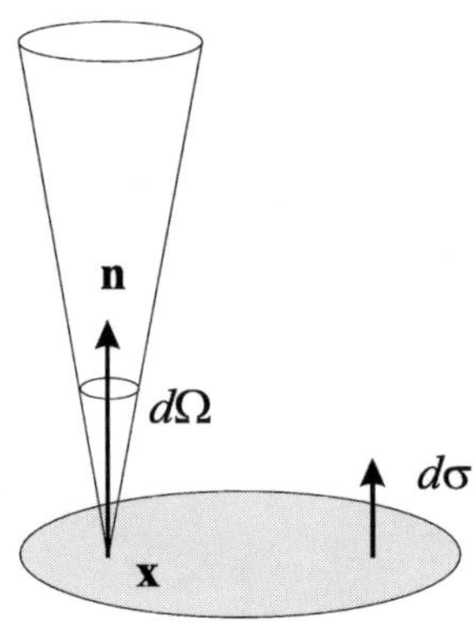

Fig. 4.6 Radiant flux dP through area element $d\sigma$.

$$dP = \varphi(\mathbf{x}, n)(\mathbf{n} \cdot d\boldsymbol{\sigma})d\Omega.$$

The quantity $\varphi(x, \mathbf{n})$ is called *the intensity* or *the brightness*, i.e.:

$$\varphi(x, \mathbf{n}) = \frac{dP}{d\Omega d\sigma \cos(\widehat{\mathbf{n}d\boldsymbol{\sigma}})}.$$

Let's consider the physical bases for the obtaining of the integro-differential *transport equation*. The variation of the radiation intensity ($\Delta\varphi$) on a small length

(Δl) in a direction $\mathbf{n}$ is connected, at the first, with an absorption, which one is proportional to an intensity (φ), secondly, with a dissipation of the radiation φ from a direction $\mathbf{n}$ on all by another directions (in Fig. 4.7 the examples of angular distributions of the dissipation of the radiation intensity by a small (a), large (b) and super-large (c) particles) are given:

$$\sim \frac{1}{4\pi}\int q(\mathbf{n}' \leftarrow \mathbf{n})\varphi(x,\mathbf{n})dn' = \varphi(x,\mathbf{n})\frac{1}{4\pi}\int q(\mathbf{n}' \leftarrow \mathbf{n})d\mathbf{n}',$$

Thirdly, with the contribution to the intensity of a flux with a direction of the radiation $\mathbf{n}$, dissipated in all other directions:

$$\sim \frac{1}{4\pi}\int q(\mathbf{n} \leftarrow \mathbf{n}')\varphi(x,\mathbf{n}')d\mathbf{n}',$$

Fourthly, with a radiation source. Therefore the transport equation takes a form

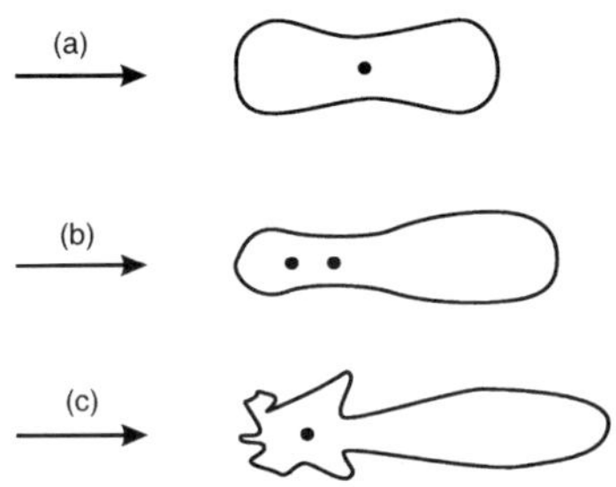

Fig. 4.7 Examples of angular distributions of a dissipation of a radiation intensity by small (a), large (b) and super-large (c) particles.

(with taking into account of $\nabla_n = (\mathbf{n}, \boldsymbol{\nabla})$ and $\alpha_s + \alpha_a \triangleq \alpha$)

$$(\mathbf{n}\cdot\boldsymbol{\nabla})\varphi(x,\mathbf{n}) + \alpha\varphi(x,\mathbf{n}) - \frac{1}{4\pi}\int q(x,\mathbf{n},\mathbf{n}')\varphi(x,\mathbf{n}')d\mathbf{n}' = s(x,\mathbf{n}). \tag{4.72}$$

And in the operator form

$$L\varphi = s, \quad L = (\mathbf{n}\cdot\boldsymbol{\nabla}) + \alpha\hat{I} - \frac{1}{4\pi} < q(\mathbf{n},\mathbf{n}')|, \tag{4.73}$$

where here symbol $< q(\mathbf{n},\mathbf{n}')|$ means $\int q(\cdot)d\mathbf{n}'$; $\varphi = \varphi(x,\mathbf{n})$ is a field of the sounding signal; $s = s(x,\mathbf{n})$ is a signal source. Substituting the operator L from the equation (4.73) as $L = L_0 - S$ (S is a scattering operator),

$$S = \alpha\hat{I} + \frac{1}{4\pi} < q(\mathbf{n},\mathbf{n}')|,$$

we write down the solution of the transport equation as an expansion on the multiple scattering:

$$\varphi = \varphi_0 + L_0^{-1}SL_0^{-1}s + L_0^{-1}SL_0^{-1}SL_0^{-1}s + \dots,$$

where

$$L_0^{-1}s = \varphi_0, \quad \varphi = \varphi_0\exp\Big\{-\int \alpha(x)dl\Big\}.$$

Let's note, that the equation of the carrying (4.72) is used for the solution of the wide range of problems, for example, at studying of the transport of monochromatic neutrons (in this case it is accepted to call it as the Boltzmann equation), in problems of X-ray and spectral tomography, at laser sounding etc.

Chapter 5

Ray theory of wave field propagation

The exact solutions of problems of sounding signals propagation are constructed only for limited number of media models. As a rule, this set of media includes uniform medium, layered homogeneous medium, uniform medium with the inclusions of high symmetry. For the interpretation of real geophysical fields (seismic, acoustic, electromagnetic) it is necessary to construct the approximate solutions for the wave propagation in non-uniform media. So, it exists in the Earth not only the interfaces, on which one the elastic properties vary by jump, but also areas, inside which there is a smoothly varying variation of elastic properties. From the physical point of view the ray theory is interpreted as follows: the waves propagate with local velocities along ray pathways and arrive in the observation point with amplitudes described by a geometrical spreading of rays from a source to the receiver point. At an enunciating of this chapter we shall follow the description introduced in (Ryzhikov and Troyan, 1994).

5.1 Basis of the Ray Theory

One of the most common method of the solution of the equations of wave propagation is the method of *geometrical optics* (Babic and Buldyrev, 1991; Babic *et al.*, 1999; Kravtsov, 2005; Bleistein *et al.*, 2000). This method is a shortwave asymptotic of a field in weak non-uniform, slow non-stationary and weak-conservative media: the sizes of the inhomogeneity are much greater than the wavelength and time intervals of the non-stationarity much more than the period of oscillation. The shortwave asymptotic allows to consider the medium locally as homogeneous and stationary and is based on the assumption of a wave field in a form (φ) a "quick" phase and a "slow" amplitude multiplier factors.

Let us consider a formal scheme of the space-time-ray method. Let a field φ satisfies homogeneous linear equations, which we shall write down, following (Kravtsov and Apresyan, 1996), in a form of the integral equation:

$$L\varphi = \int L(x, x')\varphi(x')d^4x' = 0 \tag{5.1}$$

(in this section we put $x = (x_1, x_2, x_3, t)$). Let us assume that the operator L contains a part L_0, carrying the responsibility for the description of the field φ in the conservative medium, and the part L' with a small parameter α:

$$L = L_0 + \alpha L'. \tag{5.2}$$

We represent the field φ as an asymptotic expansion (*the Debye expansion*):

$$\varphi = \mathcal{A}(y, \alpha) \exp\left\{\frac{i}{\alpha}\tau(y)\right\}, \tag{5.3}$$

where $y = \alpha x$ is " a slow" argument; τ is a phase (or eikonal); $\mathcal{A}$ is a wave amplitude. The expansion of $\mathcal{A}$ over power of α has a form

$$\mathcal{A} = \mathcal{A}_0(y) + \alpha\mathcal{A}_1(y) + O(\alpha^2).$$

Let L is a near differential operator, i.e. $L(x, x') \neq 0$ under small $\Delta x = x' - x$. Then the expansion of the phase function $\exp\{(i/\alpha)\tau(\alpha x)\}$ in the point x' accurate to $O(\alpha^2)$ can be represented as:

$$\begin{aligned}
&\exp\left\{\frac{i}{\alpha}\tau(\alpha(x + \Delta x))\right\} \\
&= \exp\left\{\frac{i}{\alpha}\left[\tau(\alpha x_0) + \alpha(\Delta x \cdot \partial_y)\tau + \frac{1}{2}\alpha^2(\Delta x \cdot \partial_y)^2\tau + O(\alpha^2)\right]\right\} \\
&= \exp\left\{\frac{i}{\alpha}\tau(y)\right\} \times \exp\{i(p \cdot \Delta x)\}\left(1 + \frac{i}{2}\alpha(\Delta x \cdot \partial_y)^2\tau\right) + O(\alpha^2),
\end{aligned}$$

where $p \triangleq \nabla_y\tau$. The expansion of the amplitude accurate to $O(\alpha^2)$ has a form:

$$\mathcal{A}(\alpha x') = \mathcal{A}(\alpha(x + \Delta x)) = (1 + \alpha(\Delta x \cdot \partial_z))\mathcal{A}(z),$$

where $z = \alpha x$. Finally, the wave field in a point x' is given by the formula

$$\begin{aligned}
\varphi(x') &= \exp\left\{\frac{i}{\alpha}\tau(y)\right\} \exp\{i(p \cdot \Delta x)\} \\
&\times \left(1 + \alpha\left[(\Delta x \cdot \partial_z) + \frac{i}{2}(\Delta x \cdot \partial_y)^2\tau(y)\right]\right) \\
&\times \mathcal{A}(z)\big|_{z=y} + O(\alpha^2).
\end{aligned} \tag{5.4}$$

Regarding this expansion and keeping linear terms on α, we can rewrite (5.1) on the next form:

$$\begin{aligned}
&\int L(x, x')\varphi(x')d^4x' \\
&= \exp\left\{\frac{i}{\alpha}\tau(\alpha x)\right\} \times \int d^4x' L(x, x') \exp\{ip(x' - x)\} \\
&\times \left(1 + \alpha)\left[(\Delta x \cdot \partial_z) + \frac{i}{2}(\Delta x \cdot \partial_y)^2\tau(y)\right]\right)\mathcal{A}(z) \\
&= \exp\left\{\frac{i}{\alpha}\tau(\alpha x)\right\}\left(1 - \alpha\left[i(\partial_p\partial_z) + \frac{i}{2}(\partial_p\partial_y)^2\tau(y)\right]\right) \\
&\times \int d^4x' L(x, x') \exp[ip(x' - x)]\mathcal{A}(z).
\end{aligned}$$

Introducing the denotation $\int d^4x' L(x,x') \exp[ip(x'-x)] \equiv \hat{L}(p)$, let us represent (5.1) as

$$\left(1 - i\alpha\left[(\partial_p\partial_z) + \frac{1}{2}(\partial_p\partial_y)^2\tau(y)\big|_{y=\alpha x}\right]\right)\hat{L}(p)\mathcal{A}(z)\big|_{z=\alpha x} = 0.$$

Finally, expanding the amplitude in the expression for $\mathcal{A}(z)$ in a point x by power α and representing $L(p)$ in the form $L = L_0 + \alpha L'$ (and equate the terms of the equal order on α), we obtain the system of recurrence equations to determine the amplitude (now we put the parameter α being equal an identity element):

$$\hat{L}_0(p)\mathcal{A}_0(x) = 0, \tag{5.5}$$

$$\hat{L}_0(p)\mathcal{A}_1(x) = \left\{i\left[(\partial_p\partial_x) + \frac{1}{2}(\partial_p\partial_y)^2\tau(y)\big|_{y=x}\right]\hat{L}_0(p) - \hat{L}'(p)\right\}A_0(x) \tag{5.6}$$

etc.

If the field is multicomponent, for example, the displacement field in the Lame equation, electromagnetic field, then the amplitude $\mathcal{A}$ is a vector function, and the operator $\hat{L}(p)$ is a matrix operator. The *resolvability condition* (5.5) is reduced to the requirement of the equality to zero of the determinant:

$$\det L_0(p) = 0. \tag{5.7}$$

If L_0 corresponds to the conservative medium, then the operator $\hat{L}_0$ is the Hermitian operator and the next canonical representation of the matrix L_0 takes place:

$$\hat{L}_0 = \sum_i \lambda_i \mathbf{e}_i \mathbf{e}_i^T, \tag{5.8}$$

where $\{\lambda_i\}$ are eigenvalues; $\{\mathbf{e}_i\}$ are orthonormal eigenvectors of the matrix L_0, i.e. $L_0\mathbf{e}_i = \lambda_i\mathbf{e}_i$, $\mathbf{e}_i^T\mathbf{e}_j = \delta_{ij}$.

The resolvability condition (5.7) subject to a canonical form (5.8) gives the *dispersion equations*:

$$\lambda_i = \lambda_i(x,p) = 0, \qquad i = 1, 2, \ldots, \tag{5.9}$$

which determine the relationship between space wave vectors $\partial\tau/\partial x_1$, $\partial\tau/\partial x_2$, $\partial\tau/\partial x_3$) and the frequency $\partial\tau/\partial x_0 \triangleq \partial\tau/\partial t$, as well as the phase (eikonal) τ space and time behavior. Because $p = \partial\tau/\partial x$, then the dispersion equation is the first-order differential equation, which is usually solved by the method of characteristics. It is usual to write the equation for the characteristic in the Hamilton form, introducing the parameter l:

$$\begin{aligned} d_l x &= \partial_p \lambda_i, \\ d_l p &= -\partial_x \lambda_i. \end{aligned} \tag{5.10}$$

The equation (5.10) describes the rays corresponding to the different types of waves, while i takes the different values.

5.2 Ray Method for the Scalar Wave Equation

The scalar wave equation represents the mathematical model underlying the description of the set of the physical processes of the signal propagation. Therefore enunciating of the concrete examples to use of the space-time ray method (Babic *et al.*, 1999) we shall begin from the wave equation

$$\Box\varphi = \Delta\varphi - \frac{1}{c^2}\frac{\partial^2}{\partial t^2}\varphi = 0. \tag{5.11}$$

In accordance with the statement form of the operator equation of the sounding signal propagation $L\varphi = s$, in this case L is

$$L = \Delta - \frac{1}{c^2}\frac{\partial^2}{\partial t^2}; \quad \varphi = \varphi(x,t); \quad s = 0$$

(we consider the wave field in an area without any sources).

Let us write the general form of an asymptotic expansion (5.3), extract the time in the explicit form, as stated below

$$\varphi(x,t) = \mathcal{A}(\alpha, \alpha x, \alpha t)\exp[i/\alpha\tau(\alpha x, \alpha t)].$$

Let's note, that the wave operator is a local operator (differential operator), i.e. in representation of the operator L as an integral operator ($L\varphi = \int L(x,t;x',t')\varphi(x',t')dx'dt'$), operator kernel is singular, i.e. the sign of integration get lost.

Approximation of the phase part of the wave field φ in the vicinity of x_0, t_0 (expansion (5.4)) accurate to $O(\alpha)$ looks like

$$\varphi(x,t) \sim \exp[i(\mathbf{p}\Delta\mathbf{x} - p_0\Delta t)],$$
$$\text{where} \quad \mathbf{p} = \frac{1}{\alpha}\boldsymbol{\nabla}\tau\big|_{x_0,t_0}, \quad p_0 = -\frac{1}{\alpha}\frac{\partial}{\partial t}\tau\big|_{x_0,t_0},$$
$$\Delta t = t - t_0, \quad \Delta\mathbf{x} = \mathbf{x} - \mathbf{x}_0,$$

i.e. the approximation in the vicinity of x_0, t_0 appears to be a plane wave. The resolvability equation (5.5) implies that

$$(\mathbf{p},\mathbf{p}) - \frac{1}{c^2(x,t)}p_0^2 = 0 \tag{5.12}$$

(here $L_0 = \Delta - [c^2(x,t)]^{-1}\partial^2/\partial t^2$), therefore

$$L_0(p) = \exp[-i(\mathbf{p}\mathbf{x} - p_0 t)]\left(\Delta - \frac{1}{c^2}\frac{\partial^2}{\partial t^2}\right)\exp[i(\mathbf{p}\mathbf{x} - p_0 t)].$$

The resolvability condition (5.12), which we have written in the form of the dispersion equation, gives a connection of the frequency $p_0 \triangleq \omega$ and the wave vector $\mathbf{p} \triangleq \mathbf{k}$ (spatial frequency):

$$\omega = \omega(\mathbf{k}) = c|\mathbf{k}|, \tag{5.13}$$

where $c = c(x,t)$; ω is the solution of the dispersion equation $|L_0(\omega(\mathbf{k}))| = 0$.

The ray approximation of the solution of (5.11) could be represented in the next form:

$$\varphi(\mathbf{k},\omega) = \mathcal{A}(\mathbf{k})\delta(L_0(\omega(\mathbf{k}))) = \mathcal{A}_{\mathbf{k}}\delta(\omega - \omega(\mathbf{k})), \tag{5.14}$$

where $\mathcal{A}_{\mathbf{k}} \triangleq \mathcal{A}|\partial\omega L(\mathbf{k},\omega)|^{-1}$ (a consequence of the formula $\delta(f(x)) = |\partial_x f|^{-1}\times$ $\times\delta(x - x')$, where x' is a root of the equation $f(x) = 0$). The presence in the representation of (5.14) the Dirac delta function $\delta(|L_0|)$ means, that the amplitude of a field φ can be differ from zero only for wave vectors $\mathbf{k}$, fitting to the dispersion equation, i.e. a constraint equation for the frequency and the modulus of the wave vector (5.13).

Let us note, that the expression (5.14) corresponds to the single type of wave (single-mode propagation in conservative medium), that means the equation $|L_0(\omega(\mathbf{k}))| = 0$ has a one real root.

In the general case of the multimode propagation, the wave field φ can be represented by the superposition of the several types of waves:

$$\varphi = \sum_{\mu} \mathcal{A}_{\mathbf{k}}\delta(\omega - \omega^{\mu}(\mathbf{k})).$$

In the space-time representation the solution φ is described as a set of the travelling waves:

$$\varphi(x,t) = \frac{1}{(2\pi)^{3/2}}\int \mathcal{A}_{\mathbf{k}} \exp[i(\mathbf{kx} - \omega(\mathbf{k})t)]d\mathbf{k}.$$

Writing the inverse Fourier transform of $\varphi(x,t)$, it is possible to find out that

$$\mathcal{A}_{\mathbf{k}}e^{i\omega(\mathbf{k})t} = \frac{1}{(2\pi)^{3/2}}\int \varphi(\mathbf{x},t)e^{-i\mathbf{kx}}d\mathbf{x},$$

i.e. the space Fourier operator of the free wave field is an oscillating one.

The resolvability equation (5.5) written for a phase τ is called *a general eikonal equation*:

$$(\nabla_x\tau, \nabla_x\tau) = \frac{1}{c^2(x,t)}\left[\frac{\partial}{\partial t}\tau\right]^2. \tag{5.15}$$

This equation is a generalisation of the eikonal equation for the known case of the stationary media ($c = c(x)$, $\tau = \tau - t$, $c^{-2}(x) \triangleq n^2(x)$ plays usually the role of the refractive index).

We may name the equation (5.15) as a characteristic equation of the wave equation (5.11). The hypersurfaces $\tau(x,t) = \text{const}$ are the characteristics of the wave equation. The equation (5.15) can be represented as the Hamilton–Jacobi equation:

$$\frac{\partial\tau}{\partial t} + H(\boldsymbol{\nabla}_x\tau, x, t) = 0, \tag{5.16}$$

where $H = c(x,t)|\boldsymbol{\nabla}\tau|$. The primary method to solve this equation is the characteristics method. Let us remind the algorithm of this method for the general form of the equation (5.16):

$$H\left[x^{\mu}, \frac{\partial\tau}{\partial x^{\mu}}\right] = 0, \quad \mu = 0,\dots,M, \tag{5.17}$$

i.e. we consider the general partial equation of the second order (index $\mu = 0$ is connected with the time parameter). Let supplement (5.17) by the initial conditions defined on hypersurface Q, we then start to solve the Cauchy problem.

We consider Q as a parameter defined value, i.e.

$$Q = \{x^\mu : x_0^\mu(\gamma_1, \dots, \gamma_m)\}, \qquad \gamma \in \Gamma,$$

where the vectors $\{\partial \mathbf{x}/\partial\gamma\}$ are linear independent (surface Q is regular, im kleinen). Let the determinant D is nonsingular on the surface Q:

$$D = \begin{Vmatrix} \partial H/\partial\tau_0 & \dots \partial H/\partial\tau_m \\ \partial x^0/\partial\gamma_1 & \dots \partial x^m/\partial\gamma_1 \\ \dots & \dots\dots \\ \partial x^0/\partial\gamma_m & \dots \partial x^m/\partial\gamma_m \end{Vmatrix} \neq 0, \tag{5.18}$$

where

$$\tau_\mu \triangleq \frac{\partial}{\partial\gamma_\mu}\tau.$$

To set the initial conditions on Q as

$$\tau\big|_Q = \tau^0(\gamma), \qquad \frac{\partial\tau}{\partial x^\mu}\big|_Q = \tau_\mu^0(\gamma).$$

The characteristic system of equations for the equation (5.17) is the system of equations

$$\begin{aligned} dx^\mu/ds &= \partial H/\partial\tau_\mu, \\ d\tau_\mu/ds &= -\partial H/\partial x^\mu, \\ d\tau/ds &= -\sum_\mu \tau_\mu \partial H/\partial\tau_\mu, \end{aligned} \tag{5.19}$$

where

$$s : \left|\frac{\partial(x^0, \dots, x^m)}{\partial(s, \gamma^1 \dots, \gamma^m)}\right|_{s=0} \neq 0$$

owing to (5.18), i.e. locally (in the vicinity of s) the coordinate frame $(s, \gamma^1, \dots, \gamma^m)$ is non-degenerate.

Let us present the method of derivation of the system of characteristic equations. Let us consider (5.17) as a constraint equation of the independent variables $\{x^\mu\}$ and $\{p^\mu \triangleq \partial\tau/\partial x^\mu\}$, i.e. let's find the solution in *phase space* $X \times \mathcal{P}$, where $X = \{x = (x^0, x^1, \dots, x^m)\}$, $\mathcal{P} = \{p = (p^0, p^1, \dots, p^m)\}$, $H(x^\mu, p^\mu) = 0$. Owing to this constraint, the differential H must be identically equal to zero:

$$dH = \sum_\mu \left(\frac{\partial H}{\partial x^\mu}dx^\mu + \frac{\partial H}{\partial p^\mu}dp^\mu\right) = 0.$$

Hence, it is obviously, the formal vectors $\partial H/\partial x$, $\partial H/\partial p$ and correspondingly vectors dx, dp are orthogonal. For the orthogonality condition to be certainly satisfied, it is necessary to put the first components of dx, dp to be proportional to the second components of $\partial H/\partial x$, $\partial H/\partial p$, whereas the relation of other components has an opposite sign of the constant of proportionality.

Introducing the formal parameter s (i.e. we let $p = p(s)$, $x = = x(s)$), we can write

$$dx^\mu/ds = \partial H/\partial p_\mu,$$
$$dp^\mu/ds = -\partial H/\partial x^\mu.$$

To exclude the parameter s we should write the third equation, which determines $d\tau/ds$ under the condition $p = \partial\tau/\partial x$:

$$\frac{d\tau}{ds} = \sum_\mu \frac{\partial\tau}{\partial x_\mu}\frac{dx^\mu}{ds} = \sum_\mu p^\mu \frac{\partial H}{\partial p^\mu}.$$

Solving the system (5.19) under initial conditions, we obtain $x^\mu = x^\mu(s, \gamma^1, \dots \gamma^m)$, $\tau_\mu = \tau_\mu(s, \gamma^1, \dots \gamma^m)$, $\tau = \tau(s, \gamma^1, \dots \gamma^m)$. Using the initial coordinates $(x^0, \dots, x^m)$, we find the functions $\tau = \tau(x)$ and $\tau_\mu = \tau_\mu(x)$, that are the solutions of the Cauchy problem, which satisfies both (5.17) and initial conditions.

Returning to the equation (5.16), let us write the characteristic equations ($x_0 \equiv t$):

$$\begin{aligned}
&\frac{dt}{ds} = 1, \quad \frac{dx^1}{ds} = \frac{\partial H}{\partial\tau_1} \dots \frac{dx^m}{ds} = \frac{\partial H}{\partial\tau_m},\\
&\frac{d\tau_1}{ds} = -\frac{\partial H}{\partial x^1} \dots \frac{d\tau_m}{ds} = -\frac{\partial H}{\partial x^m},\\
&\frac{d\tau}{ds} = \tau_0 + \sum_{\mu=1}^m \hat{\tau}_\mu \frac{\partial H}{\partial \tau_\mu}.
\end{aligned} \tag{5.20}$$

Here, in the last equation the addend $\tau_0 = \partial\tau/\partial t$ is emphasis ed and we take into account that $\partial H/\partial\tau_0 = 1$.

Let us remind an analogy between the ray in optics and the trajectory in mechanics. Writing the eikonal equations as

$$H(x,p) = \frac{1}{2}\Big[(\mathbf{p}, p) - n^2(\mathbf{x})\Big] = 0,$$

we can obtain the equations for the rays in the Hamilton form:

$$d\mathbf{x}/ds = \mathbf{p},$$
$$\frac{d\mathbf{p}}{ds} = \frac{1}{2}\boldsymbol{\nabla} n^2(\mathbf{x}).$$

Following to $\partial\tau/\partial s = (\mathbf{p}, p)$ we have the next equation for the eikonal τ

$$\tau = \tau^0 + \int_{s_0}^{s} p^2 ds = \tau^0 + \int_{s_0}^{s} n^2(x(s))ds.$$

The parameter s is connected with an arc length of the ray (dl) by virtue of

$$(dl)^2 = (d\mathbf{x})^2 = \left(\frac{\partial H}{\partial \mathbf{p}}\right)^2 ds^2$$

(in the last equality we take into account the first equations of the characteristic system). In this case

$$ds = dl|\partial H/\partial \mathbf{p}|^{-1} = dl|\mathbf{p}|^{-1} = dl/n(x),$$

and τ depends on arc length l as

$$\tau = \tau^0 + \int_A^B n(\mathbf{x})dl.$$

The system of Hamilton equations relatively to $\mathbf{x}$ and $\mathbf{p}$, excluding $\mathbf{p}$, can be written as

$$\frac{d^2\mathbf{x}}{ds^2} = \frac{1}{2}\nabla n^2(\mathbf{x}).$$

Let us compare this equation with the dynamic one for the motion of a particle in the potential field ($\mathbf{F} = -\nabla\varphi$):

$$d^2\mathbf{x}/dt^2 = -\nabla\varphi.$$

This analogy allows the results obtained in optics, to use in mechanics, and on the contrary. Essential difference of posing of mechanical and wave problems is the assignment of the initial data: by virtue of the specificity of the wave problems the initial conditions determine not an alone trajectory, as it is accepted in mechanics, but a set (congruence) of "trajectories" — rays.

For wave problems are of interest (exotic from the point of view of mechanics) the next example

$$H(k\tau_i, t, x) = kH(\tau_i, t, x),$$

where $i = 1 \div m$, $k > 0$, i.e. the last equation of the system (5.20) has a form

$$d\tau/ds = \tau_0 + H = \partial\tau/\partial t + H \equiv 0,$$

at that $\tau =$ const along the curve, that is the solution of the system (5.20) (loss of the dispersion).

Let's note, that the solution of the characteristic system (5.19) as the system of the ordinary differential equations with given initial conditions, by virtue of the uniqueness theorem represents a family of curves without intersections in the phase space.

It is accepted in geometrical optics to call a spatially-time projection of the characteristic of the eikonal equation as the *spatially-time ray*, $x^\mu = x^\mu(s, \gamma^1, \ldots, \gamma^m)$, which one can be obtained or as the solution of the system of equations (5.19), or as an extreme value of the functional

$$\tau = \int_A^B \sum_\mu \tau_\mu \frac{\partial H}{\partial \tau} ds$$

on every possible pathways pairing a dots A and B. (In Fig. 5.1 the projection of a characteristic, given in a phase space, on the physical space $x_1 x_2$ with the

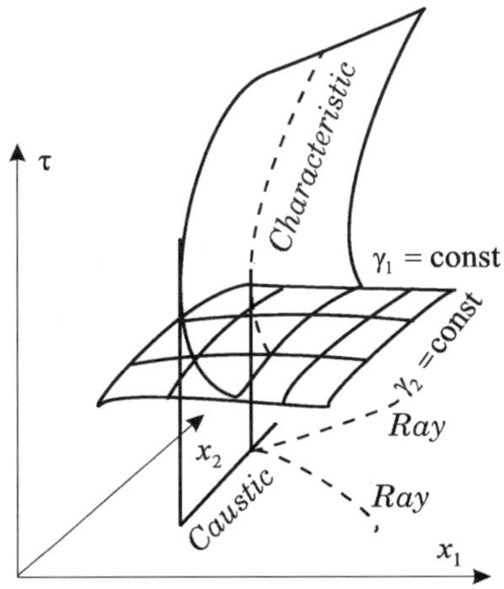

Fig. 5.1 Projection of the characteristic, given in a phase space.

formation of caustics is represented.) For the case of the simplest Hamilton function $H = [(\mathbf{p}, \mathbf{p}) - n^2(\mathbf{x})]/2$ this functional reads as

$$\tau = \int_A^B n(\mathbf{x})dl = \int_A^B \frac{dl}{c(\mathbf{x})}\,.$$

with a physical sense as a time of propagation from the point A to the point B. Here the ray corresponds to the stationary time of propagation. The stationary condition of the functional τ: $\delta\tau = 0$ is called *Fermat's principle.* The solution of the appropriate extremum problem forms the basis for the constructing algorithms of the numerical calculation of rays in inhomogeneous media.

The modern interpretation of the Fermat's principle bases on the representation concerning *interference of the wave perturbations (Huygens–Kirchhoff principle)*, propagating from the point A to the point B on every possible virtual pathways. Thus, those pathways "survive" only, for which one the variation of the eikonal (phase) has a value of $\sim \lambda/2$. In Fig. 5.2 the rays having a common point with boundary of the Fresnel zone are represented, i.e. taking into account the Fresnel zone, the contribution of all remaining pathways will be a negligible small quantity by virtue of a compensation of interfering waves with different phases (in calculus mathematics it corresponds to the stationary phase principle). According to the modern interpretation, the the ray concept lies in ray representation by a dimensional ray tube (Fig. 5.3) with a diameter about the first Fresnel zone. Let's mark, that in a case of the nondispersive medium the Fermat's principle can be interpreted as a requirement of a stationarity of the propagation time of a wave from the point A to the point B.

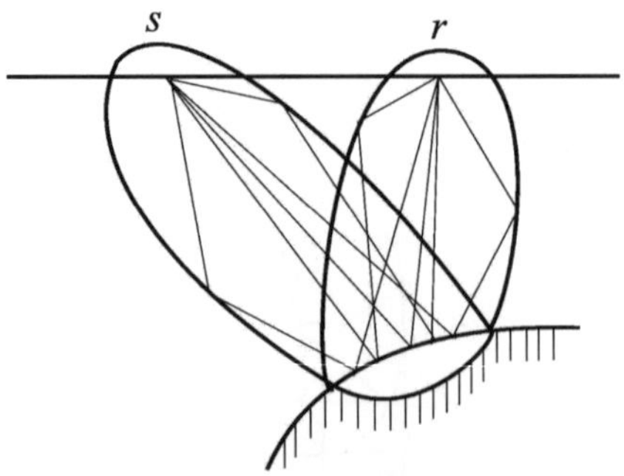

Fig. 5.2 The rays having a common point with boundary of the Fresnel zone.

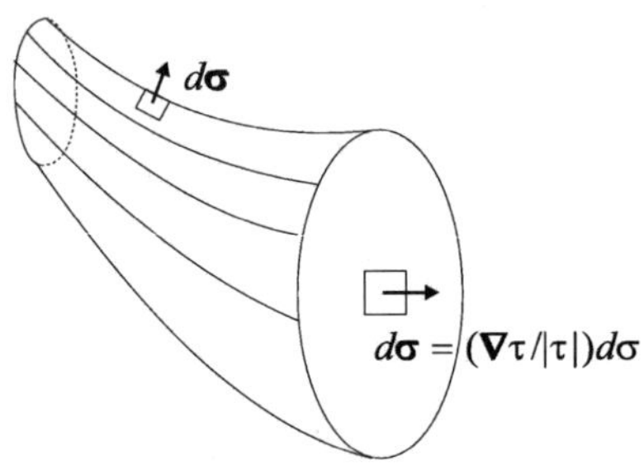

Fig. 5.3 Space ray tube.

5.3 Shortwave Asymptotic Form of the Solution of the One-Dimensional Helmholtz Equation (WKB Approximation)

Let us consider one more example of the application of the perturbation theory to the solution of the Helmholtz equation:

$$\frac{d^2}{dx^2}\varphi + V(x)\varphi = 0. \tag{5.21}$$

In accordance with the perturbation theory we write:

$$\varphi = \exp\{i\tau(x)\}. \tag{5.22}$$

Then the equation (5.21) has a form

$$-(\tau')^2 + i\tau'' + V = 0. \tag{5.23}$$

Considering τ'' being small we obtain $\tau' = \pm\sqrt{V}$, i.e.

$$\tau(x) = \pm \int \sqrt{V} dx. \tag{5.24}$$

Validity condition of the above approximation is

$$|\tau''| \approx \frac{1}{2}\left|\frac{V'}{\sqrt{V}}\right| \ll |V|.$$

With the help of the formulas (5.22) and (5.23) it is to see, that $(\sqrt{V})^{-1}$ has a physical sense of the wavelength ("exponential length", i.e. it is a character interval of the variation of V in a case of $\mathrm{Im}\tau > 0$). Thus, the approximation (5.24) is valid, if the variation of V at distance about wavelength is much less, than value of V, i.e. $V(x)$ has a small variation in a scale of the wavelength.

Following approximation we shall find by solving of the equation (5.23) at

$$\tau'' \approx \pm\frac{1}{2}\frac{V'}{\sqrt{V}}, \qquad (\tau')^2 \approx V \pm \frac{i}{2}\frac{V'}{\sqrt{V}}, \qquad \tau' \approx \pm\sqrt{V} + \frac{i}{4}\frac{V'}{V}. \tag{5.25}$$

After integration of the obtained expression and can write down

$$\tau(x) \approx \pm \int \sqrt{V} dx + \frac{i}{4} \ln V. \tag{5.26}$$

Using approximation (5.26), we can construct a superposition of the solutions corresponding to the different signs. Therefore WKB approximation is valid, if $|\Delta V(x)| \leq |V(x)|$.

In the sewing point $V(x) = 0$ is valid, and $\Delta V(x)$ is singular, then the exact solution of the equation (5.21) in this point is a regular solution, while the WKB solution has singularity in this point. Therefore we find the solution $\varphi(x)$ as a speed key $\tilde{\varphi}_+$ and $\tilde{\varphi}_-$:

$$\varphi(x) = a_+\tilde{\varphi}_+(x) + a_-\tilde{\varphi}_-(x).$$

We shall find those coefficients using the next equality for $\varphi'(x)$:

$$\varphi'(x) = a_+\tilde{\varphi}'_+(x) + a_-\tilde{\varphi}'_-(x),$$

$a_+ = \mathrm{const}$, $a_- = \mathrm{const}$. Solution of this system about a_+ and a_- is placed below

$$a_+ = \frac{\varphi\tilde{\varphi}'_- - \varphi'\tilde{\varphi}_-}{\tilde{\varphi}_+\tilde{\varphi}'_- - \tilde{\varphi}'_+\tilde{\varphi}_-},$$

and for approximation (5.26) we can write

$$\tilde{\varphi}(x) \approx (V(x))^{-1/4}\left\{c_+ \exp\left(i\int\sqrt{V}dx\right) + c_- \exp\left(-i\int\sqrt{V}dx\right)\right\}. \tag{5.27}$$

The solution (5.27) is suitable in any field, where the validity condition is satisfied, but it is obviously broken in the vicinity of the point $x^0 : V(x^0) = 0$, it is necessary to be able to join of exponential (at $V(x) < 0$) and oscillating (at $V(x) > 0$) solutions, i.e. we need *the connecting formulas.*

Let's note, that the WKB solutions (5.27), which we have written in a form $\tilde{\varphi}_+$ and $\tilde{\varphi}_-$ (correspond to $c_+ = 1$ and $c_- = 1$), are exact solutions for the equation of (5.21), where $V(x)$ is exchanged by the following expression:

$$V(x) \Rightarrow V(x) + \Delta V(x) \triangleq V(x) + \frac{1}{4}\frac{V''}{V} - \frac{5}{16}(\frac{V'}{V})^2,$$

$$a_- = \frac{\varphi\tilde{\varphi}'_+ - \varphi'\tilde{\varphi}_+}{\tilde{\varphi}_+\tilde{\varphi}'_- - \tilde{\varphi}'_+\tilde{\varphi}_-}.$$

Taking into account that determinant of the system is equal to $\tilde{\varphi}_+\tilde{\varphi}'_- - \tilde{\varphi}'_+\tilde{\varphi}_- = -2i$ and $\varphi\tilde{\varphi}''_\mp - \varphi''\tilde{\varphi}_\mp = \Delta V(x)\varphi\tilde{\varphi}_\mp$, we obtain

$$\frac{da_\pm}{dx} = \mp\frac{i}{2}\Delta V(x)\varphi\tilde{\varphi}_\mp = \mp\frac{i}{2}\frac{\Delta V(x)}{\sqrt{V(x)}}\left\{a_\pm + a_\mp \exp\mp\int_{x_0}^{x}\sqrt{V}dx\right\}.$$

The above expression is an estimation of the error being able to appear in the WKB approximation along the large interval Δx.

5.4 The Elements of Elastic Wave Ray Theory

According to the ray theory of propagation of the seismic wave, body wave propagates with local velocity along the ray (having the jogs on interfaces of elastic medium, according to a Snell's law), with the amplitude given by a geometrical spreading of the ray (Babic and Buldyrev, 1991; Goldin, 1984; Petrashen *et al.*, 1985). Using the general ray theory (see Sec. 5.1) in the stationary medium, we represent the eikonal as $\tau(x,t)$ as $t - \tau(x)$, where $\tau(x)$ is called *a wavefront.* Applicability of the ray theory is connected with the assumption of much more fast variation of the wave process in a normal direction to the wavefront ($\mathbf{n} = \boldsymbol{\nabla}\tau/|\boldsymbol{\nabla}\tau|$) in comparison with variations of the characteristics of the medium, i.e. we suppose a validity of the *shortwave asymptotic*, when the ratio of the wavelength to the typical sizes of an inhomogeneity of the medium is a small quantity.

Under the selected eikonal form the wave vector can be represented as $\mathbf{p} = (p_0; p_1, p_2, p_3) = (p_0, \mathbf{p}) = (1, \boldsymbol{\nabla}\tau)$, where vector $\mathbf{p}$ is coaxial to the vector of the phase velocity (a normal direction to the wavefront) is called *a refraction vector.*

The Lame operator in the form (4.16) has the vector structure:

$$\hat{L} = \hat{I}\rho\frac{\partial^2}{\partial t^2} - \boldsymbol{\nabla}\cdot\tilde{K}\partial_{\mathbf{x}},$$

here I is an unit operator in R^3 space. Operator $\hat{\partial}_{\mathbf{x}}$ reads as

$$\hat{\partial}_{\mathbf{x}}\,\mathbf{a} = \left\|\begin{matrix} \partial a_1/\partial x_1 & \partial a_2/\partial x_1 & \partial a_3/\partial x_1 \\ \partial a_1/\partial x_2 & \partial a_2/\partial x_2 & \partial a_3/\partial x_2 \\ \partial a_1/\partial x_3 & \partial a_2/\partial x_3 & \partial a_3/\partial x_3 \end{matrix}\right\|.$$

Let us write a concrete form of the equation (5.5) for the Lame operator

$$(\rho\hat{I} - \tilde{K}\mathbf{p}\,\mathbf{p}^T)\mathcal{A}_0 = 0. \tag{5.28}$$

Here $(\tilde{K}\mathbf{p}\,\mathbf{p}^T)_{ik} = \sum_j\sum_l K_{ijlk}p_jp_l$, $\mathbf{p} = \boldsymbol{\nabla}\tau$. The solvability condition for the equation (5.28)

$$\det(\rho\hat{I} - \tilde{K}\mathbf{p}\,\mathbf{p}^T) = 0 \tag{5.29}$$

describes the fronts existing in the elastic medium. Let us write down the tensor of the elastic modules $\tilde{K}$ for the isotropic elastic medium:

$$K_{ijkl} = \lambda\delta_{ij}\delta_{kl} + \mu(\delta_{ik}\delta_{jl} + \delta_{il}\delta_{jk}).$$

Choosing a local (ray) orthogonal coordinate frame

$$\mathbf{e}_\tau = \frac{\mathbf{p}}{|\mathbf{p}|}\,, \qquad \mathbf{e}_\beta = \mathbf{e}_\tau \times \mathbf{e}_n\,,$$

then we write down the coordinate representation of $L_0(\mathbf{p})\mathcal{A}_0 = 0$:

$$\begin{Vmatrix} (\lambda+2\mu)(\mathbf{p}\cdot\mathbf{p}^T)-\rho & 0 & 0 \\ 0 & \mu(\mathbf{p}\cdot\mathbf{p}^T)-\rho & 0 \\ 0 & 0 & \mu(\mathbf{p}\cdot\mathbf{p}^T)-\rho \end{Vmatrix} \cdot \begin{Vmatrix} \mathcal{A}_t \\ \mathcal{A}_n \\ \mathcal{A}_\beta \end{Vmatrix} = 0\,. \tag{5.30}$$

The solvability condition for the system (5.30) leads us to the conclusion that, in accordance with the ray method, there are three kind of the waves in uniform and isotropic elastic medium:

(1) quasi-longitudinal wave with the velocity $(\lambda(x)+2\mu(x))/\rho(x) \triangleq c_{\mathrm{P}}^2(x)$ and the polarization vector coinciding with $\mathbf{e}_\tau$, which describes by the eikonal equation:

$$(\mathbf{p}\cdot\mathbf{p}) = \frac{\rho(x)}{\lambda(x)+2\mu(x)};$$

(2) quasi-shear waves of two kinds with polarisation vectors $\mathbf{e}_n$ and $\mathbf{e}_\beta$, propagating with equal velocities $\mu(x)/\rho(x) \triangleq c_{\mathrm{S}}^2(x)$, which are describes by equation $(\mathbf{p}\cdot\mathbf{p}) = \rho(x)/\mu(x)$.

5.5 The Ray Description of Almost-Stratified Medium

The propagation of an acoustic wave at ocean is well approximated by the model of the almost-stratified medium, i.e. the medium with smoothly varying (in comparison with depth) velocity of the wave propagation in a horizontal direction. Following the logic of Sec. 5.1, we shall consider horizontal coordinates $\rho = (x, y)$ as "slow" arguments, and vertical coordinate z — as a "fast" argument, i.e. velocity $c = c(\alpha x, \alpha y, z)$.

The wave equation for the time component of the Fourier transform, in the spatial domain that does not contain the sources (4.41), reduces to the Helmholtz equation:

$$\left(\frac{\partial^2}{\partial x^2} + \frac{\partial^2}{\partial y^2} + \frac{\partial^2}{\partial z^2} + k^2(\alpha x, \alpha y, z)\right)\varphi = 0, \tag{5.31}$$

$$k^2 = \frac{\omega^2}{c^2}.$$

Let's in the spatial domain in which the solution of the above equation is considered for the interfaces of the water column $Z^\pm = Z^\pm(\alpha x, \alpha y)$ the next boundary condition is valid

$$[a^\pm(\alpha x, \alpha y)I \pm b^\pm(\alpha x, \alpha y)(\mathbf{n}\cdot\boldsymbol{\nabla})]\big|_{Z^\pm}\varphi = 0, \tag{5.32}$$

where $a^{\pm}$, $b^{\pm}$ are the real quantities and

$$n^{\pm} = \left[1 + \left(\frac{\partial Z^{\pm}}{\partial x}\right)^2 + \left(\frac{\partial Z^{\pm}}{\partial y}\right)\right]^{-1/2} \left(-\frac{\partial Z^{\pm}}{\partial x}, -\frac{\partial Z^{\pm}}{\partial y}, 1\right)$$

is an outer normal vector. Let us write (5.31) and boundary conditions (5.32) in the form $\mathcal{L}\varphi = 0$:

$$\mathcal{L} = \begin{Vmatrix} \hat{L} \\ \hat{\Gamma} \end{Vmatrix}$$

where operator $\hat{\Gamma}$ in an explicit form allows the argument of the asymptotic expansion.

Compressing the horizontal coordinates, we make the variation scale of the velocity be the same in the all directions $x_0 = \alpha x$, $y_0 = \alpha y$, $z_0 = z$. In the new coordinates the operator $\mathcal{L}$ looks like

$$\hat{L} => \left[\alpha^2 + \left(\frac{\partial^2}{\partial x_0^2} + \frac{\partial^2}{\partial y_0^2}\right) + \left(\frac{\partial^2}{\partial z_0^2} + k^2(x_0, y_0, z_0)\right)\right] \triangleq \alpha^2 \Delta_0 + \tilde{L}_z, \tag{5.33}$$

$$\hat{\Gamma} => \left[a^{\pm}\hat{I} + b^{\pm}\left(\frac{\partial}{\partial z_0} - \alpha^2(\mathbf{n}^{\pm} \cdot \boldsymbol{\nabla}_0)\right)\right]\Big|_{Z^{\pm}}. \tag{5.34}$$

Taking into account the asymptotic representation

$$\varphi \sim \mathcal{A}(x_0, y_0, z_0; \alpha) \exp\left\{\frac{i}{\alpha}\tau(x_0, y_0)\right\}$$

the solvability condition (5.5) (with the presence of the boundary conditions (5.34)) can be written as

$$[e^{-i(\nabla_0\tau\cdot\boldsymbol{\rho})}\alpha^2\Delta_0 e^{i(\nabla_0\tau\cdot\boldsymbol{\rho})} + \tilde{L}_z]\mathcal{A}_0 = 0,$$
$$\hat{\Gamma}_z \triangleq \left(a^{\pm}\hat{I} + b^{\pm}\frac{\partial}{\partial z}\right)\mathcal{A}_0 = 0,$$

where $\boldsymbol{\rho} = (x, y)$, i.e.

$$[(\boldsymbol{\nabla}\tau \cdot \boldsymbol{\nabla}\tau) - \tilde{L}_z]\mathcal{A}_0 = 0, \tag{5.35}$$

$$\Gamma_z\big|_{Z^{\pm}} \mathcal{A}_0 = 0. \tag{5.36}$$

The equation (5.35) leads to the problem of eigenvalues and eigenfunctions of the operator $\tilde{L}_s$ with boundary conditions (5.36):

$$\tilde{L}_z\mathcal{A}_0 = \lambda^2\mathcal{A}_0,$$
$$\hat{\Gamma}\big|_{Z^{\pm}} = 0, \tag{5.37}$$

where

$$\lambda^2 = (\boldsymbol{\nabla}\tau \cdot \boldsymbol{\nabla}\tau). \tag{5.38}$$

Let us consider the equation (5.37) (omitting the zero index):

$$\frac{\partial^2}{\partial z^2}\mathcal{A}_0^n + \left(\frac{\omega^2}{c^2_{x,y}(z)} - \lambda_n^2\right)\mathcal{A}_0^n = 0,$$

$$a^{\pm}\mathcal{A}_0^n\,|_{Z^{\pm}_{x,y}} + b^{\pm}\frac{\partial}{\partial z}\mathcal{A}_0^n\,|_{Z^{\pm}_{x,y}} = 0.$$

Introducing the notations $c_{x,y}\,(z), Z^{\pm}_{x,y}$, we underscore "slow" arguments, i.e. we consider the eigenfunction and eigenvalue problem (Sturm–Liouville problem) as parametric one, which depends on a point (x, y).

Equation (5.37) can be solved by WKB method (see. Sec. 5.3), which allows us to find the short-wave asymptotic for amplitude $\mathcal{A}_0^n$. It is obviously that eigenvalues λ_n^2 depend on the interface position $Z^{\pm}_{x,y}$, the frequency ω and the depth dependence of the velocity:

$$\lambda_n^2 = \lambda_n^2\,(Z^{\pm}_{x,y}, \omega, c_{x,y}\,(z)).$$

It is possible to solve the equation (5.38) (eikonal equation) by the method of characteristics (see Sec. 5.2):

$$\frac{d\boldsymbol{\rho}}{ds} = \mathbf{p}, \qquad \frac{d\mathbf{p}}{ds} = \frac{1}{2}\boldsymbol{\nabla}\lambda^2(\boldsymbol{\rho}),$$

$$\frac{d\tau}{ds} = (\mathbf{p}\cdot\mathbf{p}), \qquad \frac{d\tau}{ds} = \lambda^2(\boldsymbol{\rho}).$$

Let us represent the amplitude $\mathcal{A}_0^n$ in the form

$$\mathcal{A}_0^n = A_0(x, y, z) = a_0(x, y)\psi_n(x, y, z),$$

where $\{\psi_n\}$ is an orthonormal basis of the operator, $\int\limits_{Z^-}^{Z^+} \psi_n\psi_{n'}dz = \delta_{nn'}$. Therefore it is possible to obtain the transport equation for reduced amplitude a_0, which depends on horizontal coordinate only; in accordance with formula (5.6) we can write:

$$a_0\boldsymbol{\nabla}^2\tau + 2(\boldsymbol{\nabla}\tau\cdot\boldsymbol{\nabla}a_0) = 0.$$

It is possible to represent the transport equation as:

$$\boldsymbol{\nabla}\cdot(a_0^2\boldsymbol{\nabla}\tau) = 0. \tag{5.39}$$

Integrating (5.39) on the ray tube volume (see Fig. 5.3)and using the Gauss theorem, we obtain

$$\int (a_0^2\cdot\boldsymbol{\nabla}\tau)\cdot d\boldsymbol{\sigma} = 0.$$

Taking into account that the tube cross-section is represented as $d\boldsymbol{\sigma} = \boldsymbol{\nabla}\tau/|\boldsymbol{\nabla}\tau|\sigma$, and at the tube wall is valid $\boldsymbol{\nabla}\tau\cdot d\boldsymbol{\sigma} = 0$, we can write: $\lambda_n a_0^2\cdot\Delta\sigma = \text{const}$. Here $\Delta\sigma$ is a square of the tube cross-section, i.e. reduced amplitude varies as $a_0 \sim r^{-1/2}$.

For space-time ray, using asymptotic representation

$$\varphi \sim \mathcal{A}(x_0, y_0, z, t_0; \alpha)\exp\left[\frac{i}{\alpha}\tau(x_0, y_0, t_0)\right],$$

we obtain

$$\lambda_n^2 = \lambda_n^2(\omega, x, y, [c(z)], t)$$

and, respectively, the eikonal equation reads as

$$(\nabla\tau \cdot \nabla\tau) = \lambda_n^2(\omega, x, y, [c(z)], t). \tag{5.40}$$

Taking into consideration (5.40), the characteristic equations can be represented as

$$\begin{aligned}\frac{dx}{p_x} = \frac{dy}{p_y} &= \frac{dt}{\lambda_n \partial_\omega \lambda_n} = \frac{d\tau}{\lambda_n(\lambda_n - \omega\partial_\omega\lambda_n)} \\ &= \frac{dp_x}{\lambda_n \partial_x \lambda_n} = \frac{dp_y}{\lambda_n \partial_y \lambda_n} = -\frac{d\omega}{\lambda_n \partial_t \lambda_n}\end{aligned} \tag{5.41}$$

$(p_x = \partial/\partial x\, \tau,\ p_y = \partial/\partial y\, \tau)$

and define the horizontal space-time rays.

After differentiation of (5.40) on $\mathbf{p} = (p_x, p_y)$ and taking into consideration that $\omega = \omega(\mathbf{p})$, we obtain

$$\mathbf{p} = \lambda_n \partial_\omega \lambda_n \nabla_{\mathbf{p}} \omega. \tag{5.42}$$

Using the first two equalities from (5.41) and equation (5.42), we can write

$$\frac{d\boldsymbol{\rho}}{dt} = \frac{\mathbf{p}}{\lambda_n \partial_\omega \lambda_n} = \nabla_{\mathbf{p}} \omega.$$

Hence, the horizontal space-time rays determine the trajectories of points that move with the group velocity.

Let us consider representation of the exact solution of the problem (5.31), (5.32) in a form of the normal modes for a stratified ocean. Each normal mode is a solution of the equation (5.31) with separated variables:

$$\varphi(z, \boldsymbol{\rho}) = \varphi^z(z)\varphi^\rho(\boldsymbol{\rho}),$$

$$\left[\frac{\partial^2}{\partial z^2} + (k^2(z) - \lambda_n^2)\right]\varphi^z(z) = 0, \tag{5.43}$$

$$\Gamma\big|_{Z\pm}\, \varphi(z) = 0,$$

$$\left[\frac{\partial}{\partial \rho}\left(\rho\frac{\partial}{\partial \rho}\right) + \lambda_n^2\right]\varphi^\rho(\rho) = 0,$$

$$\Gamma\big|_\infty\, \varphi^\rho = 0. \tag{5.44}$$

At Fig. 5.4 the system of rays into the sound channel with bounds z', z'', which are turning points (in terms of WKB method), is represented.

The Sturm–Liouville problem (5.43) has an infinite number of simple real eigenvalues $\lambda_0^2 > \lambda_1^2 > \lambda_2^2 > > \ldots$, where $\lambda_n : \lambda_n^2 > 0$ is the finite number and $\lambda^2 < 0$ is the infinite number. We should choose $\lambda_n > 0$, if $\lambda_n^2 > 0$, and $\mathrm{Im}\lambda_n > 0$, if

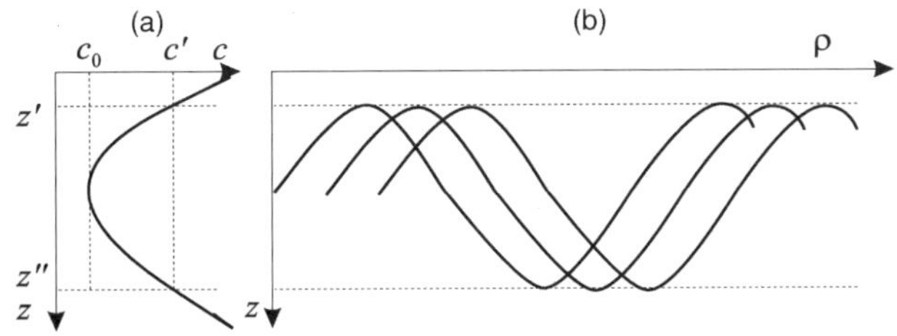

Fig. 5.4 The system of rays in the sound channel.

$\lambda_n^2 < 0$. In this case the solution of the equation (5.44) φ^ρ satisfies under $\lambda = \lambda_n$ the radiation condition:

$$\lim_{\rho \longrightarrow \infty} \rho^{1/2} \left(\frac{\partial}{\partial \rho} \varphi^\rho - i\lambda \varphi^\rho \right) = 0$$

and it is proportional to the Hankel function $H_0^{(1)}(\lambda_n \rho)$:

$$\varphi^\rho \sim H_0^{(1)}(\lambda_n \rho).$$

Thus, the exact solution $\varphi(z, \rho)$ of the homogeneous equation (5.31) with the use of the basis $\{\varphi_n(z)\}$ of the equations (5.43) can be represented as infinite series:

$$\varphi(z, \rho) = \sum_{n=0}^{\infty} \mathcal{A}_n \varphi_n^- H_0^{(1)}(\lambda_n \rho). \tag{5.45}$$

Using the above expression, let us to write the solution of the wave equation with the point source

$$(\nabla^2 + k^2)\varphi = -\delta(z - z_0)\frac{\delta(\rho)}{2\pi\rho}. \tag{5.46}$$

For this purpose we shall take the advantage of the equality

$$\left[\frac{\partial}{\partial \rho} \left(\rho \frac{\partial}{\partial \rho} \right) + \lambda_n^2 \right] H_0^{(1)}(\lambda_n \rho) = \frac{4i\delta(\rho)}{2\pi\rho}.$$

Substituting the solution (5.45) of the homogeneous equation (5.31) to the expression (5.46), we obtain the connection of the coefficients $\mathcal{A}_n$ and eigenfunctions $\varphi_n^z(z)$:

$$\sum_{n=0}^{\infty} \mathcal{A}_n \varphi_n^z = \frac{i}{4}\delta(z - z_0).$$

Hence, making the scalar product of the above equality with φ_n^z, we obtain

$$\mathcal{A}_n = \frac{i}{4}\varphi_n^z(z_0).$$

The exact solution of the acoustic equation describing propagation of the signal, generated by the point source at the depth z_0, has the next representation in a stratified ocean with a constant depth

$$\varphi(\rho, z) = \frac{i}{4}\sum_{n=0}^{\infty} \varphi_n^z(z_0)\varphi_n^z(z)H_0^{(1)}(\lambda_n \rho). \tag{5.47}$$

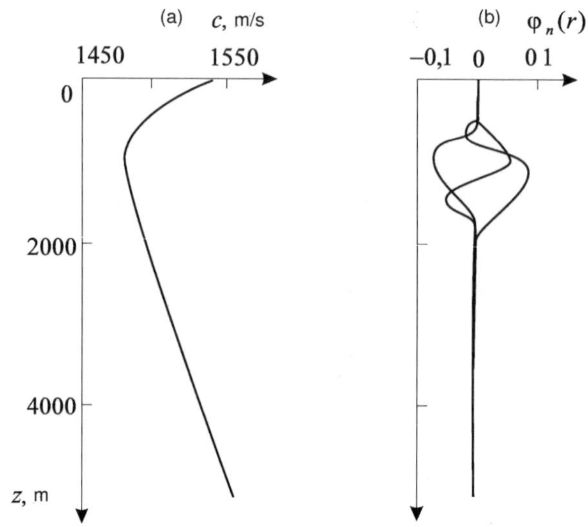

Fig. 5.5 The typical profile of the sound velocity in ocean (a) and the first tree modes (b).

At Fig. 5.5 a typical sound velocity profile in ocean (a) and the first three modes for the frequency 30 Hz (b) are represented.

The asymptotic of the exact solution (5.47) in the wave zone (ρ is great enough), using asymptotic for the Hankel function, can be written as

$$\varphi(\rho, z) \overset{\rho\to\infty}{\sim} \frac{i}{2^{3/2}\pi^{1/2}\rho^{1/2}} \sum_{n=0}^{\infty} \lambda_n^{1/2} \varphi_n^z(z_0)\varphi_n^z(z) \exp\left[i\left(\lambda_n \rho - \frac{\pi}{4}\right)\right]. \tag{5.48}$$

(We will use the expressions (5.48) for the development of the algorithms.)

5.6 Surface Wave in Vertically Inhomogeneous Medium

At study of a crystal structure and the upper earth mantle, the information contained in the surface waves which extracted on the seismograms is widely used, because their signal-to-noise merit has enough high magnitude in comparison with the body waves (an amplitude of surface waves are described by the factor $r^{-1/2}$ and for the body waves this factor is r^{-1}). Transiting through an area with a various geological feature, the waves accumulate the information on elastic properties and a geometry of these areas. This information is most concentrated in dependence of the velocity of propagation on frequency. In turn, the extraction of the surface waves is guaranteed by a presence of the polarization.

At a long-range propagation of the ground waves allows the ray method and to take advantage of the asymptotic expansion similar to the expansion of the acoustic field at a stratified ocean (see Sec. 5.5):

$$\boldsymbol{\varphi}(x,y,z,t;\alpha) = \mathcal{A}(\alpha x,\alpha y,z;\alpha)\exp\left\{\frac{i}{\alpha}\tau(\alpha x,\alpha y,\alpha t)\right\}.$$

In this representation the phase velocity is equal to $c = |\boldsymbol{\nabla}\tau|^{-1}$ and we consider the rays laying in a horizontal plane.

The Lame operator for an isotropic inhomogeneous medium reads as

$$\begin{aligned} L &= \rho\partial_t^2 - (\lambda+\mu)\boldsymbol{\nabla}(\boldsymbol{\nabla}\cdot) - \mu\Delta - (\boldsymbol{\nabla}\lambda)(\boldsymbol{\nabla}\cdot) \\ &\quad - (\boldsymbol{\nabla}\mu)\times(\boldsymbol{\nabla}\times) - 2(\boldsymbol{\nabla}\mu,\boldsymbol{\nabla}). \end{aligned} \tag{5.49}$$

In order to use the expression (5.5) to build zero approximation of geometrical optics for the sounding vector field

$$L_0(p)\mathcal{A}_0 = 0, \tag{5.50}$$

it needs to obtain the p-representation of the Lame operator (5.49):

$$\exp\{-i[(\mathbf{p}\cdot\mathbf{x}) + p_0 t]\}L\exp\{i[(\mathbf{p}\cdot\mathbf{x}) + p_0 t]\} \triangleq L(p).$$

Here $\mathbf{p} = (p_x, p_y)$ is a gradient on horizontal coordinates; $p_0 = \partial\tau/\partial t$. In Cartesian coordinates (x,y,z) the matrix operator L has the next terms:

$$\begin{aligned} L_{ii} &= \rho\partial_t^2 - \Bigg[(\lambda+\mu)\partial_i^2 + \mu\sum_{j=1}^{3}\partial_j^2 + (\partial_i\lambda)\partial_i \\ &\quad - \sum_{j=i+1}^{i+2}(\partial_j\mu)\partial_j + 2\sum_{j=1}^{3}(\partial_j\mu)\partial_j\Bigg], \end{aligned}$$

where $i = \begin{matrix}1 & 2 & 3\\ x & y & z\end{matrix}$ with cyclic permutation, for example,

$$\begin{matrix}3\\z\end{matrix} + 2 = \begin{matrix}2\\y\end{matrix}\;:$$

$$L_{ij} = -[(\lambda+\mu)\partial_i\partial_j + \partial_i\lambda\partial_j + (\partial_j\mu)\partial_i].$$

Let us write separately the appropriate components L_{ij} $(i,j \neq z)$ in p-representation, taking into account that $\tau = \tau(x,y,t)$,

$$\begin{aligned} L_{ii}(\mathbf{p}) &= -\rho p_0^2 + (\lambda+\mu)p_i^2 + \mu(\mathbf{p},p) - \mu\partial_z^2 - \partial_z\mu\partial_z \\ &= -\rho p_0^2 + (\lambda+\mu)p_i^2 + \mu(\mathbf{p},p) - \partial_z\mu\partial_z, \end{aligned}$$

$$L_{ij} = (\lambda+\mu)p_i p_j,$$

and components which contain z

$$L_{jz} = -i[(\lambda+\mu)\partial_z + (\partial_z\mu)p_j],$$

$$L_{zj} = -i[(\lambda+\mu)p_j\partial_z + (\partial_z\lambda)p_j],$$

$$L_{zz} = -\rho p_0^2 - (\lambda + 2\mu)\partial_z^2 + \mu(\mathbf{p} \cdot \mathbf{p}) - (\partial_z(\lambda + 2\mu))\partial_z.$$

Let us consider the local orthogonal coordinates (*ray coordinates*) with orts

$$\mathbf{e}_\tau = \frac{\nabla\tau}{|\nabla\tau|} \triangleq \frac{\mathbf{p}}{|\mathbf{p}|}, \quad \mathbf{e}_z = \mathbf{e}_z, \quad \mathbf{e}_\beta = [\mathbf{e}_\tau \times \mathbf{e}_z].$$

Then $\mathcal{A} = (\mathcal{A}_\tau, \mathcal{A}_\beta, \mathcal{A}_z)$; $\mathbf{p} = (p_\tau, p_\beta)$, furthermore $p_\beta \equiv 0$,

$$\begin{aligned}
L_{\tau\tau}(p) &= -\rho p_0^2 + (\lambda + 2\mu)(\mathbf{p}, \mathbf{p}) - \partial_z(\mu\partial_z),\\
L_{\beta\beta}(p) &= -\rho p_0^2 + \mu(\mathbf{p}, \mathbf{p}) - \partial_z(\mu\partial_z),\\
L_{zz}(p) &= -\rho p_0^2 + \mu(\mathbf{p}, \mathbf{p}) - \partial_z((\lambda + 2\mu)\partial_z),\\
L_{\tau z}(p) &= -i\left[(\lambda + \mu)p_\tau \frac{\partial}{\partial z} + (\partial_z\mu)p_\tau\right],\\
L_{z\tau}(p) &= -i\left[(\lambda + \mu)p_\tau \frac{\partial}{\partial z} + (\partial_z\lambda)p_\tau\right],\\
L_{\beta\tau} &= L_{\tau\beta} = L_{\beta z} = L_{z\beta} = 0.
\end{aligned}$$

So, the matrix $\mathbf{L}(p)$ has the next shape

$$\mathbf{L}(p) = \begin{Vmatrix} L_{\tau\tau} & 0 & L_{\tau z} \\ 0 & L_{\beta\beta} & 0 \\ L_{z\tau} & 0 & L_{zz} \end{Vmatrix},$$

and the vector equation (5.50) can be represented in the form

$$\mathbf{L}_{\beta\beta}\mathcal{A}_\beta = 0, \tag{5.51}$$

$$\begin{aligned}
\mathbf{L}_{\tau\tau}\mathcal{A}_\tau + \mathbf{L}_{\tau z}\mathcal{A}_z &= 0,\\
\mathbf{L}_{z\tau}\mathcal{A}_\tau + \mathbf{L}_{zz}\mathcal{A}_z &= 0.
\end{aligned} \tag{5.52}$$

The equations (5.51) and (5.52) should be supplemented with corresponding boundary conditions. If we consider the surface wave propagation inside half space with a weak horizontal inhomogeneity, then boundary conditions should define at the free surface $Z^0 = Z^0(\tau, \beta)$ and over infinity $Z \longrightarrow \infty$.

Let us write the operator Γ of the boundary condition at the free surface $\mathbf{t} = \tau\mathbf{n} = 0$:

$$\Gamma\mathcal{A} = \Gamma_{ij}\mathcal{A}_j,$$

$$\Gamma_{ii} = n_i(\lambda + 2\mu)\partial_i + \sum_{k \neq i} n_k\mu\partial_k,$$

$$\Gamma_{ij} = n_i\lambda\partial_j + n_j\mu\partial_i.$$

We remind that

$$\mathbf{n} = (1 + (\partial_\tau Z^0)^2 + (\partial_\beta Z^0)^2)^{-1/2}(\partial_\tau Z^0, \partial_\beta Z^0, -1),$$

then we can write the p-representation of the operator Γ $(\Gamma(p) = e^{-i(\mathbf{p},x)}\Gamma e^{(\mathbf{p},x)}$, $\mathbf{p} = (p_\tau, p_\beta))$:

$$\Gamma(p) = -\begin{Vmatrix} \mu\partial_z & 0 & i\mu p_\tau \\ 0 & \mu\partial_z & 0 \\ i\lambda p_\tau & 0 & (\lambda+2\mu)\partial_z \end{Vmatrix}.$$

The complete solvability condition for the equation $L\boldsymbol{\varphi} = 0$, $\Gamma\boldsymbol{\varphi} = 0$ or $\mathcal{L}\boldsymbol{\varphi} = 0$ has a form $\mathcal{L}_0(p)\mathcal{A}_0 = 0$, i.e. in this case the equation for the β-component of the amplitude $\mathcal{A}$ $(L_{\beta\beta}\mathcal{A}_\beta = 0,\ \Gamma\mathcal{A}_\beta = 0)$ transforms to the Sturm–Liouville problem

$$\begin{aligned} &\partial_z(\mu\partial_z)\mathcal{A}_\beta - (\lambda_n^2 - \rho p_0^2)\mathcal{A}_\beta = 0, \\ &\partial_z\, \mathcal{A}_\beta|_{Z^0} = 0, \qquad \mathcal{A}_\beta \overset{z\to 0}{\longrightarrow} 0. \end{aligned} \tag{5.53}$$

Here $\lambda_n^2 = \mu(\mathbf{p},\mathbf{p})$, i.e. the eikonal equation (and respectively the ray geometry of the surface wave) determines by the mode (λ_n) and oscillation of the component $\mathcal{A}_\beta$ along z axis.

We should note, that in the problem (5.53) the shear modulus μ appears only. The polarization vector is orthogonal to the propagation direction and lies in horizontal plane. The such wave is the well known Love wave (Levshin *et al.*, 1989).

The reminder of the boundary conditions at the free surface for a linear combination is described by the vectors with polarizations $\mathbf{e}_z$ and $\mathbf{e}_\tau$:

$$\mu\partial_z\mathcal{A}_\tau - i\mu p_\tau\mathcal{A}_z = 0,$$

$$i\lambda p_\tau\mathcal{A}_\tau + (\lambda+2\mu)\partial_z\mathcal{A}_z = 0.$$

Supplemented by conditions $\mathcal{A}_\tau \overset{z\to\infty}{\longrightarrow} 0$ and $\mathcal{A}_z \overset{z\to\infty}{\longrightarrow} 0$ the above equations and equations (5.52) describe the Rayleigh wave (Aki and Richards, 2002; Levshin *et al.*, 1989; Yanovskaya, 1996).

5.7 Ray Approximation of Electromagnetic Fields

The description of the processes of the propagation in the nonuniform medium can be based on the ray theory of electromagnetic fields. The ray method can be considered as a basis for obtaining the transport equations and to establish a connection between the statistical characteristics of the medium and the parameters of the phenomenological transport theory of an electromagnetic radiation.

Let us write the equations for an electric intensity vector using the formulas (4.63), (4.64):

$$\nabla\times\mathbf{E} = -\mu\frac{1}{c}\frac{\partial}{\partial t}\mathbf{H},$$

$$\nabla\times\mathbf{H} = \frac{4\pi}{c}\mathbf{j} + \frac{\varepsilon}{c}\frac{\partial}{\partial t}\mathbf{E},$$

If the medium is described by the inductivity tensor $\hat{\varepsilon}$ and the magnetic conductivity $\mu \approx 1$, and the extraneous currents ($\mathbf{j} = 0$) are missed, we can write:

$$\hat{L}\mathbf{E} = \left(\boldsymbol{\nabla} \times \boldsymbol{\nabla} x - c^{-2}\frac{\partial^2}{\partial t^2}\hat{\varepsilon}\right)\mathbf{E} = 0.$$

Using the general approach from Sec. 5.1, to write the asymptotic expansion of the vector $\mathbf{E}$ in the next form

$$\mathbf{E} = \mathcal{A}(y, \alpha)\exp\left\{\frac{i}{\alpha}\tau(y)\right\},$$

where $y = (y_0, y_1, y_2, y_3) \triangleq (\alpha t, \alpha x, \alpha y, \alpha z)$. The operator L in the p-representation ($\mathbf{p} \triangleq (p_0, p_1, p_2, p_3) \triangleq (p_0, \mathbf{p})$, $p_0 = \omega$) has the next form:

$$L(p) = e^{-i(\mathbf{p},\mathbf{x})}\hat{L}e^{i(\mathbf{p},\mathbf{x})} = -\mathbf{p} \times \mathbf{p}x - \frac{\omega^2}{c^2}\hat{\varepsilon}.$$

Let us represent the operator $L(p)$ as $L = L_0 + \alpha L'$, where $\alpha L' \triangleq \triangleq (-\omega^2/c^2)(\hat{\varepsilon} - \hat{I}\operatorname{Re}\operatorname{sp}\hat{\varepsilon}/3)$ describes an anisotropy and a nonconservativity of the media, $L_0 = -\mathbf{p} \times \mathbf{p}x - (\omega^2/c^2)\varepsilon_0\hat{I}$, ε_0 is real and equal to $\operatorname{Re}\operatorname{sp}\hat{\varepsilon}/3$. The equation for a zero approximation of the ray method in the local coordinates

$$\mathbf{e}_1 = \frac{\mathbf{p}}{|\mathbf{p}|}, \quad \mathbf{e}_2 = \frac{\partial_{\mathrm{S}}\mathbf{e}_1}{|\partial_{\mathrm{S}}\mathbf{e}_1|}, \quad \mathbf{e}_3 = \mathbf{e}_1 \times \mathbf{e}_2$$

looks as follows

$$\hat{L}_0\mathcal{A} = \left\| \begin{matrix} \frac{\omega^2}{c^2}\varepsilon_0 & 0 & 0 \\ 0 & -(\mathbf{p},\mathbf{p}) + \frac{\omega^2}{c^2}\varepsilon_0 & 0 \\ 0 & 0 & -(\mathbf{p},\mathbf{p}) + \frac{\omega^2}{c^2}\varepsilon_0 \end{matrix} \right\| \cdot \left\| \begin{matrix} \mathcal{A}_1 \\ \mathcal{A}_2 \\ \mathcal{A}_3 \end{matrix} \right\| = 0, \tag{5.54}$$

and in the canonical representation:

$$\hat{L}_0(p) = (\mathbf{e}_2\mathbf{e}_2^T + \mathbf{e}_3\mathbf{e}_3^T)\left[\frac{\omega^2}{c^2}\varepsilon_0 - (\mathbf{p},\mathbf{p})\right] + \mathbf{e}_1\mathbf{e}_1^T\frac{\omega^2}{c^2}\varepsilon_0.$$

In the case of the equality of the eigenvalues $\lambda_2 = \lambda_3 = (\omega^2/c^2)\varepsilon_0 - (\mathbf{p},\mathbf{p})$, it is possible to make the concluding about the degeneration of the polarisation of the shear waves ($(\mathbf{e}_2 \cdot \mathbf{p}) = 0$, $(\mathbf{e}_3 \cdot \mathbf{p}) = 0$). From the equation (5.9) in this case it follows $|\mathbf{p}|^2 = (\omega^2/c^2)\varepsilon_0$, i.e. the dispersion relation has the form

$$\omega - \omega(\mathbf{p},\mathbf{x}) = \omega - |\mathbf{p}|c\varepsilon_0^{-1/2} = 0.$$

The group velocity $\mathbf{v}_{\mathrm{gv}} = (\partial/\partial\mathbf{p})\,\omega(p,x)$ is equal to $\mathbf{v}_{\mathrm{gv}} = c\varepsilon_0^{-1/2}\mathbf{e}_1$ in this case. Instead of the ray parameter s we introduce an arc length l, which is connected with the ray parameter by the next relation

$$(dl)^2 = \left(\frac{\partial H}{\partial \mathbf{p}}\right)^2 ds^2,$$

where

$$H = \frac{1}{2}\left[(\mathbf{p},\mathbf{p}) - \frac{\omega^2}{c^2}\varepsilon_0\right]. \tag{5.55}$$

From here it follows that $dl = |\mathbf{p}|ds$.

Let us write the ray equations using arc length of the space projection of the ray dl:

$$\frac{d\mathbf{x}}{dl} = \frac{1}{|\mathbf{p}|}\frac{d\mathbf{x}}{ds} = \frac{1}{|\mathbf{p}|}\mathbf{p} = \frac{\mathbf{v}_{\mathrm{g}}}{|\mathbf{v}_{\mathrm{g}}|} = \mathbf{e}_1, \tag{5.56}$$

$$\frac{dt}{dl} = \frac{1}{|\mathbf{p}|}\frac{dt}{ds} = \frac{\omega}{|\mathbf{p}|} = \frac{1}{|\mathbf{v}_{\mathrm{g}}|}, \tag{5.57}$$

$$\frac{d\mathbf{p}}{dl} = \frac{1}{|\mathbf{p}|}\frac{d\mathbf{p}}{ds} = -\frac{1}{|\mathbf{p}|}\frac{\partial H}{\partial \mathbf{x}} = \frac{|\mathbf{p}|}{|\mathbf{p}|}\frac{\varepsilon_0^{1/2}}{c}\partial_{\mathbf{x}}\omega = \mathbf{v}_{\mathrm{g}}^{-1}\partial_{\mathbf{x}}\omega, \tag{5.58}$$

$$\frac{d\omega}{dl} = \frac{1}{|\mathbf{p}|}\frac{d\omega}{ds} = -\frac{1}{|\mathbf{p}|}\frac{\partial H}{\partial t} = \frac{1}{|\mathbf{p}|}\frac{\omega}{c^2}\varepsilon_0\partial_t\omega(\mathbf{p},x) = \mathbf{v}_{\mathrm{g}}^{-1}\partial_t\omega. \tag{5.59}$$

Taking into account the equations (5.56), (5.59) it is possible to write equation for kinematics of a point belonging to the ray

$$\frac{d\mathbf{x}}{dt} = \mathbf{v}_{\mathrm{gv}}.$$

So, the point motion describes by the group velocity, which for the isotropic medium is coincided with the vector $\mathbf{p}$.

The equations (5.58), (5.59) allow to obtain the connection between the wave vector and the spatial gradient ω:

$$\frac{d\mathbf{p}}{dt} = \partial_{\mathbf{x}}\omega.$$

At last, from the relations (5.56) – (5.59) follow, that total time derivative of ω is equal to the local derivative:

$$\frac{d\omega}{dt} = \partial_t\omega.$$

The ray tomography can be realized not only using phase measurements, but also using the polarization vector.

Let us consider the derivation of *the law of rotation of the polarization vector* (Kravtsov and Apresyan, 1996). For this purpose we write the relation (5.6) for the amplitude of a zero approximation of the ray method $\mathbf{E}_0$:

$$\partial_p\partial_x L_0\,(p)\mathbf{E}_0\,(x) = L'(p)\mathbf{E}_0(\mathbf{x}), \tag{5.60}$$

where

$$L'(p) = -\frac{\omega^2}{c^2}\left(\hat{\varepsilon} - \frac{1}{3}\hat{I}\mathrm{Resp}\hat{\varepsilon}\right).$$

We will use a doubled Hamilton function H (5.55) instead of $L_0(p)$ and rewrite the formula (5.60) as follows

$$\partial_{p_0}H\partial_t\mathbf{E}_0 + \partial_{\mathbf{p}}H\partial_{\mathbf{x}}\mathbf{E}_0 = \frac{1}{2}L'(p)\mathbf{E}_0(x).$$

To change the derivatives of the Hamilton function:

$$\partial_{p_0} H = \frac{dt}{ds}, \qquad \partial_{\mathbf{p}} H = \frac{d\mathbf{x}}{ds}$$

and to pass to the parameter l (arc length of the spatial ray), we obtain

$$\frac{\partial t}{\partial l}\partial_t \mathbf{E}_0 + \frac{d\mathbf{x}}{dl}\partial_{\mathbf{x}}\mathbf{E}_0 = \frac{1}{2}|\mathbf{p}|L'\mathbf{E}_0. \tag{5.61}$$

Let us note, that in the left hand side of (5.61) there is the first total derivative of $\mathbf{E}_0$ on length of the nonstationary ray, i.e.

$$\frac{d}{dl}\mathbf{E}_0 = \frac{\omega}{2l}\sqrt{\varepsilon_0}L'\mathbf{E}_0.$$

Writing the equation with zero approximation for $\mathbf{E}_0$ on the dispersion surface, in other words, taking into account that in accordance with the equation (5.54)

$$\mathbf{E}_0 = \mathcal{A}_2\mathbf{e}_2 + \mathcal{A}_3\mathbf{e}_3,$$

we project the equation (5.61) to a plane which is orthogonal to

$$P_1 d_l \mathbf{E}_0 = \frac{\omega}{2c}\sqrt{\varepsilon}P_1 L'\mathbf{E}_0, \tag{5.62}$$

where

$$P_1 = I - \mathbf{e}_1 \cdot \mathbf{e}_1^T.$$

If an anisotropy of the absorption is absent, then $L' = 0$, and the relation (5.62) can be rewritten in a form $P_1 d_l \mathbf{E}_0 = 0$, i.e. we obtain the system of two equations for $\mathcal{A}_2$ and $\mathcal{A}_3$. Denoting the first derivative on l by a dot, we write

$$\begin{aligned} \dot{\mathcal{A}}_2 + \mathcal{A}_2(\mathbf{e}_2, \mathbf{e}_3) &= 0, \\ \dot{\mathcal{A}}_3 + \mathcal{A}_3(\mathbf{e}_3, \mathbf{e}_2) &= 0. \end{aligned} \tag{5.63}$$

Multiplying the first equation of (5.63) by $\mathcal{A}_3$, the second equation by $\mathcal{A}_2$, and after summing up, we obtain

$$\mathcal{A}_3\dot{\mathcal{A}}_2 - \mathcal{A}_2\dot{\mathcal{A}}_3 = \mathcal{A}_2^2(\mathbf{e}_3, \dot{\mathbf{e}}_2) - \mathcal{A}_3^2(\mathbf{e}_2, \dot{\mathbf{e}}_3). \tag{5.64}$$

Since $d_l(\mathbf{e}_2, \mathbf{e}_3) = 0$, then the equation (5.64) is possible to rewrite in a form

$$(\mathbf{e}_2, \dot{\mathbf{e}}_3) = \frac{\mathcal{A}_3\dot{\mathcal{A}}_2 - \mathcal{A}_2\dot{\mathcal{A}}_3}{\mathcal{A}_2^2 + \mathcal{A}_3^2}.$$

Let us note, that in the right-hand side of the above equality there is the derivative of $\mathrm{arctg}\mathcal{A}_2/\mathcal{A}_3 = \theta$, and the scalar product $(\mathbf{e}_2, \dot{\mathbf{e}}_3)$ is equal to $|\dot{\mathbf{e}}_3|$. In turn, $|\dot{\mathbf{e}}_3|$ is *rotating of curve* (ray) by definition. Having designated value of rotating as T, we shall write the law of rotating of the polarization vector as

$$d_l\,\theta = T.$$

5.8 Statement of Problem of the Ray Kinematic Tomography

Reviewed in the previous sections, the ray approximation is a natural basis for the statement problem of a ray kinematic tomography. All ray approximations in an explicit form contain a phase factor $e^{i\omega\tau}$, where the value τ in a spatial point x is determined by the value of τ in the some point x_0 (for example, a source point) and by the integration of the (ray) refraction index along the ray from the point x_0 to a point x. Therefore if the experimental data allow to trace a phase of signal, then the connection of properties of the medium and the travel time of a sounding signal can be recognized. The same concept allows to institute a field of the attenuation factor of the medium.

As an example of the interpretation of the ray approximation, let as consider the case of an anisotropic reference medium. Let a scalar wave field $\varphi(x,t)$ satisfies the equation

$$L\varphi = 0, \tag{5.65}$$

where

$$L = \partial_t^2 - \sum_{ij} A_{ij}(x)\partial_i\partial_j, \quad i,j = 1 \div 3,$$

a matrix A is positive defined independently of a point x, i.e. the operator L is hyperbolic everywhere in an applicable domain. The characteristic equation is written in this case as

$$(\mathbf{p}, \hat{A}\mathbf{p}) = 1 \tag{5.66}$$

and it can be considered as an eikonal equation; the vector $\mathbf{p}$ is equal to $\boldsymbol{\nabla}\tau$. To equate gradient of the characteristic surface to zero, we obtain a differential equation (in a vector form) of the first order relatively to the derivative $\mathbf{p}$:

$$2(\boldsymbol{\nabla}\cdot\mathbf{p})^T\hat{A}\mathbf{p} + (\mathbf{p}, (\boldsymbol{\nabla}\hat{A})\mathbf{p}) = 0. \tag{5.67}$$

From this equation we obtain the characteristic equations:

$$\begin{aligned}\frac{d\mathbf{x}}{ds} &= \hat{A}\mathbf{p},\\ \frac{d\mathbf{p}}{ds} &= -\frac{1}{2}\Big(\mathbf{p}, (\nabla\cdot\hat{A})\mathbf{p}\Big).\end{aligned} \tag{5.68}$$

The last equality is the characteristic equation (5.66).

The characteristic equations (5.68) determine bicharacteristics of the wave equation (5.65), and for their building it is necessary to establish the initial conditions

$$\mathbf{x}\big|_{s=0} = \mathbf{x}^{(0)}, \qquad \mathbf{p}\big|_{s=0} = \mathbf{p}^{(0)},$$

where $\mathbf{p}^{(0)}: \ (\mathbf{p}^{(0)}, A\mathbf{p}^{(0)}) = 1$.

The ray direction at the point $s = 0$ we determine using the first equation of the system (5.68):

$$\frac{d\mathbf{x}}{ds}\Big|_{s=0} = \hat{A}_{x^0}\mathbf{p}^{(0)}.$$

Having noted, that $\mathbf{p} = \hat{A}^{-1} d\mathbf{x}/ds$, we write the equation (5.66) in a form

$$\left(\frac{d\mathbf{x}}{ds}, \hat{A}^{-1}\frac{d\mathbf{x}}{ds}\right) = 1 . \tag{5.69}$$

Parameter s accurate within additive constant is equal to eikonal τ, because

$$\frac{d\tau}{ds} = \left(\nabla_x \tau, \frac{d\mathbf{x}}{ds}\right) = (\mathbf{p}, A\mathbf{p}) = 1.$$

Therefore from the equation (5.69) we obtain

$$ds^2 = d\tau^2 = (d\mathbf{x}, \hat{A}^{-1} d\mathbf{x}) ,$$

and we can interpret the characteristic equations (5.68). Taking into account the relation

$$\tau = \tau^0 + \int\limits_{\mathcal{L}} (d\mathbf{x}, \hat{A}^{-1}\, d\mathbf{x})^{1/2} , \tag{5.70}$$

we conclude that the characteristic equations (5.68) determine the parametric determination of geodesic curves in the space with the metric $d\tau = = (d\mathbf{x}, A^{-1}\, d\mathbf{x})$. Using the equality $|d\mathbf{x}|^2 = dl^2$ (dl is a length of the ray arc), it is possible to write $d\mathbf{x} = \mathbf{n}\,(x) dl$ ($\mathbf{n}(x)$ is a unit vector with direction coinciding with the direction of the signal propagation in the point $\mathbf{x}$). The expression for the (*slowness*) can be represented as

$$\frac{d\tau}{dl} = v^{-1}(x, \mathbf{n}) = (\mathbf{n}, A^{-1}\mathbf{n}) .$$

Therefore the integral from the formula (5.70) over a geodesic curve $\mathcal{L}$

$$\int\limits_{\mathcal{L}} (d\mathbf{x}, A^{-1}\, d\mathbf{x})^{1/2} = \int\limits_{\mathcal{L}} \frac{dl}{v(x, \mathbf{n})} ,$$

has a sense of a minimum propagation time of the signal along a curve connected two spatial points. The explicit form of the velocity dependence $v(x, \mathbf{n})$ on the direction of propagation of the signal ($\mathbf{n}$) contains a specific character of the anisotropic model.

The restoration of coefficients $A_{ij}\,(x)$ using the known lengths of geodesic curves $\mathcal{L}$, which joining the pairs of arbitrary points $\mathbf{x}$ and $\mathbf{x}_0$ is a substance of the inverse kinematic problem. In the standard statements of such problems, the pairs of the points $\mathbf{x}$ and $\mathbf{x}_0$ are considered laying on the boundary of the explored area. Let's consider a statement of the linearized solution of a three-dimensional inverse kinematic problem. Let, the basic reference medium is characterized by a matrix $A_0\,(\mathbf{x})$, and the problem is to restore $A(\mathbf{x})$:

$$\hat{A}(x) = \hat{A}_0\,(x) + \alpha \hat{A}_1\,(x) . \tag{5.71}$$

Here α is a small parameter; $\hat{A}_0$ and $\hat{A}_1$ are the positive defined matrices. To represent the eikonal as the next expansion

$$\tau(x_0\,, x) = \sum_{n=0}^{\infty} \alpha^n\, \tau_n\,(x_0\,, x) . \tag{5.72}$$

By substituting the expressions (5.71) and (5.72) into the characteristic equation, we obtain

$$(\mathbf{p}_0 + \alpha \mathbf{p}_1 \, , (A_0 + \alpha A_1)(\mathbf{p}_0 + \alpha \mathbf{p}_1)) = 1 \, .$$

Here, in expansion on $\mathbf{p} = \boldsymbol{\nabla}\tau \approx \boldsymbol{\nabla}\tau_0 + \alpha\boldsymbol{\nabla}\tau_1 \stackrel{\Delta}{=} \mathbf{p}_0 + \alpha \mathbf{p}_1$ we keep only the linear on α terms. Taking into account that

$$(\mathbf{p}_0, \hat{A}_0 \, \mathbf{p}_0 \,) = 1 \, , \tag{5.73}$$

and keeping the terms of order α, we obtain the equation

$$2(\mathbf{p}_1 \, , \hat{A}_0) \, \mathbf{p}_0 + (\mathbf{p}_0 \, , \hat{A}_1 \, \mathbf{p}_0) = 0 \, . \tag{5.74}$$

The equation (5.73) determines the geodetic $\mathcal{L}_0 \, (x_0, x)$. Taking into account that $ds = d\tau_0$ we can write

$$(\mathbf{p}_1 \, , A_0 \, \mathbf{p}_0 \,) = \left(\mathbf{p}_1 \, , \frac{d\mathbf{x}_0}{d\tau_0} \right) = \frac{d\tau_1}{d\tau_0} \, ,$$

and from equation (5.74) we get the expression for the deduction τ_1:

$$\tau_1(x, x_0) = -\frac{1}{2} \int\limits_{\mathcal{L}_0(x_0,x)} (\mathbf{p}_0, \hat{A}_1 \, (x) \mathbf{p}_0 \,) d\tau_0 \, . \tag{5.75}$$

This expression is a basis for solution the linearized inverse kinematic problem. Let's note, that $\hat{A}_1 \, (x)$ should satisfies the condition $(A_1 \, (x) + A_0 \, (x) > 0)$.

At the remote sensing problem the kinematic statement, as a rule, connects with the isotropic reference medium, therefore

$$\hat{A}_0 \, (x) = a_0 \, (x) = v_0^2 \, (x) \, ,$$
$$\hat{A}_1 \, (x) = a_1 \, (x) = v_1^2 \, (x) \, ,$$
$$(\mathbf{p}_0 \, , A_1 \, \mathbf{p}_0 \,) = a_1 \, (x) (\mathbf{p}_0 \, , \mathbf{p}_0 \,) = \frac{a_1 \, (x)}{a_0} \, , \quad d\tau_0 = \frac{dl_0}{v_0} \, ,$$

and an appropriate constraint equation to correct the propagation time from the point x_0 to a point x looks like follows

$$\tau_1 \, (x, x_0) = -\frac{1}{2} \int\limits_{\mathcal{L}_0(x_0,x)} \frac{v_1^2 \, (x) dl_0}{v_0^3 \, (x)} \stackrel{\Delta}{=} \int\limits_{\mathcal{L}_0(x_0,x)} \theta(x) dl_0 \, . \tag{5.76}$$

Here dl_0 is the length of ray arc $\mathcal{L}_0(x_0, x)$, which joins together the points x_0 and x in the reference medium with the velocity $v_0 \, (x)$.

The representation (5.76) is a basis for the statement of the ray tomography problems. It will be demonstrated in Sec. 11.7, the characteristic equation is determined by higher derivatives of the propagation operator L only. Therefore the ray concept is easily transferred on restoration of the attenuation factor, which is connected with the first derivatives, which are included in the propagation operator L:

$$\ln \frac{|\Phi(\omega)|}{|\Phi_0(\omega)|} = - \int\limits_{\mathcal{L}_0} \beta_\omega \, (x) dl_0,$$

where $(\Phi(\omega)$ is the Fourier transform of the signal $\boldsymbol{\varphi}$ observed at the point x; $\Phi_0 \, (\omega)$ is the Fourier transform of a theoretical sounding signal, exited in the point x_0, which propagates inside the reference medium; the ray $\mathcal{L}_0 \, (x_0, x)$ joins together the source point and the observation point; $\beta_\omega \, (x)$ is a ray attenuation coefficient for the frequency ω.

Chapter 6

Methods for parameter estimation of geophysical objects

At a choice of method for the parameters estimation, it is necessary to implement study of properties of the estimated values. The main properties of the estimate were considered in § 2.2, they are: the consistence, unbiasedness, efficiency (minimum of variance), robustness and normality. Some of these properties can be executed asymptotically, i.e. at a great number of measurements, for example, asymptotic unbiasedness, asymptotic consistence, asymptotic normality. Let's consider methods of a point estimation most frequently used at processing of real geophysical data.

6.1 The Method of Moments

The method of moments is reduced to following. Let's we have a repeated sample $x_1, x_2, \ldots, x_n$ of a random variable ξ with the likelihood function

$$L(x_1, x_2, \ldots, x_n; \theta_1, \ldots, \theta_s) = \prod_{i=1}^{n} p(x_i; \theta_1, \ldots, \theta_s). \tag{6.1}$$

To find S sample moments

$$\tilde{\alpha}_s = \frac{1}{n} \sum_{i=1}^{n} x_i^s, \quad s = 1, 2, \ldots$$

If $\alpha_s = M[x_i^{2s}]$ exists, then at $n \to \infty$ the law of large numbers is valid, and we can write

$$P[|\tilde{\alpha}_s - \alpha_s| > \varepsilon] \to 0 \text{ if } n \to \infty,$$

i.e. at a large sample size $\tilde{\alpha}_s$ tends to α_s with probability close to 1. Therefore it is natural to construct the equations

$$\int_{-\infty}^{\infty} x^s p(x; \theta_1, \ldots, \theta_s) dx = \alpha_s, \quad s = 1, 2, \ldots, \tag{6.2}$$

which allow to find the estimates of the desired parameters $\theta_1, \ldots, \theta_s$. At $n \to \infty$ these estimates are asymptotically unbiased and asymptotically normal.

6.2 Maximum Likelihood Method

For any sample $(x_1, x_2, \ldots, x_n)$ the *likelihood function* (6.1) is a function of an unknown parameter vector $\boldsymbol{\theta}$. The maximum likelihood estimate of $\boldsymbol{\theta}$ is a quantity $\hat{\boldsymbol{\theta}}$, which corresponds to a maximum of the likelihood function $\ln L(x_1, x_2, \ldots, x_n; \boldsymbol{\theta})$. We can obtain the vector of estimates $\hat{\boldsymbol{\theta}}$ by the solution of the system of S likelihood equations:

$$\frac{\partial \ln L(x_1, \ldots, x_n; \boldsymbol{\theta})}{\partial \theta_s} = 0, \quad s = 1, \ldots, S, \tag{6.3}$$

if the likelihood function is a differentiable function.

In spite of the fact that the maximum likelihood method leads to the more complicate computation in comparison of the method of moments, the maximum likelihood estimator is preferable (especially in a case of a small sample size) due to the following theorems, which we shall give without the proof.

(1) If the effective estimate exists, it will be the unique solution $\hat{\boldsymbol{\theta}}$ of the system of likelihood equations (6.3).
(2) If the sufficient estimate exists, each solution of the system of likelihood equations (6.3) will be a function of this estimate. At sufficiently general conditions of the regularity of the function $\ln L$, that are usually executed in practice, the maximum likelihood estimator has following properties.
   (1) At a great enough number of observations the solution of the likelihood equation $\hat{\theta}$ or the system of likelihood equations $\hat{\boldsymbol{\theta}}$ (in case of a parameter vector) converges in probability to a true value θ or $\boldsymbol{\theta}$, i.e. the maximum likelihood method gives the consistent estimate.
   (2) The estimate of the maximum likelihood method is an asymptotically efficient estimate, i.e. at a great number of observations the Rao-Cramer inequality transfers to the equality, thus the variance of the estimated parameter reaches the lower edge:

$$\sigma^2_{\hat{\theta}_n} \overset{n \to \infty}{=} [I^{(F)}(\theta_n)]^{-1}$$

   in the case of one parameter θ,

$$R(\hat{\boldsymbol{\theta}}) \overset{n \to \infty}{=} [I^{(F)}(\boldsymbol{\theta}_n)]^{-1}$$

   in the case of the vector parameter $\boldsymbol{\theta}$.
   (3) The maximum likelihood estimate $\hat{\boldsymbol{\theta}}$ is an asymptotic normal estimate, i.e.

$$\sqrt{n}(\hat{\boldsymbol{\theta}}_n - \boldsymbol{\theta}) \sim N(0, [I^{(F)}(\boldsymbol{\theta}_n)]^{-1})$$

   if $n \to \infty$.
   (4) The maximum likelihood estimate is an asymptotic unbiased estimate, i.e.

$$M[\hat{\boldsymbol{\theta}}_n] = \boldsymbol{\theta}$$

   if $n \to \infty$.

For a finite sample size n there is only one case, when the maximum likelihood estimator is optimal — it is the exponential shape of the distribution function of the general population. The maximum likelihood method is widely used for a finding of the estimates of parameters of geophysical objects in the problems of the quantitative interpretation. Thus, as a rule, the additive model of the experimental data is used

$$\boldsymbol{u} = \{u_1, u_2, \ldots, u_n\}, \quad \boldsymbol{u} = \boldsymbol{f}(\boldsymbol{\theta}) + \boldsymbol{\varepsilon}, \tag{6.4}$$

where $\boldsymbol{f}(\boldsymbol{\theta})$ is the model of the geophysical field under investigation, which depends on the vector of desired parameters, $\boldsymbol{\varepsilon}$ is a random error of measurement. The normality hypothesis of the random component $\boldsymbol{\varepsilon}$ ($\boldsymbol{\varepsilon} \in N(0, R_\varepsilon)$) allows, in the case of an additive model (6.4), to write a logarithm of the likelihood function, without the terms which do not contain $\boldsymbol{\theta}$, in the form

$$l_1(\boldsymbol{u}, \boldsymbol{\theta}) = -\frac{1}{2}(\boldsymbol{u} - \boldsymbol{f}(\boldsymbol{\theta}))^T R_\varepsilon^{-1}(\boldsymbol{u} - \boldsymbol{f}(\boldsymbol{\theta})). \tag{6.5}$$

It is necessary to note, that in case of the normal distribution, the estimates of the maximum likelihood method are equivalent to the estimates of the weighted least squares method where as the weights appear the elements of an inverse covariance matrix R_ε^{-1}.

For the noncorrelated random component $\boldsymbol{\varepsilon}$ and observations with an equal accuracy, the logarithm of the likelihood function (6.5) we can rewrite in the form

$$l_1(\boldsymbol{u}, \boldsymbol{\theta}) = -\frac{1}{2\sigma_\varepsilon^2}\sum_{i=1}^{n}(u_i - f_i(\boldsymbol{\theta}))^2. \tag{6.6}$$

The maximum likelihood criterion in this case turn to the classical least square method. Substituting the relation (6.6) to the system of equations (6.3), we obtain

$$(\boldsymbol{u} - \boldsymbol{f}(\boldsymbol{\theta}))^T R_\varepsilon^{-1}\frac{\partial \boldsymbol{f}}{\partial \theta_s} = 0, \quad s = 1, 2, \ldots, S. \tag{6.7}$$

6.3 The Newton–Le Cam Method

Let's consider an approximate method of the solution of the system of equations (6.7), which one is update of the Newton iterative method of the solution of the system of algebraic equations. The idea of the method is reduced to following. The initial parameter vector $\boldsymbol{\theta}^{(0)}$ is picked from a priori data. Further, the logarithm of the likelihood function in the vicinity of the point $\boldsymbol{\theta}^{(0)}$ is expanded in the Taylor series with three first terms

$$l_1(\boldsymbol{\theta}) \approx l_1(\boldsymbol{\theta}^{(0)}) + \Delta\boldsymbol{\theta}^T\boldsymbol{d} - \frac{1}{2}\Delta\boldsymbol{\theta}^T C\Delta\boldsymbol{\theta}, \tag{6.8}$$

where $\Delta\boldsymbol{\theta}^T = (\boldsymbol{\theta} - \boldsymbol{\theta}_0)^T$ is a difference of two vectors (row vector), $\boldsymbol{d}$ is a derivative on parameters (column vector), C is a matrix of the mathematical expectations of the second derivatives

$$d_s = \left.\frac{\partial l_1(\boldsymbol{\theta})}{\partial \theta_s}\right|_{\boldsymbol{\theta}=\boldsymbol{\theta}^{(0)}}, \quad c_{ss'} = -\left.\frac{\partial^2 l_1(\boldsymbol{\theta})}{\partial \theta_s \partial \theta_{s'}}\right|_{\boldsymbol{\theta}=\boldsymbol{\theta}^{(0)}}.$$

Substituting the expression (6.8) to the system of equations (6.3), after simple transformations we obtain the system of linear likelihood equations in a form

$$C\Delta\boldsymbol{\theta} = \boldsymbol{d}. \tag{6.9}$$

This system of the linear equations can be solved, for example, by the Gauss method, if C is an improper matrix. In the case of a singularity of this matrix, it is necessary to use the singular analysis or regularization methods(see § 6.12).

By solution of the system (6.9) we obtain the estimate of the vector $\Delta\boldsymbol{\theta}$,

$$\Delta\hat{\boldsymbol{\theta}}^{(1)} = C^{-1}\boldsymbol{d},$$

and a corrected value of the parameter vector

$$\hat{\boldsymbol{\theta}}^{(1)} = \hat{\boldsymbol{\theta}}^{(0)} + \Delta\hat{\boldsymbol{\theta}}^{(1)}.$$

The value of the superscript is a number of the iteration. Further, the value of the vector $\hat{\boldsymbol{\theta}}^{(1)}$ becomes as an initial vector. We calculate the tare vector $\Delta\hat{\boldsymbol{\theta}}^{(2)}$ and the corrected vector $\hat{\boldsymbol{\theta}}^{(2)}$.

The iterative process is prolonged until the relative variation of the vector $\Delta\hat{\boldsymbol{\theta}}^{(n)}$ becomes less than a priori threshold value β:

$$\left|\frac{\Delta\hat{\theta}_s^{(n)}}{\hat{\theta}_s^{(n)}}\right| \leq \beta, \quad s = 1, \ldots, S,$$

usually β is order of $10^{-2} \div 10^{-3}$. The rate of convergence depends on the function $\boldsymbol{f}(\boldsymbol{\theta})$ and on the initial vector $\hat{\boldsymbol{\theta}}^{(0)}$.

In the considered algorithm the elements of the matrix C from (6.8) contain the second derivative of the function $\boldsymbol{f}(\boldsymbol{\theta})$ on components of vector $\boldsymbol{\theta}$. In a practical sense it is more preferable instead of the matrix C to use the matrix

$$\tilde{C} = M[C],$$

which obtains by averaging of the matrix C over the elements of the random vector $\boldsymbol{\varepsilon}$ with the next assumption $M[\boldsymbol{\varepsilon}] = 0$. In this case the elements of matrix $\tilde{C}$ do not contain the second derivatives of the function $\boldsymbol{f}(\boldsymbol{\theta})$. Let's write the expression for the elements of matrix $\tilde{C}$ belonging to n-th iteration

$$\begin{aligned}
\tilde{c}_{ss'}^{(n)} &= M\left[-(\boldsymbol{u} - \boldsymbol{f}(\boldsymbol{\theta}))^T R_\varepsilon^{-1} \frac{\partial^2 \boldsymbol{f}}{\partial\theta_s \partial\theta_{s'}}\bigg|_{\boldsymbol{\theta}=\hat{\boldsymbol{\theta}}^{(n)}}\right. \\
&\quad \left. + \frac{\partial \boldsymbol{f}^T}{\partial\theta_s}\bigg|_{\boldsymbol{\theta}=\hat{\boldsymbol{\theta}}^{(n)}} R_\varepsilon^{-1} \frac{\partial \boldsymbol{f}}{\partial\theta_{s'}}\bigg|_{\boldsymbol{\theta}=\hat{\boldsymbol{\theta}}^{(n)}}\right] \\
&= -M\left[\boldsymbol{\varepsilon}^T\right] R_\varepsilon^{-1} \frac{\partial^2 \boldsymbol{f}}{\partial\theta_s \partial\theta_{s'}} + \frac{\partial \boldsymbol{f}^T}{\partial\theta_s}\bigg|_{\boldsymbol{\theta}=\hat{\boldsymbol{\theta}}^{(n)}} R_\varepsilon^{-1} \frac{\partial \boldsymbol{f}}{\partial\theta_{s'}}\bigg|_{\boldsymbol{\theta}=\hat{\boldsymbol{\theta}}^{(n)}} \\
&= \frac{\partial \boldsymbol{f}^T}{\partial\theta_s} R_\varepsilon^{-1} \frac{\partial \boldsymbol{f}}{\partial\theta_{s'}}\bigg|_{\boldsymbol{\theta}=\hat{\boldsymbol{\theta}}^{(n)}}.
\end{aligned}$$

Let us note, that at the mathematical derivation of the components of matrix $\tilde{C}$ we take into account the next equality $M[\boldsymbol{\varepsilon}] = 0$ for the random vector $\boldsymbol{\varepsilon}$.

The matrix $\tilde{C}$ does not depend on experimental data and completely defined by the model of the determinate and random components.

Using the model of experimental data and the formula (6.8), to find an explicit form of vector $\boldsymbol{d}$:

$$d_s = (\boldsymbol{u} - \boldsymbol{f}(\boldsymbol{\theta}))^T R_\varepsilon^{-1} \frac{\partial \boldsymbol{f}}{\partial \theta_s}\bigg|_{\boldsymbol{\theta}=\hat{\boldsymbol{\theta}}^{(n)}}.$$

If the function $\boldsymbol{f}(\boldsymbol{\theta})$ is a linear one relatively to the desired parameters and we work under the model

$$\boldsymbol{u} = \psi\boldsymbol{\theta} + \boldsymbol{\varepsilon}, \tag{6.10}$$

where ψ is a plan matrix (or structure matrix) with dimension $[n \times S]$:

$$\psi = \begin{Vmatrix} \psi_{11} & \psi_{12} & \dots & \psi_{1S} \\ \psi_{21} & \psi_{22} & \dots & \psi_{2S} \\ \dots & \dots & \dots & \dots \\ \psi_{n1} & \psi_{n2} & \dots & \psi_{nS} \end{Vmatrix},$$

then the system of equations (6.7) is a linear one relative to $\boldsymbol{\theta}$ and it can be written in a matrix shape

$$(\psi^T R_\varepsilon^{-1} \psi)\boldsymbol{\theta} = \psi^T R_\varepsilon^{-1} \boldsymbol{u}.$$

It is easy to find the maximum likelihood estimate, if the matrix $\psi^T R_\varepsilon^{-1} \psi$ is non-singular:

$$\hat{\boldsymbol{\theta}} = (\psi^T R_\varepsilon^{-1} \psi)^{-1} \psi^T R_\varepsilon^{-1} \boldsymbol{u}. \tag{6.11}$$

In the case of singularity it needs to use the singular analysis or regularization (see Sec. 6.12).

We have demonstrated, that in the case of the normal distribution and at presence of correlations, the maximum likelihood method is equivalent to the weighted least squares method, and for independent components of the random vector $\boldsymbol{\varepsilon}$ it is equivalent to the least square method.

6.4 The Least Squares Method

The maximum likelihood method (MLM), as it was mentioned in § 6.2, at a normal distribution of the random component of the model is equivalent to the least squares method (LSM). The computing circuit of LSM is one of the simplest, therefore the method is used for finding the estimates even in the case when the distribution of experimental data differs from the normal distribution. The estimates, obtained by LSM in this case have smaller accuracy, but it is compensated by the simplicity of their deriving.

Let us consider a special case of LSM which has a great practical importance. This is the case of the linear model relatively to the vector of desired parameters (6.10). The desired estimates is found by minimization of the quadratic form

$$\lambda(\boldsymbol{\theta}) = (\boldsymbol{u} - \psi\boldsymbol{\theta})^T W(\boldsymbol{u} - \psi\boldsymbol{\theta}),$$
$$\hat{\boldsymbol{\theta}} = \arg\min \lambda(\boldsymbol{\theta}),$$

where W is a weight matrix. The necessary condition consists in making to zero the partial derivatives of the quadratic form on variables $\theta_1, \theta_2, \ldots, \theta_S$:

$$\frac{\partial \lambda}{\partial \theta_s} = 0, \quad s = 1, 2, \ldots, S. \tag{6.12}$$

The obtained system of equations is called the *system of normal equations*. Under accepted assumptions, the system of equations (6.12) it is possible to write as

$$(\psi^T W \psi)\boldsymbol{\theta} = \psi^T W \boldsymbol{u}.$$

Using this relation it is easy to obtain LSM estimates, if the matrix $\psi^T W \psi$ is not singular one:

$$\hat{\boldsymbol{\theta}} = (\psi^T W \psi)^{-1} \psi^T W \boldsymbol{u}. \tag{6.13}$$

The estimate (6.13) coincides with MLM estimate (6.11) if the weight W matrix is equal to the inverse covariance matrix R_ε^{-1}. In the case of an uncorrelated vector of experimental data $\boldsymbol{u}$, the main diagonal of the matrix R_ε^{-1} contains the inverse variances σ_i^2 of the measurements u_i. The analogy with MLM to clarifies a sense of the weight matrix W, which appears formally in LSM. In a case of observations with the equal accuracy

$$W = \sigma_\varepsilon^{-2} I$$

the estimate (6.13) can be written as

$$\hat{\boldsymbol{\theta}} = (\psi^T \psi)^{-1} \psi \boldsymbol{u}. \tag{6.14}$$

The covariance matrix $\hat{\boldsymbol{\theta}}$

$$R_{\hat{\boldsymbol{\theta}}} = M\left[(\hat{\boldsymbol{\theta}} - \boldsymbol{\theta})(\hat{\boldsymbol{\theta}} - \boldsymbol{\theta})^T\right] = (\psi^T \psi)^{-1} \psi^T R_\varepsilon \psi (\psi^T \psi)^{-1}$$

is obtained from (6.14) by substituting $R_\varepsilon = \sigma_\varepsilon^2 I$. Finally, we obtain

$$R_\theta = \sigma_\varepsilon^2 (\psi^T \psi)^{-1}.$$

The LSM estimates in the case of the linear model have the next optimal properties which do not depend on the distribution law of $\boldsymbol{\varepsilon}$.

(1) Unbiasness of $\hat{\boldsymbol{\theta}}$. Taking into account that $M[\boldsymbol{u}] = \psi\boldsymbol{\theta}$, the mathematical expectation of the estimate $\hat{\boldsymbol{\theta}}$

$$M[\hat{\boldsymbol{\theta}}] = (\psi^T \psi)^{-1} \psi^T M[\boldsymbol{u}]$$

is equal to the value of the desired vector parameter $\boldsymbol{\theta}$

$$M[\hat{\boldsymbol{\theta}}] = (\psi^T \psi)^{-1} \psi^T \psi \boldsymbol{\theta} = \boldsymbol{\theta}.$$

(2) Efficiency of $\hat{\boldsymbol{\theta}}$. Among the class of unbiased estimates $\boldsymbol{\theta}^*$ of the parameter vector $\boldsymbol{\theta}$, which are a linear combination of the data $\boldsymbol{u}$, LSM estimate $\hat{\boldsymbol{\theta}}$ has a minimum variance

$$R_{ss}(\hat{\boldsymbol{\theta}}) \leq R_{ss}(\boldsymbol{\theta}^*)$$

for arbitrary $s = 1, \ldots, S$.

(3) An estimate $\hat{\sigma}_\varepsilon^2$ using residual sum of squares is given by the formula:

$$\hat{\sigma}_\varepsilon^2 = (\boldsymbol{u} - \psi\hat{\boldsymbol{\theta}})^T(\boldsymbol{u} - \psi\hat{\boldsymbol{\theta}})/(n - S).$$

6.5 LSM — Nonlinear Model

The LSM estimate for the linear model has a simple computing circuit, but it also can be used as a basis for the general case of the nonlinear model:

$$\boldsymbol{u} = \varphi(\boldsymbol{\theta}) + \boldsymbol{\varepsilon}. \tag{6.15}$$

Let's represent $\boldsymbol{\varphi}(\boldsymbol{\theta})$ as an expansion in Taylor series in vicinity of some initial vector $\boldsymbol{\theta}_0$ and keeping two terms of the expansion, we rewrite the model (6.15) as:

$$\tilde{\boldsymbol{u}} = \psi\Delta\boldsymbol{\theta} + \boldsymbol{\varepsilon}, \tag{6.16}$$

where

$$\tilde{\boldsymbol{u}} = \boldsymbol{u} - \boldsymbol{\varphi}(\boldsymbol{\theta}_0), \quad \|\psi_{js}\| = \left.\frac{\partial\varphi_j}{\partial\theta_s}\right|_{\boldsymbol{\theta}=\boldsymbol{\theta}_0}, \quad \Delta\boldsymbol{\theta}^{(0)} = \boldsymbol{\theta} - \boldsymbol{\theta}_0.$$

The model (6.16) is equivalent to the model (6.10). The applying of the weighted LSM, gives the estimate

$$\Delta\hat{\boldsymbol{\theta}} = (\psi^T W\psi)^{-1}\psi^T W\tilde{\boldsymbol{u}}.$$

We repeat this procedure, but instead of $\boldsymbol{\theta}_0$ we use $\boldsymbol{\theta}_1 = \boldsymbol{\theta}_0 + \Delta\hat{\boldsymbol{\theta}}$.

The convergence of the iterative process depends on a type of the function $\boldsymbol{\varphi}(\boldsymbol{\theta})$ and depends on a choice of weights. The iterative process stops, when the relative variation of the arguments on the next step k becomes less than a given threshold value β

$$\left|\frac{\Delta\theta_s^{(k)}}{\theta_s^{(k)}}\right| \leq \beta.$$

Usually the value of β is equal to 10^{-2} – 10^{-3}.

6.6 LSM — Orthogonal Polynomials (Chebyshev Polynomials)

The inversion of the matrix $\psi^T\psi$ for a derivation of the estimate (6.14) is simplified essentially, if as the determinate function the system of the normalized orthogonal polynomials is used. In this case i-th row of matrix ψ can be represented as

$$\psi_i = \|\varphi_0(x_i), \varphi_1(x_i), \ldots, \varphi_{S-1}(x_i)\|,$$

where $\varphi_s(x_i)$ are orthogonal on a set of x:

$$\sum_{i=1}^{n} \varphi_s(x_i)\varphi_{s'}(x_i) = \begin{cases} \sum\limits_{i=1}^{n} \varphi_s^2(x_i), & \text{if } s = s', \\ 0, & \text{if } s \neq s'. \end{cases} \tag{6.17}$$

From the condition (6.17) it yields that the matrix $\psi^T\psi$ is a diagonal one

$$\psi^T\psi = \left\| \begin{matrix} \sum\limits_{i=1}^{n} \varphi_0^2(x_i) & 0 & \ldots & 0 \\ 0 & \sum\limits_{i=1}^{n} \varphi_1^2(x_i) & \ldots & 0 \\ \ldots & \ldots & \ldots & \ldots \\ 0 & 0 & \ldots & \sum\limits_{i=1}^{n} \varphi_{S-1}^2(x_i) \end{matrix} \right\|$$

and it is easy to be inverted.

We will consider the Chebyshev polynomials. The scheme of their constructing looks like the next. Let us assume that $\varphi_0(x_i) = 1$. The function $\varphi_1(x_i)$ is determined as

$$\varphi_1(x_i) = x_i + b_0\varphi_0(x_i). \tag{6.18}$$

To find the coefficient b_0 using the orthogonality condition

$$\sum_{i=1}^{n} \varphi_1(x_i)\varphi_0(x_i) = 0, \quad b_0 = \frac{\sum\limits_{i=1}^{n} x_i\varphi_0(x_i)}{\sum\limits_{i=1}^{n} \varphi_0^2(x_i)}. \tag{6.19}$$

Substituting b_0 from the formula (6.19) to the right hand side of the equality (6.18), we obtain

$$\varphi_1(x_i) = x_i - \frac{\sum\limits_{i=1}^{n} x_i\varphi_0(x_i)}{\sum\limits_{i=1}^{n} \varphi_0^2(x_i)}\varphi_0(x_i).$$

Using the expression

$$\varphi_2(x_i) = x_i^2 + b_1\varphi_1(x_i) + b_0\varphi_0(x_i),$$

to find $\varphi_2(x_i)$. Using the orthogonality condition to find b_1 and b_0:

$$\sum_{i=1}^{n} \varphi_2(x_i)\varphi_0(x_i) = 0, \quad b_0 = -\frac{\sum_{i=1}^{n} x_i^2 \varphi_0(x_i)}{\sum_{i=1}^{n} \varphi_0^2(x_i)},$$

$$\sum_{i=1}^{n} \varphi_2(x_i)\varphi_1(x_i) = 0, \quad b_1 = -\frac{\sum_{i=1}^{n} x_i^2 \varphi_1(x_i)}{\sum_{i=1}^{n} \varphi_1^2(x_i)}.$$

Finally, for $\varphi_2(x_i)$ we obtain

$$\varphi_2(x_i) = x_i^2 - \frac{\sum_{i=1}^{n} x_i^2 \varphi_1(x_i)}{\sum_{i=1}^{n} \varphi_1^2(x_i)} \varphi_1(x_i) - \frac{\sum_{i=1}^{n} x_i^2 \varphi_0(x_i)}{\sum_{i=1}^{n} \varphi_0^2(x_i)} \varphi_0(x_i).$$

By analogy we can determine $\varphi_3(x_i)$ and so on. For an arbitrary s-th function we have

$$\varphi_s(x_i) = x_i^s - \frac{\sum_{i=1}^{n} x_i^s \varphi_{s-1}(x_i)}{\sum_{i=1}^{n} \varphi_{s-1}^2(x_i)} \varphi_{s-1}(x_i) - \cdots - \frac{\sum_{i=1}^{n} x_i^s \varphi_0(x_i)}{\sum_{i=1}^{n} \varphi_0^2(x_i)} \varphi_0(x_i). \tag{6.20}$$

It is easy to show the validity of a such expression. Let's write the function $\varphi_s(x_i)$ in a following form:

$$\varphi_s(x_i) = x_i^s + b_{s-1}\varphi_{s-1}(x_i) + \cdots + b_0\varphi_0(x_i).$$

After multiplying by $\varphi_{s'}(x_i)$, $s' < s$, summing over $x_1, \ldots, x_n$ and using the orthogonality condition (6.17), we obtain

$$\sum_{i=1}^{n} x_i^s \varphi_{s'}(x_i) + b_{s'} \sum_{i=1}^{n} \varphi_{s'}^2(x_i) = 0, \quad b_{s'} = -\frac{\sum_{i=1}^{n} x_i^s \varphi_{s'}(x_i)}{\sum_{i=1}^{n} \varphi_{s'}^2(x_i)},$$

that demonstrates validity of the equality (6.20). For a case, when the abscissas $x_1, \ldots, x_n$ are preselected with an uniform sample, the recursion formula

$$\varphi_{s+1}(x_i) = \left(x_i - \frac{1}{n} \sum_{i=1}^{n} x_i \right) \varphi_s(x_i) - \frac{n^2(n^2 - i^2)}{4(4i^2 - 1)} \varphi_{s-1}(x_i)$$

for the calculation of the function $\varphi_{s+1}(x_i)$ on known $\varphi_s(x_i)$ and $\varphi_{s-1}(x_i)$ was obtained by Chebyshev.

We have considered alternatives of a representation of the matrix ψ by orthogonal polynomials for the homogeneous process.

The estimate of the coefficients of orthogonal polynomials by the LSM method, using (6.13) and orthogonality conditions (6.17) under $W = 1$, $\sum_{j=1}^{n} \varphi_{sj}^2 = 1$ has a form

$$\hat{\theta}_s = \sum_{j=1}^{n} \varphi_{sj} u_j, \quad s = 0, 1, 2, \ldots, S-1,$$

or $\hat{\boldsymbol{\theta}} = \psi^T \boldsymbol{u}$.

In this case, the covariance matrix of the estimated parameters is diagonal one

$$R_\theta = \sigma_\varepsilon^2 I,$$

and the standard deviation gives by the formula

$$\hat{\sigma}_\varepsilon^2 = \frac{(\boldsymbol{u} - \psi\hat{\boldsymbol{\theta}})^T (\boldsymbol{u} - \psi\hat{\boldsymbol{\theta}})}{n - S}.$$

6.7 LSM — Case of Linear Constraints

At the solution of applied problems a necessity frequently appears to find the estimates of desired parameters at presence of the linear constraints. It is one of the alternatives to incorporate a priori information.

Let's a vector of desired parameters $\boldsymbol{\theta}$ satisfies the system of Q equations

$$A\boldsymbol{\theta} = \boldsymbol{V}, \tag{6.21}$$

where $Q < S$. The matrix A has a dimension $[Q \times S]$ and the vector $\boldsymbol{V}$ has a length Q. To determine the desired parameters the problem can be stated as a problem on the conditional extremum, i.e. to minimize of $\lambda(\boldsymbol{\theta})$ under the condition (6.21). For the solution of this problem, as usual, we will use the Lagrange multipliers method. To build functional

$$\Phi(\boldsymbol{\theta}) = (\boldsymbol{u} - \psi\boldsymbol{\theta})^T (\boldsymbol{u} - \psi\boldsymbol{\theta}) - 2\boldsymbol{\lambda}^T (A\boldsymbol{\theta} - \boldsymbol{V}),$$

where a component of the vector $\boldsymbol{\lambda}$ is the Lagrange multiplier. The length of $\boldsymbol{\lambda}$ is equal to the length of $\boldsymbol{V}$, i.e. the lengths of the vector are equal to a number of linear constraints. As the estimate of the parameters we will choose the values, under which the functional $\Phi(\boldsymbol{\theta})$ reaches an own minimum. The necessary condition of a minimum consists in the equalling to zero of the first derivative on components of the vector $\boldsymbol{\theta}$:

$$\frac{\partial \Phi(\boldsymbol{\theta})}{\partial \theta_s} = 0, \quad s = 1, \ldots, S. \tag{6.22}$$

After the solution of the system of linear equations (6.22) relatively to θ_s, we obtain the next estimate $\hat{\boldsymbol{\theta}}$, expressed using Lagrange multipliers:

$$\tilde{\boldsymbol{\theta}}^T = \hat{\boldsymbol{\theta}}^T + \boldsymbol{\lambda}^T A (\psi^T \psi)^{-1}, \tag{6.23}$$

where $\hat{\boldsymbol{\theta}}^T$ is an estimate obtained by LSM without taking linear constraints into account.

To find $\boldsymbol{\lambda}$ we multiply both sides of the equality (6.23) by A^T from the right side and take into account the condition (6.21). As the result we obtain

$$\boldsymbol{\lambda}^T = (\boldsymbol{V}^T - \hat{\boldsymbol{\theta}}^T A^T)(A(\psi^T\psi)^{-1}A^T)^{-1}. \tag{6.24}$$

Substituting the equality (6.24) into expression (6.23), we come to final estimate for desired parameters

$$\tilde{\boldsymbol{\theta}}^T = \hat{\boldsymbol{\theta}}^T + (\boldsymbol{V}^T - \hat{\boldsymbol{\theta}}^T A^T)(A(\psi^T\psi)^{-1}A^T)^{-1}A(\psi^T\psi)^{-1}. \tag{6.25}$$

As it is possible to see from the formula (6.25), the estimate contains two addends: the first one corresponds to an estimate obtained by the least squares method without an account of the linear constraints, and second one characterizes the contribution of the linear constraints to common estimate. The covariance matrix of the estimates is expressed by the following formula:

$$R_{\tilde{\theta}} = R_{\hat{\theta}} - \sigma_\varepsilon^2(\psi^T\psi)^{-1}A^T(A(\psi^T\psi)^{-1}A^T)^{-1}A(\psi^T\psi)^{-1}, \tag{6.26}$$

$$R_{\hat{\theta}} = \hat{\sigma}_\varepsilon^2(\psi^T\psi)^{-1}. \tag{6.27}$$

On a main diagonal there are dispersions of the estimates of arguments, and the off-diagonal elements characterize a cross correlation between estimates of arguments. It is interesting to note, that the elements of the matrix consist in two terms: the first term corresponds to elements of the covariance matrix without an account of the linear constraints, and the second term characterizes influencing of the linear constraints.

To find an estimate for the variance by the formula

$$\hat{\sigma}_\varepsilon^2 = \frac{(\boldsymbol{u} - \psi\tilde{\boldsymbol{\theta}})^T(\boldsymbol{u} - \psi\tilde{\boldsymbol{\theta}})}{n - S - K}.$$

Let us consider a case of a known inverse covariance matrix of the experimental data W. In this case the estimates (6.14), (6.25) are rewritten as

$$\tilde{\boldsymbol{\theta}}^T = \hat{\boldsymbol{\theta}}^T + (\boldsymbol{V}^T - \hat{\boldsymbol{\theta}}^T A^T)(A(\psi^T W\psi)^{-1}A^T)^{-1}A(\psi^T W\psi)^{-1}, \tag{6.28}$$

$$\hat{\boldsymbol{\theta}} = (\psi^T W\psi)^{-1}\psi^T W\boldsymbol{u}. \tag{6.29}$$

The covariance matrix of the estimates $\tilde{\boldsymbol{\theta}}$, $\hat{\boldsymbol{\theta}}$ reads as

$$R_{\tilde{\theta}} = R_{\hat{\theta}} - (\psi^T W\psi)^{-1}A^T(A(\psi^T W\psi)^{-1}A^T)^{-1}A(\psi^T W\psi)^{-1}, \tag{6.30}$$

$$R_{\hat{\theta}} = (\psi^T W\psi)^{-1}. \tag{6.31}$$

It is possible to use the obtained estimates (6.28), (6.29) and covariance matrices (6.30), (6.31) at presence of a priori information about the weight of some experimental data. Such cases frequently meet, when the data under the approximation are the result of the processing and represent estimates of the desired parameters. Then the appropriate variances and covariance coefficients of the estimates of parameters can be used to create the weight matrix W.

6.8 Linear Estimation — Case of Nonstationary Model

Let us consider the methods for an estimation of the linear parameters

$$\boldsymbol{u} = \psi\boldsymbol{\theta} + \boldsymbol{\varepsilon}$$

under a condition, that random component is nonstationary one. For the simplicity we will assume that

$$\sigma_i^2 = \Sigma(f_i, \boldsymbol{\rho}) = \rho_0 + \rho_1 f_i^2,$$

where $f_i = \boldsymbol{\psi}_i^T\boldsymbol{\theta}$. In a special case $\rho_1 = 0$ we come to the stationary model. The procedure of finding of the estimates of parameters $\boldsymbol{\theta}$ and $\boldsymbol{\rho}$ is reduced to following:

(1) Computation of the zero approximation of the parameter vector $\hat{\boldsymbol{\theta}}^{(0)}$ by LSM with the unit weight coefficients

$$\hat{\boldsymbol{\theta}}^{(0)} = (\psi^T\psi)^{-1}\psi^T\boldsymbol{u}.$$

(2) Determination of the estimate of the model function

$$\hat{f}_i = \boldsymbol{\psi}_i^T\hat{\boldsymbol{\theta}}^{(0)}.$$

(3) Computation of the squared difference

$$\hat{\varepsilon}_i^2 = (u_i - \hat{f}_i)^2.$$

(4) Finding of estimate $\hat{\boldsymbol{\rho}}$:

$$\hat{\boldsymbol{\rho}} = \arg\min\,(\hat{\boldsymbol{C}} - \hat{B}\boldsymbol{\rho})^T(\hat{\boldsymbol{C}} - \hat{B}\boldsymbol{\rho}),$$
$$\hat{\boldsymbol{\rho}} = (\hat{B}^T\hat{B})^{-1}\hat{B}^T\hat{\boldsymbol{C}},$$
$$\rho_0 \geq 0, \quad \rho_1 \geq 0,$$

where

$$\hat{B} = \begin{Bmatrix} 1 \ \hat{f}_1^2 \\ 1 \ \hat{f}_2^2 \\ \dots \\ 1 \ \hat{f}_n^2 \end{Bmatrix}, \quad \hat{\boldsymbol{C}} = \begin{Bmatrix} \hat{\varepsilon}_1^2 \\ \hat{\varepsilon}_2^2 \\ \dots \\ \hat{\varepsilon}_n^2 \end{Bmatrix}.$$

(5) Computation of the variance of the random component

$$\hat{\sigma}_i^2 = \hat{\rho}_0 + \hat{\rho}_1\hat{f}_i^2.$$

(6) Determination (by weighted LSM) the estimate

$$\tilde{\boldsymbol{\theta}} = (\psi^T W\psi)^{-1}\psi^T W\boldsymbol{u},$$

where $W = \mathrm{diag}\,(\hat{\sigma}_1^{-2}, \hat{\sigma}_2^{-2}, \dots, \hat{\sigma}_n^{-2})$.

(7) Computation of the deviation $\tilde{\varepsilon}_i^2$ subject to $\tilde{\boldsymbol{\theta}}$

$$\tilde{\varepsilon}_i^2 = (u_i - \tilde{f}_i)^2,$$

where $\tilde{f}_i = \boldsymbol{\psi}_i^T\tilde{\boldsymbol{\theta}}$.

(8) Comparison of $\tilde{\varepsilon}_i^2$ with the the threshold value β_0. If $\tilde{\varepsilon}_i^2 \leq \beta_0$, then the iterative procedure is finished. If $\tilde{\varepsilon}_i^2 > \beta_0$, then $\hat{\boldsymbol{\theta}}^{(0)} = \tilde{\boldsymbol{\theta}}$, $\hat{c}_i = \tilde{\varepsilon}_i^2$, $\hat{f}_i = \tilde{f}_i$ and go to the point 4.

At the enough universal propositions about the regularity of the function Σ, which one are usually fulfilled in practice, the parametric models σ_i^2 lead to the estimates having the asymptotically normal distribution, which one at a great enough sample size coincides with the distribution of estimates $\boldsymbol{\theta}$, obtained by the weighted LSM provided that the weight coefficients are known exactly.

6.9 Bayes' Criterion and Method of Statistical Regularization

At the solution of practical problems of processing of the geophysical data the great importance has an account of the available a priori information about parameters of the explored object. The Bayes' criterion opens the ample opportunities of the use of a priori information in a statistical shape. Let's assume, that in the model (6.15) the vector $\boldsymbol{\theta}$ is random. The joint density function of the vector of experimental data $\boldsymbol{u}$ and the parameter vector $\boldsymbol{\theta}$ can be written as

$$p(\boldsymbol{u}, \boldsymbol{\theta}) = p(\boldsymbol{\theta})p(\boldsymbol{u}/\boldsymbol{\theta}) = p(\boldsymbol{u})p(\boldsymbol{\theta}/\boldsymbol{u}), \tag{6.32}$$

where $p(\boldsymbol{\theta})$ is a priori density function of the vector of desired parameters, $p(\boldsymbol{u}/\boldsymbol{\theta})$ is a conditional density function of the vector of experimental data $\boldsymbol{u}$ at the fixed parameter vector $\boldsymbol{\theta}$, $p(\boldsymbol{u})$ is a priori density function of the vector of experimental data, $p(\boldsymbol{\theta}/\boldsymbol{u})$ is a posteriori density function for the desired parameter vector $\boldsymbol{\theta}$ at fixed realization of the experimental data $\boldsymbol{u}$. Using the expression (6.32) it is possible to find a posteriori density function of the desired parameter vector

$$p(\boldsymbol{\theta}/\boldsymbol{u}) = \frac{p(\boldsymbol{\theta})p(\boldsymbol{u}/\boldsymbol{\theta})}{p(\boldsymbol{u})}. \tag{6.33}$$

In accordance with the Bayes' criterion, as the estimate of the desired parameter vector, the mathematical expectation of the desired parameter vector under a posteriori distribution is accepted

$$\hat{\boldsymbol{\theta}}_B = \int \boldsymbol{\theta}\, p(\boldsymbol{\theta}/\boldsymbol{u}) d\boldsymbol{\theta}. \tag{6.34}$$

From an analysis of the formula (6.33) we conclude, that a posteriori density function is a product of a priori density $p(\boldsymbol{\theta})$ and conditional density $p(\boldsymbol{u}/\boldsymbol{\theta})$ (the denominator does not depend on the desired parameter vector and it is introduced for normalization). Hence, a priori information about desired parameter vector consists of a priori information, which is obtained before experiment, and information extracted from the experimental data. If one of the multipliers prevails over other, then it determines a posteriori information in general. For example, if conditional density has an abrupt maximum in the parametric space ω_0 and it fast tends to zero in other space ω_1, and if a priori density $p(\boldsymbol{\theta})$ is slow varying function in ω_0, then a posteriori density practically does not depend on a shape of $p(\boldsymbol{\theta})$, which may put a constant in all space ω_0.

6.10 Method of Maximum a Posteriori Probability

Alongside with an estimate (6.34) which is the conditional mathematical expectation, wide practical applying has the method *of a maximum of a posteriori probability.* In this method as an estimate $\hat{\boldsymbol{\theta}}$, a position of a maximum of a density function $p(\boldsymbol{\theta}/\boldsymbol{u})$ in a parametric space is accepted. As the logarithm is a monotone function, the location of a maximum can be determined by the function $\ln p(\boldsymbol{\theta}/\boldsymbol{u})$, that appears by more convenient at the solution of a wide variety of the problems. If the maximum is placed inside the accessible region of variation of $\boldsymbol{\theta}$ and the density function has the first continuous derivative, then the necessary condition consist in the equality to zero of $\ln p(\boldsymbol{\theta}/\boldsymbol{u})$

$$\frac{\partial \ln p(\boldsymbol{\theta}/\boldsymbol{u})}{\partial \theta_s} = 0, \quad s = 1, 2, \ldots, S. \tag{6.35}$$

The equation (6.35) is called *the equation of the maximum a posteriori probability.* Replacing $p(\boldsymbol{\theta}/\boldsymbol{u})$ by its representation (6.33) and taking the logarithm we obtain

$$\ln p(\boldsymbol{\theta}/\boldsymbol{u}) = \ln p(\boldsymbol{\theta}) + \ln p(\boldsymbol{u}/\boldsymbol{\theta}) - \ln p(\boldsymbol{u}). \tag{6.36}$$

For finding an estimate using the maximum a posteriori probability method (MAP), let's substitute the expression (6.36) into the system of equations (6.35) and to obtain the equation of maximum a posteriori probability in a form

$$\frac{\partial \ln p(\boldsymbol{\theta})}{\partial \theta_s} + \frac{\partial \ln p(\boldsymbol{u}/\boldsymbol{\theta})}{\partial \theta_s} = 0, \quad s = 1, 2, \ldots, S. \tag{6.37}$$

The first addend from the left hand side of (6.37) characterizes a priori data and the second one is connected with an experimental data.

By analyzing of (6.37) we can conclude, that in the case of the function $\ln p(\theta)$ has a weak variation in the area of allowed values of $\boldsymbol{\theta}$, then the first term in (6.37) can be neglected, and MAP transforms to LSM. It means that LSM is a special case of MAP under the condition of an absence of a priori information, that is equivalent to a hypothesis $p(\theta) = \text{const}$ in the area of allowed values of the desired parameter vector.

At the nontrivial assignment of a priori density $p(\theta) \neq \text{const}$ MAP improves LSM estimate, and makes the solution stable. The engaging of a priori information underlies the *statistical regularization method.*

Let us consider in more detail a procedure of finding the estimate $\hat{\boldsymbol{\theta}}$ for a special case of the normal distributed random component $\boldsymbol{\varepsilon} \in N(0, R_\varepsilon)$ and normal a priori distribution $\boldsymbol{\theta} \in N(\langle\boldsymbol{\theta}\rangle, R_\theta)$ for the linear model (6.15). The logarithm of a posteriori density function looks as

$$\ln p(\boldsymbol{\theta}/\boldsymbol{u}) = -\frac{1}{2}(\boldsymbol{u} - \psi\boldsymbol{\theta})^T R_\varepsilon^{-1}(\boldsymbol{u} - \psi\boldsymbol{\theta}) - \frac{1}{2}(\boldsymbol{\theta} - \langle\boldsymbol{\theta}\rangle)^T R_\theta^{-1}$$
$$\times(\boldsymbol{\theta} - \langle\boldsymbol{\theta}\rangle) - \frac{n+S}{2}\ln(2\pi) - \frac{1}{2}(\ln|R_\theta| + \ln|R_\varepsilon|) - \ln p(\boldsymbol{u}).$$

the necessary condition of the maximum of a posteriori probability leads to the system of equations

$$(\psi^T R_\varepsilon^{-1}\psi + R_\theta^{-1})\boldsymbol{\theta} = \psi^T R_\varepsilon^{-1}\boldsymbol{u} + R_\theta^{-1}\langle\boldsymbol{\theta}\rangle,$$

and these equations give the estimate $\tilde{\boldsymbol{\theta}}$:

$$\tilde{\boldsymbol{\theta}} = (\psi^T R_\varepsilon^{-1}\psi + R_\theta^{-1})^{-1}(\psi^T R_\varepsilon^{-1}\boldsymbol{u} + R_\theta^{-1}\langle\boldsymbol{\theta}\rangle). \tag{6.38}$$

Using the equality

$$(\psi^T R_\varepsilon^{-1}\psi + R_\theta^{-1})^{-1} = R_\theta - R_\theta\psi^T(\psi R_\theta\psi^T + R_\varepsilon)^{-1},$$
$$\psi R_\theta, (\psi^T R_\varepsilon^{-1}\psi + R_\theta^{-1})^{-1}\psi^T R_\varepsilon^{-1} = R_\theta\psi^T(\psi R_\theta\psi^T + R_\varepsilon)^{-1}, \tag{6.39}$$

to represent the estimate (6.38) in a form:

$$\tilde{\boldsymbol{\theta}} = \langle\boldsymbol{\theta}\rangle + R_\theta\psi^T(\psi R_\theta\psi^T + R_\varepsilon)^{-1}(\boldsymbol{u} - \psi\langle\boldsymbol{\theta}\rangle). \tag{6.40}$$

The formula (6.40) enables a simple interpretation: the MAP estimate $\hat{\boldsymbol{\theta}}$ is represented as the sum of a priori vector $\langle\boldsymbol{\theta}\rangle$ and the correction, which one is the weighted difference between the experimental data and a priori model.

In the special case of an uncorrelated random component $R_\varepsilon = \sigma_\varepsilon^2 I_n$ and uncorrelated components of the vector $\boldsymbol{\theta}$, a priori density function contains: $R_\theta = \sigma_\theta^2 I_n$ (I_n and I_S unitary matrices with dimensions $[n \times n]$ and $[S \times S]$ respectively, σ_ε^2 is a variance of the random component, σ_θ^2 is a priori variance of the desired parameters). The formulas (6.38), (6.40) we rewrite in the next form

$$\tilde{\boldsymbol{\theta}} = (\psi^T\psi + \alpha I_S)^{-1}(\psi^T\boldsymbol{u} + \alpha\langle\boldsymbol{\theta}\rangle), \tag{6.41}$$

$$\tilde{\boldsymbol{\theta}} = \langle\boldsymbol{\theta}\rangle + \psi^T(\psi\psi^T + \alpha I_n)^{-1}(\boldsymbol{u} - \psi\langle\boldsymbol{\theta}\rangle), \tag{6.42}$$

where $\alpha = \sigma_\varepsilon^2/\sigma_\theta^2$ is a fixed regularization parameter of zero order. We should note, that regularization parameter under the statistical interpretation has a clear sense: it is the energy ratio, noise/signal.

If α tends to zero, the MAP estimate $\tilde{\boldsymbol{\theta}}$ (6.41) transforms to LSM estimate $\hat{\boldsymbol{\theta}}$ (6.11).

The covariance matrices of the estimates of the parameter vector, given by the formulas (6.38), (6.40), (6.41), (6.42), can be written as

$$R_{\tilde{\theta}} = (\psi^T R_\varepsilon^{-1}\psi + R_\theta^{-1})^{-1}, \tag{6.43}$$

$$R_{\tilde{\theta}} = R_\theta - R_\theta\psi^T(\psi R_\theta\psi^T + R_\varepsilon)^{-1}\psi R_\theta, \tag{6.44}$$

$$R_{\tilde{\theta}} = \sigma_\varepsilon^2(\psi^T\psi + \alpha I_S)^{-1}, \tag{6.45}$$

$$R_{\tilde{\theta}} = \sigma_\theta^2(I_S - \psi^T(\psi\psi^T + \alpha I_n)^{-1}\psi). \tag{6.46}$$

The interpretation quality should estimate on a basis of the analysis of the matrix $R_{\tilde{\theta}}$. The main diagonal of such matrix contains the variances of the estimates of parameters and far-diagonal elements which describe correlations between the estimates. The matrices (6.43)–(6.46) can be used for the design of the experiment. We should note, a choice of formulas for the estimation $\tilde{\boldsymbol{\theta}}$ ((6.38) or (6.40)) mainly depends on the computational capability. If the computation of the inverse covariance matrix of the high dimension $R_{\tilde{\theta}}$ and R_ε is difficult, then the form of (6.40) is preferable.

6.11 The Recursion Algorithm of MAP

At the solution the problem of an approximation of a great volume of the experimental data, the boundedness of a computer memory results in the necessity of constructing of the interpretative algorithms for the finding of the estimate of the parameter vector by a method of the maximum of a posterior probability. Let's consider a stable iterative algorithm, which is based on the representation of the estimate in the form (6.40).

As an initial value for the parameter vector estimate we choose a priori vector $\boldsymbol{\theta}^{(0)} = \langle\boldsymbol{\theta}\rangle$, and an initial covariance matrix is equal to a priori matrix $R_\theta^{(0)} = R_\theta$. The first approximation is computed by the formula

$$\boldsymbol{\theta}^{(1)} = \boldsymbol{\theta}^{(0)} + \gamma^{(1)}\Delta\boldsymbol{\theta}^{(1)},$$

where $\Delta\boldsymbol{\theta}^{(1)} = R_\theta^{(0)}\boldsymbol{\psi}^{(1)}$, $\boldsymbol{\psi}(1)$ is a column vector, with the components coinciding with the components of the first row of the matrix ψ:

$$\gamma^{(1)} = N_1/\delta_1^2, \quad N_1 = u_1 - \boldsymbol{\psi}^{(1)T}\boldsymbol{\theta}^{(0)},$$
$$\delta_1^2 = \sigma_{\varepsilon 1}^2 + \psi^{(1)T}\Delta\boldsymbol{\theta}^{(1)}.$$

The covariance matrix of the estimate $\boldsymbol{\theta}^{(1)}$ looks like

$$R_\theta^{(1)} = R_\theta^{(0)} - \Delta\boldsymbol{\theta}^{(1)}\Delta\boldsymbol{\theta}^{(1)\,T}/\delta_1^2.$$

Let's we find $(i-1)$-th approximation $\boldsymbol{\theta}^{(i-1)}$ and $R_\theta^{(i-1)}$. Then, i-th approximation we construct as the next:

$$\boldsymbol{\theta}^{(i)} = \boldsymbol{\theta}^{(i-1)} + \gamma^{(i)}\Delta\boldsymbol{\theta}^{(i)},$$

where

$$\gamma^{(i)} = N_i/\delta_i^2, \quad N_i = u_i - \boldsymbol{\psi}^{(i)T}\boldsymbol{\theta}^{(i-1)},$$
$$\delta_i^2 = \sigma_{\varepsilon i}^2 + \psi^{(i)T}\Delta\boldsymbol{\theta}^{(i)},$$
$$\Delta\boldsymbol{\theta}^{(i)} = R_\theta^{(i-1)}\boldsymbol{\psi}^{(i)T},$$
$$R_\theta^{(i)} = R_\theta^{(i-1)} - \Delta\boldsymbol{\theta}^{(i)}\Delta\boldsymbol{\theta}^{(i)\,T}/\delta_i^2.$$

6.12 Singular Analysis and Least Squares Method

By the practical application of the least squares method (LSM), using the model (6.15), we can not obtain satisfactory result in the case of a *bad condition* of the matrix $\psi^T\psi$, which is subjected to inversion. As a measure of the *condition*, it is usually used a *condition number*

$$\beta = \frac{G}{g},$$

where

$$G = \max_{\boldsymbol{\theta}} \frac{\|\psi^T\psi\boldsymbol{\theta}\|_1}{\|\boldsymbol{\theta}\|_1}, \quad g = \min_{\boldsymbol{\theta}} \frac{\|\psi^T\psi\boldsymbol{\theta}\|_1}{\|\boldsymbol{\theta}\|_1}, \quad \|\boldsymbol{\theta}\|_1 = \sum_{s=1}^{S} |\theta_s|. \tag{6.47}$$

The maximum and minimum under conditions (6.47) are determined on all non-zero parameter vectors. If the matrix $\psi^T\psi$ is singular, then $g = 0$.

The condition number β is a measure of the proximity to singularity. The condition number can be considered as an inverse value of the relative distance from a given matrix to a set of singular matrices. Let us consider the properties of the condition number β which follow from its definition.

(1) The condition number β is always greater or equal to 1 ($\beta \geq 1$), because

$$G \geq g.$$

(2) If the matrix $\psi^T\psi$ is a diagonal one, as in the case of orthogonal polynomials, then

$$\beta = \frac{\max |(\psi^T\psi)_{ii}|}{\min |(\psi^T\psi)_{ii}|}.$$

(3) If an inverse matrix exists, then

$$\beta = \|\psi^T\psi\| \, \|(\psi^T\psi)^{-1}\|.$$

(4) If the matrix $\psi^T\psi$ is a singular one, then $\beta = \infty$.

At the near singularity of the matrix $\psi^T\psi$ the condition number β has a great value and independently of the method of solution of the system of normal equations, the errors in an input information and round-off errors lead to a strong "oscillations" of the desired parameter vector. As an extreme case, when the columns of the matrix $\psi^T\psi$ are linearly dependent, the matrix $\psi^T\psi$ will be singular one with the condition number $\beta = \infty$. Hence, the high values of the condition number testify to presence of near linear the dependence of the rows of the matrix.

More stable method of computing of the parameter vector for the linear model in the case of a bad condition of the matrix is based on the singular value decomposition (SVD) of the structural matrix ψ. This approach is more affective to account of the input information, the round-off errors and the linear dependence, than classical LSM

Let us consider the singular value decomposition on an example of the linear model (6.15).

It is possible to show, that for an arbitrary real matrix ψ is valid the next representation

$$\psi = Q\Sigma P^T, \tag{6.48}$$

where $Q_{n\times n}$ and $P_{S\times S}$ are the orthogonal matrices, i.e. $QQ^T = I_{n\times n}$ and $PP^T = I_{S\times S}$ respectively, and $\Sigma_{n\times S}$ is a diagonal matrix with the elements

$$\Sigma_{ij} = \begin{cases} \sigma_i,\ i=j,\ i=1,\dots,n, \\ 0,\ \ i\neq j,\ j=1,\dots,S, \end{cases}, \quad \Sigma = \begin{bmatrix} \sigma_1 & \dots & 0 \\ \dots & \dots & \dots \\ 0 & \dots & \sigma_S \\ 0 & \dots & 0 \\ \dots & \dots & \dots \\ 0 & \dots & 0 \end{bmatrix}. \tag{6.49}$$

The matrix Q consists of n orthonormal eigenvectors of the matrix $\psi\psi^T$, and the matrix P consists of orthonormal eigenvectors of the matrix $\psi^T\psi$. The diagonal elements σ_i of the matrix Σ are non-negative quantities of square root of the eigenvalues (*singular numbers*) of the matrix $\psi^T\psi$. Assume that $\sigma_1 \geq \sigma_2 \geq \dots \geq \sigma_S \geq 0$. If a matrix rank of the matrix ψ is equal to r, then $\sigma_{r+1} = \sigma_{r+2} = \dots = \sigma_S = 0$.

For a simplicity of the analysis we suppose that the random component of the model $\boldsymbol{\varepsilon}$ is non-correlated one. Then, taking into account (6.48), the LSM criterion can be written as

$$\hat{\boldsymbol{\theta}} = \arg\min \lambda(\boldsymbol{\theta}),$$

where

$$\lambda(\boldsymbol{\theta}) = (\boldsymbol{u} - \psi\boldsymbol{\theta})^T(\boldsymbol{u} - \psi\boldsymbol{\theta}) = (\boldsymbol{u} - Q\Sigma P^T\boldsymbol{\theta})^T QQ^T$$
$$\times (\boldsymbol{u} - Q\Sigma P^T\boldsymbol{\theta})(\boldsymbol{y} - \Sigma\boldsymbol{b})^2.$$

Using the property $QQ^T = I$, we obtain the next representation for the quadratic form λ:

$$\lambda(\boldsymbol{b}) = (\boldsymbol{y} - \Sigma\boldsymbol{b})^2, \tag{6.50}$$

where

$$\boldsymbol{y} = Q^T\boldsymbol{u}, \quad \boldsymbol{b} = P^T\boldsymbol{\theta}. \tag{6.51}$$

Minimizing (6.50) on $\boldsymbol{b}$, we obtain the estimate $\hat{\boldsymbol{b}}$

$$\hat{\boldsymbol{b}} = \Sigma^+\boldsymbol{y} \tag{6.52}$$

or, come back to initial variables, we obtain

$$\hat{\boldsymbol{\theta}} = P\Sigma^+Q^T\boldsymbol{u}. \tag{6.53}$$

The matrix

$$\Sigma^+ = \begin{bmatrix} \sigma_1^{-1} & \dots & 0 & 0 & \dots & 0 \\ \dots & \dots & \dots & \dots & \dots & \dots \\ 0 & \dots & \sigma_S^{-1} & 0 & \dots & 0 \end{bmatrix}$$

is called *pseudoinverse matrix* with respect to the matrix Σ, the pseudoinverse matrix has dimension $[S \times n]$ and satisfies the next properties

(1) $\Sigma\Sigma^+\Sigma = \Sigma$;
(2) $\Sigma^+\Sigma\Sigma^+ = \Sigma^+$;
(3) $\Sigma\Sigma^+$ is a symmetric matrix;
(4) $\Sigma^+\Sigma$ is a symmetric matrix.

It is possible to demonstrate a unique existence of a such matrix. If the diagonal elements σ_{ii} of the matrix Σ do not equal to zero, then estimate b_i $(i = 1, 2, \ldots, S)$ will be written as

$$\hat{b}_i = y_i/\sigma_{ii}, \quad i = 1, 2, \ldots, S.$$

However if some σ_{ii} are small, such procedure is unwanted. In this case it is necessary to analyze the values σ_{ii} before finding of an estimate $\hat{\boldsymbol{b}}$. It is possible to show, that the singular numbers will be distinct from zero if and only if the columns of the matrix ψ are linearly independent. The analysis of the singular numbers should be implemented on a basis of the suspected accuracy of the input data and the calculation accuracy of the computer. On a basis of a priori accuracy to choose the threshold value α, to which one compare everyone σ_{ii}. If $\sigma_{ii} \geq \alpha$, then $\hat{b}_i = y_i/\sigma_{ii}$ for $i = 1, 2, \ldots, i_0$. If $\sigma_{ii} < \alpha$, then $\hat{b}_i = 0$ for $i = i_0 + 1, \ldots, S$.

As a matter of fact, at $\sigma_{ii} < \alpha$ the component of the vector b_i can be equal to an arbitrary value. This arbitrary rule is caused by the non-uniqueness of the LSM solution. To exclude an influence of unstable components of the vector $\hat{b}_i$ at $\sigma_{ii} < \alpha$, in the singular analysis their values are putted to zero. Let us consider an interpretation of the threshold value α by the example of a model with the random parameter vector $\boldsymbol{\theta}$. Let's the mathematical expectation of the parameter vector is equal to zero $\langle\boldsymbol{\theta}\rangle = 0$ and a priori covariance matrix $\langle\boldsymbol{\theta}\boldsymbol{\theta}^T\rangle = R_{\boldsymbol{\theta}}$ is given. For the simplicity we assume that the components of the parameter vector $\boldsymbol{\theta}$ and the random component $\boldsymbol{u}$ of the model have a lack of correlation i.e.

$$R_{\boldsymbol{\theta}} = \sigma_\theta^2 I, \quad R_{\boldsymbol{\varepsilon}} = \sigma_\varepsilon^2 I.$$

Moreover we assume that the random vectors $\boldsymbol{\theta}$ $\boldsymbol{\varepsilon}$ are uncorrelated among themselves.

Using the singular decomposition (6.48) of the structural matrix ψ, and notations (6.51), to rewrite the model (6.10) in a form

$$\boldsymbol{y} = \sum \boldsymbol{b} + \boldsymbol{e},$$

where $\boldsymbol{e} = Q^T\boldsymbol{\varepsilon}$. We find the solution as a linear form

$$\hat{\boldsymbol{b}} = \Gamma\boldsymbol{y}, \tag{6.54}$$

where

$$\Gamma = \begin{bmatrix} \gamma_1 & 0 & \ldots & 0 & 0 & \ldots & 0 \\ 0 & \gamma_2 & \ldots & 0 & 0 & \ldots & 0 \\ \ldots & \ldots & \ldots & \ldots & \ldots & \ldots & \ldots \\ 0 & 0 & \ldots & \gamma_S & 0 & \ldots & 0 \end{bmatrix}$$

is a diagonal rectangular matrix with the dimension $[S \times n]$. To find an estimate of the parameter vector $\boldsymbol{b}$ to determine the elements of the matrix Γ is enough. Let's to use the mean-square criterion. In this case the error of estimation and the corresponding covariance matrix have a form

$$\boldsymbol{\xi} = \hat{\boldsymbol{b}} - \boldsymbol{b} = (\Gamma\Sigma - I)\boldsymbol{b} + \Gamma\boldsymbol{e},$$

$$\langle \boldsymbol{\xi}\boldsymbol{\xi}^T \rangle = (\Gamma\Sigma - I)\langle \boldsymbol{b}\boldsymbol{b}^T \rangle (\Gamma\Sigma - I)^T + \Gamma\langle \boldsymbol{e}\boldsymbol{e}^T \rangle \Gamma^T = (\Gamma\Sigma - I)R_b(\Gamma\Sigma - I)^T + \Gamma R_e \Gamma^T.$$

The standard deviation of the desired error is equal to

$$\sigma_\xi^2 = \mathrm{Sp}(\langle \boldsymbol{\xi}\boldsymbol{\xi}^T \rangle) = \sum_{j=1}^{S} [(\gamma_j \sigma_j - 1)^2 \sigma_{b_j}^2 + \gamma_j \sigma_{e_j}^2],$$

where $\mathrm{Sp}(A)$ is the spur of the matrix A.

In the special case of $R_{\boldsymbol{\theta}} = \sigma_\theta^2 I$ $R_{\boldsymbol{\varepsilon}} = \sigma_\varepsilon^2 I$, we obtain $\sigma_b^2 = \sigma_\theta^2$, $\sigma_{e_j}^2 = \sigma_\varepsilon^2$, and introducing the notation $\lambda(\gamma_j) = \mathrm{Sp}(\langle \boldsymbol{\xi}\boldsymbol{\xi}^T \rangle)$, we can write

$$\lambda(\gamma_j) = \sum_{j=1}^{S} [(\gamma_j \sigma_j - 1)^2 \sigma_\theta^2 + \gamma_j^2 \sigma_\varepsilon^2].$$

We determine the elements γ_j using the minimum condition of the mean square error, i.e. by the solution of the system of equations

$$\frac{\partial \lambda(\gamma_j)}{\partial \gamma_j} = 0, \quad j = 1, \ldots, S.$$

The desired estimate looks like

$$\hat{\gamma}_j = \frac{\sigma_j}{\sigma_j^2 + \sigma_\varepsilon^2/\sigma_\theta^2}.$$

If for the certain random parameters $\sigma_\theta^2 \to \infty$ is valid, then $\hat{\gamma}_j = 1/\sigma_j$ and $\Gamma = \Sigma^+$, i.e. we obtain the LSM estimate (6.52) with a singular decomposition of the structural matrix.

In the case of the singular analysis the mean-square value of the desired error will be written as

$$\lambda(i_0) = \langle \|\boldsymbol{\xi}\|^2 \rangle = \sigma_\theta^2 (S - i_0) + \sigma_\varepsilon^2 \sum_{j=1}^{i_0} 1/\sigma_j^2.$$

The minimization of $\lambda(i_0)$ on i_0 leads to a choice of the number i_0, when an appropriate singular number satisfies the relations

$$\sigma_{i_0} \geq \frac{\sigma_\varepsilon}{\sigma_\theta} \quad \text{and} \quad \sigma_{i_0+1} < \frac{\sigma_\varepsilon}{\sigma_\theta}.$$

Hence, to find the boundary value i_0 it is necessary to know a priori value of the noise variance σ_ε^2 and a priori variance of the desired parameters σ_θ^2.

So, we find the matrices Γ for three estimation procedures: "usual" LSM, when the parameter $\boldsymbol{\theta}$ is unknown and nonrandom

$$\Gamma_{ii} = \gamma_i = \frac{1}{\sigma_i}, \quad i = 1, \dots, S;$$

the method of statistical regularization for a model with a random parameter vector

$$\Gamma_{ii} = \gamma_i = \frac{\sigma_i}{\sigma_i^2 + \sigma_\varepsilon^2/\sigma_\theta^2}, \quad i = 1, \dots, S;$$

the method of singular analysis

$$\gamma_j = \begin{cases} 1/\sigma_j, & j = 1, 2, \dots, i_0 \quad \sigma_{i_0} \geq \alpha, \\ 0, & j = i_0 + 1, \dots, S \; \sigma_{i_0+1} < \alpha. \end{cases}$$

At the small singular numbers σ_i in comparison with the ratio $\sigma_\varepsilon/\sigma_\theta$ ($\sigma_i \ll \sigma_\varepsilon/\sigma_\theta$), for the "usual" LSM (the first procedure), corresponding value γ_i goes up fast, and it leads to the instability of the solution. In the cases of the statistical regularization (the second procedure) and the singular analysis (the third procedure) $\gamma_i \approx 0$ and $\gamma_i = 0$ are valid correspondingly. At a great value of σ_i ($\sigma_i \gg \sigma_\varepsilon/\sigma_\theta$) all three methods lead to the same result.

It is necessary to point out, that the singular analysis is one of the expedients of the regularization of the solution, and the basic difference from the estimates obtained by the statistical regularization in a range of great values of σ_i, but a little bit larger than $\sigma_\varepsilon/\sigma_\theta$, is the next: $\gamma_i = 0$ for the singular analysis and it decreases smoothly as

$$\gamma_i = \frac{\sigma_i}{\sigma_i^2 + \sigma_\varepsilon^2/\sigma_\theta^2}, \quad i = 1, \dots, S,$$

for the statistical regularization method (Fig. 6.1). The important result of this sec-

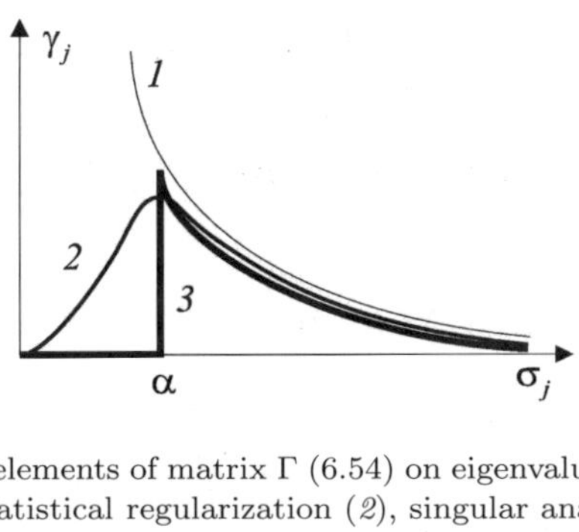

Fig. 6.1 Dependence of the elements of matrix Γ (6.54) on eigenvalues σ_j (6.49) for an estimation by the methods: LSM (*1*), statistical regularization (*2*), singular analysis (*3*).

tion consists in establishing the connection between the threshold value α, specifying in the singular analysis using a priori data, and the noise-to-signal ratio ($\sigma_\varepsilon/\sigma_\theta$), which one is customary for an interpreter and has an explicit physical sense.

The condition measure β, introduced above, can be expressed through the minimum and maximum singular numbers

$$\beta = \frac{\sigma_{\max}}{\sigma_{\min}}.$$

As it was mentioned, the value β can be used as a measure of the linear dependence of columns of the matrix $\psi^T\psi$. If the value of β is close to 1, then the columns are "practically independent". If β is sufficiently great, then the columns are "practically dependent". Introducing the threshold value α reduces the condition number β up to $\sigma_{\max}/\alpha$, that leads to a more stable determination of the desired parameters.

6.12.1 *Resolution matrix*

Let us to consider an additive model

$$\boldsymbol{u} = \psi\boldsymbol{\theta} + \boldsymbol{\varepsilon}. \tag{6.55}$$

We introduce the matrix L, which is an "inverse" matrix with respect to the matrix ψ and it defines a "rule" of the derivation of the estimate $\hat{\boldsymbol{\theta}}$ of the vector $\boldsymbol{\theta}$ using the vector of measured data $\boldsymbol{u}$,

$$\hat{\boldsymbol{\theta}} = L\boldsymbol{u}.$$

To define the resolution matrix R as a product of L and ψ

$$R = L\psi.$$

For clearing up of a sense of the resolution matrix, we multiply the left hand side and the right hand side of the equation (6.55) by the matrix L. At tending the values of the additive noise addend $\boldsymbol{\varepsilon}$ to zero, we obtain

$$\hat{\boldsymbol{\theta}} \xrightarrow{\varepsilon\to 0} R\boldsymbol{\theta}.$$

In case of the solution of the system of the linear equations (6.55) by LSM, the resolution matrix is an identity matrix:

$$L = (\psi^T\psi)^{-1}\psi^T, \quad R = L\psi = I.$$

In case of the singular decomposition (6.48) of the matrix ψ, we obtain:

$$R = P\Sigma^+Q^TQ\Sigma P^T = PP^T = I.$$

6.13 The Method of Least Modulus

Together with the LSM method, for processing of the geophysical data, the method of the least modulus is widespread. This method leads to the optimum estimates of parameters at presence in a random noise of the outliers. With the reference to the linear model (6.15) the method of least modulus is reduced to the minimization of the function

$$\lambda(\boldsymbol{\theta}) = \sum_{i=1}^{n} v_i|u_i - \sum_{s=1}^{S}\psi_{is}\theta_s|,$$

$$\hat{\boldsymbol{\theta}} = \arg\min_{\boldsymbol{\theta}} \sum_{i=1}^{n} v_i|u_i - \sum_{s=1}^{S}\psi_{is}\theta_s|. \tag{6.56}$$

Let us note, that the method of least modulus is a special case of the maximum likelihood method with the Laplace distribution of the random parameter vector $\boldsymbol{\varepsilon}$. The "tails" of the Laplace distribution fall slowly, than the "tails" of the Gaussian distribution. At the method of least squares the weight multipliers v_j have a clear physical sense, they are in inverse proportion to the standard deviations of the measurements.

To find a minimum of the function $\lambda(\boldsymbol{\theta})$ two approaches can be used. The first approach is based on the linear-programming technique. The second one is the iterative procedure based on the weighted LSM.

Let us consider the second approach in detail.

We introduce a function of two vector arguments $\boldsymbol{\theta}$ and $\boldsymbol{\rho}$

$$\lambda_1(\boldsymbol{\theta}, \boldsymbol{\rho}) = \sum_{i=1}^{n} \frac{v_i^2 |u_i - \sum\limits_{s=1}^{S} \psi_{is}\theta_s|^2}{v_i |u_i - \sum\limits_{s=1}^{S} \psi_{is}\rho_s|}. \tag{6.57}$$

If $\boldsymbol{\theta} = \boldsymbol{\rho}$, then the function (6.57) transforms to the function (6.56):

$$\lambda_1(\boldsymbol{\theta}, \boldsymbol{\theta}) = \lambda(\boldsymbol{\theta}).$$

The presentation (6.56) enables to create an iterative procedure for finding an estimate by the method of least modulus. Let's an initial approximation for the desired parameter vector $\boldsymbol{\theta}^{(0)}$ is given. We substitute the value of $\boldsymbol{\theta}^{(0)}$ to the expression (6.57) instead of the vector $\boldsymbol{\rho}$, then

$$\lambda_1(\boldsymbol{\theta}, \boldsymbol{\theta}^{(0)}) = \sum_{i=1}^{n} v_i \frac{|u_i - \sum\limits_{s=1}^{S} \psi_{is}\theta_s|^2}{|u_i - \sum\limits_{s=1}^{S} \psi_{is}\theta_s^{(0)}|}, \tag{6.58}$$

or, introducing the notation

$$w_i^0 = \frac{v_i}{|u_i - \sum\limits_{s=1}^{S} \psi_{is}\theta_s^{(0)}|}, \tag{6.59}$$

we obtain

$$\lambda_1(\boldsymbol{\theta}, \boldsymbol{\theta}^{(0)}) = \sum_{i=1}^{n} w_i^{(0)} |u_i - \sum_{s=1}^{S} \psi_{is}\theta_s|. \tag{6.60}$$

The quadratic form of (6.58), (6.60) explicitly corresponds to the quadratic form of LSM. The estimate of the desired parameter vector is determined by

$$\hat{\boldsymbol{\theta}} = (\psi^T W^{(0)} \psi)^{-1} \psi^T W^{(0)} \boldsymbol{u}, \tag{6.61}$$

where $W^{(0)} = \mathrm{diag}(w_1^0, w_2^0, \ldots, w_n^0)$.

Let's accept the estimate (6.61) as the first approximation of the estimate for the least modulus method:

$$\boldsymbol{\theta}^{(1)} = \hat{\boldsymbol{\theta}},$$

and to use it for computing the new weight multipliers. To substitute the obtained estimate to the expression (6.57) instead of $\boldsymbol{\rho}$:

$$\lambda_1(\boldsymbol{\theta}, \boldsymbol{\theta}^{(1)}) = \sum_{i=1}^{n} w_i^{(1)} (u_i - \sum_{s=1}^{S} \psi_{is}\theta_s)^2, \tag{6.62}$$

where

$$w_i^{(1)} = \frac{v_i}{|u_i - \sum\limits_{s=1}^{S} \psi_{is}\theta_s^{(1)}|}.$$

Further, the LSM estimate we find by minimizing of the quadratic form (6.62)

$$\hat{\boldsymbol{\theta}} = (\psi^T W^{(1)} \psi)^{-1} \psi^T W^{(1)} \boldsymbol{u}.$$

We accept this estimate as the second approximation of the estimate for the least modulus method:

$$\boldsymbol{\theta}^{(2)} = \hat{\boldsymbol{\theta}}.$$

Substituting $\boldsymbol{\theta}^{(2)}$ to the expression (6.57) instead of $\boldsymbol{\theta}$ we find $\boldsymbol{\theta}^{(3)}$. The iterative procedure continues until at k-th step the threshold condition

$$\frac{|\theta_s^{(k)} - \theta_s^{(k-1)}|}{|\theta_s^{(k)}|} < \delta \sim 10^{-2} \div 10^{-3}$$

is satisfied. Usually the value of δ is a order of 10^{-2}–10^{-3}.

The main element of the considered procedure is LSM. At each step we make correction of the weights W_i in accordance with the LSM estimate obtained at the previous step.

6.14 Robust Methods of Estimation

Most of the considered estimation algorithms were based on the assumption of the normal distribution of the random vector $\boldsymbol{\varepsilon}$. If this assumption is not valid, then estimates obtained by LSM lose the optimality. Besides the LSM estimates are sensitive to the great errors of observations. Therefore, the creation of the algorithms, which are stable at deviations from the initial assumptions about the model, is an actual problem. In mathematical statistics the methods of estimation, which are stable against the deviations of the supposed distributions of the random components from a true model are designed. These methods are called the *robust methods*. Three robust algorithms will be considered below.

6.14.1 *Reparametrization algorithm*

Let us consider the robust estimation algorithm for the parameter vector $\boldsymbol{\theta}$ using the model (6.15), and based on the reparametrization of an initial model (Meshalkin and Kurochkina, 1981). As the result of the reparametrization, the new parameters appear as the stable ones in wide class of the noise at a deviation of their distribution from the normal distribution. The estimation procedure looks like the next.

(1) To set an initial approach for the parameter vector $\boldsymbol{\theta}$ and a variance σ_0^2. As the vector
$\boldsymbol{\theta}_0$ an LSM estimate can be used. The variance can be estimated by the formula
$$\sigma_0^2 = \frac{(\boldsymbol{u} - \psi\boldsymbol{\theta}_0)^T(\boldsymbol{u} - \psi\boldsymbol{\theta_0})}{n - S}.$$
(2) In k-th iteration we compute the deviation
$$e_i = u_i - \boldsymbol{\psi}_i^T\hat{\boldsymbol{\theta}}^{(k)},$$
where $\boldsymbol{\psi}_i^T$ is i-th row of the matrix. To determine the weight function
$$w(\varepsilon_i) = \exp\left\{-\frac{\varepsilon_i^2\eta}{2\hat{\sigma}_k^2}\right\}.$$
At that we should determine a value of the constant η, which determines a width of the weight bell-shaped function. Such value may choose by an experimental determination. With the reference to the processing of geophysical data, this value is advisable to put $\eta = 0.2$.
(3) To find the estimate $\hat{\boldsymbol{\theta}}$ by minimizing of the quadratic form:
$$\lambda(\boldsymbol{\theta}) = \sum_{j=1}^{n}(u_j - \boldsymbol{\psi}_j^T\boldsymbol{\theta})^2 w(e_j),$$
$$\hat{\boldsymbol{\theta}} = (\psi^T W\psi)^{-1}\psi^T W\boldsymbol{u},$$
where $W = \operatorname{diag}(w_1, w_2, \ldots, w_n)$.
(4) To calculate the normalized weighted deviation
$$Y = \frac{\lambda(\boldsymbol{\theta})}{\sum\limits_{i=1}^{n}(\varepsilon_i)}.$$
(5) To find the estimate $\hat{\theta}_{k+1}$ and dispersion $\hat{\sigma}_{k+1}^2$ using the equation
$$\hat{\boldsymbol{\theta}}_{k+1} = \hat{\boldsymbol{\theta}}_k Y^{-1}\hat{\sigma}_k^2(\hat{\boldsymbol{\theta}} - \hat{\boldsymbol{\theta}}_k), \quad (\hat{\boldsymbol{\sigma}}_{k+1}^2)^{-1} = Y - \frac{\eta}{\hat{\sigma}_k^2}.$$
(6) If the difference between the estimates obtained on k-th and $(k+1)$-th iterations have an inessential difference, i.e.
$$\max_{\mu}\left|\frac{\hat{\theta}_{\mu_{k+1}} - \hat{\theta}_{\mu_k}}{\hat{\theta}_{\mu_k}}\right| \leq \delta_1, \quad \left|\frac{\hat{\sigma}_{\mu_{k+1}} - \hat{\sigma}_{\mu_k}}{\hat{\sigma}_{\mu_k}}\right| \leq \delta_2,$$
where δ_1 and δ_2 are the constants of order 10^{-2}–10^{-3}, then $\boldsymbol{\theta}_k$, σ_k^2 are the desired estimates, otherwise the iterative procedure will be repeated, starting from the point 2. As it is shown by the numerical simulation, it needs 7–10 iterations.

(7) The covariance matrix of the estimates of the desired parameters is determined by the formula

$$R_{\hat{\theta}_k} = (\psi^T W \psi)^{-1} \hat{\sigma}_k^2.$$

Described algorithm may easily be realized on a computer, because its basic element is the weighted LSM procedure.

Let us consider two robust nonlinear estimators.

6.14.2 *Huber robust method*

Let's the random components ε_i in the model (6.15) are independent and equally distributed with a symmetric density function. If the probability of a great outlier is equal to α, then

$$p(n) = (1 - \alpha)X(e) + \alpha Z(e),$$

where the supposed distribution $X(n)$ and "weed" $Z(e)$ are the symmetric densities:

$$X(e) = 1 - X(-e), \quad Z(e) = 1 - Z(-e).$$

Assuming that the random components have the normal distribution, we come to a model of the crude errors, which can be applied to the static correction at processing of seismic data. In that case instead of minimizing of a quadratic form, for the determination of the parameter vector $\boldsymbol{\theta}$, it is proposed to solve the next problem (Huber, 2004):

$$\hat{\boldsymbol{\theta}} = \arg\min_{\theta} \sum_{i=1}^{n} H(u_i - \boldsymbol{\psi}_i^T \boldsymbol{\theta}), \tag{6.63}$$

where H is the correspondingly chosen function.

In the general case the solution of the problem (6.63) reduces to the system of nonlinear equations

$$\sum_{i=1}^{n} h(u_i - \boldsymbol{\psi}_i^T \boldsymbol{\theta}) \psi_{is} = 0,$$

where $s = 1, 2, \ldots, S$, $h(\varepsilon) = H'(\varepsilon)$. As the function H we can choose the next function (Fig. 6.2)

$$H(\varepsilon) = \begin{cases} \varepsilon^2/2, & |\varepsilon| < c, \\ c|\varepsilon| - c^2/2, & |\varepsilon| \geq c. \end{cases}$$

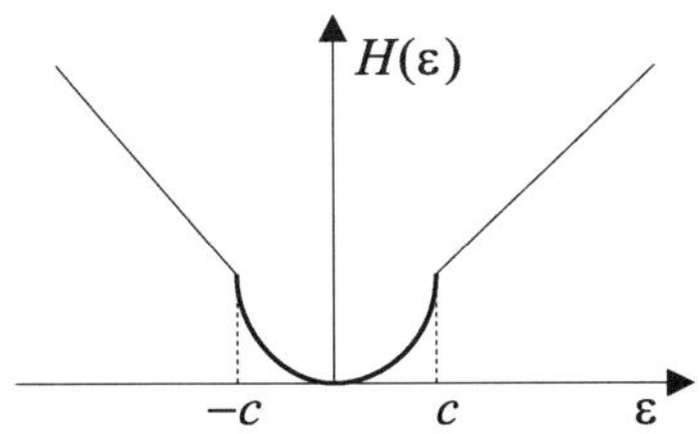

Fig. 6.2 Huber's proximity function between measured data and model data.

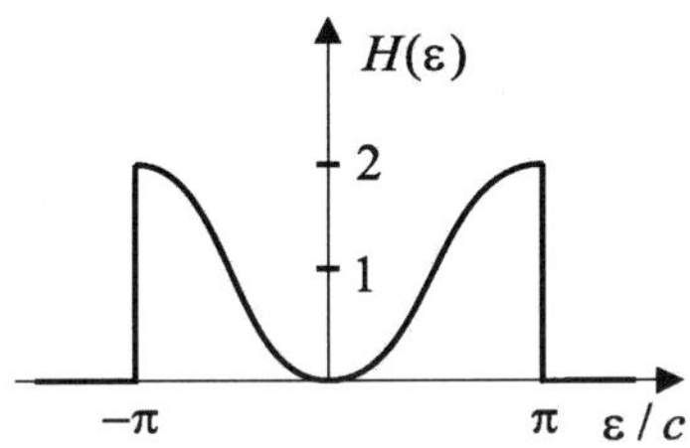

Fig. 6.3 The proximity function between measured data and model data.

6.14.3 *Andrews robust method*

Let us consider one more robust estimator (Andrews, 1974) where the weighted LSM method appears as a basic element. To find the estimate of the model parameters by minimizing the next nonlinear function:

$$\hat{\boldsymbol{\theta}} = \arg\min \sum_{i=1}^{n} H(u_i - \boldsymbol{\psi}_i^T \boldsymbol{\theta}), \tag{6.64}$$

where

$$H(\varepsilon) = \begin{cases} 1 - \cos(\varepsilon_i / c), & |\varepsilon_i| < c\pi, \\ 0, & |\varepsilon_i| \geq c\pi, \end{cases}$$

(see Fig. 6.3). As the estimate $\hat{\sigma}$ we accept a median of the absolute values:

$$\hat{\sigma} = \text{med}\, |u_i - \boldsymbol{\psi}_i^T \boldsymbol{\theta}|.$$

The problem (6.64) leads to the system of linear equations

$$\sum_{i=1}^{n} h(\varepsilon_i) \psi_{is} = 0, \quad s = 1, \ldots, S, \tag{6.65}$$

$$h(\varepsilon_i) = H'(\varepsilon_i) = \begin{cases} \sin(\varepsilon_i / c)/c, & |\varepsilon_i| < c\pi, \\ 0, & |\varepsilon_i| \geq c\pi. \end{cases}$$

The solution of the system of linear equations using the iterative procedure leads to the next steps.

(1) To establish the initial vector $\boldsymbol{\theta}_0$ or to use for its determination LSM.
Let us consider the estimator of $\hat{\boldsymbol{\theta}}_{k+1}$, if the estimates $\hat{\boldsymbol{\theta}}_k$, $k = 0, 1, \ldots$ are known.

(2) To calculate the deviation $\hat{\varepsilon}_{ik} = u_i - \boldsymbol{\psi}_i \hat{\boldsymbol{\theta}}_k$.

(3) To find the estimate $\hat{\sigma}_k = \mathrm{med}\, \hat{\varepsilon}_{ik}$.

(4) To calculate the weight function

$$w_{ik} = \begin{cases} [\sin(\hat{\varepsilon}_{ik}/c)/c][\hat{\varepsilon}_{ik}]^{-1}, & |\varepsilon_{ik}|/c < \pi, \\ 0, & |\varepsilon_{ik}|/c \geq \pi. \end{cases}$$

(5) To solve the system of equations (6.65), which can be represented in the form

$$\sum_{i=1}^{n} w_{ik} \psi_{is} \varepsilon_{ik} = 0.$$

$$\sum_{i=1}^{n} w_{ik} \psi_{is} (u_i - \boldsymbol{\psi}_i^T \boldsymbol{\theta}^{(k+1)}) = 0.$$

Using the weighted LSM to obtain the estimate

$$\hat{\boldsymbol{\theta}}_{k+1} = (\psi^T W_k \psi)^{-1} \psi^T W_k \boldsymbol{u},$$

where $W = \mathrm{diag}(w_1, w_2, \ldots w_n)$.

(6) To check an accuracy of the obtained estimates

$$\max_S \frac{|\hat{\theta}_{sk+1} - \hat{\theta}_{sk}|}{|\hat{\theta}_{sk}|} \leq \delta_1, \qquad \frac{|\hat{\sigma}_{k+1} - \hat{\sigma}_k|}{|\hat{\sigma}_k|} \leq \delta_2,$$

where δ_1 and δ_2 are the values of the order 10^{-2}–10^{-3}. If the inequalities are satisfied, then $\boldsymbol{\theta}_k$ and $\hat{\sigma}_k$ are the desired estimates. If not, we continue the procedure from the point 2. For the case of the real data the recommended value of c is 1,5.

The essential difference of this method from the reparametrization algorithm is that for a great deviation the value of the weight function puts to zero.

6.15 Interval Estimation

We have considered some examples of the point estimation. In practice the interval estimation, which is intimately connected with the point estimation, is widespread as well. The interval estimation gives not only a numerical value of parameters, but also it produces an accuracy and reliability of the estimation. So, for example, at treating experimental data the great interest is introduced by a problem of the definition of an interval, in which one the true value of parameter lays with the given probability. The problems of a such type are solved by the interval estimation method.

Let's the estimate $\hat{\theta}$ is unbiased estimate of the parameter θ. To establish enough great probability β ($\beta = 0.9, \ldots, 0.95$), we find an admissible deviation value δ of $\hat{\theta}$ and θ:

$$P(|\hat{\theta} - \theta| < \delta) = \beta, \tag{6.66}$$

Moreover, the deviations exceeding on an absolute value of δ, occur with the small probability $1 - \beta$. The expression (6.66) can be represented as

$$P(\hat{\theta} - \delta < \theta < \hat{\theta} + \delta) = \beta. \tag{6.67}$$

This inequality yields that with the probability β the random interval $I_\beta = [\hat{\theta} - \delta, \hat{\theta} + \delta]$ covers a true value of the parameter (Fig. 6.4). The probability β is usually

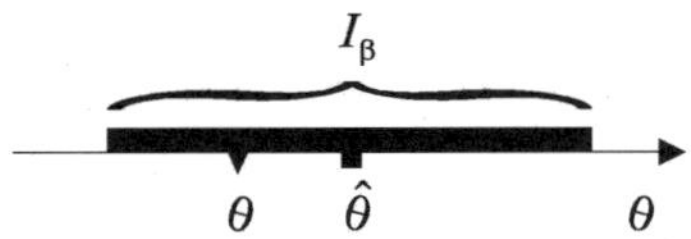

Fig. 6.4 The confidence interval I_β and estimated parameter $\hat{\theta}$.

called the *confidence probability,*, and the random interval $[\hat{\theta} - \delta, \hat{\theta} + \delta]$ is called the *confidence interval.* As it follows from the expression (6.67), to determine the confidence interval it is necessary to know a probability distribution of the estimate. As it was already mentioned, in the case of the normal distribution an error vector of the estimates obtained by LSM in case of the linear model (6.15) have the normal distribution, i.e.

$$\langle \hat{\theta}_s \rangle = \theta_s, \quad R_{\hat{\theta}_s} = \sigma_\varepsilon^2[(\psi^T W \psi)^{-1}]_{ss} \quad s = 1, 2, \ldots, S.$$

Each component of the estimating parameter vector $\hat{\theta}_s$ has the normal distribution with the mathematical expectation θ_s and the variance $\sigma_\varepsilon^2[(\psi^T \psi)^{-1}]_{ss}$:

$$\frac{\hat{\theta}_s - \theta_s}{\sigma_\varepsilon[(\psi^T W \psi)^{-1}]_{ss}^{1/2}} \in N(0, 1). \tag{6.68}$$

To introduce the deviation vector $\hat{\boldsymbol{\varepsilon}} = \boldsymbol{u} - \psi\hat{\boldsymbol{\theta}}$ and the quadratic form of deviation $\hat{\boldsymbol{\varepsilon}}^T W \hat{\boldsymbol{\varepsilon}}$. The random vector $\hat{\boldsymbol{\theta}}$ is S-dimensional normal vector, and the random vector $\hat{\boldsymbol{\varepsilon}}$ is $(n - S)$-dimensional normal vector. These vectors are independent. The random value

$$\frac{1}{\sigma^2} \hat{\boldsymbol{\varepsilon}}^T W \hat{\boldsymbol{\varepsilon}} \tag{6.69}$$

has the distribution χ^2_{n-S} and does not dependent on $\hat{\boldsymbol{\theta}}$.

Using the relations (6.68) and (6.69), to build the variable

$$t_{n-s} = \frac{\hat{\theta}_s - \theta_s}{[(\psi^T W \psi)^{-1}]^{1/2} \left[\hat{\boldsymbol{\varepsilon}}^T W \hat{\boldsymbol{\varepsilon}}/(n - S)\right]^{1/2}}$$

$$= \frac{\hat{\theta}_s - \theta_s}{\left[(\psi^T W \psi)^{-1} \hat{\boldsymbol{\varepsilon}}^T W \hat{\boldsymbol{\varepsilon}}/(n - S)\right]^{1/2}},$$

which is a ratio of a random variable with the normal distribution ($N(0,1)$) and a sum of squared normally distributed variables with a number of degrees of freedom $n-S$. Such variable has the Student's distribution with a number of degrees of freedom $n-S$ and it enables to create the confidence interval for θ_s.

To specify the confidence probability β. For given dimensions of the observed data vector and the desired parameter vector S to determine a number of degrees of freedom $k=n-S$. The values of k and β are an input data for the determination of γ, such that

$$P[|t_{n-S}| \leq \gamma] = \beta.$$

The confidence interval with the confidence probability β describes by the formula

$$I_\beta = \left[\hat{\theta}_s \pm \gamma[(\psi^T W \psi)^{-1}]_{ss}^{1/2} \left[\frac{\hat{\boldsymbol{\varepsilon}}^T W \hat{\boldsymbol{\varepsilon}}}{n-S}\right]^{1/2}\right],$$

or, taking into account, that

$$\hat{\sigma}_\varepsilon = \frac{\hat{\boldsymbol{\varepsilon}}^T W \hat{\boldsymbol{\varepsilon}}}{n-S}, \quad \hat{\sigma}_{\theta_s} = \hat{\sigma}_\varepsilon [(\psi^T W \psi)^{-1}]_{ss},$$

we obtain

$$I_\beta = [\hat{\theta}_s \pm \gamma \hat{\sigma}_{\theta_s}]. \tag{6.70}$$

From an analysis of the formula (6.70) follows that the confidence interval determines both the confidence probability and an accuracy of the estimate. Together with the interval estimations of the parameters, a practical importance possesses estimation of the confidence bounds for an approximated function. Let's deterministic component of a model with a true parameter vector is written as

$$\boldsymbol{f} = \psi \boldsymbol{\theta}, \tag{6.71}$$

where the estimate of the function f, obtained by the substitution to the right hand side of equality (6.71) the LSM estimate of the vector $\boldsymbol{\theta}$, can be written as

$$\hat{\boldsymbol{f}} = \psi \hat{\boldsymbol{\theta}}.$$

Let's note, that $\hat{\boldsymbol{f}}$ does not depend on $\hat{\boldsymbol{\varepsilon}}^T W \hat{\boldsymbol{\varepsilon}}$. To define the function t_{n-S} as a ratio of the variable

$$\frac{\hat{f}_i - f_i}{\sigma_\varepsilon [\psi(\psi^T W \psi)^{-1} \psi^T]_{ii}^{1/2}}$$

with the normal distribution and the variable

$$\frac{1}{\sigma}\sqrt{\frac{\hat{\boldsymbol{\varepsilon}}^T W \hat{\boldsymbol{\varepsilon}}}{n-S}} = \frac{1}{\sigma}\hat{\sigma}_\varepsilon.$$

The square of this variable belongs to χ^2_{n-S} distribution with $n-S$ degrees of freedom:

$$t_{n-S} = \frac{\hat{f}_i - f_i}{[\psi(\psi^T W \psi)^{-1} \psi^T]_{ii}^{1/2} [\hat{\boldsymbol{\varepsilon}}^T W \hat{\boldsymbol{\varepsilon}}/(n-S)]^{1/2}} \in \mathrm{St}\,(t_k).$$

The variable t_k belongs to the Student's distribution with $k = n - S$ degrees of freedom. For given β and (a number of degrees of freedom) k to determine (for example, using tabulated data) γ such that

$$P(|t_{n-S}| \leq \gamma) = \beta.$$

The confidence interval has the form

$$I_\beta = \left[\hat{f}_i \pm \gamma\hat{\sigma}_\varepsilon[\psi(\psi^T\psi)^{-1}\psi^T]_{ii}^{1/2}\right].$$

After calculation of I_β for $i = 1, 2, \ldots, n$, we obtain the confidence band, which covers the function $\boldsymbol{f}(\theta)$ with the probability β (Fig. 6.5).

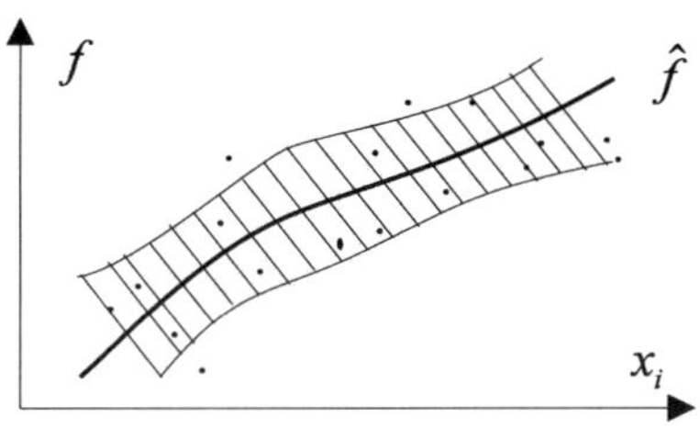

Fig. 6.5 Graphic representation of confidence band I_β.

Let us note, that the asymptotically efficient estimates and the efficient estimates under sufficiently general conditions have the lowest confidence intervals.

6.16 The Method of Backus and Gilbert of the Linear Inverse Problem Solution

At the solution of the inverse geophysical problems the model of observed data u_i $(i = 1, 2, \ldots, n)$ can be represented in the form

$$u_i = \int_0^T \varphi_i(t)\theta(t)dt + \varepsilon_i, \tag{6.72}$$

where $\varepsilon_i \in N(0, R_\varepsilon)$, $\theta(t)$ is unknown desired function of the parameters, $\varphi_i(t)$ is a given function. In a seismic case u_i is a sample of the seismogram, $\varphi_i(t)$ is a signal shape, $\theta(t)$ is a function, which characterized the reflectivity properties of a medium, $[0, T]$ is an observation interval. The desired function $\theta(t)$ in point t_0 we will describe by the mean value

$$\langle\theta(t_0)\rangle = \int_0^T C(t_0, t)\theta(t)dt, \tag{6.73}$$

where $C(t_0, t)$ is an averaging function. If the averaging function is the delta function, then $C(t_0, t) = \delta(t - t_0)$ and the mean value $\langle\theta(t_0)\rangle$ is equal to the true value

$\theta(t_0)$. Hence, under the practical calculation it is necessary to fit the averaging function close to the delta function, and to fulfil the normalization condition:

$$\int_0^T C(t_0,t)dt = 1.$$

In the method of Backus and Gilbert the averaging function (Backus and Gilbert, 1967) appears as a linear combination of the signals $\varphi_i(t)$ with unknown coefficients ρ_i:

$$C(t_0,t) = \sum_{i=1}^{n} \varphi_i(t)\rho_i(t_0). \tag{6.74}$$

Substituting the expression (6.74) to the right hand side of the equality (6.73), we obtain

$$\langle\theta(t_0)\rangle = \sum_{i=1}^{n} \rho_i(t_0)f_i, \tag{6.75}$$

where

$$f_i = \int_0^T \varphi_i(t)\theta(t)dt.$$

As it follows from the formula (6.75), the solution is represented as a linear combinations of the functionals f_i. Though, the random component ε_i brings a difference between the observed data u_i and the model function f_i. Therefore, as the estimate, it is taken a linear combination of the observed data u_i

$$\langle\hat{\theta}(t_0)\rangle = \sum_{i=1}^{n} \rho_i(t_0)u_i. \tag{6.76}$$

The problem is to find $\rho_i(t_0)$ in accordance with the stated criterion. To construct a decision rule we introduce the proximity function $s(t_0)$, which is a measure of deviation between the averaging function $C(t_0,t)$ and δ-function:

$$s(t_0) = 12 \int_{t_0-L/2}^{t_0+L/2} (t-t_0)^2 C^2(t_0,t)dt, \tag{6.77}$$

where coefficient 12 is chosen in accordance with the condition: if $C(t_0,t)$ is a function with a squared shape (Fig. 6.6)

$$C(t_0,t) = \begin{cases} 1/L, & \text{if } t \in [t_0 - L/2 \le t \le t_0 + L/2], \\ 0, & \text{if } t \notin [t_0 - L/2 \le t \le t_0 + L/2], \end{cases}$$

then $s(t_0) = L$. Substituting the formulas (6.74) to the right hand side of the equality (6.77), we obtain the proximity function

$$s(t_0) = \boldsymbol{\rho}^T(t_0)S(t_0)\boldsymbol{\rho}(t_0),$$

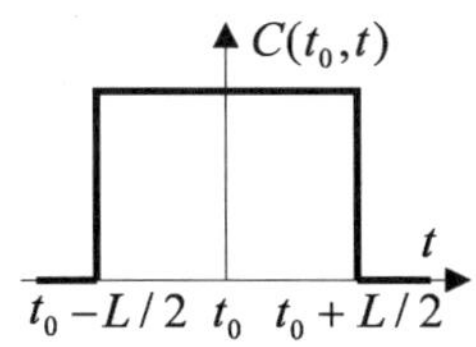

Fig. 6.6 Graphic representation of $C(t_0, t)$.

where

$$\boldsymbol{\rho}(t_0) = [\rho_1(t_0), \rho_2(t_0), \ldots, \rho_n(t_0)],$$
$$S(t_0) = \|S_{\mu\nu}(t_0)\|^n_{\mu,\nu},$$
$$S_{\mu\nu}(t_0) = 12 \int\limits_{t_0-L/2}^{t_0+L/2} \varphi_\mu(t)\varphi_\nu(t)(t-t_0)^2 dt.$$

The second component of the decision rule is the variance

$$\sigma^2_\rho(t_0) = \langle(\langle\theta(t_0)\rangle - \langle\hat{\theta}(t_0)\rangle)^2\rangle = \langle(\sum_{i=1}^{n} \rho_i(t_0)(u_i - f_i))^2\rangle$$
$$= \boldsymbol{\rho}^T(t_0)R_\varepsilon\boldsymbol{\rho}(t_0). \tag{6.78}$$

At the development of the formula (6.78) the representations (6.75) and (6.76) are used. The decision function looks usually as the following

$$g(\boldsymbol{\rho}(t_0)) = \boldsymbol{\rho}^T(t_0)G\boldsymbol{\rho}(t_0), \quad G = S\cos\theta + \beta R_\varepsilon \sin\theta,$$

where θ is a parameter belonging to the interval $0 \leq \theta \leq \pi/2$, β is a coefficient to fill in the order of the deviation and the variance. If $\theta = 0$, we get the decision function, which is entirely determined by the deviation $s(t_0)$ and the standard deviation $\sigma_\rho(t_0)$. The deviation and standard deviation characterize correspondingly the resolution and accuracy of the solution. We find the desired vector using the condition

$$\hat{\boldsymbol{\rho}} = \arg\min g(\boldsymbol{\rho}(t_0)),$$
$$\text{if} \quad \boldsymbol{B}^T\boldsymbol{\rho}(t_0) = 1,$$
$$\boldsymbol{B} = \|B_\nu\|^n_{\nu=1}, \quad B_\nu = \int\limits_0^T \varphi_\nu(t)dt. \tag{6.79}$$

Taking into account the theorem about minimum of the quadratic form with the linear constraints, we can obtain

$$\hat{\boldsymbol{\rho}}(t_0) = \frac{G^{-1}\boldsymbol{B}}{(\boldsymbol{B}^T G^{-1}\boldsymbol{B})}. \tag{6.80}$$

A linear combination of the obtained estimate $\hat{\boldsymbol{\rho}}(t_0)$ with the values of the seismic record allows to get the reflectivity property in the given point (t_0).

6.17 Genetic Algorithm

Genetic algorithms (GA), first proposed by John Holland (Holland, 1992) are based on analogies with the processes of biological evaluation. GA are very important methods for the solution of non-linear problems. The basic steps in GA are: *coding*, *selection*, *crossover*, *mutation*.

6.17.1 *Coding*

Common to any basic GA is the digitization of the list of model parameters using a binary coding scheme.

Consider, for example, the coding of the pressure wave velocity. The low velocity limit of interest may be 1500 m/s and the upper limit 1810 m/s. Assume that the desired resolution is 10 m/s. For this coding five bits is required (see Fig. 6.7). The algorithm must now determine the fitness of the individual models. This means that the binary information is decoded into the physical model parameters and the forward problem is solved. The resulting synthetic data is estimated, then compared

0	0	0	0	0	$v_{min} = 1500$ m/s
0	0	0	0	1	$v = 1510$ m/s
0	0	0	1	0	$v = 1520$ m/s
0	0	0	1	1	$v = 1530$ m/s
...	...	...	...	...	...
1	1	1	1	1	$v_{max} = 1810$ m/s

Fig. 6.7 The binary coding of the values of velocity.

with the actual observed data using the specific fitness criteria. Depending on the problem, the definition of the fitness will vary. For example, we represent the normalized correlation function, recently used in the geophysical inversion,

$$F(\boldsymbol{\theta}) = \frac{u_0 \otimes u_s(\boldsymbol{\theta})}{(u_0 \otimes u_0)^{1/2}(u_s \otimes u_s(\boldsymbol{\theta}))^{1/2}}$$

and the squared error function

$$F(\boldsymbol{\theta}) = \sum_{i=1}^{n}(u_{i0} - u_{is}(\boldsymbol{\theta}))^2,$$

where $\otimes$ is a sign of the correlation, u_0 and $u_s(\boldsymbol{\theta})$ correspond to the observed data and the model data for the parameter $\boldsymbol{\theta}$.

6.17.2 *Selection*

The selection pairs of the individual models for the reproduction is based on their fitness values. Models with the higher fitness values are more likely to get the selection than models with low fitness values. Let us consider the fitness proportionate selection.

The most basic selection method uses the ratio of each model's fitness function to the sum of the fitness of all models in the population to define its probability of the solution, i.e.

$$p_s(\theta_i) = \frac{F(\theta_i)}{\sum\limits_{j=1}^{n} F(\theta_j)},$$

where n is the number of models in the population. The selection, based on this probability, proceeds until a subset of the original models have been paired.

In a basic GA, if the population originally contained 100 models, fifty pairs are selected based on their fitness values. We shall take the models, which satisfied to the inequality

$$p_s(\theta_i) \geq \delta,$$

where δ is a given threshold value. Let we obtain L models, which form $L/2$ pairs. Each pair of the models will now produce two offsprings using the genetic operation of a crossover and a mutation. This will result in a completely new population of individuals.

6.17.3 *Crossover*

A crossover is the mechanism that allows the genetic information between the paired models to be shared. In the terminology of the geophysical inversion, the crossover caused the exchange of some information between the paired models thereby generating new models. The crossover can perform in two modes: single and multi-point. In the single point crossover, one bit positioning the binary string is selected at random from the uniform distribution. All of the bits to the down of this bit are now exchanged between the two models, generating two new models (see Fig. 6.8(a) In multi-point crossover, this operation is carried out independently for each model parameter in the string (Fig. 6.8(b)).

Example 6.1. The pair velocity model parameters v_1 and v_2 are the extreme members for the coding scheme from 1500 m/s to 1810 m/s (Fig. 6.9 and Table 6.1).

6.17.4 *Mutation*

The last genetic operation is a mutation. The mutation is a random alternation of a bit. It can be carried out during the crossover process. The mutation rate is also specified by a probability determined by the algorithm designer.

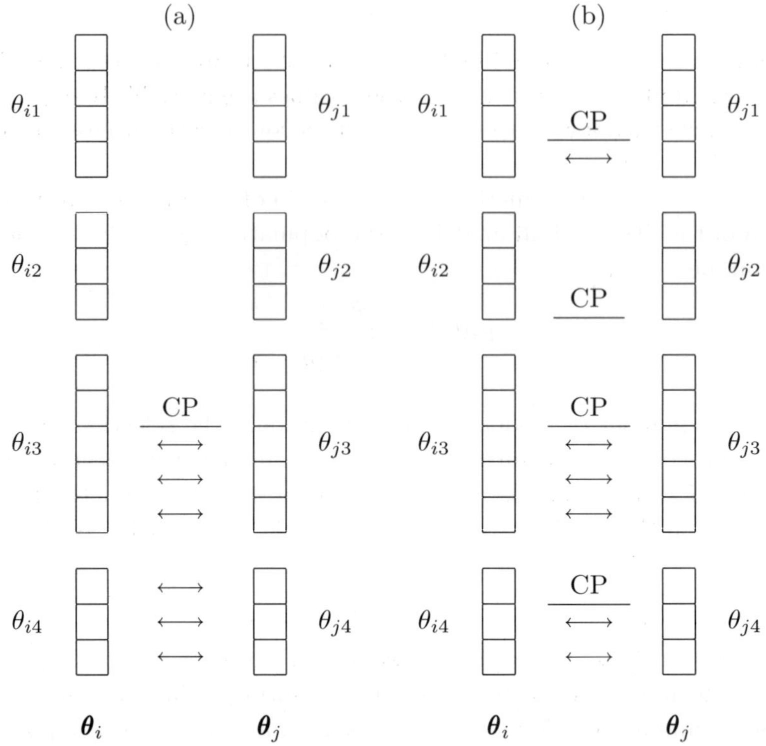

Fig. 6.8 An example of the single-point crossover (a) and the multi-point (b) crossover (CP is a crossover point).

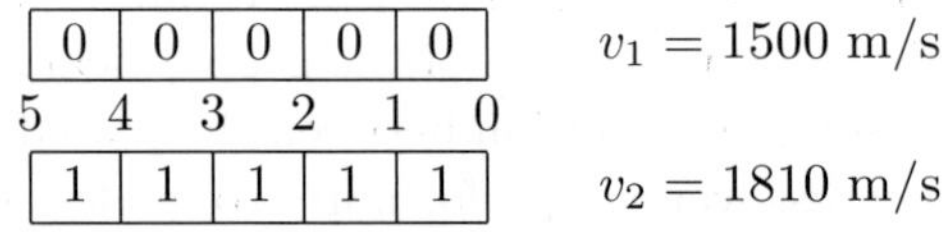

Fig. 6.9 The pair of the extreme positions for the coding scheme.

Table 6.1 An example of the crossover position.

Crossover position	v_1, m/s	v_2, m/s
0	1500	1810
1	1510	1800
2	1530	1780
3	1570	1740
4	1650	1660
5	1810	1500

Example 6.2. The process of the mutation for an initial velocity value of 1760 m/s (Fig. 6.10).

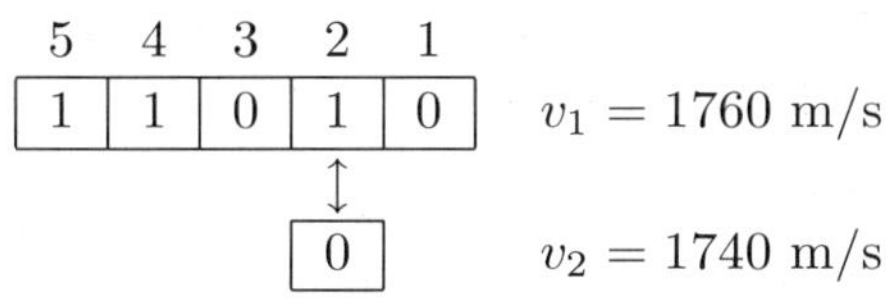

Fig. 6.10 An illustration of the mutation procedure.

If the random value ε, obtained from the random generator, less the threshold value P_m, then the mutation procedure is produced, if $\varepsilon > P_m$, then the mutation is not produced.

6.17.5 *Choice*

We choose from each of the $L/2$ pairs a model, which has the larger fitness function (Fig. 6.11). Thus we obtain $L/2$ models, which form randomly $L/4$ pairs. Then we

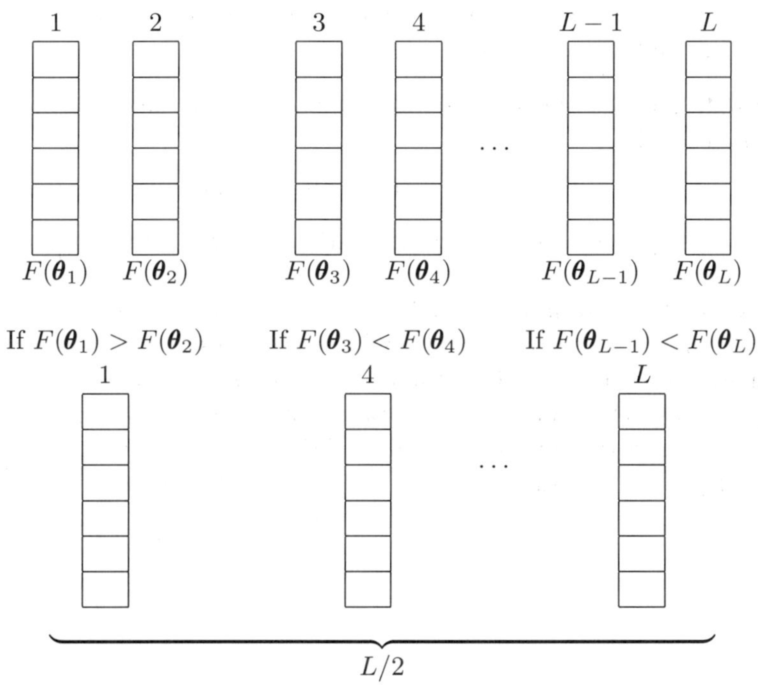

Fig. 6.11 A reduction of the models during GA.

produce the procedures: the crossover, the mutation and the choice. This procedure is continued until we obtain the optimal model. The main advantage of GA is that makes possible to find the global maximum much faster than, for example, the Monte-Carlo method. We note that the local linearization methods can give only local maximum and they do not always lead to the global maximum.

Example 6.3. Test of GA on the synthetic seismogram.

In the specific case of seismic signals the fitness function has the form

$$F(A_1, \ldots, A_M, \tau_1, \ldots, \tau_M) = \sum_{i=1}^{n} \left(u_i - \sum_{\mu=1}^{M} A_\mu \varphi(t_i - \tau_\mu) \right)^2 ,$$

where the shape of the seismic wave is

$$\varphi = \exp\{-\beta|t|\} \cos \omega t,$$

and A_μ is the amplitude, τ_μ is the arrival time of the seismic wave with the number μ. Possible bound for each of the unknown parameters (A_μ, τ_μ) and the discrete interval is determined a priori. The parameters of the attenuation β and the frequency ω are fixed also. The estimates of parameters to be find by minimization of the fitness (or objective) function $F(A_1, \ldots, A_M, \tau_1, \ldots, \tau_M)$.

During applying of GA to the optimization problem we choose the type of the bit string encoding, the size of the working population L and the values of probabilities of the crossover and the mutation.

Example 6.4. The size of the population, probabilities of the crossover and mutations are respectively:

$$L = 50, \quad P_c = 0.6, \quad P_m = 0.001.$$

Algorithm is tested under the variety population size, the mutation and crossover rates.

In a case of two interfering seismic waves the threshold of the correct wave separation is about 1/4 wave period and signal-to-noise ratio must be equal or greater than two. For the case of three interfering seismic waves, if $|\tau_1 - \tau_2| = |\tau_2 - \tau_3| \geq 0.016$ s. Three waves are separated with correct values of the estimated parameters, which correspond to the value of the wave displacement between 1/4 and 1/3 wave period. The signal-to-noise ratio must be also greater or equal to two.

Chapter 7

Statistical criteria for choice of model

Before the processing of geophysical data the interpreter faces with a choice of the suitable model of observed data. For some of these problems, for example, the extraction of a seismic signal, the resolution of the seismic signals, the determination of a number of interfering signals, the determination of the degree of the polynomial etc., in mathematical statistics a number of the methods for the solution of such problems are developed. Widely it is used the test of the parametric hypothesis with the subjective setting of the significance level. The information criterion, based on the properties of the maximum likelihood estimates and the Fisher information, is free from this imperfection. Let us start our consideration with the method of the testing of parametric hypothesis.

7.1 Test of Parametric Hypothesis

The problem statement in the case of the test of parametric hypothesis usually is the next: as a null hypothesis H_0 is regarded to the statement, that the desired parameter vector $\boldsymbol{\theta}$ is equal to $\boldsymbol{\theta}_0$; the first hypothesis (alternative hypothesis) H_1 consists in the statement, that $\boldsymbol{\theta} \neq \boldsymbol{\theta}_0$. The estimate of the parameter vector can be determined by the maximum likelihood method and it is tested: the consilience of the hypothesis H_0 with the given error probability. Let on a basis of the model of the medium with the parameter vector, the model field $\boldsymbol{u}_0$ is calculated. As the result of the observation we obtain the measured field $\boldsymbol{u}$, which is described by the model (3.2). To find the estimate $\hat{\theta}$ using the maximum likelihood method. Practical significance has a problem of adjustment of the model parameter vector $\boldsymbol{\theta}_0$ with the obtained estimate $\hat{\boldsymbol{\theta}}$. This problem is a typical one for the test of parametric hypothesis at that, the null hypothesis H_0 consists in the statement $\boldsymbol{\theta} = \boldsymbol{\theta}_0$.

As a decision criterion we shall use the method of likelihood ratio, which has at an enough great number of observations possesses the same optimal properties as the maximum likelihood method. Moreover, for enough great size of the measured data $\boldsymbol{u}$, the criterion of likelihood ratio possesses an important asymptotic property: if $\boldsymbol{\theta}$ is the S-dimensional vector and the likelihood function is regular in a sense of the first and the second derivatives on $\boldsymbol{\theta}$, then in the case of the validity of the null hypothesis and at enough great sample size, the distribution of the likelihood ratio

$$\lambda = -2\ln\frac{L(\boldsymbol{u},\boldsymbol{\theta}_0)}{L(\boldsymbol{u},\hat{\boldsymbol{\theta}})}$$

tends to χ^2_S distribution with S degrees of freedom.

The application of the criterion of the likelihood ratio corresponds in this case to the next procedure:

$$\begin{cases} \text{if } \lambda < \chi^2_{\alpha,S}, & \text{then } H_0 \text{ does not contradict } \boldsymbol{u}, \\ \text{if } \lambda > \chi^2_{\alpha,S}, & \text{then } H_0 \text{ is rejected,} \end{cases}$$

where $\chi^2_{\alpha,S}$ is the value of the function χ^2 with a significance level α and a number of degrees of freedom S.

For the model

$$\boldsymbol{u} = \begin{cases} \text{either } \boldsymbol{f}(\boldsymbol{\theta}_0) + \boldsymbol{\varepsilon}_0 : H_0\ \boldsymbol{\varepsilon}_0 \in N(0, \sigma_0^2 I), \\ \text{or} \quad\ \ \boldsymbol{f}(\boldsymbol{\theta}_1) + \boldsymbol{\varepsilon}_1 : H_1\ \boldsymbol{\varepsilon}_1 \in N(0, \sigma_1^2 I), \end{cases} \tag{7.1}$$

where $\boldsymbol{\varepsilon}_0$, $\boldsymbol{\varepsilon}_1$ are the random vectors with the normally distributed and independent components with unknown variances, σ_0^2 and σ_1^2. The criterion of the likelihood ratio can be written as

$$\lambda = -2\ln\left[\frac{L(\boldsymbol{u},\boldsymbol{\theta}_0,\hat{\sigma}_0^2)}{L(\boldsymbol{u},\boldsymbol{\theta}_1,\hat{\sigma}_1^2)}\right]. \tag{7.2}$$

The estimates $\hat{\sigma}_0^2$ and $\hat{\sigma}_1^2$ we find by the maximization of the appropriate likelihood functions $L(\boldsymbol{u},\boldsymbol{\theta}_0,\sigma_0^2)$ and $L(\boldsymbol{u},\boldsymbol{\theta}_1,\sigma_1^2)$:

$$\hat{\sigma}_0^2 = \frac{1}{n}(\boldsymbol{u} - \boldsymbol{f}(\boldsymbol{\theta}_0))^2, \quad \hat{\sigma}_1^2 = \frac{1}{n}(\boldsymbol{u} - \boldsymbol{f}(\boldsymbol{\theta}_1))^2.$$

Substituting the estimates σ_0^2, σ_1^2 in the expression (7.2), and after a simple transformation we obtain

$$\lambda = n\ln\frac{\hat{\sigma}_0^2}{\hat{\sigma}_1^2} = n\ln\left(1 + \frac{\hat{\sigma}_0^2 - \hat{\sigma}_1^2}{\hat{\sigma}_1^2}\right). \tag{7.3}$$

The criterion χ^2 for the function λ reduces to the following: if $\lambda < \chi^2_{\alpha,S}$, then obtained estimate $\hat{\boldsymbol{\theta}}$ does not contradict with the significance level α to the hypothesis H_0: $\boldsymbol{\theta} = \boldsymbol{\theta}_0$, if $\lambda > \chi^2_{\alpha,S}$, then the model $\boldsymbol{\theta} = \boldsymbol{\theta}_0$ does not fill in obtained estimates and the null hypothesis is rejected. The value of $\chi^2_{\alpha,S}$ is determined using tabulated quantities of χ^2 with a number of degrees of freedom S and a significance level α.

7.2 Criterion of a Posteriori Probability Ratio

Along with considered above solution of the problem of test of statistical hypothesis, using the properties of the criterion of likelihood ratio, at a solution of the problems of qualitative interpretation is applied the criterion of a posteriori probability ratio for the model (7.1) at given parameter vectors $\hat{\boldsymbol{\theta}}_0$ and $\hat{\boldsymbol{\theta}}_1$.

Let us consider the hypotheses H_0 and H_1 with given a priori probabilities $p(0)$ and $p(1)$ respectively. Using the Bayes theorem for the calculation of a posteriori probabilities to find the criterion function

$$\lambda = \ln \frac{P(1/\boldsymbol{u})}{P(0/\boldsymbol{u})} = \ln \frac{P(1)p(\boldsymbol{u}/1)}{P(0)p(\boldsymbol{u}/0)}.$$

An application of the criterion consists in a test of inequalities

$$\begin{cases} \text{if } \lambda < 0, \text{ then } H_0 \text{ does not contradict } \boldsymbol{u}, \\ \text{if } \lambda \geq 0, \text{ then } H_0 \text{ is rejected.} \end{cases}$$

The criterion function λ is a random one, because it depends on a random measured data. Hence, a most completely description of λ is reduced to the determination of its distribution function. For a simplicity of the analysis, we assume that $\boldsymbol{\varepsilon}_0 \in N(0, R)$, $\boldsymbol{\varepsilon}_1 \in N(0, R)$, where R is the covariance matrix of the random component, then λ will be written as

$$\lambda = \frac{1}{2}(\boldsymbol{u} - \boldsymbol{f}(\boldsymbol{\theta}_0))^T R^{-1}(\boldsymbol{u} - \boldsymbol{f}(\boldsymbol{\theta}_0)) - \frac{1}{2}(\boldsymbol{u} - \boldsymbol{f}(\boldsymbol{\theta}_1))^T R^{-1}(\boldsymbol{u} - \boldsymbol{f}(\boldsymbol{\theta}_1)). \tag{7.4}$$

To find the conditional mathematical expectations of λ_0, λ_1 and the conditional variances of $\sigma^2_{\lambda_0}$, $\sigma^2_{\lambda_1}$, using the following models

$$\boldsymbol{u} = \boldsymbol{f}(\boldsymbol{\theta}_0) + \boldsymbol{\varepsilon}, \quad \boldsymbol{u} = \boldsymbol{f}(\boldsymbol{\theta}_1) + \boldsymbol{\varepsilon},$$

$$\begin{aligned}
\langle\lambda_0\rangle &= \frac{1}{2}\langle\boldsymbol{\varepsilon}^T R^{-1}\boldsymbol{\varepsilon}\rangle - \frac{1}{2}(\boldsymbol{f}(\boldsymbol{\theta}_0) - \boldsymbol{f}(\boldsymbol{\theta}_1))^T R^{-1}(\boldsymbol{f}(\boldsymbol{\theta}_0) - \boldsymbol{f}(\boldsymbol{\theta}_1)) \\
&\quad - \frac{1}{2}\langle\boldsymbol{\varepsilon}^T R^{-1}\boldsymbol{\varepsilon}\rangle - \langle(\boldsymbol{f}(\boldsymbol{\theta}_0) - \boldsymbol{f}(\boldsymbol{\theta}_1))^T R^{-1}\boldsymbol{\varepsilon}\rangle \\
&= -\frac{1}{2}(\boldsymbol{f}(\boldsymbol{\theta}_0) - \boldsymbol{f}(\boldsymbol{\theta}_1))^T R^{-1}(\boldsymbol{f}(\boldsymbol{\theta}_0) - \boldsymbol{f}(\boldsymbol{\theta}_1)) = -\langle\lambda_1\rangle,
\end{aligned} \tag{7.5}$$

$$\begin{aligned}
\sigma^2_{\lambda_0} &= \langle(\lambda_0 - \langle\lambda_0\rangle)^2\rangle = \langle[(-\frac{1}{2}(\boldsymbol{f}(\boldsymbol{\theta}_0) - \boldsymbol{f}(\boldsymbol{\theta}_1))^T R^{-1} \\
&\quad \times (\boldsymbol{f}(\boldsymbol{\theta}_0) - \boldsymbol{f}(\boldsymbol{\theta}_1))) - (\boldsymbol{f}(\boldsymbol{\theta}_0) - \boldsymbol{f}(\boldsymbol{\theta}_1))^T R^{-1}\boldsymbol{\varepsilon} \\
&\quad + \frac{1}{2}(\boldsymbol{f}(\boldsymbol{\theta}_0) - \boldsymbol{f}(\boldsymbol{\theta}_1))^T R^{-1}(\boldsymbol{f}(\boldsymbol{\theta}_0) - \boldsymbol{f}(\boldsymbol{\theta}_1))]^2\rangle \\
&= (\boldsymbol{f}(\boldsymbol{\theta}_0) - \boldsymbol{f}(\boldsymbol{\theta}_1))^T R^{-1}\langle\boldsymbol{\varepsilon}\boldsymbol{\varepsilon}^T\rangle R^{-1}(\boldsymbol{f}(\boldsymbol{\theta}_0) - \boldsymbol{f}(\boldsymbol{\theta}_1)) \\
&= (\boldsymbol{f}(\boldsymbol{\theta}_0) - \boldsymbol{f}(\boldsymbol{\theta}_1))^T R^{-1}(\boldsymbol{f}(\boldsymbol{\theta}_0) - \boldsymbol{f}(\boldsymbol{\theta}_1)) = \sigma^2_{\lambda_1}.
\end{aligned} \tag{7.6}$$

From an analysis of the formula (7.4) leads that λ is a linear function of the measured data $\boldsymbol{u}$. Hence, at mentioned above assumptions, λ has the normal distribution

$$p(\lambda/0) = \frac{1}{\sqrt{2\pi}\sigma_{\lambda_0}} \exp\left\{-\frac{(\lambda - \langle\lambda_0\rangle)^2}{2\sigma^2_{\lambda_0}}\right\},$$

for the null hypothesis H_0 and

$$p(\lambda/1) = \frac{1}{\sqrt{2\pi}\sigma_{\lambda_1}} \exp\left\{-\frac{(\lambda - \langle\lambda_1\rangle)^2}{2\sigma^2_{\lambda_1}}\right\}$$

for the hypothesis H_1. In accordance with the criterion, if λ belongs to the interval $[-\infty, 0]$, then the hypothesis H_0 is not rejected. If λ belongs to the interval $[0, \infty]$ it is not rejected the alternative hypothesis H_1. The next four special cases can appear (Fig. 7.1):

(1) the hypothesis H_0 is true and it is not rejected;
(2) the hypothesis H_0 is true, but it is rejected (error of the first kind, alpha error);
(3) the hypothesis H_0 is false and it is rejected;
(4) the hypothesis H_0 is false, but it is not rejected (error of the second kind, beta error).

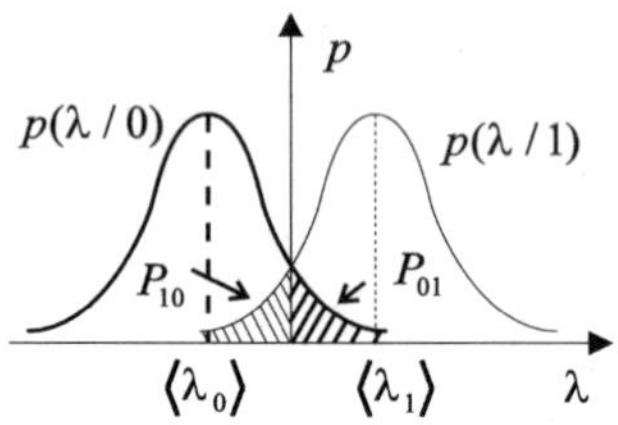

Fig. 7.1 The density functions $p(\lambda/0)$ (hypothesis H_0) and $p(\lambda/1)$ (hypothesis H_1) and decision errors.

Let denote a probability of the error of the first kind as P_{01} and a probability of the error of the second kind as P_{10}. To calculate these probabilities:

$$P_{01} = \int_0^\infty p(\lambda/0)d\lambda = \frac{1}{\sqrt{2\pi}\sigma_{\lambda_0}} \int_0^\infty \exp\left\{-\frac{(\lambda - \langle\lambda_0\rangle)^2}{2\sigma^2_{\lambda_0}}\right\}d\lambda$$
$$= \left[1 - \Phi\left(-\frac{\langle\lambda_0\rangle}{\sigma_{\lambda_0}}\right)\right],$$
$$P_{10} = \int_{-\infty}^0 p(\lambda/1)d\lambda = \frac{1}{\sqrt{2\pi}\sigma_{\lambda_1}} \int_{-\infty}^0 \exp\left\{-\frac{(\lambda - \langle\lambda_1\rangle)^2}{2\sigma^2_{\lambda_1}}\right\}d\lambda$$
$$= \Phi\left(-\frac{\langle\lambda_1\rangle}{\sigma_{\lambda_1}}\right),$$
$$\Phi(x) = \frac{1}{\sqrt{2\pi}} \int_{-\infty}^x \exp\{-t^2/2\}dt, \quad \Phi(-x) = 1 - \Phi(x),$$

where $\Phi(x)$ is the normal density function. To estimate a decision quality, probability of the total error is introduced

$$P_e = P(0)P_{01} + P(1)P_{10}.$$

For the considered special case of the equality of a priori probabilities: $P(0) = = P(1) = 1/2$ by $\langle\lambda_0\rangle = -\langle\lambda_1\rangle$, $\sigma_{\lambda_0} = \sigma_{\lambda_1}$ we obtain

$$P_e = \frac{1}{2}\left[1 - \Phi\left(-\frac{\langle\lambda_0\rangle}{\sigma_{\lambda_0}}\right)\right] + \frac{1}{2}\Phi\left(\frac{\langle\lambda_0\rangle}{\sigma_{\lambda_0}}\right) = \Phi\left(\frac{\langle\lambda_0\rangle}{\sigma_{\lambda_0}}\right).$$

Substituting (7.5) (λ_0) and (7.6) (σ_{λ_0}) in above expression, we obtain

$$P_e = \Phi(-\alpha/2) = 1 - \Phi(\alpha/2) \tag{7.7}$$

$$\alpha = [(\boldsymbol{f}(\boldsymbol{\theta}_0) - \boldsymbol{f}(\boldsymbol{\theta}_1))^T R^{-1}(\boldsymbol{f}(\boldsymbol{\theta}_0) - \boldsymbol{f}(\boldsymbol{\theta}_1))]^{1/2}. \qquad (7.8)$$

At the calculation we should set an allowed value of the error probability P_e (unreliability of the problem) and to find a value of α_0 (Fig. 7.2). If we know this

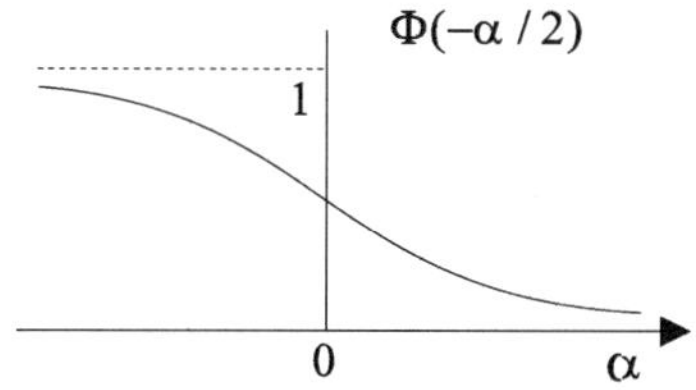

Fig. 7.2 Graphic representation of the function $\Phi(-\alpha/2)$.

value, then in all cases of the validity of the inequality

$$[(\boldsymbol{f}(\boldsymbol{\theta}_0) - \boldsymbol{f}(\boldsymbol{\theta}_1))^T R^{-1}(\boldsymbol{f}(\boldsymbol{\theta}_0) - \boldsymbol{f}(\boldsymbol{\theta}_1))]^{1/2} \geq \alpha_0,$$

the unreliability can not be greater than P_e. So, a value of α_0 determines a threshold condition for the decision with the given efficiency.

Specifically, at $P_e = 0.05$, $\alpha_0 = 3.3$ the threshold condition looks like

$$\alpha = [(\boldsymbol{f}(\boldsymbol{\theta}_0) - \boldsymbol{f}(\boldsymbol{\theta}_1))^T R^{-1}(\boldsymbol{f}(\boldsymbol{\theta}_0) - \boldsymbol{f}(\boldsymbol{\theta}_1))]^{1/2} \geq \alpha_0 = 3.3.$$

The inequalities obtained above allow us to determine the analytical threshold conditions between the parameters of the considered problem in the case if the unreliability is smaller than 0.05.

Example 7.1. Let us consider an unreliability of the extraction of the single regular wave on a noise background (Fig. 7.3):

$$u_{ki} = \begin{cases} \text{either } A\varphi(t_i - \tau - k\Delta x\gamma) + \varepsilon_{ki}, \\ \text{either } \varepsilon_{ki}, \end{cases}$$

where $k = 1, 2, \ldots, K$ (K is a number of geophones); Δx is a distance between geophones; A, τ, γ are the amplitude, the arrival time, the apparent velocity of a seismic wave respectively. The unreliability P_e is determined by the expression

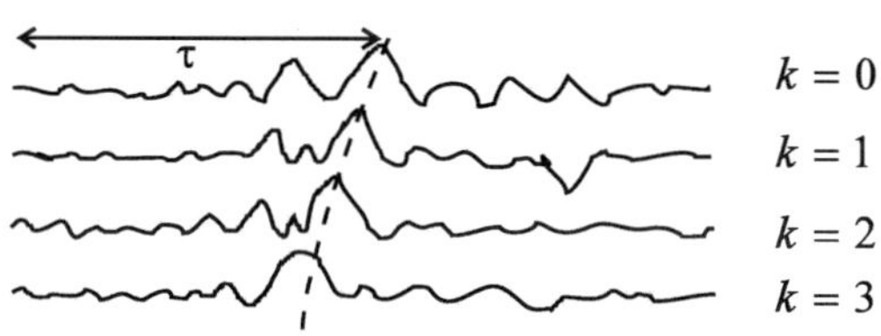

Fig. 7.3 The regular wave.

(7.7), in addition, α has the next form:

$$\alpha^2 = \frac{A^2}{\sigma^2} \sum_k \sum_i \sum_{i'} r_{ii'}^{-1} \varphi(t_i - \tau - k\Delta x\gamma)\varphi(t_{i'} - \tau - k\Delta x\gamma),$$

where $r_{ii'}^{-1}$ are the elements of the inverse covariance noise matrix, σ^2 is the noise variance. In the specific case of the uncorrelated noise component we obtain

$$\alpha^2 = \frac{A^2}{\sigma^2} \sum_{i=1}^{n} \sum_{k=1}^{K} \varphi^2(t_i - \tau - k\Delta x \gamma).$$

7.3 The Signal Resolution Problem

Let us consider the model for the signal resolution

$$u_{ki} = \begin{cases} \text{either } A_0\varphi(t_i - \tau_0 - k\Delta x\gamma_0) + \varepsilon_{ki}, \\ \text{or } \quad A_1\varphi(t_i - \tau_1 - k\Delta x\gamma_1) + A_2\varphi(t_i - \tau_2 - k\Delta x\gamma_2) + \varepsilon_{ki}. \end{cases}$$

The unreliability of the waves separation P_e (or probability of erroneous decision) is described by the formula (7.7), where

$$\begin{aligned} \alpha^2 = \frac{1}{\sigma^2} \sum_k \sum_i \sum_{i'} r_{ii'}^{-1} (A_0\varphi(t_i - \tau_0 - k\Delta x\gamma_0) - A_1\varphi(t_i - \tau_1 - k\Delta x\gamma_1) \\ - A_2\varphi(t_i - \tau_2 - k\Delta x\gamma_2))(A_0\varphi(t_{i'} - \tau_0 - k\Delta x\gamma_0) \\ - A_1\varphi(t_{i'} - \tau_1 - k\Delta x\gamma_1) - A_2\varphi(t_{i'} - \tau_2 - k\Delta x\gamma_2)). \end{aligned}$$

For the case of the uncorrelated random component, at the symmetric arrangement of the waves f_1 and f_2 with respect to f_0, i.e.

$$\tau_1 = \tau_0 - \Delta\tau, \quad \gamma_1 = \gamma_0 - \Delta\gamma,$$
$$\tau_2 = \tau_0 + \Delta\tau, \quad \gamma_2 = \gamma_0 + \Delta\gamma,$$

and under the condition $A_0 = A_1/2 = A_2/2$ we have

$$\begin{aligned} \alpha^2 = \frac{A_0^2}{\sigma^2} \sum_k \sum_i (\varphi(t_i - \tau_0 - k\Delta x\gamma_0) - \frac{1}{2}\varphi(t_i - (\tau_0 - \Delta\tau) \\ - k\Delta x(\gamma_0 - \Delta\gamma)) - \frac{1}{2}\varphi(t_i - (\tau_0 + \Delta\tau) - k\Delta x(\gamma_0 + \Delta\gamma)))^2. \end{aligned}$$

The analysis of α^2 can lead that the probability of erroneous decision (unreliability) depends on: a choice of the observation scheme (Δx and K), the signal-to-noise ratio (A/σ), the signal shape $\varphi(t)$, the kinematic parameters of waves (τ_0, γ_0, τ_1, γ_1, τ_2, γ_2). Studying a dependence of α^2 on these parameters it is possible to solve the next problems.

(1) To determine source-to-receiver offset distance, provided the given kinematic resolution of waves with the assigned probability error P_e (at the fixed signal-to-noise ratio).
(2) To estimate the minimum acceptable signal-to-noise ratio, which gives a needed resolution of the waves using the kinematic parameters with the assigned unreliability for the chosen observation scheme.

Example 7.2. We consider the function $\alpha^2(\Delta\tau)$ corresponds to the signals $\varphi(t) = \exp\{-\beta t\}cos(\Omega t)$ under the condition $\gamma_1 = \gamma_2 = \gamma_0$, i.e. resolution is studied only for the time differences $\tau_1 = \tau_0 - \Delta\tau$, $\tau_2 = \tau_0 + \Delta\tau$. An example of a such function is represented on Fig. 7.4. In the initial part of the function there is a maximum at the time $\tau \sim T/2$ and after several oscillations the curve tends to an asymptote, which corresponds to a waves reception without an interference.

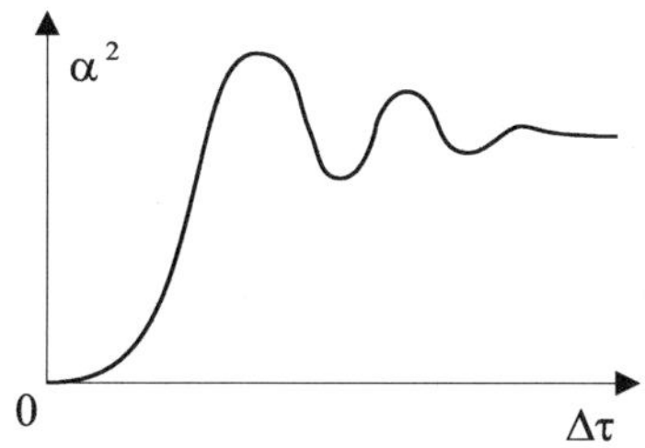

Fig. 7.4 Graphic representation of the function α^2.

Example 7.3. Let us consider an example of the resolution of magnetized bodies. We will find the threshold condition for the discrimination of the vertical thin layer, with the infinite embedded lower edge and located under the coordinate origin transversely to the observation axis x, and two similar layers located symmetrically relative to the coordinate origin at the distances $\pm\xi$. The magnetization I of the single object is equal to the doubled magnetization of each of the paired layer. So a total magnetization of the pair of layers is equal to the magnetization of the single layer. If the random component is uncorrelated at the observation points, then

$$f_{0k} = \frac{2I}{h}\left[\frac{1}{1+(k\Delta x/h)^2}\right], \quad f_{1k} = \frac{I}{h}\left[\frac{1}{1+(k\Delta x/h-\xi/h)^2}\right],$$
$$f_{2k} = \frac{I}{h}\left[\frac{1}{1+(k\Delta x/h+\xi/h)^2}\right]. \tag{7.9}$$

An example of the magnetic fields produced by the considered objects are represented on Fig. 7.5. The threshold condition can be written as

$$\sum_{k=-K}^{K}\left[\left(2\left(1+\left(\frac{k\Delta x}{h}-\frac{\xi}{h}\right)^2\right)\right)^{-1}+\left(2\left(1+\left(\frac{k\Delta x}{h}+\frac{\xi}{h}\right)^2\right)\right)^{-1}\right.$$
$$\left.-\left(1+\left(\frac{k\Delta x}{h}\right)^2\right)^{-1}\right]^2 \geq \left(\frac{\sigma h}{2I}\right)^2\alpha_0^2.$$

The obtained inequality determines the minimum value of ξ, which still makes possible to distinguish the interference field of the pairs of layers from the field of the single layer.

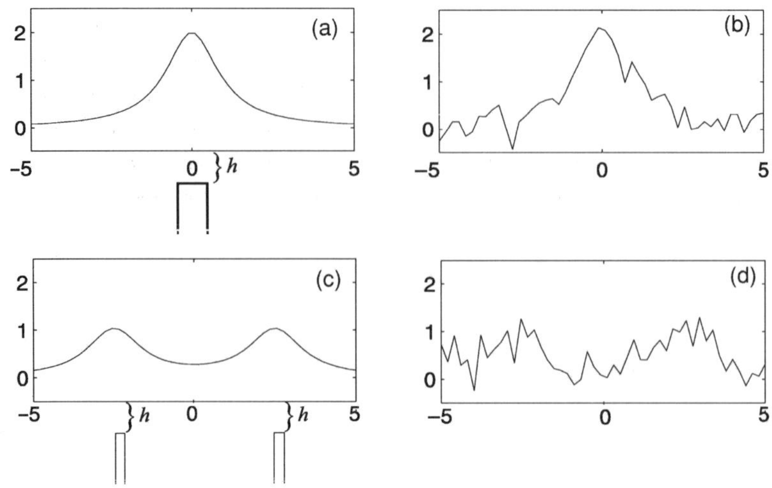

Fig. 7.5 The magnetic field produced by the vertical layers ($I = 1$, $\xi = 2.5$, $h = 1.0$): (a), (c) — without the noise; (b), (d) — with uncorrelated Gaussian noise $N(0, 0.25)$; (a), (c) — one layer; (b), (d) — two layers.

7.4 Information Criterion for the Choice of the Model

Let us consider the information criterion (Akaike, 1974). The application of the maximum likelihood method for the problem of the choice of the model at the given set of allowable models with a different dimension of the parameter vectors, leads to the choice of the model with greater dimension of the parameter vector. Therefore this method can not be used for the intuitive chose of the "right" model. However, the analysis of the maximum likelihood estimates results in that, taking into account the asymptotic properties of the maximum likelihood estimate, the likelihood function is possible to consider as the keen value relatively to the estimated parameters in a vicinity of their true quantities. This property can be used for the development of the fitting criterion of the model with the probability structure described by the density function $p_1(\boldsymbol{u}, \boldsymbol{\theta})$ with respect to the probability structure described by the density function $p_0(\boldsymbol{u})$. In the case of the classical maximum likelihood method, the estimation is implemented in the framework of one family of distributions. The information criterion allows belonging of $p_1(\boldsymbol{u}, \boldsymbol{\theta})$ and $p_2(\boldsymbol{u})$ to the family with various dimensions of the desired parameter vectors or to various families. At this case the problem of the choice of the model is stated as the estimation problem and we are free from the subjective decisions.

Let us consider the construction of the information criterion. Let the values of the observed geophysical field $u_1, u_2, \ldots, u_n$ are independent random values with a priori unknown density function $p_0(\boldsymbol{u})$. The parametric family $p_1(\boldsymbol{u}, \boldsymbol{\theta})$ has an unknown parameter vector, moreover, we do not know both the components of the vector $\boldsymbol{\theta}$, and its dimension. The estimate of the mean value of the logarithm of likelihood function has the form

$$\frac{1}{n}\sum_{i=1}^{n}\ln p_1(u_i,\boldsymbol{\theta}). \tag{7.10}$$

Passing on to the limit in the formula (7.10) under $n \to \infty$, we obtain the mean value of the logarithmic likelihood function

$$B[p_0(\boldsymbol{u}),p_1(\boldsymbol{u},\boldsymbol{\theta})] = \lim \frac{1}{n}\sum_{i=1}^{n}\ln p_1(u_i,\boldsymbol{\theta})$$
$$= \int_{-\infty}^{\infty} p_0(\boldsymbol{u})\ln p_1(\boldsymbol{u},\boldsymbol{\theta})du_1\dots du_n.$$

Taking into account the asymptotic efficiency of the maximum likelihood estimate, we can conclude that the mean value of the logarithmic likelihood function $B[p_0(\boldsymbol{u}),p_1(\boldsymbol{u},\boldsymbol{\theta})]$ is a most sensitive measure with respect to the small deviations of $p_1(\boldsymbol{u},\boldsymbol{\theta})$ from $p_0(\boldsymbol{u})$. The difference

$$I_k(p_0(\boldsymbol{u}),p_1(\boldsymbol{u},\boldsymbol{\theta})) = B[p_0(\boldsymbol{u}),p_0(\boldsymbol{u})] - B[p_0(\boldsymbol{u}),p_1(\boldsymbol{u},\boldsymbol{\theta})]$$

is known as *the Kullback information* (Kullback, 1978), which is a measure of the deviation of the actual density function $p_0(\boldsymbol{u})$ from the expected one $p_1(\boldsymbol{u},\boldsymbol{\theta})$. If $p_1(\boldsymbol{u},\boldsymbol{\theta})$ everywhere is equal to $p_0(\boldsymbol{u})$, then the information I_k is equal to zero.

Let us consider a special case. Let $p_0(\boldsymbol{u}) = p_0(\boldsymbol{u},\boldsymbol{\theta}_0)$, where $\boldsymbol{\theta}_0$ is the actual parameter vector. The mean value of the logarithmic likelihood function and the Kullback information depend only on the parameter vectors $\boldsymbol{\theta}_0$ and $\boldsymbol{\theta}$:

$$I_k(\boldsymbol{\theta}_0,\boldsymbol{\theta}) = B(\boldsymbol{\theta}_0,\boldsymbol{\theta}_0) - B(\boldsymbol{\theta}_0,\boldsymbol{\theta}),$$

where

$$B(\boldsymbol{\theta}_0,\boldsymbol{\theta}_0) = \int_{-\infty}^{\infty} p_0(\boldsymbol{u}),\boldsymbol{\theta})\ln p(\boldsymbol{u}),\boldsymbol{\theta}_0)du_1\dots du_n.$$

If the vector $\boldsymbol{\theta}_0$ is sufficiently close to the vector $\boldsymbol{\theta}$ ($\boldsymbol{\theta} = \boldsymbol{\theta}_0 + \Delta\boldsymbol{\theta}$), then it is possible the next approximation

$$I_k(\boldsymbol{\theta}_0,\boldsymbol{\theta}_0+\Delta\boldsymbol{\theta}) = \frac{1}{2}\Delta\boldsymbol{\theta}^T I_F\Delta\boldsymbol{\theta}, \tag{7.11}$$

where

$$\|I_F\|_{ss'} = \left\langle \frac{\partial\ln p(\boldsymbol{u},\boldsymbol{\theta})}{\partial\theta_s}\frac{\partial\ln p(\boldsymbol{u},\boldsymbol{\theta})}{\partial\theta_s}\right\rangle$$

the elements of the Fisher matrix. After the analysis of the formula (7.11) we can conclude, that if the estimate of the parameter $\hat{\boldsymbol{\theta}}$ is located in a small neighborhood of the parameter $\boldsymbol{\theta}_0$, then the deviation of the density $p_1(\boldsymbol{u},\boldsymbol{\theta})$ from the actual density $p_0(\boldsymbol{u},\boldsymbol{\theta}_0)$ is measured by the mean square deviation

$$\frac{1}{2}(\hat{\boldsymbol{\theta}}-\boldsymbol{\theta}_0)^T I_F(\hat{\boldsymbol{\theta}}-\boldsymbol{\theta}_0).$$

If $\boldsymbol{\theta}$ is closed to $\boldsymbol{\theta}_0$ in the maximum point of $B(\boldsymbol{\theta}_0, \boldsymbol{\theta})$:

$$\hat{\boldsymbol{\theta}} = \arg\max B(\boldsymbol{\theta}_0, \boldsymbol{\theta}),$$

it is possible to proof, that under some regularity condition the maximum likelihood estimate $\hat{\boldsymbol{\theta}}$ has an asymptotically normal distribution, i.e.

$$\sqrt{n}(\hat{\boldsymbol{\theta}} - \boldsymbol{\theta}) \in N(0, I_F^{-1}), \tag{7.12}$$

$$n(\hat{\boldsymbol{\theta}} - \boldsymbol{\theta})^T I_F (\hat{\boldsymbol{\theta}} - \boldsymbol{\theta}) \in \chi^2_{Sm}, \tag{7.13}$$

where χ^2_{Sm} is χ^2-distribution with a number of degrees of freedom equated to a length the vectors $\hat{\boldsymbol{\theta}}$ and $\boldsymbol{\theta}$, which belongs to the same parametric space S_m. But the vector $\boldsymbol{\theta}_0$ has the length S_{m_0}, which is not equal to the length of the vectors $\hat{\boldsymbol{\theta}}$ and $\boldsymbol{\theta}$, in general.

Let us evaluate the mean Kullback information for the maximum likelihood estimate $\hat{\boldsymbol{\theta}}$. To rewrite the formula (7.11) in the form

$$I_k(\boldsymbol{\theta}_0, \hat{\boldsymbol{\theta}}) = \frac{1}{2}(\hat{\boldsymbol{\theta}} - \boldsymbol{\theta})^T I_F (\hat{\boldsymbol{\theta}} - \boldsymbol{\theta}). \tag{7.14}$$

To use the condition (7.13), we multiply the left hand side and the right hand side of (7.14) by $2n$. After simple reduction we will have

$$\begin{aligned} 2nI_k(\boldsymbol{\theta}_0, \hat{\boldsymbol{\theta}}) &= n(\hat{\boldsymbol{\theta}} - \boldsymbol{\theta})^T I_F (\hat{\boldsymbol{\theta}} - \boldsymbol{\theta}) + n(\boldsymbol{\theta} - \boldsymbol{\theta}_0)^T I_F (\boldsymbol{\theta} - \boldsymbol{\theta}_0) \\ &+ 2n(\hat{\boldsymbol{\theta}} - \boldsymbol{\theta})^T I_F (\boldsymbol{\theta} - \boldsymbol{\theta}_0). \end{aligned} \tag{7.15}$$

In accordance with the property (7.13), the first term from the right hand side of the equality (7.15) belongs to the $\chi^2_{S_m}$ distribution asymptotically, and the third term has an asymptotically normal distribution. The calculation of the mean value of the left hand side and the right hand side of the equation (7.15) on the ensemble of observed data, we obtain

$$\langle 2nI_k(\boldsymbol{\theta}_0, \hat{\boldsymbol{\theta}}) \rangle = n(\boldsymbol{\theta} - \boldsymbol{\theta}_0)^T I_F (\boldsymbol{\theta} - \boldsymbol{\theta}_0) + S_m.$$

To find $n(\boldsymbol{\theta} - \boldsymbol{\theta}_0)^T I_F (\boldsymbol{\theta} - \boldsymbol{\theta}_0)$ we use the estimate $(\boldsymbol{\theta}_0, \boldsymbol{\theta}_0)$ and $(\boldsymbol{\theta}, \boldsymbol{\theta}_0)$ in the form of the sample mean of the logarithmic likelihood function. Substituting instead of $\boldsymbol{\theta}$ its estimate $\hat{\boldsymbol{\theta}}$ and to correct an arising bias by above way, we obtain:

$$n(\boldsymbol{\theta} - \boldsymbol{\theta}_0)^T I_F (\boldsymbol{\theta} - \boldsymbol{\theta}_0) = 2\sum_{i=1}^{n} \ln p(u_i, \boldsymbol{\theta}_0) - \sum_{i=1}^{n} \ln p(u_i, \boldsymbol{\theta}) + S_m. \tag{7.16}$$

Substituting the expression (7.16) to the right hand side of the equality (7.15) and taking into account that

$$I_k(\boldsymbol{\theta}, \boldsymbol{\theta}) = B(\boldsymbol{\theta}_0, \boldsymbol{\theta}_0) - B(\boldsymbol{\theta}_0, \boldsymbol{\theta}),$$

$$B(\boldsymbol{\theta}_0, \boldsymbol{\theta}_0) = \frac{1}{n}\sum_{i=1}^{n} \ln p(u_i, \boldsymbol{\theta}_0),$$

we obtain the final expression for the function of the information criterion

$$IC(S_m, \hat{\boldsymbol{\theta}}) = -2\langle nB(\boldsymbol{\theta}_0, \hat{\boldsymbol{\theta}})\rangle = -2\ln p(n, \hat{\boldsymbol{\theta}}) + 2S_m. \tag{7.17}$$

The first term from the right hand side of this expression is equal to the maximum value of the likelihood function with an opposite sign. The second term is equal to a doubled length of the vector $\boldsymbol{\theta}$. The function $IC(S_m, \hat{\boldsymbol{\theta}})$ can be considered as discord measure of the identifiable model.

The criterion application is reduced to following. To choose the competing models $m = 1, 2, \ldots, M$, with various lengths of the parameter vectors $S_1, S_2, \ldots, S_M$. We calculate the estimates by the maximum likelihood method $\hat{\boldsymbol{\theta}}_m$ and using the formula (7.17) to find $IC(S_m, \hat{\boldsymbol{\theta}})$. In accordance with the criterion, we choose the model connected with a maximum value of IC, i.e.

$$S_m = \arg\max_m [IC(S_m, \hat{\boldsymbol{\theta}}_m)]. \tag{7.18}$$

The relevant advantage of the considered criterion is the actuation of the choice of dimensionality of the model during the estimation. A quality of the choice is estimated by an absolute value of the function IC in the maximum point on m. The function IC is equal, within a constant factor, to the negative entropy. The information criterion can be considered as a modification of the maximum entropy method, which has a wide applying at the processing of geophysical data.

7.5 The Method of the Separation of Interfering Signals

The important step of the processing of geophysical data is the problem of the separation of a geophysical field on its components. For example, one of the objectives of the analysis of a seismic field consists in the extraction of the reflected waves which carry an information about parameters of the layered medium. In the magnetometry the actual problem is the extraction of the fields produced by some anomalies.

Let us consider the principles of the development of the iterative algorithm for the separation of the interfering signals. We will find not only the signal parameters, but to find also a number of the signals for the case of the representation of the field model in the form

$$\boldsymbol{u}_k = \sum_{\mu=1}^{M} \boldsymbol{f}_{\mu k}(\boldsymbol{\theta}_\mu) + \boldsymbol{\varepsilon}_k,$$

where $\boldsymbol{\theta}_\mu$ is the unknown parameter vector for μ-th signal, $\boldsymbol{\varepsilon}_k \in N(0, R)$. For example, in the case of seismic signals, we can write

$$\boldsymbol{f}_{\mu k} = \|A_\mu \varphi(t_i - \tau_\mu - k\Delta x \gamma_\mu)\|_{i=1}^{n}.$$

To find the parameters we use the maximum likelihood method (see Sec. 6.2). By analogy with the formula (6.5) we write the logarithmic likelihood function, excluding the terms independent of $\boldsymbol{\theta}$,

$$l(\boldsymbol{u}_k, \boldsymbol{\theta}) = \frac{1}{2}\sum_k (\boldsymbol{u}_k - \sum_{\mu=1}^{M} \boldsymbol{f}_{\mu k}(\boldsymbol{\theta}_\mu))^T R^{-1}(\boldsymbol{u}_k - \sum_{\mu=1}^{M} \boldsymbol{f}_{\mu k}(\boldsymbol{\theta}_\mu)),$$

where $\boldsymbol{\theta} = \|\theta_\mu\|_{\mu=1}^M$ is a full set of the vectors $\boldsymbol{\theta}_\mu$. Because $\boldsymbol{u}_k$ does not depend on the desired parameters $\boldsymbol{\theta}_\mu$, then it is enough to maximize the function

$$g(\boldsymbol{u}_k, \boldsymbol{\theta}) = \sum_{\mu=1}^{M}\sum_k \Big[\boldsymbol{u}_k^T R^{-1}\boldsymbol{f}_{\mu k}(\boldsymbol{\theta}_\mu) - \frac{1}{2}\sum_{\mu' \neq \mu} \boldsymbol{f}_{\mu' k}^T(\boldsymbol{\theta}) R^{-1}\boldsymbol{f}_{\mu k}(\boldsymbol{\theta})$$
$$- \frac{1}{2}\boldsymbol{f}_{\mu k}^T(\boldsymbol{\theta}_\mu) R^{-1}\boldsymbol{f}_{\mu k}(\boldsymbol{\theta}_\mu).\Big], \tag{7.19}$$

This function is a sufficient statistic and we will call it *the sufficient reception function.* The first terms in the square brackets of the formula (7.19) is a correlation between the signal $\boldsymbol{f}_{\mu k}(\boldsymbol{\theta}_\mu)$ of a number μ with measured data, the second term determines a mutual correlation of the various signals and the third term is an energy of μ-th signal. If the signals do not interact (by the reason of the orthogonality or there is no overlapping between them in the space-time domain), then the second term in square brackets is equal to zero and we can rewrite the formula (7.19) as

$$g(\boldsymbol{u}_k, \boldsymbol{\theta}) = \sum_{\mu=1}^{M}\sum_k \left[\boldsymbol{u}_k^T R^{-1}\boldsymbol{f}_{\mu k}(\boldsymbol{\theta}_\mu) - \frac{1}{2}\boldsymbol{f}_{\mu k}^T(\boldsymbol{\theta}_\mu) R^{-1}\boldsymbol{f}_{\mu k}(\boldsymbol{\theta}_\mu)\right]. \tag{7.20}$$

As far as an each term of the sum over μ in the formula (7.20) depends only on the desired parameter $\boldsymbol{\theta}_\mu$, the maximum of $g(\boldsymbol{u}_k, \boldsymbol{\theta})$ coincides with the maxima of the terms of the sum over μ. It allows to find the parameters $\boldsymbol{\theta}_\mu$ separately (independently for each μ). In the case of correlated signals we should depart from the above procedure and to vary the parameters simultaneously. In this case the various iterative multi-step procedures can be offered with the variation only parameters of one signal at each step.

For the algorithm development we rewrite the sufficient reception function (7.19) and write down separately the terms corresponding to the m-th signal:

$$g(\boldsymbol{u}_k, \boldsymbol{\theta}) = \Big[\sum_k \boldsymbol{u}_k^T R^{-1}\boldsymbol{f}_{mk}(\boldsymbol{\theta}_m) - \sum_{\substack{\mu=1\\ \mu\neq m}}^{M}\sum_k \boldsymbol{f}_{\mu k}(\boldsymbol{\theta}_\mu) R^{-1}\boldsymbol{f}_{mk}(\boldsymbol{\theta}_m)$$
$$-\frac{1}{2}\sum_k \boldsymbol{f}_{mk}^T(\boldsymbol{\theta}_m) R^{-1}\boldsymbol{f}_{mk}(\boldsymbol{\theta}_m)\Big] + \Big[\sum_{\substack{\mu=1\\ \mu\neq m}}^{M}\sum_k \boldsymbol{u}_k^T R^{-1}\boldsymbol{f}_{\mu k}(\boldsymbol{\theta}_\mu)$$
$$-\frac{1}{2}\sum_{\substack{\mu=1\\ \mu\neq m}}^{M}\sum_{\substack{\mu'=1\\ \mu'\neq m}}^{M}\sum_k \boldsymbol{f}_{\mu' k}^T(\boldsymbol{\theta}_{\mu'}) R^{-1}\boldsymbol{f}_{\mu k}^T(\boldsymbol{\theta}_\mu)$$
$$-\frac{1}{2}\sum_{\substack{\mu=1\\ \mu\neq m}}^{M}\sum_k \boldsymbol{f}_{\mu k}^T(\boldsymbol{\theta}_\mu) R^{-1}\boldsymbol{f}_{\mu k}(\boldsymbol{\theta}_\mu)\Big]. \tag{7.21}$$

In the formula (7.21) the terms in the second square brackets do not depend on $\boldsymbol{\theta}_m$. From the above formula we can conclude that if all signals $\boldsymbol{f}_{\mu k}$ are known, except only one signal $\boldsymbol{\theta}_m$, then for the determination of $\boldsymbol{\theta}_m$ we can maximize the expression in the first square brackets of the formula (7.21). This expression we denote as $g_m(\boldsymbol{u}_k, \boldsymbol{\theta}_m)$ and write down in the form

$$g_m(\boldsymbol{u}_k, \boldsymbol{\theta}_m) = \sum_k \boldsymbol{Y}_{km}^T R^{-1} \boldsymbol{f}_{mk}(\boldsymbol{\theta}_m) - \frac{1}{2} \sum_k \boldsymbol{f}_{mk}^T(\boldsymbol{\theta}_m) R^{-1} \boldsymbol{f}_{mk}(\boldsymbol{\theta}_m), \tag{7.22}$$

where

$$\boldsymbol{Y}_{km}^T = \boldsymbol{u}_k^T - \sum_{\substack{\mu=1 \\ \mu \neq m}}^{M} \boldsymbol{f}_{k\mu}^T(\boldsymbol{\theta}_\mu). \tag{7.23}$$

The expression $g_m(\boldsymbol{u}_k, \boldsymbol{\theta}_m)$ makes a sense of the sufficient reception function for a single signal with a number m, but there instead of the measured data $\boldsymbol{u}_k$ the model data $\boldsymbol{Y}_{mk}$ is used. The model data is obtained by the subtraction from the measured data of the signals $\boldsymbol{f}_{\mu k}$ with indices $\mu \neq m$.

The idea of the iterative process consists in reiterated and by turn look-up of the vectors $\boldsymbol{\theta}$ ($m = 1, 2, \ldots, M$) with the help of the sufficient reception function (7.22) and with the use of the values of others parameters which were found in the previous iterations of θ_μ ($\mu \neq m$). If we do not know the number of signals M, then the computing circuit looks as following.

At the beginning it is assumed that we have only one signal and we determine its parameters by minimizing of the sufficient reception function (7.22) at $m = 1$:

$$g_1(\boldsymbol{u}_k, \boldsymbol{\theta}_1) = \sum_k \left[\boldsymbol{u}_k^T R^{-1} \boldsymbol{f}_k(\boldsymbol{\theta}_1) - \frac{1}{2} \boldsymbol{f}_k^T(\boldsymbol{\theta}_1) R^{-1} \boldsymbol{f}_k(\boldsymbol{\theta}_1) \right]. \tag{7.24}$$

We find a maximum of the function $g_m(\boldsymbol{u}_k, \boldsymbol{\theta}_1)$, using, for example, one of the gradient methods:

$$\hat{\boldsymbol{\theta}}_1^{(1)} = \max_{\boldsymbol{\theta}_1} g_1(\boldsymbol{u}_k, \boldsymbol{\theta}_1).$$

Then, we suppose that there are two signals and we find the second vector $\hat{\boldsymbol{\theta}}_2^{(1)}$ by maximizing of

$$g_2(\boldsymbol{u}_k, \hat{\boldsymbol{\theta}}_1, \boldsymbol{\theta}_2) = \sum_k \boldsymbol{Y}_{k2}^T R^{-1} \boldsymbol{f}_{k2}(\boldsymbol{\theta}_2) - \frac{1}{2} \sum_k \boldsymbol{f}_{k2}^T(\boldsymbol{\theta}_2) R^{-1} \boldsymbol{f}_{k2}(\boldsymbol{\theta}_2),$$

where

$$\boldsymbol{Y}_{k2}^T = \boldsymbol{u}_k^T - \boldsymbol{f}_{k1}(\hat{\boldsymbol{\theta}}_1).$$

After that, taking into account that we have two signals, we again find the vector $\hat{\boldsymbol{\theta}}_1^{(1)}$ by maximizing of

$$g_2(\boldsymbol{u}_k, \boldsymbol{\theta}_1, \hat{\boldsymbol{\theta}}_2^{(1)}) = \sum_k \boldsymbol{Y}_{k1}^T R^{-1} \boldsymbol{f}_{k1}(\boldsymbol{\theta}_1) - \frac{1}{2} \sum_k \boldsymbol{f}_{k1}^T(\boldsymbol{\theta}_1) R^{-1} \boldsymbol{f}_{k1}(\boldsymbol{\theta}_1),$$

$$\boldsymbol{Y}_{k1} = \boldsymbol{u}_k^T - \boldsymbol{f}_{k2}(\hat{\boldsymbol{\theta}}_2),$$

at that we use the obtained estimates $\hat{\boldsymbol{\theta}}_2^{(1)}$. Then we find $\hat{\boldsymbol{\theta}}_2^{(2)}$, supposing that $\hat{\boldsymbol{\theta}}_1^{(1)}$ is equal to the obtained values, etc. The values of the maxima of the sufficient reception function, obtained in the process of the parameters refinement, are an increasing number sequence. We stop the procedure if the values of the maxima become stable.

If

$$\frac{g_2(\hat{\boldsymbol{\theta}}_1^{(n)}, \hat{\boldsymbol{\theta}}_2^{(n)}) - g_2(\hat{\boldsymbol{\theta}}_1^{(n-1)}, \hat{\boldsymbol{\theta}}_2^{(n-1)})}{g_2(\hat{\boldsymbol{\theta}}_1^{(n-1)}, \hat{\boldsymbol{\theta}}_2^{(n-1)})} < \delta,$$

then the refinement can be stopped at $\delta \sim 10^{-2} \div 10^{-3}$. Further, we suppose that we have three signals and we find the vector $\hat{\boldsymbol{\theta}}_3^{(1)}$ by maximizing

$$g_3(\boldsymbol{u}_k, \hat{\boldsymbol{\theta}}_1, \hat{\boldsymbol{\theta}}_2, \boldsymbol{\theta}_3) = \sum_k \boldsymbol{Y}_{k3}^T R^{-1} \boldsymbol{f}_{k3}(\boldsymbol{\theta}_3) - \frac{1}{2} \sum_k \boldsymbol{f}_{k3}^T(\boldsymbol{\theta}_3) R^{-1} \boldsymbol{f}_{k3}(\boldsymbol{\theta}_3),$$

$$\boldsymbol{Y}_{k3}^T = \boldsymbol{u}_k^T - \boldsymbol{f}_{k1}^T(\hat{\boldsymbol{\theta}}_1) - \boldsymbol{f}_{k2}^T(\hat{\boldsymbol{\theta}}_2),$$

where $\hat{\boldsymbol{\theta}}_1$, $\hat{\boldsymbol{\theta}}_2$ are the estimates obtained after refinement. Then we find again of $\hat{\boldsymbol{\theta}}_1$, with fixing of $\hat{\boldsymbol{\theta}}_2$ and $\hat{\boldsymbol{\theta}}_3$ by obtained values. After that we find $\boldsymbol{\theta}_2^{(1)}$ under fixed values of $\hat{\boldsymbol{\theta}}_1^{(1)}$ and $\hat{\boldsymbol{\theta}}_3^{(1)}$, and etc. During the implementation of such calculations we suppose that we have three signals and we make the cyclic rearrangement of the indices μ. We stop the refinement if an increment of the value of the maximum becomes small enough:

$$\frac{g_3(\hat{\boldsymbol{\theta}}_1^{(n)}, \hat{\boldsymbol{\theta}}_2^{(n)}, \hat{\boldsymbol{\theta}}_3^{(n)}) - g_3(\hat{\boldsymbol{\theta}}_1^{(n-1)}, \hat{\boldsymbol{\theta}}_2^{(n-1)}, \hat{\boldsymbol{\theta}}_3^{(n-1)})}{g_3(\hat{\boldsymbol{\theta}}_1^{(n-1)}, \hat{\boldsymbol{\theta}}_2^{(n-1)}, \hat{\boldsymbol{\theta}}_3^{(n-1)})} < \delta.$$

Further we suppose that there are four signals and again we refine the parameters of the signals $\boldsymbol{\theta}_m$ by the cyclic rearrangement of the indices $\mu = 1, 2, 3, 4$. In each step we maximize the sufficient reception function (7.22) with the use of the appropriate modified data (7.23). Each time at the determination of the signal parameters for new value of $M = m$, the threshold condition is tested

$$\alpha_m \geq \alpha_0,$$

where α_m corresponds to the extraction of the single regular wave with the given probability of the error (7.7):

$$P_e = \Phi(-\alpha_m/2), \quad \alpha_m = \left[\sum_k \boldsymbol{f}_{mk}^T(\hat{\boldsymbol{\theta}}_m) R^{-1} \boldsymbol{f}_{mk}(\hat{\boldsymbol{\theta}}_m) \right]^{1/2}.$$

In the special case of the uncorrelated random component $R = \sigma^2 I$ with the probability of the error of 0.05 the threshold condition for the seismic signal

$$\boldsymbol{f}_k(\boldsymbol{\theta}_m) = \|A_m \varphi(t_i - \tau_m - k\Delta x \gamma_m)\|_{i=1}^n$$

has the form

$$\frac{A_m}{\sigma}\left[\sum_k \varphi^2(t_i - \tau_m - k\Delta x\gamma_m)\right]^{1/2} \geq \alpha_m = 3.3,$$

at that, the value of α_m is the signal-to-noise ratio. The sufficient reception function (7.22) for a seismic signal at given assumptions reads as

$$g(A_m, \tau_m, \gamma_m) = \frac{A_m}{\sigma^2}\sum_i \varphi(t_i)\sum_k y(t_i + \tau_{mk}) - \frac{A_m^2}{\sigma^2}K\sum_i \varphi^2(t_i), \tag{7.25}$$

$$y(t_i + \tau_{mk}) = u(t_i + \tau_{mk}) - \sum_{\mu' \neq m} A_{\mu'}\varphi(t_i - \tau_{\mu' k} - \tau_{\mu k}). \tag{7.26}$$

Maximizing the function (7.25) over the amplitude A_m, we obtain the estimate in the explicit form

$$\hat{A}_m = \frac{\sum\limits_i \varphi(t_i)\sum\limits_k y(t_i + \tau_{mk})}{\sum\limits_i \varphi^2(t_i)}. \tag{7.27}$$

Substituting the formula (7.27) to the expression (7.25) we obtain the function which depends on kinematic parameters only

$$g_m(\tau_m, \gamma_m) = \frac{1}{\sigma^2}\frac{\left[\sum\limits_i \varphi(t_i)\sum\limits_k y(t_i + \tau_{mk})\right]^2}{K\sum\limits_i \varphi^2(t_i)}. \tag{7.28}$$

After analysis of the formula (7.28) we can conclude that the sufficient reception function in the parameter space (the arrival time and the apparent slowness) corresponds to a summogram of the controlled directional reception (CDR) $\sum_k Y(t_i + \tau_{mk})$ with the subsequent optimum filter. Let us note, that CDR summogram is the discrete analog of the Radon transform in a seismogram plane.

Chapter 8

Algorithms of approximation of geophysical data

In areas with a compound subsurface geology at the constructing of the model of a geophysical field and at the solution of the inverse geophysical problems the approximation and interpolation of geophysical data by *the spline functions* and a piecewise-polynomial approximation are widely used (Ahlberg *et al.*, 1967; Schumaker, 2007; Aronov, 1990; Sudvarg, 1974). Such approach enables to describe the geophysical fields of the various smoothness, that is typical for geophysical data (Volkov, 1981; Troyan, 1981b). We shall consider univariate cubic splines, parametric splines, splines of the fourth order together with the examples of their application for the processing of geophysical data.

For the case of an areal spread, the problem of imaging of a spatial seismic section subject to borehole data is actual. We shall consider an example of the seismic horizon imaging with the use of an areal spread and borehole data.

8.1 The Algorithm of Univariate Approximation by Cubic Splines

Let's on the abscissa axis in the points

$$x_1 < x_2 < \cdots < x_K \tag{8.1}$$

the values of the observed function are given

$$u_1, u_2, \ldots, u_K. \tag{8.2}$$

Here we suppose a monotone increasing of the argument x that is easily satisfied in the practice.

From the beginning we consider an application of the spline functions for the interpolation problem (Schumaker, 2007). The interpolating cubic spline is a function $\varphi(x)$, which in the nodes x_k is equal to

$$\varphi(x_k) = u_k, \tag{8.3}$$

and inside the intervals (x_k, x_{k+1}) it is determined by the cubic polynomials $f_k(x)$, which in the points x_k $(k = 2, 3, \ldots, K-1)$ are supposed doubly differentiable:

$$f_k(x) = a_k(x - x_k)^3 + b_k(x - x_k)^2 + c_k(x - x_k) + d_k. \tag{8.4}$$

For the determination of such interpolation function we should find $4(k-1)$ of the coefficients a_k, b_k, c_k, d_k.

Let us consider an arbitrary interval x_k, x_{k+1}. The spline function in the boundary points of the interval x_k and x_{k+1} must satisfy to the next conditions

$$\begin{aligned}\varphi(x_k) &= f_k(x_k) = d_k,\\ \varphi(x_{k+1}) &= f_k(x_{k+1}) = a_k\Delta x_k^3 + b_k\Delta x_k^2 + c_k\Delta x_k + d_k,\end{aligned} \tag{8.5}$$

$$\begin{aligned}\varphi'(x_k) &= f_k'(x_k) = c_k,\\ \varphi'(x_{k+1}) &= f_k'(x_{k+1}) = 3a_k\Delta x_k^2 + 2b_k\Delta x_k + c_k,\end{aligned} \tag{8.6}$$

$$\begin{aligned}\varphi''(x_k) &= f_k''(x_k) = 2b_k,\\ \varphi''(x_{k+1}) &= f_k''(x_{k+1}) = 6a_k\Delta x_k + 2b_k.\end{aligned} \tag{8.7}$$

The values of desired coefficients can be found using the conditions (8.5), (8.7) for the first and the second derivatives, and relation (8.3). After the analytic transformation we obtain

$$\begin{aligned}a_k &= \frac{1}{6\Delta x_k}(\varphi''(x_{k+1}) - \varphi''(x_k)),\\ b_k &= \frac{1}{2}\varphi''(x_k),\\ c_k &= \frac{\Delta u_k}{\Delta x_k} - \frac{1}{6}\Delta x_k(\varphi''(x_{k+1}) + \varphi''(x_k)),\\ d_k &= u_k,\end{aligned} \tag{8.8}$$

where $\Delta u_k = u_{k+1} - u_k$. Using the condition (8.6) and formulas for the coefficients (8.8), we obtain the formulas for the first derivatives of the spline functions

$$\begin{aligned}\varphi'(x_k) &= \frac{\Delta u_k}{\Delta x_k} - \frac{1}{6}\Delta x_k(\varphi''(x_{k+1}) + 2\varphi''(x_k)),\\ \varphi'(x_k) &= \frac{\Delta u_{k-1}}{\Delta x_{k-1}} - \frac{1}{6}\Delta x_{k-1}(\varphi''(x_k) + 2\varphi''(x_{k-1}),\end{aligned}$$

$k = 1, 2, \ldots, K-1$. For the determination of the second derivatives $\varphi''(x_k)$ of the spline functions we require a continuity of the first derivative in the boundary of two intervals

$$f'(x_k) = f_k'(x_k), \quad k = 2, 3, \ldots, K-1. \tag{8.9}$$

Substituting the condition (8.6) into the expression (8.9), we obtain

$$3a_{k-1}\Delta x_{k-1}^2 + 2b_{k-1}\Delta_x k - 1c_{k-1} = c_k.$$

Using the representation (8.8) for the coefficients, we obtain the system of equations

$$\begin{aligned}\Delta_{k-1}\varphi''(x_{k-1}) &+ 2(\Delta x_{k-1} + \Delta x_k)\varphi''(x_k) + \Delta x_k\varphi''(x_{k-1})\\ &= 6\left(\frac{\Delta u_k}{\Delta x_k} - \frac{\Delta u_{k-1}}{\Delta x_{k-1}}\right).\end{aligned} \tag{8.10}$$

Using the system of $K-2$ equations (8.10) we can determine the unknown $\varphi''(x_k)$, then, taking into account the values of u_k, we find the coefficients a_k, b_k, c_k, d_k with the help of (8.8). Before calculation we must designate the values of $\varphi''(x_1)$ and $\varphi''(x_k)$, which in the case of cubic spline usually are equal to zero:

$$\varphi''(x_1) = \varphi''(x_k) = 0. \tag{8.11}$$

The formula (8.10) can be written down in a matrix form

$$A\boldsymbol{\varphi}'' = \boldsymbol{y}, \tag{8.12}$$

where

$$A = \begin{Vmatrix} 2(\Delta x_1 + \Delta x_2) \dots & \dots & 0 \\ \Delta x_2 & \dots & \dots & 0 \\ \dots & \dots & \dots & \dots \\ 0 & \dots & \Delta x_{K-2} & 2(\Delta x_{K-2} + \Delta x_{K-1}) \end{Vmatrix}.$$

The matrix A of the system (8.12) is a three-diagonal symmetric matrix with the positive diagonal elements, consequently, it is a positive-defined matrix. Thus, an unique existence of the solution of the system (8.12) is ensured. The solution can be found, for example, by the Gauss elimination method (Voevodin and Kuznetsov, 1984).

Once it is more convenient to represent the conditions for the second derivative in the form

$$\varphi''(x_1) = \alpha\varphi''(x_2), \quad \varphi''(x_{k-1}) = \alpha\varphi''(x_k), \tag{8.13}$$

where α, β are the arbitrary numbers. In the practice enough often to choose $\alpha = \beta = 1$, so far the second derivatives in the points x_1 and x_k are equal to the second derivatives in the points x_2 and x_{k-1} respectively. Not infrequently the equalities $\alpha = \beta = 1/2$ are used.

Substituting the conditions (8.13) in the system of equations (8.10), and, for $k = 2$ and $k = K-1$, we obtain

$$\begin{gathered} ((2+\alpha)\Delta x_1 + 2\Delta x_2)\varphi''(x_2) + \Delta x_2\varphi''(x_3) = 6\left(\frac{\Delta u_2}{\Delta x_2} - \frac{\Delta u_1}{\Delta x_1}\right), \\ \Delta x_{K-2}\varphi''(x_{K-2}) + (2\Delta x_{K-2} + (2+\beta)\Delta x_{K-1})\varphi''(x_{K-1}) \\ = 6\left(\frac{\Delta u_{K-1}}{\Delta x_{K-1}} - \frac{\Delta u_{K-2}}{\Delta x_{K-2}}\right). \end{gathered}$$

Other equations remain without changing. This implies that in the matrix of the system (8.12) only two elements in the first and last rows are varied. The first and last components of the vector $\boldsymbol{y}$ of absolute terms are varied also. In this case the unique existence of the cubic splines is supplied by the conditions

$$-\left(\frac{\Delta x_2}{\Delta x_1} + 2\right) \le \alpha < \infty, \quad -\left(\frac{\Delta x_{K-2}}{\Delta x_{K-1}} + 2\right) \le \beta < \infty.$$

Thus far we was considered the next scheme of the determination of the cubic splines: for given x_k, u_k and for the boundary conditions (8.11) or (8.13) we determine the vector of the second derivatives φ'' by the solution of the system of linear equations (8.12), then, using x_k, y_k, $\varphi''(x_k)$ and formula (8.8) we calculate the coefficients a_k, b_k, c_k, d_k, and using the expression (8.4), the cubic splines are calculated. By analogous way the cubic splines can be determined using the next three consequences $\{x_k\}$, $\{y_k\}$, $\{\varphi'(x_k)\}$. The conditions (8.5) and (8.6) allow the expressions for the coefficients

$$\begin{aligned} a_k &= \frac{1}{\Delta x_k^2}\left(-2\frac{\Delta u_k}{\Delta x_k} + \varphi'(x_k) + \varphi'(x_{k+1})\right), \\ b_k &= \frac{1}{\Delta x_k}\left(3\frac{\Delta u_k}{\Delta x_k}\varphi'(x_k) - \varphi'(x_{k+1})\right), \\ c_k &= \varphi'(x_k), \quad d_k = u_k, \quad k = 1, 2, \dots, K-1. \end{aligned} \tag{8.14}$$

The second derivative, represented through the first derivative, are determined by the formulas

$$\begin{aligned} \varphi''(x_k) &= \frac{2}{\Delta x_k}\left(\frac{\Delta u_k}{\Delta x_k}2\varphi'(x_k) - \varphi'(x_{k+1})\right), \\ \varphi''(x_K) &= \frac{2}{\Delta x_{K-1}}\left(-3\frac{\Delta u_{K-1}}{\Delta x_{K-1}} + \varphi'(x_{K-1}) + 2\varphi'(x_K)\right). \end{aligned}$$

The continuity of the spline functions in the nodes for $k = 2, 3, \dots, K-1$ is formalized as

$$f''_{k-1}(x_k) = f''_k(x_k). \tag{8.15}$$

Substituting the condition (8.7) to the formula (8.15), we obtain

$$6a_{k-1}\Delta x_{k-1} + 2b_{k-1} = 2b_k. \tag{8.16}$$

Using the expression (8.14) and the formula (8.16), we obtain the system of $K-2$ linear equations with unknown $\varphi'(x_2)$, $\dots$, $\varphi'(x_{k-1})$,

$$\begin{aligned} \frac{1}{\Delta x_k}\varphi'(x_{k-1}) + 2\left(\frac{1}{\Delta x_{k-1}} + \frac{1}{\Delta x_k}\right)\varphi'(x_k) + \frac{1}{\Delta x_k}\varphi'(x_{k+1}) \\ = \frac{3}{\Delta x_k}\frac{\Delta u_{k-1}}{\Delta x_{k-1}} + \frac{3}{\Delta x_k}\frac{\Delta u_k}{\Delta x_k}, \end{aligned} \tag{8.17}$$

where $\varphi'(x_1)$ and $\varphi'(x_K)$ are given beforehand. We should note, that the matrix of coefficients of the system (8.17) has the properties, which are analogous to the properties of the matrix connected with the system (8.12). Thus the unique existence of the solution of the system (8.17) is satisfied. As before, for the solution of the system, for example, the Gauss method can be used (Voevodin and Kuznetsov, 1984).

Let us consider the smoothing of experimental data u_k with the use of splines.

Let's the observations and the observation points are described by the relations (8.1) and (8.2). The edge conditions we set in the points x_1 and x_K by the values of the first or the second derivatives of the desired function $\varphi(x)$, i.e. $\varphi'(x_1)$ and $\varphi'(x_K)$, or $\varphi''(x_1)$ and $\varphi''(x_K)$. For the calculation of the approximating function $\varphi(x)$, let require a proportionality of the difference $(u_k - \varphi(x_k))$ with a jump of the third derivative in the points $(k = 2, 3, \ldots, K-1)$ (Ahlberg *et al.*, 1967):

$$g_k(u_k - \varphi(x_k)) = \alpha_k, \quad k = 1, 2, \ldots, K, \tag{8.18}$$

where

$$g_k > 0, \quad \alpha_1 = f_1'''(x_1),$$
$$\alpha_k = f_k'''(x_k) - f_{k-1}'''(x_k), \quad \alpha_n = f_1'''(x_n),$$

f_k are the polynomials, given by the formula (8.4), g_k are the coefficients of the proportionality (must be positive). Using the conditions (8.18) and equations (8.8) for the coefficients, we obtain the system of K equations with $K-2$ unknown $\varphi''(x_2)$ (for $k = 2, \ldots, K-1$) and K unknown $\varphi(x_k)$ $(k = 1, 2, \ldots, K)$ in the form

$$\frac{1}{\Delta x_1}(\varphi''(x_1) - \varphi''(x_2)) = g_1(u_1 - \varphi(x_1)),$$
$$\frac{1}{\Delta x_{k-1}}(\varphi''(x_{k-1}) - \varphi''(x_k)) - \frac{1}{\Delta x_k}(\varphi''(x_k) - \varphi''(x_{k+1})) = g_k(u_k - \varphi(x_k)),$$
$$\frac{1}{\Delta x_{K-1}}(\varphi''(x_{K-1}) - \varphi''(x_K)) = g_K(u_K - \varphi(x_K)).$$

Moreover, we have $K-2$ equations else relative to the same unknowns, which can be obtained by substituting u_k for $\varphi(x_k)$. In the case of the edge conditions for the second derivative in the form of (8.11), we can write the mentioned systems in the matrix form

$$-B^T\boldsymbol{\varphi}'' + G\boldsymbol{\varphi} = G\boldsymbol{u}, \quad A_1\boldsymbol{\varphi}'' + B\boldsymbol{\varphi} = 0, \tag{8.19}$$

where

$$G = \begin{Vmatrix} g_1 & & & \\ & g_2 & & \\ & & \ddots & \\ & & & g_K \end{Vmatrix},$$

$$B = \begin{Vmatrix} -\frac{1}{\Delta x_1} & \left(\frac{1}{\Delta x_1} + \frac{1}{\Delta x_2}\right) & \cdots & 0 & 0 \\ 0 & -\frac{1}{\Delta x_1} & \cdots & \cdots & 0 \\ \cdots & \cdots & \cdots & \cdots & \cdots \\ 0 & 0 & \cdots & -\frac{1}{\Delta x_2} & -\frac{1}{\Delta x_{K-1}} \end{Vmatrix},$$

$$\boldsymbol{\varphi}'' = [\varphi''(x_2), \ldots, \varphi''(x_2)], \quad \boldsymbol{\varphi} = [\varphi(x_1), \ldots, \varphi(x_k)], \quad \boldsymbol{u} = [u-1, \ldots, u_k].$$

The matrix A_1 can be obtained from the matrix A of the system (8.12) by multiplying all its elements by 1/6. Because the matrix A is positive defined, and for the positive values of g_k the diagonal matrix G is positive defined, then the system (8.12) has the solution for all g_k, which satisfies the inequality $0 < g_k < \infty$.

Let's consider an influence of the values of g_k on the smoothing. We multiply the first system of equation (8.19) by matrix G^{-1}:

$$-G^{-1}B^T\boldsymbol{\varphi}'' + I\boldsymbol{\varphi} = I\boldsymbol{u}_b,$$

where I is an identity matrix. At tending g_k to the infinity, the first term of the matrix equation tends to zero, and $\varphi(x_k)$ tends to u_k. At that, the second system of (8.19) transforms to the system of (8.12), and we come to the problem of the interpolation by cubic splines. At $g_k \to 0$ the values $\varphi(x_k)$ move away from the observed data u_k. In this case the approximation by the spline functions corresponds to a stright line approximation with the use of the method of least squares. So far, the values of g_k connected with a degree of smoothing of the experimental data.

We consider the solution of the system of equations (8.19). Using the second system we determine the vector $\boldsymbol{\varphi}$, because the matrix G is diagonal one and it is easy to invert:

$$\boldsymbol{\varphi} = G^{-1}(B^T\boldsymbol{\varphi}'' + G\boldsymbol{u}).$$

Substituting this equation to the first system, we obtain the estimate for the vector of the second derivative

$$\boldsymbol{\varphi}'' = -(A_1 + BG^{-1}B^T)^{-1}B\boldsymbol{u}.$$

The matrix $A_1 + BG^{-1}B^T$ is a positive defined and fifth diagonal matrix, therefore the solution can be found by the Gauss method.

Let us consider the results of smoothing of stacking velocity along a profile. At Fig. 8.1 the initial data and smoothing data at various values of the weighted multipliers g_k (the same for all points $g_k = DP$) are represented. From the analysis of the curves it follows that at decreasing of the weighted multipliers from 1 to 0,0001 a transition from smoothing of high frequencies to low frequencies occurs.

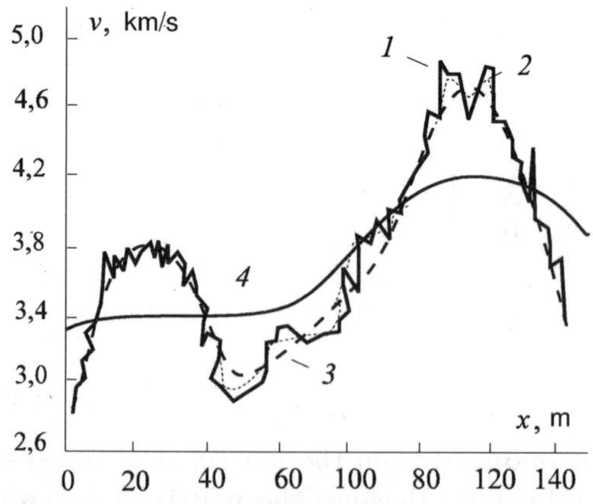

Fig. 8.1 Smoothing of the effective velocity along profile by the cubic spline functions: *1* is initial data, *2* — $DP = 10^{-1}$, *3* — $DP = 10^{-2}$, *4* — $DP = 10^{-5}$.

At Fig. 8.2 an example of smoothing of gravity $\Delta g(x)$ along a profile by the cubic splines is represented. Obtained curves can be used in subsequent interpretation steps.

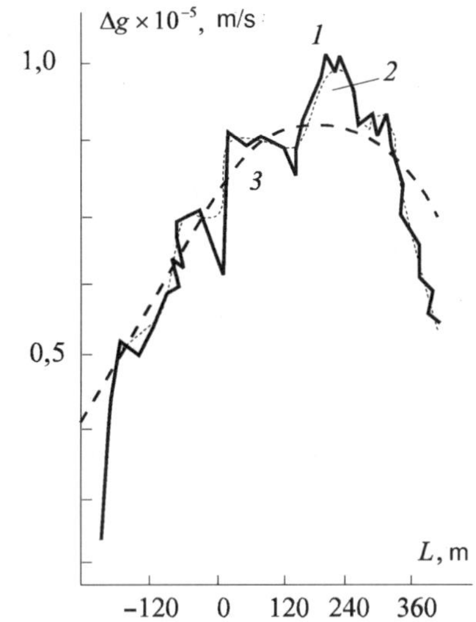

Fig. 8.2 Smoothing of gravity along profile: *1* is initial data, *2* — $DP = 1$, *3* — $DP = 10^{-4}$.

The spline approximation is widely used for the approximation of the seismic interfaces. In the special case of the solution of a direct problem by the ray method, the stability of the iterative processes depends on the smoothness of the interfaces and their first derivative. The cubic splines can provide the high smoothness and the continuity of the first derivative.

8.2 Periodic and Parametric Spline Functions

The interpolation and approximation of the closed and periodic curves can be implemented using cubic splines with the periodic edge condition (Späth, 1973):

$$f_1(x_1) = f_{K-1}(x_K), \quad f_1'(x_1) = f_{K-1}'(x_K), \quad f_1''(x_1) = f_{K-1}''(x_K). \tag{8.20}$$

Relations (8.20) are the conditions for the function $\varphi(x_k)$:

$$\varphi(x_1) = \varphi(x_K), \quad \varphi'(x_1) = \varphi'(x_K), \quad \varphi''(x_1) = \varphi''(x_K). \tag{8.21}$$

Let us consider the interpolation problem. Using the second (join of the first derivatives) condition from (8.20), to write down the equation for the system (8.10) ($k = 2$) in the form

$$\begin{aligned} 2(\Delta x_{K-1} + \Delta x_1)\varphi''(x_1) + \Delta x_1 \varphi''(x_2) + \Delta x_{K-1}\varphi''(x_{K-1}) \\ = 6\left(\frac{\Delta u_1}{\Delta x_1} - \frac{\Delta u_{K-1}}{\Delta x_{K-1}}\right). \end{aligned} \tag{8.22}$$

If $k = K - 1$ we obtain from the system of equations (8.10) with the use of the condition (8.21) the following equation

$$\Delta x_{K-1}\varphi''(x_1) + \Delta x_{K-2}\varphi''(x_{K-2}) + 2(\Delta x_{K-2} + \Delta x_{K-1}) \times\varphi''(x_{K-1}) = 6\left(\frac{\Delta u_{K-1}}{\Delta x_{K-1}} - \frac{\Delta u_{K-2}}{\Delta x_{K-2}}\right). \tag{8.23}$$

Taking into account the equations (8.22) and (8.23), we rewrite the system of equation (8.12) in the form

$$A_2\boldsymbol{\varphi}'' = \boldsymbol{y_2},$$

where $\boldsymbol{\varphi}'' = [\varphi''(x_1), \varphi''(x_2), \ldots, \varphi''(x_{K-1})]^T$ is a column vector;

$$\boldsymbol{y}_2 = \left[6\left(\frac{\Delta u_1}{\Delta x_1} - \frac{\Delta u_{K-1}}{\Delta x_{K-1}}\right), 6\left(\frac{\Delta u_2}{\Delta x_2} - \frac{\Delta u_1}{\Delta x_1}\right), \ldots \right.$$
$$\left. \ldots, 6\left(\frac{\Delta u_{K-1}}{\Delta x_{K-1}} - \frac{\Delta u_{K-2}}{\Delta x_{K-2}}\right)\right]$$

is a column vector;

$$A_2 = \left\| \begin{matrix} 2(\Delta x_{K-1} + \Delta x_1) & \ldots & \Delta x_{K-1} \\ \ldots & \ldots & \ldots \\ \Delta x_{K-1} & \ldots & 2(\Delta x_{K-2} + \Delta x_{K-1}) \end{matrix} \right\|.$$

The matrix A_2 is a cyclic three-diagonal matrix, and it is possible to prove its positive definiteness. The system of equations can be solved by the Gauss method.

At a solution of the interpolation problem of the closed curves we must depart from the strict monotonicity of the data on the argument x_k (see condition (8.1)). We represents the curve by a set of arbitrary points (x_k, u_k), $k = 1, 2, \ldots, K$, using a parametric representation by two functions $x = x(v)$ and $u = u(v)$. At that, we assume a strict monotonicity for the parameter v ($v_1 < v_2 < \cdots < \cdots < v_n$).

We shall describe the curve using the cubic splines of two modes: (v_k, x_k) and (v_k, u_k), $k = 1, 2, \ldots, K$. We suppose that the values of first (second) derivatives x'_1, φ'_1 (x''_1, φ''_1) by τ in the points v_1 and v_K are given. In the practice, the parameter v usually is determined in the following way:

$$v_1 = 0, \quad v_k = v_{k-1} + \Gamma_{k-1}, \quad k = 1, 2, \ldots, K, \tag{8.24}$$

where

$$\Gamma_k = \sqrt{\Delta x_k^2 + \Delta u_k^2}.$$

This method can be generalized to a three-dimensional case (x_k, u_k, v_k). The parameter v is represented as

$$v_1 = 0, \quad v_k = v_{k-1} + \sqrt{\Delta x_k^2 + \Delta u_k^2 + \Delta v_k^2}, \quad k = 2, \ldots, K,$$

and for an each component the spline function is calculated. At the interpolation of the closed curve in a plane or in a space it is necessary to specify the boundary conditions: $x_1 = x_K$, $\varphi_1 = \varphi_K$ and to use the parametric splines.

Let us consider the approximation problem of the non-monotone and closed curves. We will approximate the corves given by the coordinates u_k, v_k, containing, in the general case, the random component, with the help of the cubic splines v_k, ψ_k, v_k, φ_k. The values of v_k are the arguments and they calculated by the formula (8.24). The coefficient of proportionality between a jump of the third derivative and a deviation of the observed and approximated data is chosen, in general case for each coordinates separately:

$$\psi'''(t_k) = g_k(v_k - \psi(t_k)), \quad \psi'''(t_k) = h_k(u_k - \varphi(t_k)).$$

For the approximation of the closed curves it is possible to use the periodic spline functions for the variables u and v respectively. As the parameter we will consider t_k, which is determined by the formula (8.24) (instead of Δx_k we use Δv_k). Let's the next boundary condition is specified:

$$u_1 = u_k, \quad p_1 = p_k, \quad v_1 = v_k, \quad g_1 = g_k, \tag{8.25}$$

as well as

$$\varphi_1 = \varphi_K, \quad \varphi_1'' = \varphi_K'', \quad \psi_1 = \psi_K, \quad \psi_1'' = \psi_K. \tag{8.26}$$

Using the conditions (8.25) and (8.26), we obtain the system the same type as (8.19) for the variables $\varphi_1, \ldots, \varphi_{k-1}$, $\varphi_1'', \ldots, \varphi_{k-1}''$, $\psi_1, \ldots, \psi_{k-1}$, $\psi_1'', \ldots, \psi_{k-1}''$. We find the solution of the system by the method described above.

Let us consider the results of the approximation of isolines of the complete magnetic-field vector (Fig. 8.3). From the analysis of Fig. 8.3 we can conclude that

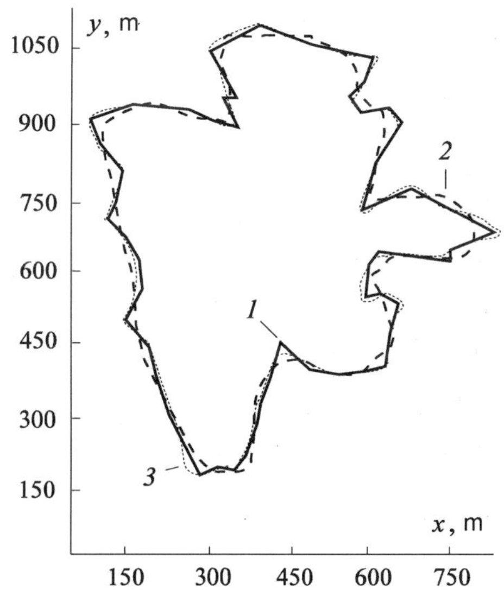

Fig. 8.3 Smoothing of isolines of the complete magnetic-field vector by the parametric splines at the various weight coefficients: *1* is initial data; *2* — $DP = 10^{-3}$; *3* — $DP = 10^{-5}$.

the smoothness degree is increased together with decreasing of the values of the weight coefficients.

8.3 Application of the Spline Functions for Histogram Smoothing

One of the tasks of mass processing of the results of geophysical observation is the task of a histogram smoothing with the purpose to obtain a differentiable density function.

Let us consider that histogram is described by the abscissas $x_1 < x_2 < \dots x_K$ and the corresponding staircase function, having in the interval $[x_k, x_{k+1}]$ the ordinate H_k, $k = 1, 2, \dots, K-1$. The problem consists in the construction of the smooth (differentiable) curve, if the integral of this curve at the interval $[x_k, x_{k+1}]$ is equal to a square of the appropriate rectangle of the histogram (Späth, 1973). Let us suppose also, that for the each abscissa x_k we specify the ordinate $\varphi(x_k)$, which is crossed by the curve. To determine the coefficients of the fourth order polynomial in the interval $[x_k, x_{k+1}]$:

$$f_k(x) = a_k(x - x_k)^4 + b_k(x - x_k)^3 + c_k(x - x_k)^2 + d_k(x - x_k) + l_k \tag{8.27}$$

under the following conditions

$$f_k(x_k) = \varphi(x_k), \quad f_k(x_{k+1}) = \varphi(x_{k+1}), \quad f_k'(x_k) = \varphi'(x_k), \quad f_k'(x_{k+1}) = \varphi'(x_{k+1}),$$

$$\int\limits_{x_k}^{x_{k+1}} f_k(x)dx = H_k \Delta x_k.$$

Using these expressions to find the polynomial coefficients

$$\begin{aligned} a_k = {} & \frac{5}{\Delta x_k^4}\left[\frac{1}{2}\Delta x_k(\varphi'(x_{k+1}) - \varphi'(x_k)) \right. \\ & \left. - 3(\varphi(x_{k+1}) + \varphi'(x_k)) - 2H_k\right], \\ b_k = {} & -\frac{4}{\Delta x_k^3}\left[\frac{1}{2}\Delta x_k(\varphi'(x_{k+1}) - 3\varphi'(x_k)) \right. \\ & \left. - 7\varphi(x_{k+1}) - 8\varphi(x_k) + 15H_k\right], \\ c_k = {} & \frac{3}{2\Delta x_k^2}[\Delta x_k(\varphi'(x_{k+1}) - 3\varphi'(x_k) - 8\varphi(x_{k+1}) \\ & - 12\varphi(x_k)) + 20H_k], \quad d_k = \varphi'(x_k), \quad e_k = \varphi(x_k). \end{aligned}$$

The continuity condition for the second derivative is written down as $f_{k-1}''(x_k) = f_k''(x_k)$. Involving the representation (8.27) as well, we obtain the system of equations for determination of the first derivatives $\varphi'(x_k)$:

$$\begin{aligned} & \frac{1}{\Delta x_{k-1}}\varphi'(x_{k-1}) + 3\left(\frac{1}{\Delta x_{k-1}} + \frac{1}{\Delta x_k}\right)\varphi'(x_k) \\ & -\frac{1}{\Delta x_{k+1}} = 20\left[\frac{H_k}{\Delta x_k^2} - \frac{H_{k-1}}{\Delta x_{k+1}^2}\right] + 8\frac{\varphi(x_{k-1})}{\Delta x_{k-1}^2} \\ & +12\left[\frac{1}{\Delta x_{k-1}^2} + \frac{1}{\Delta x_k^2}\right] - 8\frac{\varphi(x_{k+1})}{\Delta x_k^2}. \end{aligned} \tag{8.28}$$

Now it is necessary to determine the edge conditions for the first derivatives in the points x_1, x_K: $\varphi'(x_1)$, $\varphi'(x_K)$. The system of equations (8.28) has a three-diagonal and positive defined matrix of the coefficients. So, the solution of the system possesses the unique existence properties.

For the calculation of $\varphi(x_k)$ can propose the next expedient:

$$\varphi(x_1) = g_1 H_1, \quad \varphi(x_k) = g_k H_(k-1) + (1-g_k)H_k, \quad \varphi(x_K) = g_K H_{K-1},$$

where g_k $(k = 1, 2, \ldots, K)$ are the arbitrary coefficients from the interval $[0, 1]$. Sometimes the values g_k are chosen equal to $1/2$ for all intervals.

Let us consider an example of smoothing of the velocity histograms, obtained by the seismic data for two beds (Fig. 8.4). The obtained curves can be used for

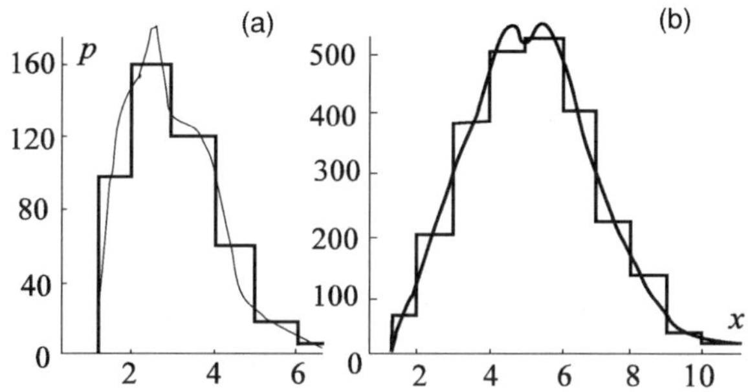

Fig. 8.4 Histogram smoothing using the spline functions. (a) corresponds to the bed number one 1, $g_k = 0,3,\ 0,3,\ 0,5,\ 0,6,\ 0,3,\ 0,3,\ 0,2$; (b) corresponds to the bed number two, $g_k = 0,\ 0,6,\ 0,5,\ 0,4,\ 0,1,\ 0,7,\ 0,4,\ 0,6,\ 0,3,\ 0,2,\ 0,5,\ 0,3,\ 0,3$.

the calculation of moments of the distribution. The histogram smoothing can be applied to other processing stages as well.

8.4 Algorithms for Approximation of Seismic Horizon Subject to Borehole Observations

Let the area under the study is covered by an arbitrary set of profiles registered by the multifold coverage scheme. (Troyan and Sokolov, 1982). In addition, there are a few deep boreholes with boring data. As the result of the processing of the profiles we obtain the depth profiles. Let us consider that we have the interlocked profiles, i.e. in the cross points of the profiles the depth deviations do not exceed the threshold values. As the initial observed data $\boldsymbol{U} = [\boldsymbol{u}_m(x_k, y_k)]$ we consider the bed depths corresponding to the depth profile coordinates: x_k, y_k; k are the points of the area under investigation, $k = 1, 2, \ldots, K$; m is a number of a bed, $m = 1, 2, \ldots, M$. In this case the model of observations can be written down as

$$\boldsymbol{U} = \psi \boldsymbol{\rho} + \boldsymbol{n}, \tag{8.29}$$

where

$$\boldsymbol{U} = \begin{Vmatrix} \boldsymbol{u}_1 \\ \boldsymbol{u}_2 \\ \dots \\ \boldsymbol{u}_M \end{Vmatrix}, \quad \boldsymbol{\rho} = \begin{Vmatrix} \boldsymbol{\rho}_1 \\ \boldsymbol{\rho}_2 \\ \dots \\ \boldsymbol{\rho}_M \end{Vmatrix}, \quad \boldsymbol{n} = \begin{Vmatrix} \boldsymbol{n}_1 \\ \boldsymbol{n}_2 \\ \dots \\ \boldsymbol{n}_M \end{Vmatrix},$$

$$\psi = [\psi_1, \psi_2, \dots, \psi_M],$$

$\boldsymbol{\rho}_m$ is the parameter vector, ψ_m is the structural matrix of m-th bed. For example, in the case of the polynomial of power two, we have

$$\psi_m = \begin{Vmatrix} 1 & x_1 & y_1 & x_1^2 & x_1 y_1 & y_1^2 \\ 1 & x_2 & y_1 & x_2^2 & x_2 y_1 & y_2^2 \\ \dots & \dots & \dots & \dots & \dots & \dots \\ 1 & x_K & y_1 & x_K^2 & x_K y_K & y_K^2 \end{Vmatrix}.$$

The random component $\boldsymbol{n}_m = [n_m(x_k, y_k)]$ in the points x_k, y_k is centralized, and the correlation between the beds is determined by the covariance matrix

$$r_n = \begin{Vmatrix} R_{11} & R_{12} & \dots & R_{1M} \\ R_{21} & R_{22} & \dots & R_{2M} \\ \dots & \dots & \dots & \dots \\ R_{M1} & R_{M2} & \dots & R_{MM} \end{Vmatrix}. \tag{8.30}$$

The values of the bed depth $\boldsymbol{V}$ in the boreholes are assumed to be under linear constraints, which determine the parameters of approximating planes:

$$a\boldsymbol{\rho} = \boldsymbol{V},$$

where

$$A = [A_1, A_2, \dots, A_M], \quad \boldsymbol{V} = \begin{Vmatrix} \boldsymbol{V_1} \\ \boldsymbol{V_2} \\ \dots \\ \boldsymbol{V_M} \end{Vmatrix},$$

$\boldsymbol{V}_m$ is a set of the depth for m-th bed obtained using borehole data; A_m is the structural matrix for m-th bed, which one, for example, for the polynomial of power to has the following shape

$$A_m = \begin{Vmatrix} 1 & x_1 & y_1 & x_1^2 & x_1 y_1 & y_1^2 \\ 1 & x_2 & y_1 & x_2^2 & x_2 y_1 & y_2^2 \\ \dots & \dots & \dots & \dots & \dots & \dots \\ 1 & x_K & y_1 & x_K^2 & x_K y_K & y_K^2 \end{Vmatrix}.$$

The estimate $\tilde{\boldsymbol{\rho}}$ is determined by the formula (see expressions (6.23), (6.25))

$$\tilde{\boldsymbol{\rho}}^T = \hat{\boldsymbol{\rho}}^T + (\boldsymbol{V}^T - \tilde{\boldsymbol{\rho}} A^T)(AB^{-1}A^T)^{-1}AB^{-1}, \tag{8.31}$$

where

$$\hat{\boldsymbol{\rho}} = B^{-1}\psi^T R_n^{-1}\boldsymbol{U}, \quad B = \psi^T R_n^{-1}\psi, \tag{8.32}$$

and it is the estimate of the $\boldsymbol{\rho}$ by the least squares method without linear constraints.

The covariance matrix of the estimate of parameters $\tilde{\boldsymbol{\rho}}$ can be represented as (see formulas (6.26), (6.27))

$$R_{\tilde{\rho}} = R_{\hat{\rho}} - B^{-1}A^T(AB^{-1}A^T)^{-1}AB^{-1}, \quad R_{\hat{\rho}} = B^{-1},$$

where $R_{\hat{\rho}}$ is the covariance matrix of the estimate of parameters $\hat{\boldsymbol{\rho}}$, obtained by the least squares method without linear constraints.

Let us consider the variants of introducing of the correlation dependence between the random components of the model (8.29).

8.4.1 *The Markovian type of correlation along the beds and no correlation between beds*

The matrix R_n (8.30) in this case is a diagonal one and its inverse matrix R_n^{-1} reads as:

$$R_n^{-1} = \begin{Vmatrix} R_{11}^{-1} & 0 & \dots & 0 \\ 0 & R_{22}^{-1} & \dots & 0 \\ \dots & \dots & \dots & \dots \\ 0 & 0 & \dots & R_{nn}^{-1} \end{Vmatrix}.$$

The matrices R_{mm}^{-1} are the inverse matrices with respect to the matrix of a type of (3.22):

$$R_{mm} = \sigma^2 \begin{Vmatrix} 1 & \beta_m & \beta_m^2 & \dots & \beta_m^K \\ \beta_m & 1 & \beta_m & \dots & \beta_m^{K-1} \\ \dots & \dots & \dots & \dots & \dots \\ \beta_m^K & \beta_m^{K-1} & \beta_m^{K-2} & \dots & 1 \end{Vmatrix}.$$

Here, the value of β_m have a sense of correlation coefficients between the points with the distance of $\Delta r = \sqrt{(\Delta x)^2 + (\Delta y)^2}$ for the bed with the number m. The inverse matrix R_{mm}^{-1} reads as

$$R_{mm}^{-1} = \frac{1}{\sigma_m^2(1-\beta_m^2)} \begin{Vmatrix} 1 & -\beta_m & 0 & \dots & 0 & 0 \\ -\beta_m & 1+\beta_m^2 & -\beta_m^2 & \dots & 0 & 0 \\ 0 & -\beta_m & 1+\beta_m^2 & \dots & 0 & 0 \\ \dots & \dots & \dots & \dots & \dots & \dots \\ 0 & 0 & 0 & \dots & -\beta_m & 1 \end{Vmatrix}. \tag{8.33}$$

Using the representation (8.33), we find the matrix B:

$$\begin{aligned} B_{ss'} = \sum_m \frac{1}{\sigma_m^2(1-\beta_m^2)} \Big[& (1+\beta_m^2) \sum_k \psi_{mks}\psi_{mks'} \\ & - \beta_m \sum_k (\psi_{mks}\psi_{m(k+1)s'} + \psi_{m(k+1)s}\psi_{mks'}) \\ & - \beta_m^2 (\psi_{mKs}\psi_{mKs'} + \psi_{m1s}\psi_{m1s'}) \Big]. \end{aligned} \tag{8.34}$$

8.4.2 Markovian correlation between the beds and no correlation along bed

By repeating above mentioned arguments, we pass to the next representation of the matrix B, which is similar to (8.34):

$$B_{ss'} = \sum_k \Big[\sum_m \frac{1}{\sigma_m^2(1-\beta_m^2)}[(1+\beta_m^2)\psi_{mks}\psi_{mks'} - \beta_m(\psi_{mks}\psi_{(m+1)ks} + \psi_{(m+1)ks}\psi_{mks'}) - \beta_m^2(\psi_{1ks}\psi_{1ks'} + \psi_{Mks}\psi_{Mks'})]\Big].$$

8.4.3 Conformance inspection of seismic observation to borehole data concerning bed depth

The next problem represents great practical significance. Let using the maximum likelihood method, the estimation of the parameters $\tilde{\boldsymbol{\rho}}$ of approximated surface with taking into account and without taking into account the borehole observation are found. And it is necessary to solve the problem about the conformance of seismic observations to the borehole data concerning the bed depth. This problem can be solved by the test of the parametric hypothesis method (see Sec. 7.1). We consider the linear model (8.29) (we put $M = 1$ and $R_n = \sigma^2 I$ for the simplicity sake) and we test the hypothesis H_0: $A\boldsymbol{\rho} = \boldsymbol{V}$. For the hypothesis checking we shall use the method of the likelihood ratio

$$q = \frac{L(\tilde{\boldsymbol{\rho}}_{ML}, \tilde{\sigma}^2)}{L(\hat{\boldsymbol{\rho}}_{ML}, \hat{\sigma}^2)}. \tag{8.35}$$

As it was demonstrated in Sec. 6.2, the estimates $\hat{\boldsymbol{\rho}}$ obtained by the least squares method and the estimates $\hat{\boldsymbol{\rho}}_{\mathrm{ML}}$ obtained by the maximum likelihood method coincide in the case of the normal distribution of the random component $\boldsymbol{n}$.

To find the estimates $\boldsymbol{\rho}$ under the condition of the validity of the hypothesis H_0: $A\boldsymbol{\rho} = \boldsymbol{V}$. In this case the maximum likelihood method reduces to the conditional extremum problem, which can be solved by the Lagrange multipliers method. To construct the function

$$F(\boldsymbol{\rho}) = \ln L + 2\lambda^T(A\boldsymbol{\rho} - \boldsymbol{V}), \tag{8.36}$$

where $\boldsymbol{\lambda}$ is a column vector contained undetermined Lagrange multipliers. The estimate $\hat{\boldsymbol{\rho}}_{\mathrm{ML}}$ we find by the minimization of (8.36) under the condition, that the equality $A\boldsymbol{\rho} = \boldsymbol{V}$ is valid:

$$\partial F/\partial \boldsymbol{\rho} = 0. \tag{8.37}$$

After simple transforms we obtain

$$\tilde{\boldsymbol{\rho}}_{\mathrm{ML}} = \hat{\boldsymbol{\rho}}_{\mathrm{ML}} + 2\sigma^2\lambda^T A(\psi^T\psi)^{-1}. \tag{8.38}$$

To postmultiply the left hand side and right hand side of the equality (8.38) by A^T and to use the equality $\tilde{\boldsymbol{\rho}}^T A^T = \boldsymbol{V}^T$, we obtain

$$\boldsymbol{\lambda}^T = \frac{1}{2\sigma^2}(\boldsymbol{V}^T - \boldsymbol{\rho}_{\mathrm{ML}}^T A^T)(A(\psi^T\psi)^{-1}A^T)^{-1}. \tag{8.39}$$

Substituting $\boldsymbol{\lambda}^T$ from the formula (8.39) to the expression (8.38), we obtain

$$\tilde{\boldsymbol{\rho}}_{\mathrm{ML}}^T = \hat{\boldsymbol{\rho}}_{\mathrm{ML}}^T + (\boldsymbol{V}^T - \boldsymbol{\rho}_{\mathrm{ML}} A^T)(A(\psi^T\psi)^{-1}A^T)^{-1}A(\psi^T\psi)^{-1}.$$

The obtained estimate $\tilde{\boldsymbol{\rho}}_{\mathrm{ML}}$ is equivalent to the estimate $\tilde{\boldsymbol{\rho}}$, given by the formula (6.25). The estimate for the standard deviation can be written as

$$\tilde{\sigma}^2 = \frac{1}{K}(\boldsymbol{u} - \psi\tilde{\boldsymbol{\rho}})^T(\boldsymbol{u} - \psi\tilde{\boldsymbol{\rho}}).$$

Substituting the obtained estimates to the likelihood function we calculate its maximum value for the hypothesis H_0 and its alternative:

$$L_{\max}(\tilde{\sigma^2}, \tilde{\boldsymbol{\rho}}) = (2\pi\tilde{\sigma^2})^{-K/2}q^{-K/2}, \tag{8.40}$$

$$L_{\max}(\hat{\sigma^2}, \hat{\boldsymbol{\rho}}) = (2\pi\hat{\sigma^2})^{-K/2}q^{-K/2}. \tag{8.41}$$

Using the expressions (8.40), (8.41), to rewrite the formula (8.35) in the form

$$q = \frac{L_{\max}(\tilde{\sigma^2}, \tilde{\boldsymbol{\rho}})}{L_{\max}(\hat{\sigma^2}, \hat{\boldsymbol{\rho}})} = \left(\frac{\hat{\sigma}^2}{\tilde{\sigma^2}}\right)^{K/2}. \tag{8.42}$$

We use the criterion as follows: if $q < q_{\mathrm{t}}$, then the hypothesis H_0 is rejected, if $q \geq q_{\mathrm{t}}$, then the hypothesis H_0 is valid (and in our case the depth values calculated using borehole data does not contradict to the depth values calculated using seismic observations. To determine the threshold value of q_{t} it is necessary to determine the probability distribution of q, but the most simple way consists in the determination of the constraint of the considered criterion with the Fisher information criterion. If the hypothesis H_0 is correct one, then the relation

$$F = \frac{[(\boldsymbol{u} - \psi\tilde{\boldsymbol{\rho}})^2 - (\boldsymbol{u} - \psi\hat{\boldsymbol{\rho}})^2]/Q}{(\boldsymbol{u} - \psi\hat{\boldsymbol{\rho}})^2/(K - S)} \tag{8.43}$$

belongs to Fisher distribution (see Sec. 1.8.6) with Q and $K - S$ degrees of freedom ($F \in \Phi_{Q,K-S}$). By analyzing the formula (8.42) and (8.43), we can conclude that the function F can be expressed through q:

$$F = \frac{K - S}{Q}[q^{-2/K} - 1].$$

The likelihood ratio criterion is reduced to the Fisher criterion: if $F > \Phi_{Q,K-S,\alpha}$ (the value of $\Phi_{Q,K-S,\alpha}$ is calculated subject to given confidence level α), then the hypothesis H_0 is rejected and we conclude that with the given probability of error, there is an inadequacy between the depth values obtained using borehole data and depth values obtained using seismic data. The value $\Phi_{Q,K-S,\alpha}$ can be determined using, for example, a tabular information on the Fisher distribution at given $Q, K - S$ and the probability of the error (confidence level).

8.4.4 *Incorporation of random nature of depth measurement using borehole data*

The above considered problem statement has been limited by an exact setting of the bed depth using borehole data. Such assumption is justified, if the accuracy of the seismic observation is more rough in comparison with the borehole measurement. However, the consideration of the random nature of the depth measurements using the borehole data has a practical significance.

The model of observations in this case can be represented as: $\boldsymbol{V} = A\boldsymbol{\rho}+\boldsymbol{\varepsilon}$. Here $\boldsymbol{V}$, A, $\boldsymbol{\rho}$ has the former sense, $\boldsymbol{\varepsilon} = [\boldsymbol{\varepsilon}(x_k, y_k)]$ is a random component of the model, measured at the points (x_k, y_k) with the covariance matrix P_ε.

The estimate of the parameter $\boldsymbol{\rho}$ we find using minimizing of the following square form

$$l(\boldsymbol{\rho}) = (\boldsymbol{u} - \psi\boldsymbol{\rho})^T R_n^{-1}(\boldsymbol{u} - \psi\boldsymbol{\rho}) + (\boldsymbol{V} - A\boldsymbol{\rho})^T P_\varepsilon^{-1}(\boldsymbol{V} - A\boldsymbol{\rho}). \tag{8.44}$$

The formula (8.44) is the objective function in the parameter space $\boldsymbol{\rho}$, jointly for seismic and borehole observations. The estimate of the parameter vector $\boldsymbol{\rho}$ we find by minimizing of (8.44) with respect to $\boldsymbol{\rho}$:

$$\overline{\boldsymbol{\rho}}^T = (\boldsymbol{u}^T R_n^{-1}\psi + \boldsymbol{V}^T P\varepsilon^{-1}A)(\psi^T R_n^{-1}\psi + A^T P\varepsilon^{-1}A)^{-1.}. \tag{8.45}$$

The formula (8.45) can be transformed to the following shape

$$\overline{\boldsymbol{\rho}}^T = (\hat{\boldsymbol{\rho}}^T B_1 + \hat{\boldsymbol{\rho}}_{\text{bh}}^T)(E + B_1)^{-1}, \tag{8.46}$$

where $\hat{\boldsymbol{\rho}}$ is determined by the formula (6.11):

$$B_1 = \Psi^T R_n^{-1}\Psi(A^T P_\varepsilon^{-1}A)^{-1},$$
$$\hat{\boldsymbol{\rho}}_{\text{bh}}^T = \boldsymbol{V}^T P_\varepsilon^{-1}A(A^T P_\varepsilon^{-1}A)^{-1}.$$

The formula (8.46) can be simplified, if the following condition is valid

$$\lambda_{max}(B_1) < 1, \tag{8.47}$$

where λ_{max} is the maximum eigenvalue of the matrix B_1.

In the most practical cases the accuracy of borehole data concerning the bed depth is higher than the bed depth obtained with the use of seismic data. Analysis of the matrix B_1 follows that in these cases the inequality (8.47) is valid and the formula (8.46) can be rewritten in the approximation form:

$$\overline{\boldsymbol{\rho}}^T \approx (\hat{\boldsymbol{\rho}}^T B_1 + \hat{\boldsymbol{\rho}}_{\text{bh}}^T)(E - B_1). \tag{8.48}$$

In practice it is often to suppose the random components $\boldsymbol{n}$ and $\boldsymbol{\varepsilon}$ are uncorrelated. In this case the formulas (8.46), (8.48) for the estimate $\boldsymbol{\rho}$ become simpler:

$$\overline{\boldsymbol{\rho}}^T = \left(\frac{\boldsymbol{u}^T\Psi}{\sigma_n^2} + \frac{\boldsymbol{V}^T A}{\sigma_\varepsilon^2}\right)\left(\frac{\Psi^T\Psi}{\sigma_n^2} + \frac{A^T A}{\sigma_\varepsilon^2}\right)^{-1}, \tag{8.49}$$

$$\overline{\boldsymbol{\rho}} \sim \left(\frac{\sigma_\varepsilon^2}{\sigma_n^2}\hat{\boldsymbol{\rho}}^T(\Psi^T\Psi)(A^T A)^{-1} + \hat{\boldsymbol{\rho}}_{\text{bh}}^T\right)$$
$$\times \left(E - \frac{\sigma_\varepsilon^2}{\sigma_n^2}\hat{\boldsymbol{\rho}}^T(\Psi^T\Psi)(A^T A)^{-1}\right).$$

Let's, for simplicity, the variances of the random components σ_n^2 and σ_ε^2 are the same for all beds.

We consider some limit values for the formula (8.49):

$$1)\ \lim_{\sigma_\varepsilon \to 0} \overline{\boldsymbol{\rho}} = \hat{\boldsymbol{\rho}}_{\text{bh}};\ 2)\ \lim_{\sigma_n \to 0} \overline{\boldsymbol{\rho}} = \hat{\boldsymbol{\rho}};$$
$$3)\ \lim_{\sigma_n \to 0} \overline{\boldsymbol{\rho}} = \hat{\boldsymbol{\rho}}_{\text{bh}};\ 4)\ \lim_{\sigma_\varepsilon \to 0} \overline{\boldsymbol{\rho}} = \hat{\boldsymbol{\rho}}.$$

The first and third limiting cases, the second and fourth limiting cases are equivalent. This result agrees with the statement: at estimation of the parameters using two kinds of experimental data, preferable data have a minimal variance.

Considered above description of the borehole bed depth as a random value contains one special case of the depth description by the nonrandom values with the help of the linear constraints.

The depths obtained by the seismic data $\boldsymbol{u}$ and the borehole observations $\boldsymbol{V}$ are independent, therefore the likelihood function can be written as

$$p(\boldsymbol{\rho}, \boldsymbol{u}, \boldsymbol{V}) = p(\boldsymbol{\rho}, \boldsymbol{u}) p(\boldsymbol{\rho}, \boldsymbol{V}), \tag{8.50}$$

where $p(\boldsymbol{\rho}, \boldsymbol{u})$ and $p(\boldsymbol{\rho}, \boldsymbol{V})$ belong the normal distribution with mathematical expectations and covariance matrices respectively $\boldsymbol{\Psi}_\rho$, R_n and $\boldsymbol{A}\rho$, P_ε. For simplicity we assume that the random component is uncorrelated (matrix P_ε is diagonal) with the equal variances σ_ε^2. The transform from the random depths V to fixed ones is implemented by tending σ_ε to zero:

$$p_\varepsilon(\boldsymbol{\rho}, \boldsymbol{V}) = \lim_{\sigma_\varepsilon \to 0} p(\boldsymbol{\rho}, V).$$

This limit process we implement using the characteristic functions. The characteristic function of the normal distribution (Cramer, 1946) with a reference to our model looks like

$$Z(\varepsilon) = \exp\left\{-\frac{\varepsilon^2 \sigma_\varepsilon^2}{2}\right\}, \quad Z_0(\varepsilon) = \lim_{\sigma_\varepsilon \to 0} Z(\varepsilon) = 1.$$

After the inverse Fourier transform of $Z_0(\varepsilon)$, we obtain

$$p_0(\boldsymbol{\rho}, \boldsymbol{V}) = \delta(\boldsymbol{V} - A\boldsymbol{\rho}),$$

and the formula (8.50) can be rewritten as

$$p(\boldsymbol{\rho}, \boldsymbol{u}, \boldsymbol{V}) = p(\boldsymbol{\rho}, \boldsymbol{u}) \delta(\boldsymbol{V} - A\boldsymbol{\rho}). \tag{8.51}$$

The determination of the desired parameter vector $\boldsymbol{\rho}$ by the maximum likelihood method on the basis of the expression (8.51) is equivalent to maximizing of the function $p(\boldsymbol{\rho}, \boldsymbol{u})$ under the condition $\boldsymbol{V} = A\boldsymbol{\rho}$.

Thus, we come to using of the borehole data in a form of the linear constraints.

8.4.5 *Application of a posteriori probability method to approximation of seismic horizon*

Let us consider one more modification of the defined problem, based on the models similar to (8.29), (6.21) with random parameter vector $\boldsymbol{\rho}$. A priori information about the parameter vector $\boldsymbol{\rho}$ is introduced in a probability form. In the most practical cases we can assume a priori distribution to be normal with the mathematical expectation $\langle\rho\rangle$ and the covariance matrix D. As the objective function for the parameter estimation we use a posteriori distribution of the vector $\boldsymbol{\rho}$ with given observation data $\boldsymbol{u}$, $\boldsymbol{V}$. In accordance with the Bayes theorem (see Sec. 6.9) we obtain

$$p(\boldsymbol{\rho}/\boldsymbol{u},\boldsymbol{V}) = \frac{1}{p(\boldsymbol{u},\boldsymbol{V})}p(\boldsymbol{\rho})p(\boldsymbol{u},\boldsymbol{V}/\boldsymbol{\rho}). \tag{8.52}$$

To find the vector of estimates $\hat{\boldsymbol{\rho}}$ we use a posteriori probability method:

$$\tilde{\boldsymbol{\rho}}_{\text{bh}} = \max_{\boldsymbol{\rho}} \ln p(\boldsymbol{\rho}/\boldsymbol{u},\boldsymbol{V}).$$

If the depths calculated using seismic data $\boldsymbol{u}$ and using borehole observations $\boldsymbol{V}$ are independent, then the density function $p(\boldsymbol{u},\boldsymbol{V}/\boldsymbol{\rho})$ can be rewritten as

$$p(\boldsymbol{u},\boldsymbol{V}/\boldsymbol{\rho}) = p(\boldsymbol{u}/\boldsymbol{\rho})p(\boldsymbol{V}/\boldsymbol{\rho}). \tag{8.53}$$

Substituting the formula (8.53) to the expression (8.52) and discard the terms without a dependence on parameters, we obtain the function

$$\begin{aligned} F(\boldsymbol{\rho}) = -\{&(\boldsymbol{\rho} - \langle\boldsymbol{\rho}\rangle)^T D^{-1}(\boldsymbol{\rho} - \langle\boldsymbol{\rho}\rangle) + (\boldsymbol{u} - \Psi\boldsymbol{\rho})^T R_n^{-1}(\boldsymbol{u} - \Psi\boldsymbol{\rho}) \\ &+ (\boldsymbol{V} - A\boldsymbol{\rho})^T P_\varepsilon^{-1}(\boldsymbol{V} - A\boldsymbol{\rho})\}. \end{aligned} \tag{8.54}$$

Maximizing the function (8.54) over the parameter vector $\boldsymbol{\rho}$, we obtain the desired estimate:

$$\begin{aligned} \tilde{\boldsymbol{\rho}}_{\text{MAP}} &= [\langle\boldsymbol{\rho}^T\rangle D^{-1} + \boldsymbol{u}^T R_n^{-1}\Psi + \boldsymbol{V}^T P_\varepsilon^{-1} A] \\ &\times [D^{-1} + \Psi^T R_n^{-1}\Psi + A^T P_\varepsilon^{-1} A]^{-1}. \end{aligned} \tag{8.55}$$

The elements of the Fisher information matrix for the considered models can be written as

$$I_{ss'} = -\left\langle \frac{\partial \ln 2p(\boldsymbol{u}/\boldsymbol{\rho})}{\partial\rho_s\partial\rho_{s'}} \right\rangle - \left\langle \frac{\partial \ln 2p(\boldsymbol{V}/\boldsymbol{\rho})}{\partial\rho_s\partial\rho_{s'}} \right\rangle - \left\langle \frac{\partial \ln 2p(\boldsymbol{\rho})}{\partial\rho_s\partial\rho_{s'}} \right\rangle. \tag{8.56}$$

Substituting the expressions $p(\boldsymbol{u}/\boldsymbol{\rho})$, $p(\boldsymbol{V}/\boldsymbol{\rho})$, $p(\boldsymbol{\rho})$ to the formula (8.56) we obtain the following expression:

$$I = \Psi^T R_n^{-1}\Psi + A^T P_\varepsilon^{-1} A + D^{-1}. \tag{8.57}$$

For the considered linear models the covariance matrix of estimates is computed by the inversion of the information matrix I:

$$R_{\tilde{\boldsymbol{\rho}}_{\text{MAP}}} = [D^{-1} + \Psi^T R_n^{-1}\Psi + A^T P_\varepsilon^{-1} A]^{-1}. \tag{8.58}$$

8.4.6 *Case of uncorrelated components of random vector*

Let us consider special, but important from the practical standpoint case of the uncorrelated random components n, ε and parameter vector $\boldsymbol{\rho}$, i.e. $D = \sigma_\rho^2 E$, $R_n = \sigma_n^2 E$, $P_\varepsilon = \sigma_\varepsilon^2 E$ are valid. Under considered assumptions the formulas (8.55) and (8.58) can be rewritten in the following form

$$\tilde{\boldsymbol{\rho}}_{\mathrm{MAP}}^T = \left[\frac{\langle\boldsymbol{\rho}\rangle}{\sigma_\rho^2} + \frac{\boldsymbol{u}^T\Psi}{\sigma_n^2} + \frac{\boldsymbol{V}^T A}{\sigma_\varepsilon^2}\right] \times \left[\frac{1}{\sigma_\rho^2}E + \frac{\Psi^T\Psi}{\sigma_n^2}\frac{A^T A}{\sigma_\varepsilon^2}\right], \tag{8.59}$$

$$R_{\tilde{\boldsymbol{\rho}}_{\mathrm{MAP}}} = \left[\frac{1}{\sigma_\rho^2}E + \frac{\Psi^T\Psi}{\sigma_n^2}\frac{A^T A}{\sigma_\varepsilon^2}\right]. \tag{8.60}$$

For the further analysis we rewrite the expressions (8.59), (8.60) in the form:

$$\tilde{\boldsymbol{\rho}}_{\mathrm{MAP}}^T = [\sigma_n^2\sigma_\varepsilon^2\langle\boldsymbol{\rho}\rangle^T + \sigma_\varepsilon^2\sigma_\rho^2\boldsymbol{u}^T\Psi + \sigma_\rho^2\sigma_n^2\boldsymbol{V}^T A] \times R_{\tilde{\boldsymbol{\rho}}_{\mathrm{MAP}}}/(\sigma_\rho^2\sigma_\varepsilon^2\sigma_n^2), \tag{8.61}$$

$$R_{\tilde{\boldsymbol{\rho}}_{\mathrm{MAP}}} = \sigma_\rho^2\sigma_\varepsilon^2\sigma_n^2[\sigma_n^2\sigma_\varepsilon^2 E + \sigma_\rho^2\sigma_\varepsilon^2\Psi^T\Psi + \sigma_n^2\sigma_\rho^2 A^T A]^{-1}. \tag{8.62}$$

Let's consider a few proceedings to limit on a basis of the formulas (8.59)–(8.62):

$$\lim_{\sigma_n\to 0}\tilde{\boldsymbol{\rho}}_{\mathrm{MAP}}^T = \boldsymbol{u}^T\Psi(\Psi^T\Psi)^{-1} = \tilde{\boldsymbol{\rho}}^T,$$

$$\lim_{\sigma_\varepsilon\to 0}\tilde{\boldsymbol{\rho}}_{\mathrm{MAP}}^T = \boldsymbol{V}^T A(A^T A)^{-1} = \hat{\boldsymbol{\rho}}_{\mathrm{MAP}}^T,$$

$$\lim_{\sigma_\rho\to 0}\tilde{\boldsymbol{\rho}}_{\mathrm{MAP}}^T = \langle\boldsymbol{\rho}\rangle^T,$$

$$\lim_{\sigma_n\to 0} R_{\tilde{\boldsymbol{\rho}}_{\mathrm{MAP}}} = \lim_{\sigma_\varepsilon\to 0} R_{\tilde{\boldsymbol{\rho}}_{\mathrm{MAP}}} = \lim_{\sigma_\rho\to 0} R_{\tilde{\boldsymbol{\rho}}_{\mathrm{MAP}}} = 0.$$

So far, if the variance of the depth random component obtained by the seismic data tends to zero, then the estimate of parameters is fully determined by the seismic data. If the borehole observations have zero value of the variance, then the estimate $\tilde{\boldsymbol{\rho}}_{\mathrm{MAP}}$ is determined by the borehole data only. And, at last, if a priori data is reliable ($\sigma_\rho \to 0$), then the solution is equal to the given mean vector $\langle\boldsymbol{\rho}\rangle$. The variance $\tilde{\boldsymbol{\rho}}_{\mathrm{MAP}}$ in all considered cases tends to zero:

$$\lim_{\sigma_n\to 0}\tilde{\boldsymbol{\rho}}_{\mathrm{MAP}}^T = \left[\frac{\langle\boldsymbol{\rho}\rangle^T}{\sigma_\rho^2} + \frac{\boldsymbol{V}^T A}{\sigma_\varepsilon^2}\right]\left[\frac{1}{\sigma_\rho^2}E + \frac{A^T A}{\sigma_\varepsilon^2}\right]^{-1},$$

$$\lim_{\sigma_n\to 0} R_{\tilde{\boldsymbol{\rho}}_{\mathrm{MAP}}} = \left[\frac{1}{\sigma_\rho^2}E + \frac{A^T A}{\sigma_\varepsilon^2}\right]^{-1},$$

$$\lim_{\sigma_\varepsilon\to 0}\tilde{\boldsymbol{\rho}}_{\mathrm{MAP}}^T = \left[\frac{\langle\boldsymbol{\rho}\rangle^T}{\sigma_\rho^2} + \frac{\boldsymbol{u}^T\Psi}{\sigma_n^2}\right]\left[\frac{1}{\sigma_\rho^2}E + \frac{\Psi^T\Psi}{\sigma_n^2}\right]^{-1},$$

$$\lim_{\sigma_\varepsilon\to 0} R_{\tilde{\boldsymbol{\rho}}_{\mathrm{MAP}}} = \left[\frac{1}{\sigma_\rho^2}E + \frac{\Psi^T\Psi}{\sigma_n^2}\right]^{-1},$$

$$\lim_{\sigma_n\to 0}\tilde{\boldsymbol{\rho}}^T_{\mathrm{MAP}} = \left[\frac{\boldsymbol{u}^T\Psi}{\sigma_n^2} + \frac{V^TA}{\sigma_\varepsilon^2}\right]\left[\frac{\Psi^T\Psi}{\sigma_n^2} + \frac{A^TA}{\sigma_\varepsilon^2}\right]^{-1},$$

$$\lim_{\sigma_n\to 0} R_{\tilde{\boldsymbol{\rho}}\mathrm{MAP}} = \left[\frac{\Psi^T\Psi}{\sigma_n^2} + \frac{A^TA}{\sigma_\varepsilon^2}\right]^{-1}.$$

If the variance of the random component σ_n^2 tends to infinity, it implies that a quality of the seismic data is worse than the borehole and a priori data, therefore the estimation excludes processing of the seismic data $\boldsymbol{u}$. Similar results can be obtained under conditions $\sigma_\varepsilon^2 \to \infty$ and $\sigma_\rho^2 \to \infty$. In these cases the estimate $\tilde{\boldsymbol{\rho}}$ does not intend processing of the borehole data and a priori data respectively.

Let us consider a special case of the model (8.29) with an identity matrix Ψ and a vector $\boldsymbol{\rho}$ dimension coinciding with a number of a priori data. Taking into account the borehole observation in this case is implemented by introducing it to the mean vector of a priori data together with appropriate variances. The estimate $\tilde{\boldsymbol{\rho}}^T_{\mathrm{MAP}}$ with $D = \sigma_\rho^2 E$, $R_n = \sigma_n^2 E$ can be written as

$$\tilde{\boldsymbol{\rho}}^T_{\mathrm{MAP}} = \left[\frac{\langle\boldsymbol{\rho}\rangle^T}{\sigma_\rho^2} + \frac{\boldsymbol{u}^T}{\sigma_n^2}\right]\left[\frac{1}{\sigma_\rho^2}E + \frac{1}{\sigma_n^2}E\right]^{-1},$$

and covariance matrix is written as

$$R_{\tilde{\boldsymbol{\rho}}\mathrm{MAP}} = \frac{\sigma_n^2}{[1+\sigma_n^2/\sigma_\rho^2]}E.$$

In this approach we do not need to know the structural matrix Ψ. But increasing of the dimension of the parameter vector $\boldsymbol{\rho}$ can lead to a loss of accuracy. Considered above models and appropriate solutions can be used for design of the algorithms of construction of the seismic profiles with taking into account the borehole observations.

Together with the construction of the approximating surface, the problem of construction of the confidence areas which characterize a degree of proximity of the constructed approximating surface and its parameters to the appropriate theoretical values (see Sec. 6.15) has a great practical importance. From the beginning we find the variance of a deviation of the approximating surface from the theoretical one taking into account and without taking into account the borehole observations:

$$\tilde{\sigma}^2 = \boldsymbol{\varphi}^T R_{\tilde{\rho}}\boldsymbol{\varphi}, \tag{8.63}$$

$$\hat{\sigma}^2 = \boldsymbol{\varphi}^T R_{\hat{\rho}}\boldsymbol{\varphi}. \tag{8.64}$$

Where $\boldsymbol{\varphi}$ is a vector with the components from a row of the matrix Ψ, for example, in case of the polynomial of second power we have $\boldsymbol{\varphi} = [1, x, y, x^2, xy, y^2]$. The values $\tilde{\sigma}^2(x,y)$ and $\hat{\sigma}^2(x,y)$ can be calculated in an arbitrary point (x,y) of the considered square using the formulas (8.63), (8.64). The confidence intervals are estimated using the following formulas (see Sec. 6.15):

$$P[\tilde{f}(x,y) - \gamma_{K-S+Q}\tilde{\sigma}(x,y) < f(x,y) < \tilde{f}(x,y) + \gamma_{K-S+Q}\tilde{\sigma}(x,y)] = \beta, \tag{8.65}$$

$$P[\hat{f}(x,y) - \gamma_{K-S+Q}\hat{\sigma}(x,y) < f(x,y) < \hat{f}(x,y) + \gamma_{K-S+Q}\hat{\sigma}(x,y)] = \beta, \tag{8.66}$$

$$P[\tilde{\rho}_s - \gamma_{K-S+Q}\tilde{\sigma}_{\rho_s} < \rho_s < \tilde{\rho}_s + \gamma_{K-S+Q}\tilde{\sigma}_{\rho_s}] = \beta, \tag{8.67}$$

$$P[\hat{\rho}_s - \gamma_{K-S+Q}\hat{\sigma}_{\rho_s} < \rho_s < \hat{\rho}_s + \gamma_{K-S+Q}\hat{\sigma}_{\rho_s}] = \beta, \tag{8.68}$$

where β is the confidence probability; the value γ_N is connected with the confidence probability β and can be determined using the Student's distribution with N degrees of freedom, which is equal to $K - S$ in the case without taking into account the borehole observations and it is equal to $K - S + Q$ in the case of taking into account the borehole observations. The formulas (8.65), (8.67) and (8.66), (8.68) can be used to obtain the confidence intervals with and without taking into account the borehole observations correspondingly.

8.4.7 *Approximation of parameters of approximation horizon by the orthogonal polynomials*

Together with the considered polynomial approximation it is convenient in a computational sense to use the orthogonal functions for the purpose of the approximation (see Sec. 6.6). At that the formulas (8.32) and (8.31) for the estimation of the parameters $\hat{\boldsymbol{\rho}}$ and $\tilde{\boldsymbol{\rho}}$ are reduced. In this case observation model can be represented as

$$\boldsymbol{u} = \Phi\boldsymbol{\rho} + \boldsymbol{n},$$

where Φ is a structural matrix with the rows represented by the orthogonal functions for fixed values of (x, y): $[\varphi_0(x,y), \varphi_1(x,y), \ldots, \varphi_{S-1}(x,y)]$; $\boldsymbol{\rho}$ is the vector of desired coefficients for given orthogonal functions; $\boldsymbol{n}$ is a random deviation between observed data and model values.

Let us consider the system of orthogonal functions from Sec. 6.6, then the formula for the parameter estimation $\hat{\boldsymbol{\rho}}$, $\tilde{\boldsymbol{\rho}}$ and the appropriate covariance matrices $R_{\hat{\rho}}$, $R_{\tilde{\rho}}$ can be written as

$$\hat{\boldsymbol{\rho}} = T^{-1}\Phi^T\boldsymbol{u}, \tag{8.69}$$

$$\tilde{\boldsymbol{\rho}}^T = \hat{\boldsymbol{\rho}}^T + (\boldsymbol{V} - \hat{\boldsymbol{\rho}}^T A^T)(AT^{-1}A^T)AT^{-1}, \tag{8.70}$$

$$R_{\hat{\rho}} = T^{-1}\hat{\sigma}^2, \tag{8.71}$$

$$R_{\tilde{\rho}} = R_{\hat{\rho}} - \hat{\sigma}^2 T^{-1}A^T(AT^{-1}A^T)^{-1}AT^{-1}, \tag{8.72}$$

where T is a diagonal matrix ($\Phi^T\Phi = T$).

To find the estimate $\hat{\boldsymbol{\rho}}$ without taking into account the borehole data and its correlation matrix we do not need a matrix inversion. To estimate the value $\tilde{\boldsymbol{\rho}}$ we invert the matrix with dimension $[Q \times Q]$, which determines by a number of boreholes. The formulas (8.69) – (8.72) are suitable for calculations. In special case, when we have only one borehole ($Q = 1$), the matrix A becomes a row vector and formulas (8.70), (8.72) can be written as

$$\tilde{\rho}^T = \hat{\rho}^T + \frac{1}{A_0}(\mathbf{V} - \hat{\rho}^T A^T)AT^{-1},$$

$$R_{\tilde{\rho}} = R_{\hat{\rho}} - \frac{\hat{\sigma}^2}{A_0}T^{-1}A^T A^T,$$

where $A_0 = AT^{-1}A^T$.

During obtaining the considered above estimates, we supposed that the matrix $\Psi^T R_n^{-1}\Psi$ is a nonsingular matrix, if not, we may use the estimates, obtained by the singular analysis represented in Sec. 6.12.

8.4.8 *Numerical examples of application of approximation algorithms*

Let us consider numerical simulation of the horizon approximation. As the model we use a trough. Its depth and thickness are decreased in y direction (see Fig. 8.5). From analysis of the borehole data leads that the horizon has a sinking in the direction of x axis, from the profile *II* to the profile *IV* for a quantity of 100 m (see Fig. 8.6). Approximation is implemented using the polynomials of third power. At the profiles *I* and *III* a maximum depth deviation with taking into account and without taking into account the borehole data is observed near the boreholes *5* and *3*. The approximating curve without taking into account the borehole data gives a satisfactory description of the horizon.

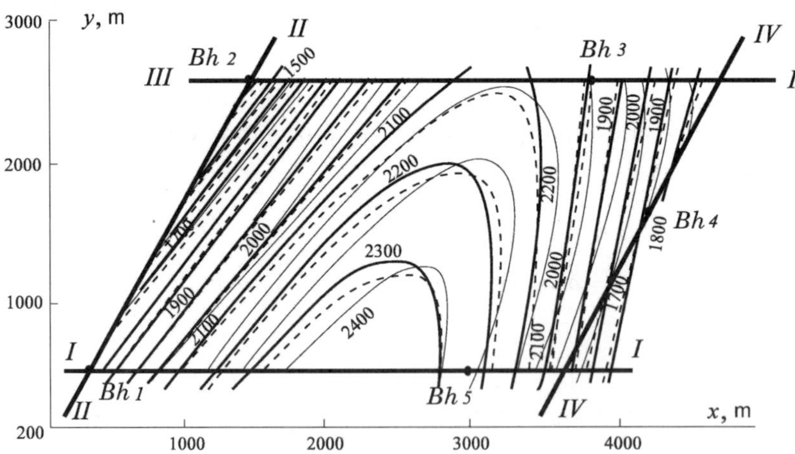

Fig. 8.5 Numerical simulation of the horizon approximation. Scheme of profiles (4 profiles (*I–IV*), 5 boreholes) and isolines of the horizon: solid line is the numerical simulation without taking into account the borehole data, dashed lined is the numerical simulation with taking into account the borehole data.

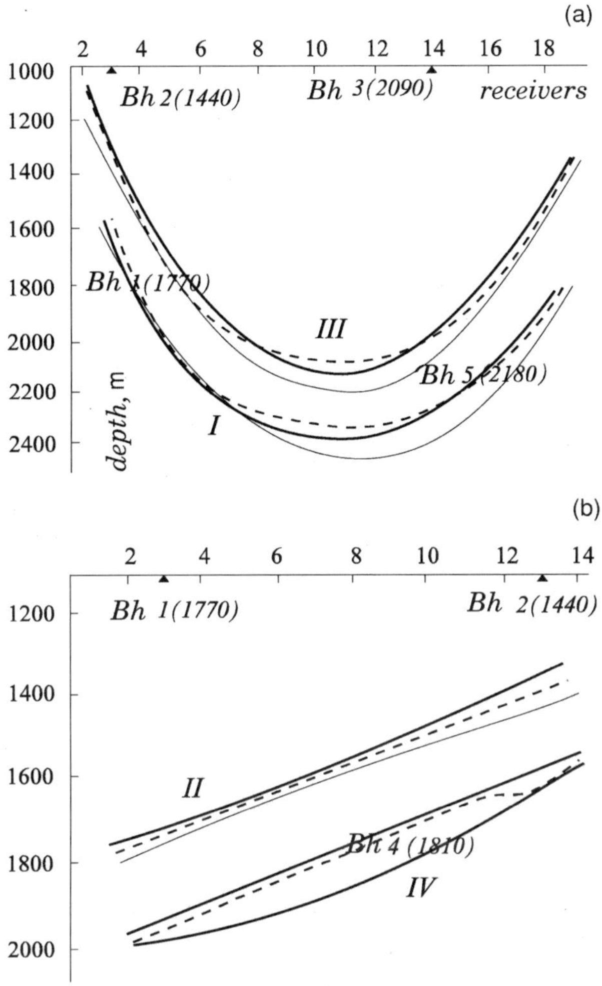

Fig. 8.6 Numerical simulation of the horizon approximation (see Fig. 8.5). Depth section along profiles: profiles *I* and *III* (*a*); profiles *II* and *IV* (*b*) ; dashed line is the model, bold line is without taking into account the borehole data, thin line is with taking into account the borehole data.

We should note that the approximated horizon, obtained with taking into account the borehole data deviates from the horizon, obtained without taking into account the borehole data, in accordance with the borehole data. Near profile *II* such deviations are insignificant. At profile *IV* the deviations have maximum values near borehole with number *4* and at the extremities of the profile the deviations tend to the values of the approximated horizon, calculated without taking into account the borehole data.

In the considered above examples the degree of the polynomial is a set using a priori information about the desired smoothness of the horizon. Let us analyze an applicability of the information criterion (7.17), (7.18) for choosing of the degree of polynomial (Akaike, 1974). Under assumptions concerning normality and noncorrelatedness of the random component, the least squares estimate $\hat{\boldsymbol{\rho}}$ coincides with the maximum likelihood estimate, which is used in the function $IC(S_m, \hat{\boldsymbol{\rho}})$. For simplicity we consider an one-dimensional approximation of the horizon along the profile and two-dimensional generalization present no difficulties. Let us find an explicit form of the function of the information criterion for the polynomial of degree m. We substitute to the formula (7.17) the normal density function and with the help of the estimate, obtained by the maximum likelihood method for the variance of the random component:

$$IC(m, \hat{\boldsymbol{\rho}}_m) = J \log 2\pi + J + J \log \hat{\sigma}^2 + Km,$$

where unlike the criterion (7.17), before a number of parameters instead of term 2 the term K is introduced. This term can be chosen empirically using some model experiments, for a better reflection of the specificity of the solved problem. As it will be clear below, the algorithm is stable in a wide range of K deviation. Because the two initial terms in the expression $IC(m, \hat{\boldsymbol{\rho}}_m)$ do not change at changing of degree of the polynomial, then for the practical using the function $IC(m, \hat{\boldsymbol{\rho}}_m)$ can be written down as

$$IC(m, \hat{\boldsymbol{\rho}}_m) = J \log \hat{\sigma}^2 + Km.$$

Let us consider an application of the criterion. Taking into account a priori information we determine a maximum degree of polynomial M, calculate the estimates $\hat{\boldsymbol{\rho}}_m$, $\hat{\sigma}^2$ and the values $IC(m, \hat{\boldsymbol{\rho}}_m)$, $m = 1, 2, \ldots, M$. As an estimate of the degree of a polynomial we choose

$$\hat{m} = \arg\min[IC(m, \hat{\boldsymbol{\rho}}_m)].$$

The mean values of $IC(m)$ are represented in Fig. 8.7, a). These values are obtained on the base of the model experiments with the various realization of the random component. The criterion function $IC(m)$ rises sharply from the left hand side of its true value and from the right hand side the function rises not so sharply, that, in the case of high noise level, can leads to a bias of estimate of the degree of the polynomial. The study of the influence of the dispersion on the criterion implementation is represented at Fig. 8.8. Let's note, that beginning from the variance σ_{bd}, the estimate of the polynomial degree $\hat{m}$ has a systematic bias, i.e. its value one unit smaller than the true value. Let us consider the results of model experiments for studying of the influence of the nonstationarity of the noise. For various values of K criterion gives the estimate of the polynomial degree, which coincides with the true value (see Fig. 8.7(b)), that illustrates the robustness of the criterion with the presence of the nonstationary noise.

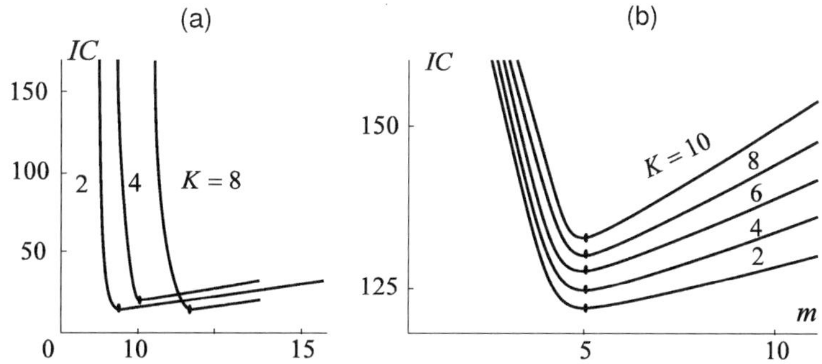

Fig. 8.7 Dependence of the criterion function IC on the degree of polynomial m, obtained by the model data with various initial degrees (a) and for nonstationary noise (b).

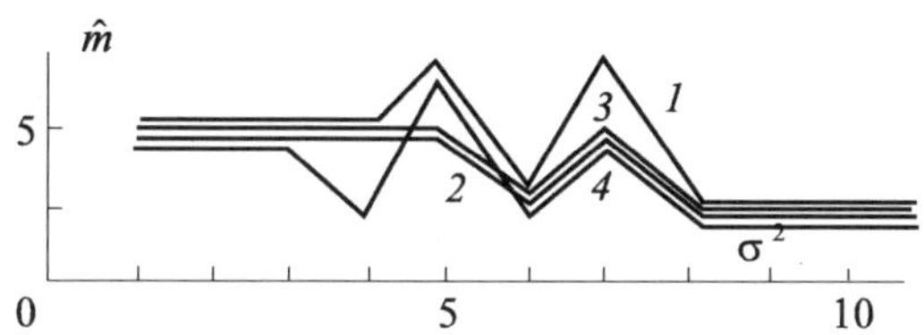

Fig. 8.8 Dependence of estimates of the degree of polynomial on noise variance for the different values of K.

8.5 Algorithm of Approximation of Formation Velocity with the Use of Areal Observations with Borehole Data

The validity of the results obtained by the seismic exploration method in regions with complex geological conditions depends on the accuracy of the determination of the formation velocity. The formation velocity can be computed using CDP data on the linear profiles and extended for an areal under interpretation together with an addition information obtained by the borehole observations. Such algorithm is considered in (Troyan and Sokolov, 1983). The solution is constructed on the basis of SVD (singular value decomposition) method (see Sec. 6.12).

Let's in the linear profiles of the investigated region the values of velocities $\boldsymbol{u}$ for an interesting horizon are calculated. The observation model can be represented as follows

$$\boldsymbol{u} = \Psi \boldsymbol{\rho} + \boldsymbol{n}, \tag{8.73}$$

where Ψ is the structural matrix; $\boldsymbol{\rho}$ is the vector of the desired parameters; $\boldsymbol{n}$ is the random component with the zero mathematical expectation and a given covariance matrix R_n. As the approximation function we will use the quadratic expressions on variables x, y, h, Δh, where (x, y) are the coordinates of a point in the profile; h is a formation top; Δh is a formation thickness. The row of the matrix Ψ is written as

$$\Psi_i = [1, x_i, y_i, h_i, x_i^2, x_i y_i, x_i h_i, x_i \Delta h_i, y_i^2, \quad y_i h_i, y_i \Delta h_i, h_i^2, h_i \Delta h_i, \Delta h_i^2]. \quad (8.74)$$

Where $i = 1, 2, \ldots, K$ is a number of initial points. The borehole data are given as the linear conditions

$$A\boldsymbol{\rho} = \boldsymbol{V}, \quad (8.75)$$

where $\boldsymbol{V}$ are the values of the velocity in the boreholes, Q is a number of dimensions of the vector $\boldsymbol{V}$, which determined by a number of the boreholes; $\boldsymbol{\rho}$ is the vector of unknown parameters; A is the matrix with the rows analogous to the rows of the matrix Ψ, but they determined in the points of the location of the boreholes.

Let's represent the velocity in the borehole in the form

$$\boldsymbol{V} = A\boldsymbol{\rho} + \boldsymbol{\varepsilon}. \quad (8.76)$$

Here $\boldsymbol{\varepsilon}$ is the random component with the zero mathematical expectation and a given covariance matrix. The solution of the such problem under the conditions (8.75) leads to the next estimate (Troyan, 1981c; Troyan and Windgassen, 1979):

$$\tilde{\boldsymbol{\rho}}^T = \hat{\boldsymbol{\rho}}^T + (\boldsymbol{V}^T - \hat{\boldsymbol{\rho}}^T A^T)(A(\Psi^T\Psi)^{-1}A^T)^{-1}A(\Psi^T\Psi)^{-1},$$
$$\hat{\boldsymbol{\rho}} = (\Psi^T\Psi)^{-1}\Psi^T\boldsymbol{u}. \quad (8.77)$$

A quality of the estimate is determined by the covariance matrix:

$$R_{\tilde{\boldsymbol{\rho}}} = \hat{\sigma}^2[(\Psi^T\Psi)^{-1} - (\Psi^T\Psi)^{-1}A^T(A(\Psi^T\Psi)^{-1}A^T)^{-1}A(\Psi^T\Psi)^{-1}].$$

If additional data about the velocity are given in the form (8.76), then the solution, obtained by the maximum likelihood method, can be written as

$$\tilde{\boldsymbol{\rho}}^T = (\boldsymbol{u}^T R_n^{-1}\Psi + \boldsymbol{V}^T P_\varepsilon A)(\Psi^T R_n^{-1}\Psi + A^T P_\varepsilon^{-1} A)^{-1}. \quad (8.78)$$

The numerical simulation with the application of the estimates (8.77), (8.78) was implemented. As the result of the such simulation it is shown, that in some cases we can not obtain a satisfactory solution if a rank of the matrix $\Psi^T\Psi$ is smaller than its dimension. If the profiles in the region under the investigation are placed in a such way that the rows of the matrix Ψ become a linear dependent or 'almost' linear dependent, then it leads to the singular matrix and the solution with the use of considered above formulas can not be obtained. In this case the singular value decomposition of the structural matrix can be used, that allows us to find a satisfactory solution in the presence of the roundoff error and linear dependence of the basis functions. In Sec. 6.12 it was shown, that an arbitrary real matrix Ψ can be represented as

$$\Psi = T\Sigma Y^T, \quad (8.79)$$

where $T[K \times K]$, $Y[S \times S]$ are the orthogonal matrices; $\Sigma[K \times S]$ is a diagonal matrix

$$\sigma_{ij} = \begin{cases} \sigma_t, & \text{for } i = j, \\ 0, & \text{for } i \neq j. \end{cases}$$

The least squares estimate, with the use of the expansion (8.79), is written as

$$\hat{\boldsymbol{b}} = \Sigma^{+}\boldsymbol{Z}, \quad \text{where} \quad \boldsymbol{b} = Y^{+}\boldsymbol{\rho}, \quad \boldsymbol{Z} = T^{T}\boldsymbol{u}.$$

If the diagonal elements σ_i of the matrix Σ do not equal to zero, then the estimate $\hat{\boldsymbol{b}}$ is $\hat{b}_i = Z_i/\sigma_i$. But if some σ_i are small, then such procedure is undesirable. The values σ_i should be analysed before obtaining the estimate $\hat{\boldsymbol{b}}$. It is proved, that the singular numbers non equal zero if and only if the columns of the matrix Ψ are linear independent. Study of the singular numbers should implement taking into account an accuracy of the input data and the calculation accuracy. Taking into account a priori accuracy we choose the threshold condition σ_{t} for the comparison of each σ_i (see Sec. 6.12): if $\sigma_i \geq \alpha$, then $\hat{b}_i = Z_i/\sigma_i$; if $\sigma_i < \alpha$, then $\hat{b}_i = 0$. In general, at $\sigma_i < \alpha$ the vector component b_i can set an arbitrary value. Such arbitrariness is a reason of the non-uniqueness of the least squares solution. To exclude an influence of the non-stable components of the vector b_i, in the case of $\sigma_i < a$, usually, these components put to zero.

Under linear constraints (8.75) in addition to the expansion (8.79) the expansion

$$a = GWF^{T},$$

can be used, where

$$w_{ij} = \begin{cases} 0, & i \neq j, \\ w_{ii} > 0, & i = j,\ i \leq Q', \\ w_{ii} = 0, & i = j,\ i > Q', \quad Q' \leq Q; \end{cases}$$

G and F are the orthogonal matrices; Q' is a number of linearly independent conditions.

Let us consider the results of the model example of a determination of the velocity taking into account the borehole data. At Fig. 8.9 the synthetic region of observations is represented. The velocities at the profiles are calculated depending on the thickness and the depth of the horizon, besides the normal noise is simulated and the velocity gradient in x direction is introduced. At Fig. 8.10 the results of processing with the use of four profiles are represented. The approximating velocity surface is computed using the borehole data (two variants of the velocity dependence) coincides with the velocity values in the boreholes and in other points has the properties peculiar to the least squares method.

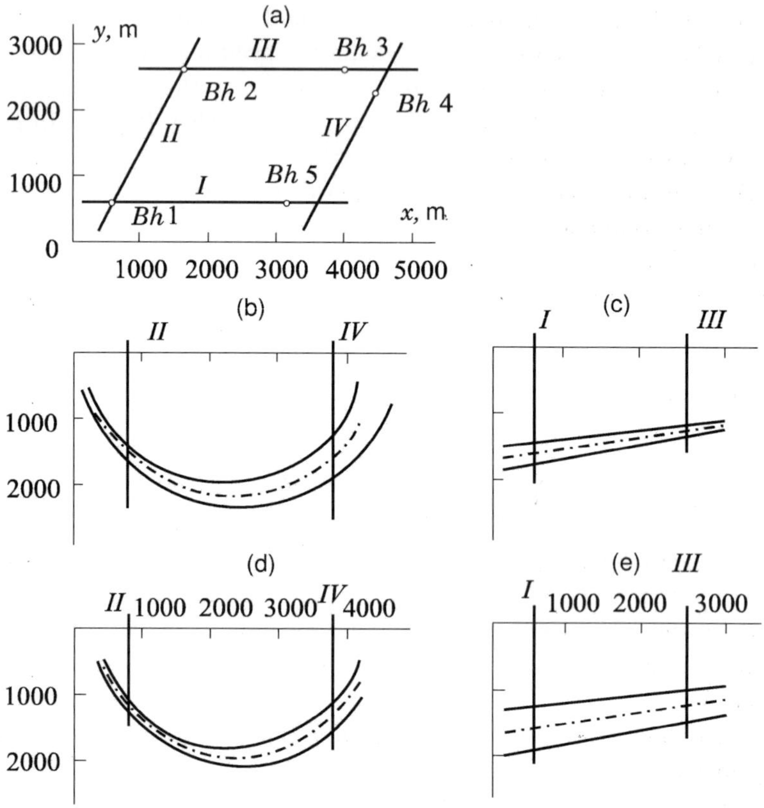

Fig. 8.9 Synthetic seismic observations (a) with location of the profiles (I – IV), boreholes and depth profiles: (b) is profile I, (c) is profile II, (d) is profile III, e is profile IV. Dash-dot lines is a model location of the horizon, solid lines are the border of deviation area of the horizon.

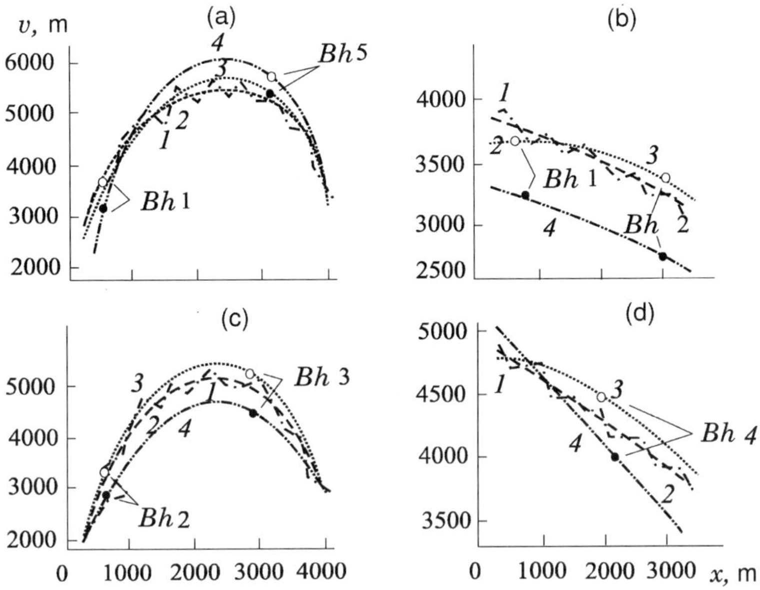

Fig. 8.10 The results of processing of the model example from Fig. 8.9.
Lines *1* are the initial velocity data along the profiles; *2* is the results of the restoration of the velocity by the seismic data without taking into account the borehole observations; *3*, *4* are the results of restoration of the velocity taking into account the borehole observations for two types of the velocity profiles in the boreholes:
(a) is the profile *I;* for borehole 1 white circle: $v = 3650$ m/s, $\Delta v = -15$ m/s, dark circle: $v = 3250$ m/s, $\Delta v = -550$ m/s; for borehole 5 white circle: $v = 5750$ m/s, $\Delta v = 470$ m/s, dark circle: $v = 5500$ m/s, $\Delta v = 220$ m/s;
(b) is profile *II;* for borehole 1 white circle: $v = 3650$ m/s, $\Delta v = -150$ m/s, dark circle: $v = 3250$ m/s, $\Delta v = -550$ m/s; for borehole 2 white circle: $v = 3375$ m/s, $\Delta v = 125$ m/s, dark circle: $v = 2750$ m/s, $\Delta v = -500$ m/s;
(c) is profile *III;* for borehole 2 white circle: $v = 3375$ m/s, $\Delta v = 120$ m/s, dark circle: $v = 2750$ m/s, $\Delta v = -50$ m/s; for borehole 3 white circle: $v = 5250$ m/s, $\Delta v = 230$ m/s, dark circle: $v = 4500$ m/s, $\Delta v = -520$ m/s;
(d) is profile *IV;* for borehole 4 white circle: $v = 4350$ m/s, $\Delta v = 190$ m/s, dark circle: $v = 3900$ m/s, $\Delta v = -260$ m/s.
Δv is a deviation between data, obtained by the seismic observations and in the boreholes.

Chapter 9

Elements of applied functional analysis for problem of estimation of the parameters of geophysical objects

A numerical solution of the estimation problem and, generally, the methods of the computational mathematics are based on the applied functional analysis. An operational language is very helpful for a compact representation of the algorithms of the inverse problem solution, to estimate of the accuracy of the inverse problem solution with taking into account the most general properties of the considered operators, to trace the modifications of some elements of the algorithm in the dependence of the concrete functional binding of the considered inverse problem. A solution of the first-kind operator equation, having the compact operator is an independent mathematical problem: the problem of the regularization of the ill-posed tasks.

The solution of any intepretating problem, connected with the processing of the experimental data, should contain the random component ("noise") as a part of the input data of the inverse problem. We should note that in this case the classical methods of the functional analysis are inadmissible, because in general case the data and noise do not belong to the image of the operator. This fact alone leads to using the methods of the mathematical statistics for the construction of the algorithms and the interpretation of the solution of the inverse problems.

The existing literature on these problems: the applied functional analysis, the solution of the ill-posed problems and the methods of mathematical statistics are very extensive and a presentation style usually is far from the "physical style", when the proof if ever produces, but without details. It is unconditionally, at such enunciating the mathematical rigor is lost, but it allows to exchange the mathematical erudition by the physical intuition, which facilitates the physical interpretation of the solution.

9.1 Elements of Applied Functional Analysis

Let us consider the basic elements of the functional analysis.

As we set the problem of the approximate solution, for example, finding of the function φ, with the definite accuracy grade satisfies the equation $L\boldsymbol{\varphi} = s$, it is appeared the problem of the definition of the measure of the accuracy, i.e. we should find the measure of the functions proximity. The measure may be only a number which corresponds to the representation of the functional space elements $\varphi \in \Phi$ on the real axis R^1 or in other words it is necessary to define a functional $N : \Phi \longrightarrow R^1$, and to determine the requirements, which must satisfy this functional as the measure of the proximity of two functions e.g. φ_1 and φ_2:

(1) a functions deviation is non-negative: $N(\varphi_1 - \varphi_2) \geq 0$, or $N(\varphi_1 - \varphi_2) = 0 \Leftrightarrow \varphi_1 = \varphi_2$;
(2) a multiplication of the both functions on a number $\lambda > 0$ leads to a change of the deviation in $|\lambda|$ times: $N(\lambda(\varphi_1 - \varphi_2)) = |\lambda| N(\varphi_1 - \varphi_2)$;
(3) by analogy with the Euclidean geometry a sum of the deviations φ_0 from φ_1 and φ_2 from φ_0 is greater than or equal to the deviation φ_1 from φ_2: $N(\varphi_1 - \varphi_0) + N(\varphi_2 - \varphi_0) \geq N((\varphi_1 - \varphi_0) + (\varphi_0 - \varphi_2))$.

It is not difficult to see that the above mentioned requirements (a, b, c) give the definition of a *norm of a function*. We rewrite the conditions a)–c), replacing $(\varphi_1 - \varphi_2) \Rightarrow \varphi$ in a) and b); $\varphi_1 - \varphi_0 \Rightarrow \psi$ and $\varphi_2 - \varphi_0 \Rightarrow \varphi$ in c) and $N(\varphi) \Rightarrow \|\varphi\|$, in the following form:

(1) $\|\varphi\| \geq 0, \quad \|\varphi\| \Leftrightarrow \varphi = 0$ is a non-negativity of a norm;
(2) $\|\lambda\varphi\| = |\lambda| \, \|\varphi\|$ is a homogeneity of a norm;
(3) $\|\varphi + \psi\| \leq \|\varphi\| + \|\psi\|$ is the triangle inequality.

Let's mark, that by the genesis of the term norm is obliged to the property of the homogeneity, which one allows at any given function φ by multiplying it on the factor λ to make the magnitude of $\|\lambda\varphi\|$ equal to unity (to normalize function).

The conditions, defined the norm, are satisfied for the following presentations:

(1) $\|\varphi\| = \int\limits_a^b |\varphi(x)| dx$. This norm defines the space of modulus integrable functions L_1 on $[a, b]$.
(2) $\|\varphi\| = \max\limits_{x \in [a,b]} |\varphi(x)|$. The norm defines the space C on $[a, b]$ of continuous functions.
(3) $\|\varphi\| = \left[\int\limits_a^b |\varphi(x)|^2 dx \right]^{1/2}$. The norm defines the space L_2 on $[a, b]$ of square integrable functions.
(4) $\|\varphi\| = \left[\int\limits_a^b |\varphi(x)|^2 \, dx + \int\limits_a^b |\varphi'(x)|^2 \, dx \right]^{1/2}$. The norm in the Sobolev's space W_2^1 on $[a, b]$.
(5) $\|\varphi\| = \left[\sum\limits_{i=1}^n \varphi_i^2 \right]^{1/2}$. The norm in the space R^n.

It should be noted that the norm notion coincides with the notion of a distance metric $\rho(\varphi_1, \varphi_2) = \|\varphi_1 - \varphi_2\|$.

A metric space (i.e. the space with the norm) is called *a complete* space, if any fundamental sequence of the elements of this space converges to an element of the space. We should remind that the *fundamental sequence* is such that for an arbitrary $\varepsilon > 0$ is found a number $n(\varepsilon)$, that for any elements of the sequence $n', n'' > 0$ are satisfied the inequality $\rho(\varphi_{n'}, \varphi_{n''}) < \varepsilon$. The complete normalized space is called *Banach space.*

By analogy with the vector analysis it is introduced a scalar product for elements of the functional space with the next properties:

(1) $S(\varphi, \varphi) \geq 0$, $S(\varphi, \varphi) = 0 \Leftrightarrow \varphi = 0$ is a non-negativity of the vector modulus;
(2) $S(\varphi, \alpha\psi_1 + \beta\psi_2) = \alpha S(\varphi, \psi_1) + \beta S(\varphi, \psi_2)$ is a linearity on the second element;
(3) $S(\varphi, \psi) = S^*(\psi, \varphi)$, if φ, ψ are complex valued.

Hence, it follows

$$S(\lambda\varphi, \psi) = \lambda^* S(\varphi, \psi).$$

We shall use a different designation for the scalar product:

$$S(\boldsymbol{\varphi}, \boldsymbol{\psi}) \triangleq (\varphi, \psi),$$

$$S(\boldsymbol{\varphi}, \boldsymbol{\psi}) \triangleq \langle\varphi|\psi\rangle$$

(the designation of Dirac: $\langle\boldsymbol{\varphi}|$ is a bra vector, $|\boldsymbol{\psi}\rangle$ is a ket vector ("*bra* + *ket* = *bracket*": $\langle|\rangle$). An example of a scalar product is

$$S(\varphi, \psi) = \int \varphi^*(x)\psi(x)dx.$$

The Banach space with a scalar product is called *the Hilbert space* H. Let us show that the construction of a scalar product permits to introduce a concordant norm. Actually, $S(\varphi, \varphi) \geq 0$, and in this sense the first property of a scalar product has analogy with the property of the norm:

$$S(\lambda\varphi, \lambda\varphi) = \lambda^*\lambda S(\varphi, \varphi) = |\lambda|^2 S(\varphi, \varphi),$$

i.e. in the contrast to the functional N, which is a homogeneous function of the first order $N(\lambda\varphi) = |\lambda| N(\varphi)$, the scalar product is a homogeneous function of the second order.

Let us show that a norm can be introduced as:

$$N(\varphi) = (S(\varphi, \varphi))^{1/2},$$

i.e. we will show that such functional satisfies the triangle inequality, and, hence, it satisfies for all three properties, which have been introduced for the norm. We write *the Shwarz's inequality*:

$$|(\psi, \varphi)| \leq (\varphi, \varphi)^{1/2} (\psi, \psi)^{1/2},$$

$$0 \leq (\varphi + \alpha\psi, \varphi + \alpha\psi) = (\varphi, \varphi) + \alpha(\varphi, \psi) + \alpha^*[(\psi, \varphi) + \alpha(\psi, \psi)].$$

So far as a scalar product is grater than or equal zero for any α, we choose α such as the term in square brackets was equal zero, i.e. $\alpha = -(\psi, \varphi) / (\psi, \psi)$ we obtain the Schwarz's inequality. We write down a triangle inequality $||\varphi + \psi|| \le ||\varphi|| + +||\psi||$, using a scalar product

$$(\varphi + \psi, \varphi + \psi)^{1/2} \le (\varphi, \varphi)^{1/2} + (\psi, \psi)^{1/2}.$$

It is really,

$$(\varphi + \psi, \varphi + \psi) = (\varphi, \varphi + \psi) + (\psi, \varphi + \psi) \le$$
$$\le (\varphi, \varphi)^{1/2}(\varphi + \psi, \varphi + \psi)^{1/2} + (\psi, \psi)^{1/2}(\varphi + \psi, \varphi + \psi)^{1/2},$$

where the last inequality is the consequence of the Schwarz inequality.

We have shown that the space with the scalar product is defined by the norm (the norm in the Hilbert space)

$$\|\varphi\| = (\varphi, \varphi)^{1/2}.$$

For instance, if the scalar product is defined as $\int \varphi^*(x)\psi(x)\,dx$, so the norm of that space reads as

$$\|\varphi\| = \left[\int \varphi^*(x)\varphi(x)dx\right]^{1/2},$$

and the corresponding Schwarz's inequality

$$\left|\int \psi^*(x)\varphi(x)dx\right| \le \left[\int \varphi^*(x)\varphi(x)dx\right]^{1/2} \left[\int \psi^*(x)\psi(x)dx\right]^{1/2}$$

takes the name *the Cauchy–Bunyakovskii inequality.* We have simultaneously shown that the functional $\left[\int |\varphi(x)|^2\,dx\right]^{1/2}$ satisfies the triangle inequality

$$\left[\int (\psi(x) + \varphi(x))^* \, (\psi(x) + \varphi(x)\,)dx\right]^{1/2}$$
$$\le \left[\int |\psi(x)|^2 dx\right]^{1/2} + \left[\int |\varphi(x)|^2 dx\right]^{1/2}.$$

The scalar product is

$$(\varphi, \psi) = \sum_{i=1}^{n} \varphi_i \, \psi_i = \boldsymbol{\varphi}^T \boldsymbol{\psi}$$

in R^n, and in the Sobolev space W_2^1:

$$(\boldsymbol{\varphi}, \boldsymbol{\psi}) = \int \varphi^*(x)\psi(x)dx + \int \varphi'^*(x)\psi'(x)dx.$$

The orthogonality condition is $(\varphi, \psi) = 0$.

We would remind the definition of the linear independence of elements $\{\varphi_i\}$, $i = 1 \div n$. The elements $\{\varphi_i\}$ are linearly independent, if from the equality $\sum_i \alpha_i \, \varphi_i = 0$ follows that $\alpha_i = 0$ for all $i = 1 \div n$.

The *basis of the Hilbert space* H is called the set $\Phi \subseteq H$ that for an arbitrary element of H there is a convergent sequence, which is a linear combination of the elements $\varphi \subseteq \Phi$, in other words, the set Φ is the basis H, if its elements are linearly independent and linear closure of Φ coincides with H. A basis is called orthonormal, if all elements of the basis are mutually orthogonal and its norm is equal to 1. The dimension of the basis is the *dimension of the Hilbert space.*

Let us show that in any basis the presentation $\varphi = \sum_i \alpha_i \varphi_i$ is unique. We assume that φ has an another presentation $\varphi = \sum_i \beta_i \varphi_i$, then $0 = \varphi - \varphi = \sum_i (\alpha_i - \beta_i)\varphi_i$. But from a condition of a linear independence of the basis it follows, that $\alpha_i = \beta_i$ for any i.

The analogue of the vector presentation in the co-ordinate system given by the orthonormal basis e_i, $\boldsymbol{\varphi} = \sum_i (\boldsymbol{\varphi}, e_i)e_i$, in the Hilbert space is determined by generalized Fourier series

$$\varphi(x) = \sum_i \alpha_i \varphi_i(x),$$

where $\varphi_i(x)$ are the basis functions; α_i are the coefficients of the expansion, which are equal to (φ, φ_i). It is a familiar example of a countable basis in the infinite dimensional Hilbert space $L_2[-\pi, \pi]$, the basis of an trigonometrical function: $(2\pi)^{-1/2}$, $(\pi)^{-1/2}\sin x$, $(\pi)^{-1/2}\cos x$, $(\pi)^{-1/2}\sin(2x)$, $(\pi)^{-1/2}\cos(2x)$, $\ldots, (\pi)^{-1/2}\sin(nx)$, $(\pi)^{-1/2}\cos(nx), \ldots$ The expansion of a function φ into Fourier series in terms of trigonometrical functions:

$$\varphi(x) = \frac{1}{\sqrt{2\pi}} \int_{-\pi}^{\pi} \varphi(x)dx + \frac{1}{\sqrt{\pi}} \sum_{k=1}^{\infty} \left[\cos(kx) \int_{-\pi}^{\pi} \varphi(x)\cos(kx)dx \right.$$
$$\left. + \sin(kx) \int_{-\pi}^{\pi} \varphi(x)\sin(kx)dx \right].$$

The possibility of the function presentation as a coefficient set in a given basis defines the effectiveness of the introduced scalar product in a functional space.

We would remind a necessary information from the operator theory. The function L, acting from the space H_1 into the space H_2, is called an *operator*:

$$H_1 \xrightarrow{L} H_2.$$

If $H_2 = R^1$, then the operator is called a *functional*. The operator L is the *linear operator*, if the condition

$$L(\alpha\varphi + \beta\psi) = \alpha L\varphi + \beta L\psi$$

is satisfied. The example of an integral operator in the space L_2:

$$L\varphi \triangleq \int \mathcal{L}(x, x')\varphi(x')dx'.$$

If the integral kernel L is such that $\int\int \mathcal{L}^2(x,x')dxdx' < \infty$, then L is the Hilbert–Schmidt operator. The *norm of the linear operator* is the functional

$$||L|| = \sup \frac{||L\varphi||}{||\varphi||}\,.$$

Let us compare the norm of the functional with the case of the finite dimensional vector space, when the matrix $\hat{L}$ transforms the vector $\boldsymbol{\varphi}$ into the vector $\boldsymbol{\psi}$, i.e. $L\varphi \triangleq \hat{L}\boldsymbol{\varphi} = \boldsymbol{\psi}$. In this case the norm of the operator L is a maximum possible length $|\boldsymbol{\psi}|$ of the vector $\boldsymbol{\psi}$, lying on the surface of the sphere ($|\boldsymbol{\varphi}| = 1$). If the norm L is less than infinity, the operator is called *limited.* The example of *an unlimited operator* is $L = d^2/dx^2$, which acting in the space of the square integrable functions on the interval $[-\pi, \pi]$:

$$\left\|\frac{d^2}{dx^2}\right\| = \sup \frac{||(d^2/dx^2)\varphi||}{||\varphi||}.$$

Choosing as a function φ the orthonormal trigonometric basis, we see that

$$\frac{||(d^2/dx^2)\varphi_k(x)||}{||\varphi_k(x)||} = k^2 \frac{||\varphi_k(x)||}{||\varphi_k(x)||} \overset{k\to\infty}{\longrightarrow} \infty.$$

From the definition of the operator norm L it follows the inequality

$$||L\varphi|| \le ||L|| \cdot ||\varphi||.$$

The operator L is called *compact,* if this operator transforms an each weakly convergent sequence $\{\varphi_n\}$ from a range of its definition into a strong convergent sequence, i.e.

$$L : (f, \varphi_n - \varphi_0) \overset{\forall f}{\longrightarrow} 0 \Rightarrow ||L(\varphi_n - \varphi_0)|| \to 0.$$

As example of the compact operator in the space L_2 the integral Hilbert–Schmidt operator

$$L\varphi \triangleq \int \mathcal{L}(x,x')\varphi(x')dx'$$

can be considered. A specificity of the infinite dimensional Hilbert space in comparison with the finite dimensional space consists in that from the weak convergence the strong convergence does not follow. Let's compare with a vector space: if

$$\forall \mathbf{f}\, (\mathbf{f}, \boldsymbol{\varphi}_n - \boldsymbol{\varphi}_0) \Rightarrow 0,$$

then $\boldsymbol{\varphi}_n \longrightarrow \boldsymbol{\varphi}_0$. This statement is equivalent to the fact that in the infinite-dimensional space an unit operator is the noncompact operator. The sequence of elements of an infinite-dimensional basis $\{\varphi_n, n = 1 \div N\}$ converges to zero weakly:

$$f_n = (f, \varphi_n) \overset{n\to\infty}{\longrightarrow} 0,$$

so far as from the Parceval equality follows $\sum f_n^2 = ||f||^2 = C < \infty$. The sequence $\{I\varphi_n = \varphi_n\}$ does not have a limit:

$$\begin{aligned} ||I(\varphi_{n'} - \varphi_{n''})|| &= (\varphi_{n'} - \varphi_{n''}, \varphi_{n'} - \varphi_{n''})^{1/2} \\ &= [(\varphi_{n'}, \varphi_{n'}) + (\varphi_{n''}, \varphi_{n''})]^{1/2} = \sqrt{2}\,. \end{aligned}$$

It would remind some definitions from the theory of operators. The operator L^* is called *the Lagrange conjugate operator*, if $(\psi, L\varphi) = (L^*\psi, \varphi)$ holds true. The operator H is called *a self-conjugate operator*, if $H^* = H$. The self-conjugate operator H, connected the spaces of complex-valued functions is called *the Hermite operator* ($H = H^+$). Let's compare with acting of the matrix $\hat{H}$ on vectors $\boldsymbol{\varphi}, \boldsymbol{\psi}$. The matrix $\hat{H}$ is a self-conjugate matrix, if the equality $(\hat{H}\boldsymbol{\psi}, \boldsymbol{\varphi}) = (\boldsymbol{\psi}, \hat{H}\boldsymbol{\varphi})$ is valid, i.e. the matrix H is symmetric ($H_{ij} = H_{ji}$)).

The operator U is called *an orthogonal*, if $(U\varphi, U\psi) = (\varphi, \psi)$ is satisfied for real φ and ψ. The equivalent determination is an operator equality

$$U^*U = I.$$

If φ and ψ are complex-valued, then the operator U, which satisfies the condition $U^+U = I$, is called *an unitary operator*. (Let's compare: $(\hat{U}\boldsymbol{\varphi}, \hat{U}\boldsymbol{\psi}) == (\boldsymbol{\varphi}, \boldsymbol{\psi})$, $\hat{U}$ is a rotation matrix, preserved the angles between vectors: $\cos(\widehat{\boldsymbol{\varphi}, \boldsymbol{\psi}}) = (\boldsymbol{\varphi}, \boldsymbol{\psi})(|\boldsymbol{\varphi}|\cdot|\boldsymbol{\psi}|)^{-1}$ and the vector lengths $|\boldsymbol{\varphi}|$, $|\boldsymbol{\psi}|$.)

The functions φ_α are called *eigenfunctions* of the operator H, if the equality

$$H\varphi_\alpha = \lambda_\alpha \varphi_\alpha,$$

is satisfied, and λ_α are called *eigenvalues*. The analogue for vectors

$$\hat{H}\boldsymbol{\varphi}_\alpha = \lambda_\alpha \boldsymbol{\varphi}_\alpha$$

is meaning, that eigenvector keeps the direction after acting of the matrix $\hat{H}$ and changes the length in λ_α times. The eigenfunctions of the operator H are linearly independent and make a basis in the Hilbert space. This basis can be made orthonormal.

We assume here a procedure of the Gramm–Schmidt orthogonalization. Let $\{\varphi_\alpha\}$ be the set of eigenvectors, then we choose as the first vector of the basis, e.g. $e_1 = \varphi_1||\varphi_1||^{-1}$. The second one is determined by $e_2 = \Delta\varphi_2||\Delta\varphi_2||^{-1}$, where $\Delta\varphi_2 = \varphi_2 - (\varphi_2, e_1)e_1$, and $(\Delta\varphi_2, e_1) = 0$. By analogy $e_3 = \Delta\varphi_3||\Delta\varphi_3||^{-1}$, where $\Delta\varphi_3 = \varphi_3 - (\varphi_3, e_2) - (\varphi_3, e_1)e_1$ and $(\Delta\varphi_3, e_2) = 0$, $(\Delta\varphi_3, e_1) = 0$. We get for an arbitrary n

$$e_n = \Delta\varphi_n||\Delta\varphi_n||^{-1},$$

where $\Delta\varphi_n = \varphi_n - \sum\limits_{\alpha=1}^{n-1}(\varphi_n, e_\alpha)e_\alpha$ and $(\Delta\varphi_n, e_\alpha) = 0$ for $\alpha = 1 \div n-1$.

We consider the differential in the Hilbert space. The linear bounded functional $p' \triangleq (\delta/\delta\varphi)p$ is called *the Frechet derivative* or *a functional gradient* $p(\varphi)$, if an equality

$$p(\varphi_0 + \delta\varphi) - p(\varphi_0) = p'\big|_{\varphi_0}(\delta\varphi) + r(\varphi_0, \delta\varphi),$$

$$\text{where } r(\varphi_0, \delta\varphi): \lim_{||\delta\varphi||\to 0} \frac{r(\varphi_0, \delta\varphi)}{||\delta\varphi||} = 0$$

is satisfied. On the Riesz theorem a linear bounded functional $l(\varphi)$ in the Hilbert space H can be presented uniquely in the form $\hat{l}(\varphi) = (l, \varphi)$. At that the equality $||\hat{l}|| = ||l||_H$ is valid. In other words, the conjugate space with the elements $\{\hat{l}\}$ can be identify with the Hilbert space. The expression, that is presented in the definition of the Frechet derivative, can be written as

$$p(\varphi_0 + \delta\varphi) = p(\varphi_0) + \left(\frac{\delta}{\delta\varphi}\Big|_{\varphi_0} p, \delta\varphi\right) + r(\varphi_0, \delta\varphi).$$

The analogue of the Frechet derivative in a finite-dimensional space for a function f of arguments $(x_1, \ldots, x_n) \triangleq x$, can be presented as

$$f(x_0 + \Delta x) = f(x_0) + (\nabla\big|_{x_0} f, \Delta x) + r(x_0, \Delta x).$$

Any measurement is a set of numbers or set of functionals $\{l_n\}$, such that the measurement data u_n, defined by a response of the device on the variation of the state φ of the system under investigation:

$$u_n = l_n(\varphi) + \varepsilon_n \quad (n = 1 \div N,$$

where ε_n is a random noise. Taking into account that the response of the device is known and using the Frechet derivative definition, the model may be written as

$$U_n = l_n(\varphi_0) + l_n'(\delta\varphi) + r_n(\varphi_0, \delta\varphi) + \varepsilon_n \triangleq u_n^0 + L_n\delta\varphi + \tilde{\varepsilon_n},$$

where $\tilde{\varepsilon_n} = r_n(\varphi_0, \delta\varphi) + \varepsilon_n$, L_n is n-th component of the linear operator L, presented N linear functionals l_n', and effected on $\delta\varphi \in \Phi$, i.e. $L : \Phi \longrightarrow R^N$. Therefore we discuss further the operator equation of the first order.

We consider the self-conjugate operator H, effected on the space of real functions, and we introduce the *Rayleigh functional*

$$\lambda(\varphi) = \frac{(\varphi, H\varphi)}{(\varphi, \varphi)}.$$

The extreme points of this functional are determined by condition $(\delta\lambda(\varphi) = 0)$ or equalling to zero of the gradient of the functional $(\lambda' = 0)$. The last equation takes the name of *the Euler equation.* Let's write the variation of the functional $\lambda(\varphi)$, taking into account that $\lambda(\varphi) = F(\varphi)/\Phi(\varphi)$

$$\delta\frac{F(\varphi)}{\Phi(\varphi)} = \frac{\delta F}{\Phi} - \frac{F\delta\Phi}{\Phi^2} = \frac{\delta F}{\Phi} - \lambda(\varphi)\frac{\delta\Phi}{\Phi},$$

i.e. the equation $\delta\lambda(\varphi) = 0$ can be represented as $\delta F - \lambda(\varphi)\delta\Phi = 0$. For the concrete form of the Rayleigh functional we have

$$(\delta\varphi, H\varphi) - \lambda(\varphi)(\delta\varphi, \varphi) = 0, \tag{9.1}$$

so

$$\begin{aligned}\Delta(\varphi, H\varphi) &= (\varphi + \Delta\varphi, H(\varphi + \Delta\varphi)) - (\varphi, H\varphi) \\ &= (\delta\varphi, H\varphi) + (\varphi, H\Delta\varphi) + (\Delta\varphi, H\Delta\varphi) = 2(\Delta\varphi, H\varphi) + O(||\Delta\varphi||^2),\end{aligned}$$

where in the last equation we used the self-conjugacy of the operator $H((\Delta\varphi, H\varphi) = (H\Delta\varphi, \varphi))$ and the reality of the function φ $((H\Delta\varphi, \varphi) = (\varphi, H\Delta\varphi))$, and by analogy

$$\Delta(\varphi, I\varphi) = 2(\Delta\varphi, \varphi) + O(||\Delta\varphi||^2).$$

The equation (9.1) can be rewritten, using a linearity of the scalar product, as

$$(\delta\varphi, H\varphi - \lambda\varphi) = 0,$$

i.e. the Euler equation for the Rayleigh functional, which determines a stationary point of the functional, reads as

$$H\varphi = \lambda\varphi,$$

and it coincides with the equation on the eigenvalues of the operator. Note, that we determined the gradient of the square functional $(\varphi, H\varphi)$ as

$$\delta\lambda = \delta(\varphi, H\varphi) = 2(H\varphi, \delta\varphi),$$

and gradient λ' as

$$\lambda' = \frac{\delta}{\delta\varphi}(\varphi, H\varphi) = 2H\varphi.$$

Remind, that the self-conjugate operator $H = H^+$ has the real eigenvalues $\lambda_\alpha^* = \lambda_\alpha$:

$$\lambda_\alpha(\varphi_\alpha, \varphi_\alpha)(\varphi_\alpha, H\varphi_\alpha) = (H\varphi_\alpha, \varphi_\alpha)^* = \lambda_\alpha^*(\varphi_\alpha, \varphi_\alpha).$$

At that, the eigenfunctions $\{\varphi_\alpha\}$ form orthonormal basis

$$0 = (\varphi_\alpha, H\varphi_\beta) - (H\varphi_\alpha, \varphi_\beta) = (\lambda_\alpha - \lambda_\beta)(\varphi_\alpha, \varphi_\beta) \Rightarrow (\varphi_\alpha, \varphi_\beta) = 0$$

by the condition $\lambda_\alpha - \lambda_\beta \neq 0$. From the stationarity of the Rayleigh function in the points of eigenfunctions it follows a condition

$$\sup \frac{(\varphi, H\varphi)}{(\varphi, \varphi)} = \lambda_{\max},$$

where $\lambda_{\max}$ is the maximum eigenvalue of the operator H. By analogy

$$\inf \frac{(\varphi, H\varphi)}{(\varphi, \varphi)} = \lambda_{\min},$$

where $\lambda_{\min}$ is a minimum eigenvalue of the operator H, i.e. for the operator $H = H^*$ the inequality

$$\lambda_{\min}(\varphi, \varphi) \leq (\varphi, H\varphi) \leq \lambda_{\max}(\varphi, \varphi)$$

is satisfied.

Defining *a non-negative operator* H by a condition $(\varphi, H\varphi) \geq 0$ for any arbitrary φ, we see that it is equivalent to the assertion $\lambda_\alpha \geq 0$. The operator H is *a positive operator* $(H > 0)$, if the inequality $(\varphi, H\varphi) > 0$ is satisfied. The operator H is *a positive defined operator*, if the condition $(\varphi, H\varphi) \geq C(\varphi, \varphi)$ is satisfied.

We bring examples of self-conjugate operators.

1. For any operator L the operator L^*L is self-conjugate, i.e. $L^*L = (L^*L)^* \geq 0$, as

$$(\psi, L^*L\varphi) = (L\psi, L\varphi) = (L^*L\psi, \varphi) = ((L^*L)^*\psi, \varphi) \geq 0.$$

2. Consider now the operator

$$L = -\frac{\partial}{\partial x}\left(q(x)\frac{\partial}{\partial x}\right)$$

in the space L_2 on $[a, b]$ at $q(x) > 0$. We shall seek for conditions for the realization of the equality $L = L^*$. We write down the scalar product and integrating it by parts:

$$\begin{aligned}
(\psi, L\varphi) &= -\int_a^b \psi(x)\frac{\partial}{\partial x}\left[q(x)\frac{\partial}{\partial x}\varphi(x)\right]dx \\
&= -\psi(x)q(x)\frac{\partial}{\partial x}\varphi(x)\Big|_a^b + \int_a^b \frac{\partial}{\partial x}\psi\left[q(x)\frac{\partial}{\partial x}\varphi(x)\right]dx \\
&= -\psi(x)q\frac{\partial}{\partial x}\varphi(x)\Big|_a^b + \varphi(x)q(x)\frac{\partial}{\partial x}\psi(x)\Big|_a^b \\
&\quad - \int_a^b \frac{\partial}{\partial x}\left[q(x)\frac{\partial}{\partial x}\psi(x)\right]\varphi(x)dx = (L\psi, \varphi) \\
&\quad + \left\{q(x)\left[\varphi(x)\frac{\partial}{\partial x}\psi(x) - \psi(x)\frac{\partial}{\partial x}\varphi(x)\right]\right\}\Big|_a^b .
\end{aligned}$$

The operator L in L_2 on $[a, b]$ is a self-conjugate operator, if the class of functions, on which it acts, is such that the following equality

$$\left\{q(x)\left[\varphi(x)\frac{\partial}{\partial x}\psi(x) - \psi(x)\frac{\partial}{\partial x}\varphi(x)\right]\right\}\Big|_a^b = 0$$

is satisfied. Using the presentation for $(\psi, L\varphi)$ after the first integration by parts we get

$$(\varphi, L\varphi) = -\varphi(x)q(x)\frac{\partial}{\partial x}\varphi(x)\Big|_a^b + \int_a^b \left|\frac{\partial}{\partial x}\varphi\right|^2 q(x)dx.$$

We can see that on the function class

$$\left\{\varphi : \varphi(x)q(x)\frac{\partial}{\partial x}\varphi(x)\Big|_a^b = 0\right\},$$

for example $\varphi\big|_a = 0$, $\frac{\partial}{\partial x}\varphi\big|_b = 0$, if the operator L is non-negative, then

$$-\frac{\partial}{\partial x}\left(q(x)\frac{\partial}{\partial x}\right) \geq 0.$$

3. We shall bring the conditions, under which the operator $L = -\Delta$ is a self-conjugate operator. Using the Gauss theorem

$$\int\limits_V \boldsymbol{\nabla}\cdot\mathbf{F}(\mathbf{x})dV = \int\limits_{\partial V}(\mathbf{F}, d\boldsymbol{\sigma})$$

and choosing as $\mathbf{F}(\mathbf{x})$ the product $\psi\boldsymbol{\nabla}\varphi$, we obtain

$$\int\limits_V \Big[(\boldsymbol{\nabla}\psi, \boldsymbol{\nabla}\varphi)\Big] + \psi\Delta\varphi dV = \int\limits_{\partial V}(d\boldsymbol{\sigma}, \psi\boldsymbol{\nabla}\varphi)$$

$$= \int\limits_{\partial V}(\psi d\boldsymbol{\sigma}\cdot\boldsymbol{\nabla})\varphi = \int\limits_{\partial V}\psi\frac{\partial\varphi}{\partial n}d\boldsymbol{\sigma}$$

the first Green's theorem. Subtraction from the right and left hand sides the analogous expressions in which the replacement by $\psi \longrightarrow \varphi$, $\varphi \longrightarrow \psi$ are produced, we obtain the presentation of the second Green's theorem

$$\int\limits_V [\psi\Delta\varphi - \varphi\Delta\psi]dV = \int\limits_{\partial V}\left[\psi\frac{\partial\varphi}{\partial n} - \varphi\frac{\partial\psi}{\partial n}\right]d\sigma.$$

Thus, if the operator $L = -\Delta$ is determined on the class of functions such that

$$\int\limits_{\partial V}\left[\psi\frac{\partial\varphi}{\partial n} - \varphi\frac{\partial\psi}{\partial n}\right]d\sigma = 0,$$

then it will be a self-conjugate operator. From the first Green's theorem we find a class of functions on which the operator $L = -\Delta$ is non-conjugate:

$$-\int\limits_V \varphi\Delta\varphi dV + \int\limits_{\partial V}\varphi\frac{\partial\varphi}{\partial n}d\sigma = \int\limits_V |\nabla\varphi|^2 dV \geq 0.$$

Thus, $(\varphi, -\Delta\varphi) \geq 0$, i.e. $-\Delta \geq 0$, if

$$\psi \in \overset{\sqcap}{\Phi} = \left\{\varphi : \int\limits_{\partial V}\varphi\frac{\partial\varphi}{\partial n}d\sigma = 0\right\}.$$

Note, that the equality $\nabla^* = -\nabla$ is satisfied on the same class of functions.

4. The integral operator with the symmetrical kernel. for example, the operator

$$L\varphi = \int\sum_j \mathcal{L}_{ij}(x, x')\varphi_j(x')dx',$$

where

$$\mathcal{L}_{ij}(x, x') = \mathcal{L}_{ji}(x', x),$$

is non-negative on L_2, if the integral kernel $\mathcal{L}_{ij}(x, x')$ is the Hilbert–Schmidt kernel.

The positive operators $H > 0$ allow to construct the Hilbert space, in which the scalar product defined as

$$(\psi, \varphi)_H \triangleq (\psi, H\varphi).$$

Accordingly the norm is

$$||\varphi||_H \triangleq (\varphi, H\varphi)^{1/2}$$

in this Friedrichs space. To examine that the form $(\psi, H\varphi)$ is a scalar product:

(1)

$$(\psi, \psi)_H = (\psi, H\psi) > 0 \quad \text{if} \quad \psi \neq 0$$

in accordance with the definition of the positive operator H;

(2)

$$(\psi, \alpha_1\varphi_1 + \alpha_2\varphi_2)_H = (\psi, H(\alpha_1\varphi_1 + \alpha_2\varphi_2)) = \alpha_1(\psi, \varphi_1)_H + \alpha_2(\psi, \varphi_2)_H$$

from behind the linearity of the operator H;

(3)

$$(\psi, \varphi)_H = (\psi, H\varphi) = (H\varphi, \psi)^* = (\varphi, H\psi)^* = (\varphi, \psi)^*_H,$$

where it is used that the operator H is the Hermitian operator ($H = H^+$).

The particular case of the Friedrichs space is the Sobolev space W_1^2, with a norm

$$||\varphi||_{W_2^1} = ||\varphi||_H,$$

where $H = I - \Delta > 0$, i.e.

$$\begin{aligned} ||\varphi||_{W_2^1} &= (\varphi, (I - \Delta)\varphi)^{1/2} \\ &= [(\varphi, \varphi)_{L_2} + (\nabla\varphi, \nabla\varphi)_{L_2}]^{1/2} = \left[\int \varphi^2 dx + \int |\nabla\varphi|^2 dx\right]^{1/2}. \end{aligned}$$

The positivity of the operator $(I - \Delta)$ is a consequence of the general expression $\sum_i H_i > 0$, if at least on the value $H_i > 0$ is positive, and other are non-negative: $(\varphi, \sum_i H_i\varphi) = \sum_i (\varphi, H_i\,\varphi) > 0$.

We consider the problems, arisen at the solution of the operator equation of the first kind

$$L\varphi = s, \tag{9.2}$$

where the operator L is a linear compact operator. We remind that the equation of the second kind has the form $\varphi = L\varphi + s$. The matrix presentation of the operator equation can be obtained using the countable basis of the Hilbert space $\{\psi_\alpha\}$. We expand the unit operator on projection operators:

$$I = \sum_\alpha P_\alpha = \sum_\alpha |\psi_\alpha\rangle\langle\psi_\alpha|,$$

$$I\varphi = \varphi = \sum_\alpha |\psi_\alpha\rangle\langle\psi_\alpha|\varphi\rangle = \sum_\alpha \varphi_\alpha|\psi_\alpha\rangle,$$

where $\langle\psi_\beta|\psi_\alpha\rangle = \delta_{\beta\alpha}$. One can rewrite the equation (9.2) as

$$LI\varphi = L\sum_\alpha |\psi_\alpha\rangle\langle\psi_\alpha|\varphi\rangle = \sum_\alpha L|\psi_\alpha\rangle\varphi_\alpha = s.$$

Projecting the both parts of the last equality on ψ_β, we obtain

$$\sum_\alpha \langle\psi_\beta|L|\psi_\alpha\rangle\varphi_\alpha = \langle\psi_\beta|s = s_\beta.$$

Introducing the designation $\langle\psi_\beta|L|\psi_\alpha\rangle = L_{\beta\alpha}$, we write a matrix expression of the operator equation: $\sum\limits_\alpha L_{\beta\alpha}\varphi_\alpha = s_\beta$. It should be noted, that an operator equation passes on a matrix one by numerical methods as the result of the discretization. Therefore we consider at first the problem of the stability of the operator L in the finite-dimensional space. Using the linearity of the operator, we obtain

$$L(\varphi + \delta\varphi) = L\varphi + L\delta\varphi = s + \delta s,$$

from which it follows

$$L\delta\varphi = \delta s. \tag{9.3}$$

Let us carry out the analysis of the errors δs on the accuracy of the restoration of the function φ, determined by the error $\delta\varphi$. Consider the relation of the relative error of the solution $||\delta\varphi||/||\varphi||$ to the relative error $||\delta s||/||s||$ in the right hand side of (9.3):

$$\Delta = \frac{||\delta\varphi||}{||\varphi||}\left(\frac{||\delta s||}{||s||}\right)^{-1} = \frac{||\delta\varphi||}{||\delta s||}\frac{||s||}{||\varphi||}.$$

We write an inequality for the upper bound Δ:

$$\sup\Delta \le \sup\frac{||\delta\varphi||}{||\delta s||}\sup\frac{||s||}{||\varphi||} = \sup\frac{||\delta\varphi||}{||L\delta\varphi||}\sup\frac{||L\varphi||}{||\varphi||}.$$

We assume that the operator L^{-1} exists and the equality $L\,L^{-1} = I$ is valid, after the substitution of variable $\delta\varphi = L^{-1}\psi$ we can write:

$$\sup\frac{||\delta\varphi||}{||L\delta\varphi||} = \sup\frac{||L^{-1}\psi||}{||\psi||} = ||L^{-1}||.$$

Then the initial inequality for $\sup\Delta$ takes the form

$$\sup\Delta \le ||L^{-1}||\,||L|| \triangleq \mathrm{cond}L,$$

where $\mathrm{cond}L$ is a *condition number* of the operator L. One can write the relation of the relative errors as

$$\frac{||\delta\varphi||}{||\varphi||} \le \mathrm{cond}L\,\frac{||\delta s||}{||s||}.$$

Taking into account that

$$||L|| = \sup\frac{||L\varphi||}{||\varphi||} = \sup\frac{(\varphi, L^*L\varphi)^{1/2}}{(\varphi,\varphi)^{1/2}} = (\lambda_{\max}^{L^*L})^{1/2},$$

$$||L^{-1}|| = \sup\frac{||L^{-1}\varphi||}{||\varphi||} = \sup\frac{||\psi||}{||L\psi||} = \left(\inf\frac{||L\psi||}{||\psi||}\right)^{-1} = (\lambda_{\min}^{L^*L})^{-1/2},$$

and the following Rayleigh relation

$$\lambda_{\min}^H(\varphi,\varphi) \le (\varphi, H\varphi) \le \lambda_{\max}^H(\varphi,\varphi),$$

for an arbitrary operator $H = H^*$ the condition number can be represented as

$$\mathrm{cond}L = \left(\lambda_{\max}^{L^*L}/\lambda_{\min}^{L^*L}\right)^{1/2}$$

for an arbitrary operator $H = H^*$. For *a symmetrical* operator L we have: $L = L^*$, $\lambda^{L^*L} = (\lambda^L)^2$, and the condition number one can write down using the eigenvectors λ of the operator L:

$$\mathrm{cond}L = \lambda_{\max}/\lambda_{\min}.$$

We consider now the influence of errors in the presentation of the operator δL on the errors of the solution $\delta\varphi$. Let the exact equation be presented as $L\varphi = s$, and the perturbed equation is $(L + \delta L)(\varphi + \delta\varphi) = s$. We assume that the operator L^{-1} exists. Taking into account the exact equation we obtain

$$L\delta\varphi + \delta L(\varphi + \delta\varphi) = 0,$$

or

$$\delta\varphi = -L^{-1}\delta L(\varphi + \delta\varphi).$$

Using these equalities one can write the inequality for the norm

$$||\delta\varphi|| \leq ||L^{-1}||\,||\delta L||\,||\varphi + \delta\varphi||. \tag{9.4}$$

Introducing to the right hand side of (9.4) the multiplier $||L||/||L||$, we get for the relative error of the field $\delta\varphi$ the expression:

$$\frac{||\delta\varphi||}{||\varphi + \delta\varphi||} \leq \mathrm{cond}L\frac{||\delta L||}{||L||}.$$

Thus the condition number is the amplification factor of the error in the case of the operator presentation. Although the condition number is the maximum amplification factor of the errors, one can show, that this upper boundary are achieved practically always. First of all we note that the unstrict inequality

$$||L^{-1}\varphi|| \leq ||L^{-1}||\,||\varphi||$$

transforms to the strict equality, if the function φ is the eigenfunction of the operator L and it corresponds to the minimum eigenvalue. We show that the condition number of the operator L is increased and the solution error is increased too by the successive increasing of the number of knots of the finite difference procedure. It should be noted, that just this fact stimulates the beginning of the active development of the methods for the solution of ill-posed problems. We use the expressions obtaining by the introduction of the Rayleigh functional. In the practical case if the self-conjugate operator $H = H^*$ we have

$$\inf\frac{(\varphi, H\varphi)}{(\varphi, \varphi)} = \lambda_{\min}.$$

Evidently, that just on that functions φ, on which the Rayleigh functional is closed to the lower boundary, the condition number is maximal. We consider the behavior of the lower boundary of the Rayleigh functional, if the operator L is the integral Hilbert–Schmidt operator, i.e. the operator with the square integrability of the integral kernel. This operator is limited (i.e. $||L|| < \infty$):

$$\frac{(L\varphi, L\varphi)}{(\varphi, \varphi)} = \frac{\int dx[\int \mathcal{L}(x, x')\varphi(x')dx']^2}{(\varphi, \varphi)} \leq \int dx \int \mathcal{L}^2(x, x')dx' = c < \infty,$$

where the last integral is limited by the definition.

We shall see the presentation of the lower boundary of the Rayleigh functional

$$\lambda_{\min}^{L^*L} \leq \frac{(L\varphi, L\varphi)}{(\varphi, \varphi)} = \frac{\int dx[\int \mathcal{L}(x, x')\varphi(x')dx']^2}{(\varphi, \varphi)}.$$

The expression in the square brackets presents the scalar product, depending on the point x as a parameter. Without a loss of the generality one can consider, that the function φ is normalized to 1 and the scalar product coincides with the projection of the scalar kernel $\mathcal{L}$ on the function φ. As it was shown, the basis of the trigonometrical functions converges weakly to zero. In this case one can consider far Fourier's components of the integrable kernel $\mathcal{L}(x, x')$ are infinitesimal values (more strictly — exponentially small values), if $\mathcal{L}$ does not have singularities on the real axis, i.e. $(\mathcal{L}, \varphi) \sim a(x)e^{-\omega x_1}$, where x_1 is a characteristic length of the function $\mathcal{L}$, i.e. the distance from the real axis to a singular point. Thus, we have obtained

$$\lambda_{\min}^{L^*L} \leq \frac{(L\varphi_n, L\varphi_n)}{(\varphi_n, \varphi_n)} \stackrel{n\to\infty}{\longrightarrow} 0,$$

that allows to make the following conclusions concerning the operator L, which satisfies only one requirement: the integral kernel must be square integrable.

(1) the lower bond of the spectrum $\lambda_{\min}^{L^*L}$ is equal to zero independently of where is zero point an eigenvalue of the operator L^*L;
(2) lower boundary of the Rayleigh functional reaches on the functions with high frequency components.

From the first condition it follows that the notion of the condition number for the case of the spaces with an infinite dimension, in the strict sense, is inapplicable even in the case when 0 is not the eigenvalue of the operator, L^*L then

$$\ker L = \{\varphi : L\varphi = 0\} = \emptyset.$$

The numerical methods are based on the finite-dimension approximation. Let us consider the behavior of the condition number under increasing of the dimension of the approximation. The lower and upper boundaries

$$\inf \frac{(L\varphi, L\varphi)}{(\varphi, \varphi)}, \qquad \sup \frac{(L\varphi, L\varphi)}{(\varphi, \varphi)},$$

which determine the condition number, are depended on the set of functions using for the creation of the Rayleigh functional. The finite difference approximation of the integral operator L can be considered as a result of the action of the operator on a set of step functions or piece linear functions. In these cases the representation of the matrix element $L_{mn} = \langle \psi_m | L | \psi_n \rangle$ of the operator L corresponds to the integration by the rectangular formula or trapezoid formula. To estimate the condition number of the operator L on the set of the piece linear functions we shall use, from the beginning, the functions $\varphi \in \overset{\sqcap}{\Phi}_n$ given by n nodes on a unit interval, then the function $\varphi \in \overset{\sqcap}{\Phi}_{2n}$ is used for more exact approximation. Then we can write the next inequalities

$$\inf_{\varphi \in \overset{\sqcap}{\Phi}_n} \frac{(L\varphi, L\varphi)}{(\varphi, \varphi)} \leq \inf_{\varphi \in \overset{\sqcap}{\Phi}_{2n}} \frac{(L\varphi, L\varphi)}{(\varphi, \varphi)},$$

$$\sup_{\varphi \in \overset{\sqcap}{\Phi}_{2n}} \frac{(L\varphi, L\varphi)}{(\varphi, \varphi)} \geq \sup_{\varphi \in \overset{\sqcap}{\Phi}} \frac{(L\varphi, L\varphi)}{(\varphi, \varphi)}.$$

These inequalities are a consequence of the embedding $\Phi_n \subset \Phi_{2n}$ (a function with n nodes is a special case of the function with $2n$ nodes). Evidently, that on a more wide set the lower boundary is decreased and upper boundary is increased. Finally, we obtain

$$\operatorname{cond}_{\varphi \in \overset{\sqcap}{\Phi}_{2n}} L \geq \operatorname{cond}_{\varphi \in \overset{\sqcap}{\Phi}_n} L. \tag{9.5}$$

In other words, more exact approximation is more unstable one. It is possible to assert a priori, that the unstable solution is oscillating one. The finite-dimension approximation can be created with the use of not only mentioned above the sets of functions (step functions or piece linear functions). The most natural basis of the problem $L\varphi = s$ is the basis of eigenfunctions of the operator L (for example, $L = A^*A$):

$$L\varphi_\alpha = \lambda_\alpha \varphi_\alpha, \qquad (\varphi_\alpha, \varphi_\beta) = \delta_{\alpha\beta}.$$

Then the Rayleigh functional can be represented as

$$\begin{aligned}
\lambda\varphi = \frac{(\varphi, L\varphi)}{(\varphi, \varphi)} &= \frac{\left(\sum\limits_\alpha (\varphi, \varphi_\alpha)\varphi_\alpha, L\sum\limits_\beta (\varphi, \varphi_\beta)\varphi_\beta\right)}{\left(\sum\limits_\alpha (\varphi, \varphi_\alpha)\varphi_\alpha, \sum\limits_\beta (\varphi, \varphi_\beta)\varphi_\beta\right)} \\
&= \frac{\sum\limits_\alpha \lambda_\alpha |(\varphi, \varphi_\alpha)|^2 \, (\varphi_\alpha, \varphi_\alpha)}{\sum\limits_\alpha |(\varphi, \varphi_a)|^2} = \lambda_1 \frac{\sum\limits_\alpha |(\varphi, \varphi_\alpha)|^2}{\sum\limits_\alpha |(\varphi, \varphi_a)|^2} \\
&+ \frac{\sum\limits_\alpha (\lambda_\alpha - \lambda_1)|(\varphi, \varphi_\alpha)|^2}{\sum\limits_\alpha |(\varphi, \varphi_a)|^2} \triangleq \lambda_1 + \Delta(\varphi),
\end{aligned}$$

where λ_1 is the minimum eigenvalue of the operator L, $\Delta(\varphi)$ is greater than zero. If we choose the next set of the function $\Phi_1 = \{\varphi : (\varphi, \varphi_1) = 0\}$, then

$$\inf_{\varphi \in \Phi_1} \lambda(\varphi) = \lambda_2,$$

if $\varphi \in \Phi_2 = \{\varphi : (\varphi, \varphi_1) = 0 \text{ and } (\varphi, \varphi_2) = 0\}$, then

$$\inf_{\varphi \in \Phi_2} \lambda(\varphi) = \lambda_3.$$

Correspondingly

$$\text{cond}_{\varphi \in \Phi} L > \text{cond}_{\varphi \in \Phi_1} L > \text{cond}_{\varphi \in \Phi_2} L > \dots,$$

if $\lambda_1 < \lambda_2 < \lambda_3 \dots$ Let's note that the obtained chain of inequalities relatively to cond$L$ can be interpreted by analogy with the case considered above, when the increasing of the instability it was connected with the set-theoretic embedding of the function spaces ((9.5): $\Phi \supset \Phi_1 \supset \Phi_2 \supset \dots$

The suppression in the solution of the projections onto eigenfunctions connected with the small eigenvalues of the operator $L = A^* A$, stabilize the solution and it is a basis of the *singular analysis*. The analogous methods are used at the numerical solution of the direct problems, when *unbounded operators* are appeared. In this case the formal criterion of the instability is behavior of cond$L$. This condition number is connected with the maximum eigenvalue appeared on the same set of the fast oscillating functions. The filtering of the high frequency component can be implemented, for example, with the use of: the low level approximation with a small number of nodes; limited set of the basis functions; the "intuitive" regularization.

Let's try to remove the reviewed above difficulties at the solution of the operator equation $L\varphi = s$, arising at the solution of the appropriate extremum problem. Let us consider the extremum problem. The finite-dimensional vector $\boldsymbol{\varphi} = \{\varphi_n, n = 1 \div N\}$ is considered as a solution of the problem $L\boldsymbol{\varphi} = s$, if the sum of the square deviations $\sum\limits_m (\sum\limits_n L_{mn}\varphi_n - s_m)^2$ is minimum one (*least squares method*), i.e.

$$\hat{\boldsymbol{\varphi}} = \arg\min_{\boldsymbol{\varphi}} (L\boldsymbol{\varphi} - s, L\boldsymbol{\varphi} - s).$$

At the limit, at $m, n \longrightarrow \infty$ the scalar product transforms to a square norm in L_2

$$\hat{\varphi} = \arg\inf ||L\varphi - s||^2_{L_2} \,. \tag{9.6}$$

The solution of this extremum problem can be obtained by the Euler equation $(\delta/\delta\varphi)\, J = 0$ for the functional

$$J(\varphi) \triangleq ||L\varphi - s||^2 = (\varphi, L^* L\varphi) - 2(L^* s, \varphi) + (s, s).$$

The gradient for the quadratic form $(\varphi, H\varphi)$ can be written as

$$\frac{\delta}{\delta\varphi}(\varphi, H\varphi) = 2H\varphi.$$

The gradient $(L^* s, \varphi)$ is equal

$$\frac{\delta}{\delta\varphi}(L^* s, \varphi) = L^* s.$$

Finally, the Euler equation for $J(\varphi)$ reads as

$$L^* L\varphi - L^* s = 0\,. \tag{9.7}$$

We obtain that the function φ, which is the solution of the extremum problem (9.6), must be simultaneously a solution of the first-order operator equation (9.7). If the operator L is compact, for example, integral operator, then the operator L^*L is compact and nonnegative. Greatest low bound of the operator L spectrum is equal to zero, and it means that the inverse operator $(L^*L)^{-1}$ is an unbounded operator or it does not exist, if 0 is eigenvalue. The application of the least squares method is possible to consider as a finite-dimension approximation in L_2 space (9.6). The instability under using the least squares method is increased together with the dimension N under the nesting condition of the finite difference approximations (9.2).

Let us note, that for physicists the assumption concerning the the continuity of the mathematical models is quite natural, i.e. small deviations of the system parameters are correspond to the small deviations of the measured data. For the case of inverse problem (and even linear inverse problem) we have an opposite situation. A small deviations in the initial data can lead to the arbitrary large oscillations of the solution. The conviction in the continuity of the mathematical models of the physical processes led to the verbiage of concept of *the well-posed problem* (Tikhonov and Arsenin, 1977).

9.2 Ill-Posed Problems

According to G. Hadamard, problem $L\varphi = s$, $\varphi \in \Phi$, $s \in S$ is called *the well-posed problem*, if the conditions are fulfilled

(1) solvability:

$$\forall\; s \in R(L)\; \exists \varphi : L\varphi = s;$$

(2) uniqueness:

$$\dim \ker(L) = 0;$$

(3) stability:

$$||L^{-1}|| = c < \infty.$$

Problems, which are not satisfied these conditions, are called *ill-posed.* As we mentioned already, all interpretating problems are ill-posed, if we do not assume $\varphi \in R^N$, where N is enough small: these problems do not satisfy at least to the condition of stability.

The problem is called *the well-posed by Tikhonov*, if it is possible to point out a such set of functions $\overset{\sqcap}{\Phi} \subset \Phi$, that function determined on these set satisfies the conditions 2) and 3). The essence of all regularization methods consists in the construction of a such set. The logical basis of the selection from the space Φ to the set $\overset{\sqcap}{\Phi}$ serves *a priori* information about the solution φ. As it was shown in Sec. 9.1, the solution of the problem

$$\hat{\varphi} = \arg \inf_{\varphi \in \Phi} ||L\varphi - s||_{L_2}$$

is ill-posed, that is connected with the compactness of the operator L^*L, i.e. with the nonexistence or unboundedness of the operator $(L^*L)^{-1}$, while the regularization is connected with the solution of the problem

$$\tilde{\varphi} = \arg \inf_{\varphi \in \overset{\sqcap}{\Phi} \subset \Phi} ||L\varphi - s||_{L_2}.$$

We consider more general problem of the minimization of one functional with constraints, which determine the set $\overset{\sqcap}{\Phi}$ through other functionals $\Delta J_i(\varphi)$, $i = 1 \div m$:

$$\tilde{\varphi} = \arg \inf_{\varphi \in \overset{\sqcap}{\Phi}} J(\varphi),$$

$$\overset{\sqcap}{\Phi} = \cap_i \varphi_i \neq \emptyset, \quad \Phi_i = \{\varphi : \Delta J_i(\varphi) \leq 0\}.$$

We show that the solution of this problem is equivalent to the search of a stationary point of the Lagrange's function $\mathcal{L}(\varphi, \boldsymbol{\lambda})$, $\boldsymbol{\lambda} \in R^m$, with $\boldsymbol{\lambda} \geq 0$. For the simplification we shall consider that all functionals incoming to the problem are differentiable. Let's $\tilde{\varphi}$ is a solution of this problem. The class $\{\delta\varphi\}$ of admissible variations is determined by conditions

$$\delta \Delta J_j(\varphi) \leq 0,$$

$$j : \Delta J_j(\tilde{\varphi}) = 0, \qquad j = 1 \div m', \ m' \leq m,$$

or

$$\left(\delta\varphi, \frac{\delta}{\delta\varphi} \Delta J_j|_{\varphi=\tilde{\varphi}} \right) \leq 0. \tag{9.8}$$

The statement, that $\tilde{\varphi}$ is a solution, is equivalent to the statement: *it does not exist* $\delta\varphi$ from the class of admissible variations, reducing to the reduction of the values $J(\varphi)$:

$$\nexists \delta\varphi : J(\tilde{\varphi} + \delta\varphi) < J(\tilde{\varphi})$$

or does not exist $\delta\varphi$ such that

$$\delta J = \left(\delta\varphi, \frac{\delta}{\delta\varphi} J\big|_{\varphi=\tilde{\varphi}}(\varphi) \right) < 0. \tag{9.9}$$

The inequality (9.9) remains valid, if we add to it the inequalities (9.8), multiplied to arbitrary positive constants λ_j, and formally include analogous expressions for constraints with the multipliers $\lambda_i = 0$:

$$\left(\delta\varphi, \frac{\delta}{\delta\varphi} J\big|_{\varphi=\tilde{\varphi}} + \sum_{i=1}^{m} \lambda_i \frac{\delta}{\delta\varphi} \Delta J_i\big|_{\varphi=\tilde{\varphi}} \right) < 0.$$

The condition of the nonexistence of the variation $\delta\varphi$ leads to the requirement

$$\frac{\delta}{\delta\varphi}J\big|_{\varphi=\tilde{\varphi}} + \sum_{i=1}^{m} \lambda_i \frac{\delta}{\delta\varphi}\Delta J_i\big|_{\varphi=\tilde{\varphi}} = 0,$$

where $H > 0$. By the Tikhonov's regularization, as a regularizing functional it is used the square of the norm in the Sobolev's space W_2^n.

As we saw previously, the extremum problem with constraints are reduced to the extremum problem with the Lagrange's function:

$$\mathcal{L}(\varphi, \alpha) = J(\varphi) + \alpha\Delta J(\varphi),$$

where α is a Lagrange's multiplier. The solution of the problem with Lagrange's function it is possible to consider as a family α-parametric problem with an indefinite parameter α, i.e. it is solved a sequence of problems in the following way: it is fixed an enough great value of α_0, at which it is certainly satisfied an initial constraint and it is found the solution relevant to the unconditional extremum problem in the space Φ. This solution corresponds to the Euler's equation for Lagrange's functional with fixed coefficients λ_i:

$$\mathcal{L}(\varphi, \lambda) = J(\varphi) + \sum_{i=1}^{m} \lambda_i \Delta J_i(\varphi).$$

The stationary point of Lagrange's functional gives the following conditions:

$$\begin{aligned} &\frac{\delta\mathcal{L}}{\delta\lambda_i}\big|_{\tilde{\lambda}_i\,\tilde{\varphi}} = 0 = \Delta J_i\,(\tilde{\varphi}), \quad i:\, \tilde{\lambda}_i > 0, \\ &\frac{\delta\mathcal{L}}{\delta\varphi}\big|_{\tilde{\varphi}\tilde{\lambda}} = 0 = \frac{\delta}{\delta\varphi}J\big|_{\varphi=\tilde{\varphi}} + \sum_{i=1}^{m} \tilde{\lambda}_i \frac{\delta}{\delta\varphi}\Delta J_i\big|_{\varphi=\tilde{\varphi}}, \end{aligned} \tag{9.10}$$

i.e. the search of the stationary point (9.10) is a problem that is equivalent to the problem (9.7), and put together the basis of a nonlinear programming and problem of an optimal control. We mention, that the stationary point of Lagrange's functional must be only a saddle point in the space $\Phi \times R_+^m$:

$$\mathcal{L}(\tilde{\varphi}, \lambda) \leq \mathcal{L}(\tilde{\varphi}, \tilde{\lambda}) \leq \mathcal{L}(\varphi, \tilde{\lambda}).$$

Therefore the search of the stationary point of the Lagrange's functional is reduced to a minimax problem

$$\mathcal{L}(\tilde{\varphi}, \tilde{\lambda}) = \min_{\varphi\in\Phi}\ \max_{\lambda\in R_+^m}\ \mathcal{L}(\varphi, \lambda)$$

or

$$\mathcal{L}(\tilde{\varphi}, \tilde{\lambda}) = \max_{\lambda\in R_+^m}\ \min_{\varphi\in\Phi}\ \mathcal{L}(\varphi, \lambda).$$

Thus we reduced the solution of the operator equation to the solution of the extremum problem. We remind some methods for the solution of the extremum (optimization) problems. We consider as a basis problem the problem of a search of the extremum point $J(\varphi)$ without constraints on φ:

$$\tilde{\varphi} = \arg \inf_{\varphi \in \Phi} J(\varphi)$$

under suggestion that the functional $J(\varphi)$ is bounded below.

The basis methods for the solution of this problem in the case of the differentiable functional are the methods which are used for the calculation of the gradient $\delta/\delta\varphi\, J(\varphi)$. Then it is possible the general structure of computational approaches to represent in the form of an explicit scheme of an iterative (relaxational) process:

$$\varphi_{n+1} = \varphi_n - H_n J'_n, \qquad H_n > 0, \quad H_n : \Phi^* \longrightarrow \Phi,$$

where $\delta/\delta\varphi|_{\varphi=\varphi_n} J$ is the gradient of the functional $J(\varphi)$; H_n determines different optimization problem.

The sense of the choice of the direction of the motion $p_n = H_n J'_n$ in the functional space Φ it is possible to illustrate on the simplest scheme of the linearization:

$$J(\varphi_{n+1}) = J(\varphi_n - H_n J'_n) \approx J(\varphi_n) - (J'_n, H_n J'_n) < J(\varphi_n),$$

where the last inequality is written with regard to the condition $J'_n \neq 0$, indicated that on the n-th step we are not found the point of the local minimum.

The simplest choice $H_n = \alpha_n I$ ($\alpha_n \in R^1_+$) corresponds to the *steepest descent method* (the motion along antigradient). One of the methods for the determination α amounts to:

$$\alpha_n = \arg \min_{\alpha \geq 0} J(\varphi_n - \alpha J').$$

The choice of the operator H_n in the form

$$H_n = \left[\frac{\delta^2}{\delta\varphi^2}\Big|_{\varphi_n} J \right]^{-1}$$

leads to the *generalized Newton's scheme*, which corresponds to the quadratic approximation of the minimizing functional $J(\varphi)$ on the functional values and its first derivatives on the n-th step:

$$J(\varphi) = J(\varphi_n) + (J', \varphi - \varphi_n) + \frac{1}{2}(\varphi - \varphi_n, J''(\varphi - \varphi_n)).$$

The function $\tilde{\varphi}$, minimizing $J(\varphi)$, is found from Euler's equation

$$J''(\varphi - \varphi_n) + J' = 0,$$

whence

$$\varphi_{n+1} = \varphi_n - [J''_n]^{-1} J'_n.$$

It should be noted that the applicability condition of the Newton's method is the existence of the inverse functional $[J'']^{-1}$, which is provided by the strict convexity of the functional $J(\varphi)$. In this connection we remind the properties of the convex functionals. A functional $J(\varphi)$ is called *a convex functional*, if for each pair φ_1 and φ_2 of the domain of the definition is realized the inequality

$$J\left((1-\lambda)\varphi_1 + \lambda\varphi_2\right) \leq (1-\lambda)J(\varphi_1) + \lambda J(\varphi_2),$$

where $\lambda \in (0,1)$ is an argument of the functional in the left hand side of the inequality, which describes an interval of a straight line $\varphi_1 + \lambda(\varphi_2 - \varphi_1)$, connected the points φ_1 and φ_2, and a value of the functional on the right hand side represents a linear interpolation of the functional values between the points φ_1 and φ_2. A linear functional is, for example, a convex functional. It is not difficult to see, that the norm is a convex functional as well:

$$||(1-\lambda)\varphi_1 + \lambda\varphi_2|| \leq ||(1-\lambda)\varphi_1|| + ||\lambda\varphi_2||.$$

We have from the triangle inequality and from the homogeneity condition

$$||(1-\lambda)\varphi_1 + \lambda\varphi_2|| \leq (1-\lambda)||\varphi_1|| + \lambda||\varphi_2||.$$

The *strictly convex functional* is called a functional satisfying to the inequality:

$$J\left((1-\lambda)\varphi_1 + \lambda\varphi_2\right) \leq (1-\lambda)J(\varphi_1) + \lambda J(\varphi_2),$$

$$\lambda \in (0,1).$$

For example, the square of the norm is a strictly convex functional:

$$\begin{aligned}&||(1-\lambda)\varphi_1 + \lambda\varphi_2||^2 \leq (1-\lambda)^2||\varphi_1||^2 + \lambda^2||\varphi_2||^2\\ &+2\lambda(1-\lambda)||\varphi_1||\,||\varphi_2|| < (1-\lambda)||\varphi_1||^2 + \lambda||\varphi_2||^2,\end{aligned}$$

and the norm $(\varphi, H\varphi) \triangleq ||\varphi||^2_H$, where $H > 0$, is a strictly convex functional as well.

The convex functionals have the following differential properties. By definition of the convexity, for an arbitrary $\delta\varphi$:

$$J(\varphi + \lambda\delta\varphi) \leq (1-\lambda)J(\varphi) + \lambda J(\varphi + \delta\varphi)$$

we obtain

$$\frac{J(\varphi + \lambda\delta\varphi) - J(\varphi)}{\lambda} \leq J(\varphi + \delta\varphi) - J(\varphi).$$

By λ tending to zero, the inequality is written in the form

$$J(\varphi + \delta\varphi) - J(\varphi) \geq \left(J'\big|_{\varphi}, \delta\varphi\right). \qquad (9.11)$$

We show that the second derivative of the strictly convex functional J'' is positive defined. We write an expansion of $J(\varphi + \delta\varphi)$ in the Taylor's series

$$J(\varphi + \delta\varphi) = J(\varphi) + (J', \delta\varphi) + \frac{1}{2}(\delta\varphi, J''\delta\varphi) + 0(||\varphi||^2)$$

and taking into account the inequality (9.11), valid for an arbitrary $\delta\varphi$, we obtain

$$J(\varphi + \delta\varphi) - J(\varphi) - (J', \delta\varphi) = \frac{1}{2}(\delta\varphi, J''\delta\varphi) + 0(||\delta\varphi||^2) \geq 0,$$

from this it follows $J'' \geq 0$. For the strictly convex functional, the inequality (9.11) passes to the strict inequality $J'' > 0$.

For example, we write a scheme for the solution of the extremum problem on the Newton's method for the functional

$$J(\theta) = ||\mathcal{P}(\theta) - u||_B^2 + \alpha||H(\theta_n - \theta_0)||^2\,, \tag{9.12}$$

where $\mathcal{P}(\theta)$ is a nonlinear operator; $H = H^* > 0$; $B = B^* > 0$. we write the first derivative of the functional in the point θ_n:

$$\left.\frac{\partial}{\partial\theta}\right|_{\theta_0} J = 2[P_n^* B(u_n - u) + \alpha H(\theta_n - \theta_0)],$$

where $P_n = \left.\frac{\partial}{\partial\theta}\right|_{\theta_0} \mathcal{P}$, $u_n = \mathcal{P}(\theta_0)$, and the second derivative:

$$\left.\frac{\partial^2}{\partial\theta^2}\right|_{\theta_n} J = 2[P_n^* BP_n + \alpha H].$$

We represent a calculation scheme for derivatives:

$$\begin{aligned}\Delta J &= J(\theta_n + \delta\theta) - J(\theta_n) = [\mathcal{P}(\theta_n + \delta\theta), B\mathcal{P}(\theta_n + \delta\theta)] \\ &+ \alpha[\theta_n + \delta\theta - \theta_0, H(\theta_n + \delta\theta - \theta_0)] - [\mathcal{P}(\theta_n), B\mathcal{P}(\theta_n)] \\ &- \alpha[\theta_n - \theta_0, H(\theta_n - \theta_0)] \approx [\mathcal{P}(\theta_n) + \mathcal{P}\delta\theta, B(\mathcal{P}(\theta) + p\delta\theta)] \\ &- [\mathcal{P}(\theta_n), B\mathcal{P}(\theta_n)] + \alpha[(\delta\theta, H(\theta_n - \theta_0)) + (\delta\theta, H\delta\theta)] \\ &\approx 2 \cdot (\delta\theta P_n^* B(u_n - u) + \alpha H(\theta_n - \theta_0)) \\ &+ [\delta\theta, (P_n^* BP_n + \alpha H)\delta\theta] \triangleq (\delta\theta, J_n') + \frac{1}{2}(\delta\theta, J_n''\delta\theta).\end{aligned}$$

We write, finally, the iteration algorithm for the solution of the nonlinear extremum problem (9.12):

$$\theta_{n+1} = \theta_n - [P_n^* BP_n + \alpha H]^{-1}\,[P_n^* B(u_n - u) + \alpha H(\theta_n - \theta_0)].$$

We represent separately the first iteration of the generalized Newton's method, when as an initial approximation it is used the value θ_0 from the equation (9.12):

$$\theta_1 = \theta_0 + [P_0^* BP_0 + \alpha H]^{-1}\,[P_0^* B(u - \mathcal{P}(\theta_0))].$$

We mention, that the input functional $J(\theta)$ is a strictly convex functional due to the strict convexity of $||H(\theta_n - \theta_0)||^2$. The Newton's method is more preferable in comparison with the gradient method due to the higher degree of the convergence.

We consider now a set $\overset{\sqcap}{\Phi}$, that it is possible to represent as quadratic on φ constraint. This is connected with the fact that Euler's equation for these problems is linear, and the algorithm may be represented formally in the operator: $\varphi = Rs$.

First of all we mention, that the manifestation of the instability was expressed in unlimited errors of the solution. Therefore it is naturally to expect that, selected as the set $\overset{\sqcap}{\Phi}$ the sphere with a finite radius in the space Φ, we get the function $\tilde{\varphi}$, which does not have certainly large errors. These concepts are an heuristic basis for *the Ivanov–Lavrentiev's regularization*:

$$\tilde{\varphi} = \arg\inf \left\{||L\varphi - s||^2_{L_2}\right\},$$
$$\tilde{\varphi} \in \overline{\Phi} = \left\{||\varphi||^2_{L_2} \leq C\right\} . \qquad (9.13)$$

The Ivanov–Lavrentiev's regularization method does not make filtering of the high-frequency components because the constraint $||\varphi||^2 \leq C$ has an integral character for the function φ. The filtering of the high-frequency components will take the place, if a constraint is imposed upon the behavior not only for an amplitude of the function but for its derivatives. These conditions are determined *the Tikhonov's regularization*:

$$\tilde{\varphi} = \arg\inf \left\{||L\varphi - s||^2_{L_2}\right\},$$

$$\tilde{\varphi} \in \overline{\Phi} = \left\{(||\varphi||^2_{L_2} + ||\nabla\varphi||^2_{L_2}) \leq C\right\},$$

or in more general form

$$\tilde{\varphi} = \arg\inf \left\{||L\varphi - s||^2_{L_2}\right\},$$

$$\tilde{\varphi} \in \overline{\Phi} = \left\{\left(\varphi, \sum_{r=0}^{n}(-1)^r \frac{\partial^r}{\partial x^r}\left[q(x)\frac{\partial^r}{\partial x^r}\right]\varphi\right) \leq C\right\}, \qquad (9.14)$$

where $q(x) > 0$. The regularization with the differential operator (9.14), represented in the last constraint is called the *regularization of n-th order* (zero, first, second, ..., n-th). Therefore constraint $(||\varphi||^2_{L_2} + ||\nabla\varphi||^2_{L_2}) \leq C$ corresponds to the first order Tikhonov's regularization ($q(x) \equiv 1$ $(\varphi, (I - \partial^2/\partial x^2)\varphi)$, and constraint $||\varphi||^2_{L_2} \leq C$ corresponds to the zero order Tikhonov's regularization.

The general scheme of the regularization can be represented as a solution of the following extremum problem:

$$\tilde{\varphi} = \arg\inf ||L\varphi - s||^2_B,$$

$$\tilde{\varphi} \in \overline{\Phi} = \left\{||\varphi||^2_H \triangleq (\varphi, H\varphi) \leq C\right\} . \qquad (9.15)$$

Then we reduce the parameter $\alpha_1 = q\alpha_0$, $q < 1$, and again we solve an unconditional extremum problem (9.15). This procedure is continued until received solution will not satisfy to the constraint

$$\tilde{\varphi}_\alpha = \arg\ \inf\ \left\{||L\varphi - s||^2_B + \alpha||\varphi||^2_H\right\},$$

$$\alpha\ :\ ||\tilde{\varphi}_\alpha||^2_H \leq C . \qquad (9.16)$$

We note, that for any value $\alpha > 0$ the extremum problem on the unconditional extremum is a problem of the convex analysis and due to the strict convexity $\alpha||\varphi||^2_H$ has an unique solution, which may be obtained by any numerical method, for example, the gradient descent method, the Newton's method with regularization, the conjugate gradient method and etc. (Marchuk, 1982). In the regularization method for ill-posed problems (9.16) α is called a *regularization parameter* and $||\varphi||_H^2$, stipulated a strict convexity of the minimizing functional is called a *stabilizing functional.*

An algorithm for the solution of an ill-posed problem is particularly simple one if the operator in the problem $L\varphi = s$ is linear and constraints are quadratic on φ. The extremum point, in this case, should correspond the Euler's linear equation, that for the problem (9.16) reads as

$$(L^*BL + \alpha H)\varphi = L^*Bs,$$

where the operator equation of the first kind is correctly solvable equation, since the operator $(L^*BL + \alpha H)^{-1}$ is bounded operator: in fact

$$(\varphi, (L^*BL + \alpha H)\varphi) = (\varphi, L^*BL\varphi) + \alpha(\varphi, H\varphi) \geq \alpha\lambda_{\min}^H(\varphi, \varphi) > 0,$$

the derivative with respect to an regularization parameter α is nonpositive:

$$\frac{\partial}{\partial\alpha}||\tilde{\varphi}_\alpha||_H^2 \leq 0,$$

because the derivative is a quadratic form with the symmetric nonpositive operator. It means, that it is possible to write $(\partial/\partial\alpha)\, R_\alpha \leq 0$ with respect to a sphere of the radius R_α, to which a solution φ_α belongs. The extremum problem

$$J(\varphi) + \alpha\Delta J(\varphi) \longrightarrow \inf$$

it is possible to interpret as an extreme problem of such kind

$$\frac{1}{2}J(\varphi) + \Delta J(\varphi) \longrightarrow \inf,$$

that corresponds to the search of the function $\tilde{\varphi}$:

$$\tilde{\varphi} = \arg\inf \Delta J(\varphi), \tag{9.17}$$

whence

$$||(L^*BL + \alpha H)^{-1}|| \leq (\alpha\lambda_{\min}^H)^{-1} = C < \infty.$$

In practice, the compact radius, which is determined by the restriction $(||\varphi||_H^2 \leq C)$, it is not known and the solution of an ill-posed problem is reduced to the following. The solution $\tilde{\varphi}$ of the ill-posed problem $L\varphi = s$ is declared the limit of the sequence $\tilde{\varphi}_\alpha$ by tending of the regularization parameter to zero:

$$\tilde{\varphi}_\alpha \overset{\alpha\to+0}{\longrightarrow} \tilde{\varphi}.$$

We shall show that the decrease of the regularization parameter corresponds to the increase of the compact radius, that is equivalent to the removal of the restriction: in fact,

$$||\tilde{\varphi}_\alpha^2||_H^2 = \left(L^*Bs, [(L^*BL + \alpha H)^{-1}\, H(L^*BL + \alpha H)^{-1}] \cdot L^*Bs\right)$$

provided that $J(\varphi) \leq C$, where the decrease of the regularization parameter α corresponds to the decrease of the square norm of the residual (Δ):

$$||\Delta||_B^2 \triangleq ||L\varphi - s||_B^2.$$

By known level of errors at the right hand side of s such approach gives the practical method for the choice of the regularization parameter (principle of the residual) (Morozov, 1993)).

We note that the regularization by the solution of operator equations is reduced to the change of the compact operator L^*BL to the strictly positive operator $(L^*BL + \alpha H)$ with $\alpha > \alpha_0 > 0$.

If it is used the Ivanov–Lavrentiev's method, then $H = I$ and by the interpretation (9.17) $\tilde{\varphi}$ is called *a quasisolution* ($\alpha \longrightarrow +0$). If the Tikhonov's method is used, then we have

$$H = \sum_{r=0}^{n} (-1)^r \frac{\partial^r}{\partial x^r}\left[q(x)\frac{\partial^r}{\partial x^r}\right],$$

that corresponds to the canonical regularization of ill-posed problems by the solution of the integral equations of the first kind, for example, a convolution equation.

9.3 Statistical Estimation in the Terms of the Functional Analysis

In the development of the regularization of ill-posed problems, the essential place is presented in the introduction of constraints either for the norm $||\varphi||_H^2 \leq C$, or for the squared of the residual $||L\varphi - s||_B^2 \leq C$. In practice, it is impossible to determine accurately the boundaries of these inequalities. We attempt to weaken hard requirements the boundary indication. We shall assume that the squared norm can take any values, but the mean value of the squared norm is equal C: $\overline{||\varphi||_H^2} = C$. To formalize the mean value notion it is necessary to give the each squared norm value the specific weight such that

$$p(\varphi): \quad (p, F(\varphi)) = C,$$

$$(p, 1) = 1,$$

$$p \geq 0.$$

In our case it will be $F(\varphi) = ||\varphi||_H^2$. We shall attempt to determine this weight introducing minimum of the arbitrariness to the construction procedure.

We shall require, that the Shannon's information functional

$$H = -(p, \ln p) \tag{9.18}$$

(see Sec. 1.9) reaches the maximum value or $(p, \ln p) \longrightarrow \min$, provided that $(P, F) = C$ taking into account of the normalization condition $(p, \mathbf{1}) = = 1$. By using Lagrange multipliers, the extremum problem is written as

$$\hat{p} = \arg\inf\left\{(p, \ln p) + \alpha(p, F) + \beta(p, 1)\right\}.$$

The corresponding Euler's equation has the form

$$\ln p + \alpha F + \beta 1 = 0,$$

from which

$$\hat{p} = p_0 e^{-\alpha F},$$

where $p_0 = e^{-\beta 1} = \text{const}$. Substituting $F(\varphi)$ into $||\varphi||_H^2$, we obtain the finally form of the weight function $p(\varphi)$, which corresponds to the minimal arbitrariness in the choice of this function

$$\hat{p}(\varphi) = p_0 e^{-\alpha ||\varphi||_H^2} = p_0 e^{-\alpha(\varphi, H\varphi)},$$

where $\alpha > 0$.

The obtained function $\hat{p}(\varphi)$ has all properties of the probability density, therefore we have the possibility to make use of the probability measure properties and to implement the statistical interpretation of all constructions including the weight function p, for example, the mathematical expectation and variance:

$$\mathrm{E}\varphi \triangleq (\varphi, p),$$

$$\mathrm{E}[(\varphi - \mathrm{E}\varphi)^2] \triangleq ((\varphi - (\varphi, p))^2, p).$$

The operation of the mathematical expectation E is linear in the explicit form:

$$\mathrm{E}(\lambda\varphi_1 + \mu\varphi_2) = \lambda \mathrm{E}(\varphi_1) + \mu \mathrm{E}(\varphi_2),$$

therefore

$$\mathrm{E}L\varphi = LE\varphi,$$

i.e. the operator of the mathematical expectation E computes with linear operators.

We shall introduce a *correlation operator (correlator)* by $\mathrm{E}\varphi = 0$:

$$K_{\varphi\varphi} : \mathrm{E}(\eta, \varphi)(\xi, \varphi)^* = (\eta, K_{\varphi\varphi}\xi) \ \forall\ \xi, \eta.$$

It follows from the definition, that the correlation operator, which can be written as $K_{\varphi\varphi} = \mathrm{E}\varphi\varphi^*$, is nonnegative:

$$(\eta, K_{\varphi\varphi}\eta) = \mathrm{E}(\eta, \varphi)(\eta, \varphi^*) = \mathrm{E}|(\eta, \varphi)|^2 \geq 0.$$

It is possible to write the correlator for the linear transformed functions $A\varphi$ and $(B\varphi)^*$:

$$\mathrm{E}(A\varphi)(B\varphi)^* = AK_{\varphi\varphi}B^*,$$

as far as

$$\begin{aligned} \mathrm{E}(\eta, A\varphi)(\xi, B\varphi)^* &= \mathrm{E}(A^*\eta, \varphi)(B^*\xi, \varphi)^* \\ &= (A^*\eta, K_{\varphi\varphi}\, B^*\xi) = (\eta, (AK_{\varphi\varphi}\, B^*)\, \xi). \end{aligned}$$

In particular, if the correlator of the function φ is equal to $K_{\varphi\varphi}$, then the correlator of the linear transform $A\varphi$ is equal to $AK_{\varphi\varphi}A^*$:

$$\mathrm{E}\psi\psi^* \triangleq \mathrm{E}(A\varphi)(A\varphi)^* = AK_{\varphi\varphi}A^*.$$

The *correlation function* $k_{\varphi\varphi}(x,x')$, which is an integral kernel of the correlation operator $K_{\varphi\varphi}$, it is possible to represent as:

$$k_{\varphi\varphi}(x,x') = \mathrm{E}\varphi(x)\varphi^*(x'),$$

since the function $\varphi(x)$ in the points x and x' are determined by projection operators (projectors) P_x and $P_{x'}$:

$$P_x(\varphi) = \int \varphi(y)\delta(x-y)dy,$$

$$P_{x'}(\varphi) = \int \varphi(y)\delta(x'-y)dy.$$

Then we write:

$$\mathrm{E}(P_x\,\varphi)(P_{x'}\,\varphi)^* = P_x\,K_{\varphi\varphi}\,P_{x'} = k_{\varphi\varphi}(x,x').$$

We introduce the Hilbert space with the scalar product:

$$(\varphi,\psi) \triangleq \mathrm{spE}\varphi\psi^* \triangleq \mathrm{sp}K_{\varphi\psi}.$$

Here the operator E is given by the measure $d\mu_{\varphi\psi} = p(\varphi,\psi)d\varphi d\psi$; $K_{\varphi\psi}$ is called *the cross-correlator* of the functions φ,ψ. We remember that the correlator $K_{\varphi\varphi}$ is an auto-correlator. We shall check that the form (φ,ψ) is a scalar product in fact:

(1) $(\varphi,\varphi) = \mathrm{sp}K_{\varphi\varphi} > 0$;
(2) $(\varphi,\psi) = \mathrm{E}\varphi\psi^*\mathrm{sp}K_{\varphi\psi} = \int k_{\varphi\psi}(x,x)\,dx =$
$= \int k^*_{\psi\varphi}(x,x)\,dx = \mathrm{sp}K^*_{\psi\varphi} = \mathrm{sp}(\mathrm{E}\psi\varphi^*)^* = (\psi,\varphi)^*$;
(3) $(\varphi,\alpha_1\psi_1+\alpha_2\psi_2) = \alpha_1(\varphi,\psi_1)+\alpha_2(\varphi,\psi_2)$.

The orthogonal functions φ and ψ are called *uncorrelated functions*, if $(\varphi,\psi) = 0$.

We note, that it is possible to introduce the Hilbert space determining the scalar product with the measure $d\mu_\varphi = p(\varphi)d\varphi$. Then the scalar product for the elements $\varphi(x)$ and $\varphi(x')$ of the Hilbert space will be given by the correlation function at the point (x,x'):

$$(\varphi|_x,\varphi|_{x'}) = k_{\varphi\varphi}(x,x').$$

Based on the scalar product properties, it is possible to write the property of the correlation function: we obtain from Schwarz's inequality

$$|(\varphi|_x,\varphi|_{x'})| \le ||\varphi|_x||^2||\varphi|_{x'}||^2,$$

i.e.

$$|k(x,x')|^2 \le \sigma^2_\varphi(x)\sigma^2_\varphi(x'), \quad \sigma^2_\varphi(x) = \mathrm{E}|\varphi(x)|^2.$$

Introducing a *correlation coefficient* r for a real function φ:

$$r = \frac{k_{\varphi\varphi}(x,x')}{\sigma_\varphi(x)\sigma_\varphi(x')},$$

we obtain the relation for the correlation coefficient: $-1 \le r \le 1$. We define random fields φ as *homogeneous*, if the following condition $k_{\varphi\varphi}(x,x') = k_{\varphi\varphi}(x-x')$

is satisfied. This condition leads, in particular, to a constancy of the variance: $\sigma_{\varphi\varphi}(x) = \text{const}$.

As mentioned above, the maximum entropy function $p(\varphi)$ is such that the function $p \sim e^{-\alpha(\varphi, H\varphi)}$ satisfies to the condition $(\varphi, H\varphi) = C$.

The function $p(\varphi)$ it is possible to interpret as the Gaussian density function, which for the $\varphi \in R^n$ has the form

$$p(\varphi) = (2\pi)^{-n/2}|K_{\varphi\varphi}|^{-1/2}\exp\Big\{ - \frac{1}{2}(\varphi, K_{\varphi\varphi}^{-1}\varphi)\Big\}.$$

We shall remember certain properties of the Gaussian integrals.

1. The Gaussian integral it is possible to represent in the following form:

$$\begin{aligned}
&\int \exp\Big[- \frac{1}{2}(\varphi, H\varphi)\Big] D\varphi \\
&= \int \exp\Big[- \frac{1}{2}(U\varphi, UHU^*U\varphi)\Big] \left|\frac{D\varphi}{DU\varphi}\right| DU\varphi \\
&= |\text{Det}\, U| \int \exp\Big[- \frac{1}{2}(\psi, \lambda\psi)\Big] D\psi \\
&= \int \exp\Big[- \frac{1}{2}\sum_{\alpha} \lambda_\alpha \psi_\alpha^2\Big] \prod_{\alpha} d\psi_\alpha.
\end{aligned} \tag{9.19}$$

Here U is an unitary matrix, its rows are eigenvectors of the matrix $H : H\varphi_\alpha = \lambda_\alpha\varphi_\alpha$, $H\varphi_\beta = \lambda_\beta\varphi_\beta$ $(\varphi_\alpha\varphi_\beta = \delta_{\alpha\beta})$;

$$U^*U = I, \quad \text{Det}U = 1, \quad UHU^* = \Lambda;$$

Λ is a diagonal matrix; $\psi = U\varphi$.

We shall remember the single-valued integral (it will be necessary for the integral (9.19) calculation):

$$I = \int \exp\Big[- \frac{1}{2}\lambda x^2\Big] dx.$$

We note that

$$I^2 = \int \exp\Big[- \frac{\lambda}{2}(x^2 + y^2)\Big] dxdy = \int\limits_0^\infty \exp\Big(- \frac{\lambda}{2}r^2\Big) 2\pi r dr$$

$$= \frac{2\pi}{\lambda}\int\limits_{-\infty}^{0} e^\rho\, d\rho = \frac{2\pi}{\lambda}, \quad I = \sqrt{\frac{2\pi}{\lambda}}.$$

From here

$$\int \exp\Big[- \frac{1}{2}(\varphi, H\varphi)\Big] d\varphi = (2\pi)^{n/2}(\text{Det}H)^{-1/2},$$

and the normalization factor $A: \int A \exp\left[-\frac{1}{2}(\varphi, H\varphi)\right] d\varphi = 1$ is equal

$$A = (2\pi)^{-n/2}(\text{Det}H)^{1/2}.$$

2. We shall find the relation between the operator H and the correlator of the function φ:

$$\mathrm{E}\varphi\varphi^* = \frac{\int \varphi\varphi^* \exp[-(\varphi, H\varphi)/2]d\varphi}{\int \exp[-(\varphi, H\varphi)/2]d\varphi} = \frac{\int \psi\psi^* \exp[-(\psi, \Sigma\psi)/2]D\psi}{\int \exp[-(\psi, \Sigma\psi)/2]d\psi}.$$

We write the matrix element $K_{\gamma\delta}$ of the correlator K:

$$K_{\gamma\delta} = \frac{\int \psi_\gamma \psi_\delta \exp[-(1/2)\sum\limits_{\alpha\beta} \Sigma_{\alpha\beta}\psi_\alpha\psi_\beta] \prod_\alpha d\psi_\alpha}{(2\pi)^{-n/2}(\mathrm{Det}\Sigma)^{-1/2}}$$
$$= -2(2\pi)^{-n/2}(\mathrm{Det}\Sigma)^{1/2}\frac{\partial}{\partial\Sigma_{\gamma\delta}}$$
$$\times \int \exp\left[-\frac{1}{2}\sum_{\alpha,\beta}\Sigma_{\alpha\beta}\psi_\alpha\psi_\beta\right]\prod_\alpha d\psi_\alpha.$$

Taking into account that

$$\frac{\partial}{\partial\Sigma_{\gamma\delta}}\int \exp[-(\psi, \Sigma\psi)/2]D\psi$$
$$= \frac{\partial}{\partial\Sigma_{\gamma\delta}}[(2\pi)^{n/2}(\mathrm{Det}\Sigma)^{-1/2}] = (2\pi)^{n/2}\left(-\frac{1}{2}\right)(\mathrm{Det}\Sigma)^{-3/2}\bar{\Sigma}_{\gamma\delta},$$

where $\bar{\Sigma}_{\gamma\delta}$ is an algebraic cofactor of the element $\Sigma_{\gamma\delta}$ of the matrix Σ, we obtain

$$K_{\gamma\delta} = \frac{\bar{\Sigma}_{\gamma\delta}}{|\mathrm{Det}\Sigma|} = [\Sigma^{-1}]_{\gamma\delta},$$

i.e. $K_{\varphi\varphi} = H^{-1}$.

The obtained relation makes possible to write the Gaussian density using the correlator $K_{\varphi\varphi}$, in the form

$$p(\varphi) = \frac{\exp[-(\varphi, K_{\varphi\varphi}^{-1}\varphi)/2]}{\int \exp[-(\varphi, K_{\varphi\varphi}^{-1}\varphi)/2]d\varphi}. \tag{9.20}$$

The distribution is called a *nondegenerate* one, if $K_{\varphi\varphi} > 0$.

We calculate the Shannon's functional for the Gaussian distribution

$$-(p, \ln p) = -\int A\exp[-(\varphi, K_{\varphi\varphi}^{-1}\varphi)/2]\left(\ln A - \frac{1}{2}(\varphi, K_{\varphi\varphi}^{-1}\varphi)\right)d\varphi$$
$$= -\ln A + \frac{1}{2}A\int(\varphi, K^{-1}\varphi)\exp[-(\varphi, K^{-1}\varphi)/2]d\varphi.$$

The second term, using change $\psi = K^{-1/2}\varphi$, it is possible to represent in the form

$$A\int(\varphi, K^{-1}\varphi)\exp[-(\varphi, K^{-1}\varphi)/2]d\varphi$$
$$= A\int(\psi, \psi)\exp[-(\psi, \psi)/2](\mathrm{Det}K)^{1/2}d\psi$$
$$= (\mathrm{Det}K)^{-1/2}(\mathrm{Det}K)^{1/2}f(n) = f(n).$$

Here the function $f(n)$ depends only on the space dimension. Finally, we write:

$$H = -(p, \ln p) = \frac{1}{2} \ln \mathrm{Det} K + f(n). \tag{9.21}$$

If the density function $p(\varphi)$ is given in the form (9.20), then $\mathrm{E}\varphi = 0$ due to the $p(\varphi)(p(\varphi) = p(-\varphi))$ evenness. For the density function

$$p(\varphi) = A \exp\left[-\frac{1}{2}(\varphi - \varphi_0, K_{\varphi\varphi}^{-1}(\varphi - \varphi_0))\right],$$

using change of the variables $\psi = \varphi - \varphi_0$, we obtain $\mathrm{E}\varphi = \varphi_0$.

We remember the probability characteristics of the description of the system of random functions φ and ψ. Let the joint density function φ and ψ is $p(\varphi, \psi)$. The projection of this density on the space Φ is called a *marginal probability density* :

$$p(\psi) = \int p(\varphi, \psi) d\varphi.$$

Respectively the projection of $p(\varphi, \psi)$ on the space ψ gives the marginal density

$$p(\varphi) = \int p(\varphi, \psi) d\psi.$$

The normalized cross-sections of the density $p(\varphi, \psi)$ determine *the conditional density* — the probability density φ by fixed ψ:

$$p(\varphi|\psi) = \frac{p(\varphi, \psi)}{p(\psi)},$$

the probability density ψ by fixed φ:

$$p(\psi|\varphi) = \frac{p(\varphi, \psi)}{p(\varphi)}.$$

The random fields φ and ψ are called *independent*, if the following conditions are fulfilled

$$p(\varphi|\psi) = p(\varphi), \qquad p(\psi|\varphi) = p(\psi).$$

As an example we consider the model $u = L\varphi + \varepsilon$, where ε and φ are described by the Gaussian distribution $\varphi \in N(0, K_\varphi)$, i.e.

$$p(\varphi) \sim A \exp\left[-\frac{1}{2}(\varphi, K_\varphi^{-1}\varphi)\right].$$

This distribution is called *a priori distribution* with respect to the desired field φ and $\varepsilon \in N(0, K_\varepsilon)$, i.e.

$$p(\varepsilon) \sim A \exp\left[-\frac{1}{2}(\varepsilon, K_\varepsilon^{-1}\varepsilon)\right].$$

We assume that φ and ε are statistically independent: $p(\varphi, \varepsilon) = p(\varphi)p(\varepsilon)$. The distributions $p(\varphi, \varepsilon)$ and $p(u)$ are Gaussian distributions, since

$$u = ||L \vdots I|| \left\| \begin{matrix} \varphi \\ \varepsilon \end{matrix} \right\|.$$

The joint probability density $p(\varphi, u)$ is determined by the correlator

$$K\begin{pmatrix} u \\ \varphi \end{pmatrix} = \mathrm{E}\begin{Vmatrix} u \\ \varphi \end{Vmatrix} ||u^* \varphi^*||$$
$$= \begin{Vmatrix} K_{uu} & K_{u\varphi} \\ K_{\varphi u} & K_{\varphi\varphi} \end{Vmatrix} = \begin{Vmatrix} L^* K_\varphi L + K_\varepsilon & LK_\varphi \\ K_\varphi L^* & K_\varphi \end{Vmatrix}.$$

It is easy to check, that the inverse operator is

$$K^{-1}\begin{pmatrix} u \\ \varphi \end{pmatrix}$$

it is possible to represent in the form

$$K^{-1}\begin{pmatrix} u \\ \varphi \end{pmatrix} = \begin{Vmatrix} K_\varepsilon^{-1} & -K_\varepsilon^{-1} L \\ -L^* K_\varepsilon^{-1} & K_{\delta\varphi}^{-1} \end{Vmatrix},$$
$$K_{\delta\varphi} = K_\varphi - K_\varphi L^* (LK_\varphi L^* + K_\varepsilon)^{-1} LK_\varphi.$$

If the inverse operator K_φ^{-1} exists, then the correlator $K_{\delta\varphi}$ can be represented in the form

$$K_{\delta\varphi} = (L^* K_\varepsilon^{-1} L + K_\varphi^{-1})^{-1},$$

using the identity

$$(A + BDB^*)^{-1} = A^{-1} - A^{-1}B(B^*A^{-1}B + D^{-1})B^*A^{-1}.$$

We write the conditional probability density φ at prescribed value of $u = L\varphi + \varepsilon$, which is called *a posteriori density*:

$$p(\varphi|u) = \frac{p(u|\varphi)}{p(u)} \sim \exp\left\{-\frac{1}{2}(\varphi - \hat{\varphi}, K_{\delta\varphi}^{-1}(\varphi - \hat{\varphi}))\right\}, \tag{9.22}$$

where

$$\hat{\varphi} = \int \varphi p(\varphi|u) d\varphi \triangleq \mathrm{E}_u \varphi = K_\varphi L^* (LK_\varphi L^* + K_\varepsilon)^{-1} u; \tag{9.23}$$

$\mathrm{E}_u\varphi$— conditional average of φ.

We represent the conditional probability density u at prescribed value of φ:

$$p(u|\varphi) = \frac{p(u, \varphi)}{p(\varphi)} \sim \exp\left\{-\frac{1}{2}(u - \hat{u}, K_\varepsilon^{-1}(u - \hat{u}))\right\}, \tag{9.24}$$

where $\hat{u} = \mathrm{E}_\varphi u = L\varphi$.

It is easily to see that the conditional probability density $p(u|\varphi)$ coincides with the distribution p_ε independently of the type of the joint distribution for the model $u = L\varphi + \varepsilon$ i.e.:

$$p(u|\varphi) = p_\varepsilon(u - L\varphi).$$

We clarify, how it was obtained the expressions (9.22) and (9.24); for that we write the quadratic form standing at the exponent of the joint distribution $p(u, \varphi)$:

$$||u\ \varphi|| K^{-1}\begin{Vmatrix} u \\ \varphi \end{Vmatrix} = (u, K_\varepsilon^{-1} u) - (u, K_\varepsilon^{-1} L\varphi) - (\varphi, L^* K_\varepsilon^{-1} u) + (\varphi, K_{\delta\varphi}^{-1}\varphi).$$

At prescribed value of u it is possible to extract the total square for the function φ:

$$(\varphi - K_{\delta\varphi}L^*K_\varepsilon^{-1}u,\ K_{\delta\varphi}^{-1}(\varphi - K_{\delta\varphi}L^*K_\varepsilon^{-1}u)),$$

and for the fixed value of φ the total square for u is written as

$$(u - K_\varepsilon K_\varepsilon^{-1}L\varphi,\ K_\varepsilon^{-1}(u - K_\varepsilon K_\varepsilon^{-1}L\varphi)).$$

It should be noted that the operator $K_{\delta\varphi}$ represents a difference of the two positive operators: $K_\varphi > 0$ and $K_\varphi L^*(LK_\varphi L^* + +K_\varepsilon)^{-1}LK_\varphi > 0$, and in this case $K_{\delta\varphi} > 0$.

We consider statistical criteria for searching of the field φ estimate for the model $u = L\varphi + \varepsilon$, using obtained expression for the conditional probability densities $p(\varphi|u)$ and $p(u|\varphi)$.

1. Let it is known only a distribution of the random component (noise) ε, i.e. it is given the density $p(\varepsilon)$. This distribution can be obtained a priori by means of the statistical analysis of a measurement error of the experiment. Then it is reasonable as a solution to take a such function $\hat{\varphi}$, that the difference $u - L\hat{\varphi}$ is maximum "similar" to the noise ε, i.e. the function $u - L\hat{\varphi} = \hat{\varepsilon}$ should be such that the probability $p_{\hat{\varepsilon}}$ would be maximal:

$$\hat{\varphi} = \arg\sup p_\varepsilon(u - L\varphi) = \arg\sup p(u|\varphi) = \arg\sup \ln p_\varepsilon(u - L\varphi). \tag{9.25}$$

Such approach of the estimation is called the *maximum likelihood method* (Rao, 1972), in this case the function $\ln p(u|\varphi)$ is called the *likelihood function.* The monotone nondecreasing function (logarithm) is chosen in connection with that many empirical distributions are exponential, in particular this concerns to the Gaussian distribution, which in accordance with the central limit theorem (Rao, 1972) is the most prevailing approximation of the distributions. The extremum problem (9.25) solution can be obtained by one of the numerical methods. Here we extract the Gaussian distribution. Since the likelihood function is a quadratic functional in this case, and it is possible to write Euler's equation in the explicit form

$$\hat{\varphi} = \arg\inf(u - L\varphi,\ K_\varepsilon^{-1}(u - L\varphi)),$$
$$(L^*K_\varepsilon^{-1}L)\varphi = L^*K_\varepsilon^{-1}u\,. \tag{9.26}$$

The analysis of the equation (9.26) shows that the maximum likelihood method is identical to the method of least squares in the case of the Gaussian distribution and has the same disadvantage: the solution instability since the operator $L^*K_\varepsilon^{-1}L$ is compact operator. In the case of the Laplace distribution, the maximum likelihood method leads to the extremum problem solution in the norm L_1:

$$\hat{\varphi} = \arg\inf ||u - L\varphi||_{L_1},$$

that is equivalent to the method of the least modulus:

$$\hat{\varphi} = \arg\inf |u - L\varphi|.$$

2. We can obtain the regularized solution $\tilde{\varphi}$ only in the case if we have a priori information about the solution. Let along with distribution $p(\varepsilon)$ it is given a priori distribution $p(\varphi)$. It seems advisable to choose as a solution that it provides the validity of two conditions simultaneously: in the first place the function $\tilde{\varphi}$ should be such that the difference $u - L\tilde{\varphi}$ would be as much as possible similar to the noise ε (by analogy with the condition of the maximum likelihood method); in the second place the function $\tilde{\varphi}$ should correspond as well as possible to a priori representation of the distribution $p(\varphi)$. Taking into account the suggestion about the statistical independence φ and ε, the validity of these conditions is reduced to the requirements:

$$\tilde{\varphi} = \arg\sup p_\varepsilon(u - L\varphi)p(\varphi) = \arg\sup p(u|\varphi)p(\varphi). \tag{9.27}$$

Accordingly to the Bayes formula

$$p(\varphi|u) = \frac{p(\varphi)\,p(u|\varphi)}{p(u)},$$

it is possible to represent the estimate $\tilde{\varphi}$ (9.27) as

$$\tilde{\varphi} = \arg\sup p(\varphi|u). \tag{9.28}$$

The obtained solution (9.28) is called an *estimate of the maximum a posteriori probability.*

The Bayes strategy determines the solution φ in the following way:

$$\tilde{\varphi} = \mathrm{E}_u\varphi = \int \varphi p(\varphi|u)d\varphi, \tag{9.29}$$

which is identical to the estimate $\tilde{\varphi}$ (9.28) in the case of the symmetric a posteriori distribution $p(\varphi|u)$.

In the case of the normal distributions $p(\varphi)$, $p(\varepsilon)$ it is possible to write the explicit form of the regularized solution (Turchin *et al.*, 1971; Fedotov, 1990):

$$\tilde{\varphi} = \arg\inf\left[||u - L\varphi||^2_{K_\varepsilon^{-1}} + ||\varphi||^2_{K_\varphi^{-1}}\right] = K_\varphi L^*(LK_\varphi L^* + K_\varepsilon)^{-1}u. \tag{9.30}$$

The analysis of the formula (9.30) shows, that by using a priori information about the fields φ, the extremum problem contains the stabilization functional $||\varphi||^2_{K_\varphi^{-1}}$. In this case the correlator can be determined to within a multiplier and then the extremum problem (9.30) has the form coinciding with the general regularization scheme (9.16):

$$\hat{\varphi} = \arg\inf[||u - L\varphi||^2_{K_\varphi^{-1}} + \alpha||\varphi - \varphi_0||^2_{K_\varphi^{-1}}]. \tag{9.31}$$

The solution $\tilde{\varphi}$ (see the formula (9.30)) is called the solution on the *statistical regularization method.* How it is seen from the equality (9.23), the solution, obtained by using the Bayes strategy, is interpreted as a conditional mean value. The statistical approach gives a possibility together with the solution to obtain the distribution of the solution errors:

$$\mathrm{E}_u((\varphi - \tilde{\varphi})(\varphi - \tilde{\varphi})^*) = K_{\delta\varphi},$$
$$K_{\delta\varphi} = K_\varphi - K_\varphi L^*(LK_\varphi L^* + K_\varepsilon)^{-1}LK_\varphi.$$

We consider a particular case of the correlators K_φ and K_ε representation:

$$K_\varphi = \sigma_\varphi^2 I, \qquad K_\varepsilon = \sigma_\varepsilon^2 I.$$

In this case the solution $\tilde{\varphi}$ (9.31) is found from the extremum problem

$$\tilde{\varphi} = \arg\inf[||u - L\varphi||^2 + \alpha_0||\varphi||^2],$$

where the coefficient $\alpha_0 = \sigma_\varepsilon^2/\sigma_\varphi^2$ is a regularization parameter, that has in this case the obvious physical sense: in the first place $\alpha_0 \longrightarrow 0$ in the case of $\sigma_\varphi^2 \to \infty$, i.e. the convergence of the regularization parameter to zero is connected identically with the removal of a priori restrictions; in the second place, the same degree of the stability (regularization) of the extremum problem is obtained by the relation $\sigma_\varphi^2 = 0(\sigma_\varepsilon^2)$. Taking into account these results it is justified physically a degree of the convergence of the regularization parameter to zero in the methods of the functional regularization: if

$$J(\varphi) = ||L\varphi - u||_{L_2}^2 + \alpha||\varphi||_H^2$$

and if $||L\varphi - u||_{L_2} = \delta$, then $\alpha = O(\delta^2)$.

We turn to the maximum likelihood method again. At the point of the solution $\hat{\varphi}$, the first derivative of the functional becomes zero:

$$\frac{\delta}{\delta\varphi}[\ln p(u|\varphi)] = 0,$$

and a measure of the measurement sensitivity to variations of $\delta\varphi$ serves the second derivative incoming to the quadratic form:

$$-\frac{1}{2}\left(\delta\varphi, \left[\frac{\delta^2}{\delta\varphi^2}\ln p(u|\varphi)\right]\delta\varphi\right).$$

Conducting an average on all set of the possible values u, we obtain the *Fisher's information operator* (Stratonovich, 1975):

$$F = -\mathrm{E}\left(\frac{\delta^2}{\delta\varphi^2}\ln p(u|\varphi)\right).$$

For the Gaussian distribution p and for our problem $u = L\varphi + \varepsilon$, we will have the following expression for the Fisher's operator:

$$F = \mathrm{E}\frac{\delta^2}{\delta\varphi^2}(L\varphi - u, K_\varepsilon^{-1}(L\varphi - u)) = L^* K_\varepsilon^{-1} L.$$

We note that the Fisher's operator determines the correspondence between distances in the decision space φ and in the measurement space:

$$(\delta\varphi, F\delta\varphi) = (L\delta\varphi, K_\varepsilon^{-1} L\delta\varphi) = (\delta u, K_\varepsilon^{-1}\delta u),$$

i.e. it characterizes the discriminant metric in the decision space

$$||\varphi_1 - \varphi_2||_F^2 = ||u_1 - u_2||_{K_\varepsilon^{-1}}^2.$$

The behavior of the likelihood functional depends on the Fisher's operator at the extremum point. If the presentation of the operator F in the eigenbasis has the form

$$F = \sum_{\alpha} \lambda_\alpha \psi_\alpha \psi_\alpha^*,$$

where $(\psi_\alpha, \psi_\beta) = \delta_{\alpha\beta}$, $\lambda = \lambda_1 \geq \lambda_2 \geq \ldots$, then directions in the functional space, defined ψ_α at big values of α, are such directions along that the likelihood functional is closed to the stationary, i.e. the functions φ_1 and φ_2 are practically indistinguishable:

$$||\varphi_1 - \varphi_2||_F^2 \approx 0,$$

if

$$\sum_{\alpha=1}^{\alpha_0} |(\varphi_1 - \varphi_2, \psi_\alpha)|^2 = ||\varphi_1 - \varphi_2||^2,$$

where the coefficient α_0 is the threshold value, at which the ratio $\lambda_{\alpha 0}/\lambda_{\max}$ is enough small.

The proper basis of the Fisher's operator is called the *Korhunen–Loeve basis.* The normal distribution, which is written in this basis, decomposes on the products of the probability densities (projects on the basis vectors):

$$p(\varphi) \sim \exp\left[-\frac{1}{2}(\varphi, F\varphi)\right] = \exp\left[-\frac{1}{2}\sum_{\alpha} \lambda_\alpha |\varphi_\alpha|^2\right]$$
$$= \prod_{\alpha} \exp\left[-\frac{1}{2}\lambda_\alpha |\varphi_\alpha|^2\right] \sim \prod_{\alpha} p(\varphi_\alpha),$$

where $\varphi_\alpha = (\psi_\alpha, \varphi)$. Writing the Fisher's operator decomposition in the form

$$F = \sum_{\alpha=1}^{\alpha_0} \lambda_\alpha \psi_\alpha \psi_\alpha^* + \sum_{\alpha=\alpha_0+1} \lambda_\alpha \psi_\alpha \psi_\alpha^*,$$

the probability density $p(\varphi)$ can be represented approximately as

$$p(\varphi) \approx \tilde{p}(\varphi) \sim \prod_{\alpha=1}^{\alpha_0} \exp\left[-\frac{1}{2}\lambda_\alpha |\varphi_\alpha|^2\right],$$

i.e. the statistical assembly is described by the ort projection, corresponding to the big eigenvalues $\lambda_\alpha (\alpha = 1 \div \alpha_0)$ of the Fisher's operator, completely enough in the probability sense

We give additional interpretation of the statistical regularization and measurement integration using the Fisher's operator.

Let along with the model of indirect measurement of the field

$$u = L\varphi + \varepsilon, \qquad \varepsilon \in N(0, K_\varepsilon),$$

is given a priori information $\varphi \in N(\varphi_0, K_\varphi)$; this information can be presented formally in the form of the direct measurement of the field φ:

$$\varphi_0 = I\varphi + \eta, \qquad \eta \in N(0, K_\varphi).$$

We note that in the general case the direct measurement model is written as $u_i = P_i\varphi + \varepsilon_i$, where P_i is a projector, which cuts out φ values at the isolated points of the parametric space, for example, at the isolated space-time points.

As we demonstrated above, the indirect measurement $u = L\varphi + \varepsilon$ generates the normal distribution in the φ space

$$P_L(\varphi) \sim \exp\left[-\frac{1}{2}(||\varphi - \hat{\varphi}||^2_{L^*K_\varepsilon^{-1}L})\right],$$

$$\hat{\varphi} = (L^*K_\varepsilon^{-1}L)^{-1}L^*u,$$

and the direct (a priori) measurement are described by the normal distribution density

$$P_I(\varphi) \sim \exp\left[-\frac{1}{2}||\varphi - \varphi_0||^2_{K_\varphi^{-1}}\right].$$

The information, obtained from the indirect and direct measurement, is considered as independent, therefore the joint estimate φ is determined by the product of the probability densities. Then

$$\tilde{\varphi} = \arg\inf(||\varphi - \hat{\varphi}||^2_{F_1} + ||\varphi - \varphi_0||^2_{F_2}), \tag{9.32}$$

here the Fisher's operators F_1 and F_2 determine the amount of information, obtained accordingly in the indirect measurement ($F_1 = = L^*K_\varepsilon^{-1}L$) and the amount of a priori information ("direct" measurement): $F_2 = K_\varphi^{-1}$. Euler equation for the extremum problem (9.32) has the form

$$(F_1 + F_2)\tilde{\varphi} = F_1\hat{\varphi} + F_2\varphi_0,$$

$$\begin{aligned}\tilde{\varphi} &= (F_1 + F_2)^{-1}\,(F_1\hat{\varphi} + F_2\varphi_0)\\ &= (L^*K_\varepsilon^{-1}L + K_\varphi^{-1})^{-1}\,(L^*K_\varepsilon^{-1}u + K_\varphi^{-1}\varphi_0)\\ &= \varphi_0 + K_\varphi L^*(LK_\varphi L^* + K_\varepsilon)^{-1}(u - L\varphi_0),\end{aligned}$$

or

$$\tilde{\varphi} - \varphi_0 = K_\varphi L^*(LK_\varphi L^* + K_\varepsilon)^{-1}(u - L\varphi_0),$$

i.e. we obtain the estimate of the standard statistical regularization for the model

$$L(\varphi - \varphi_0) + \varepsilon = u - L\varphi_0.$$

Let us consider the general scheme of the integration of the M measurements (indirect and direct) with using a priori information

$$\begin{aligned}&L_1\varphi + \varepsilon_1 = u_1,\\ &L_2\varphi + \varepsilon_2 = u_2,\\ &\dots\\ &L_M\varphi + \varepsilon_M = u_M,\\ &L_{M+1}\varphi + \varepsilon_{M+1} = u_{M+1} \equiv \varphi_0,\end{aligned}$$

where for the indirect measurement the L_i is an integral operator, for the direct measurement the $L_i = P_i$ is an projector, such that $P = P^*, P = P^2$. If $\varepsilon_i \in N(0, K_{\varepsilon_i})$, $i = 1 \div M+1$, then the optimal estimate of the field φ, has the form

$$\tilde{\varphi} = F^{-1}\left(\sum_{i=1}^{M+1} F_i u_i\right).$$

Here the operator F^{-1}, is an inverse one in relation to the Fisher operator

$$F = \sum_{i=1}^{M+1} F_i = \sum_{i=1}^{M} (L_i^* K_{\varepsilon_i}^{-1} L_i) + I K_{\varphi}^{-1} I.$$

It determines the error correlator of the solution $\tilde{\varphi}$:

$$\mathrm{E}(\delta\varphi\delta\varphi^*) = F^{-1}.$$

Thus, the integration procedure should include a priori information or direct measurement as far as only in this case it is possible to get the stable solution. It should be noted that the direct measurement is able to substitute a priori information if they envelop the measurement of the whole field φ, but not the measurement at isolated points.

The considered statistical interpretation of the measurement integration permits to write the problem of the recovery of the non-stationary field φ_t, described of the system state. In this case the role of a priori information can fulfill the equation, described the dynamic of the field φ_t:

$$D_t\varphi_t + \eta_t = g, \qquad \eta_t \in N(0, K_\eta),$$

$$K_\eta = \sigma_\eta^2 \delta(t - t')$$

(g is a source function) provided that the boundedness of the inverse dynamic operator D_t.

We note that in the particular case, if the dynamic operator has the form $D_t = (\partial/\partial t + A(t))$, were $A(t)$ is a positive operator, we obtain the dynamic model which is used in the dynamic Kalman–Bucy filtering (Albert, 1972).

Replacing the integration of the indirect and direct measurement by the operator L, we write the system which determines the solution φ taking into account a priori information (dynamic model):

$$L\varphi_t + \varepsilon = u,$$
$$D_t\varphi_t + \eta_t = g.$$

The formal solution can be represented using the Fisher operator and writing the Euler equation:

$$(L^* K_\varepsilon^{-1} L + D_t^* K_\eta^{-1} D_t)\tilde{\varphi}_t = (L^* K_\varepsilon^{-1} u + D_t^* K_\eta^{-1}(g + \varphi_t^0),$$

$$\varphi_t^0 : \ D_t\varphi_t^0 = 0, \ \varphi_t^0|_{t=0} = \varphi_0.$$

The correlation operator of the solution errors is the inverse operator in relation to the Fisher operator $F = F_t$:

$$F_t = (L^* K_\varepsilon^{-1} L + D_t^* K_\eta^{-1} D_t) > 0$$

due to the positiveness of the operator $D_t^* K_\eta^{-1} D_t$.

We analyze the solution, obtained by using the mathematical statistics principles.

Let the solution of the linear problem $u = L\varphi + \varepsilon$ is obtained by the linear procedure R:

$$\hat{\varphi} = Ru = RL\varphi + R\varepsilon.$$

Then the error of the obtained solution is described by the solution

$$\delta\varphi \triangleq \hat{\varphi} - \varphi = (RL - I)\varphi + R\varepsilon.$$

The error $\delta\varphi$ does not depend on the true value φ, if $RL = I$. We obtain this operator R by using the method of least squares, if the inverse operator exists $(L^* K_\varepsilon^{-1} L)^{-1}$. In fact:

$$R = (L^* K_\varepsilon^{-1} L)^{-1} L^* K_\varepsilon^{-1},$$

$$RL = (L^* K_\varepsilon^{-1} L)^{-1} (L^* K_\varepsilon^{-1} L) = I.$$

The quantity of the conditional mean value of the error $\delta\varphi$ is called *bias* (b):

$$b \triangleq \mathrm{E}_\varphi \delta\varphi = (RL - I)\varphi.$$

The characteristic of the solution accuracy is the error correlation:

$$K_{\delta\varphi} \triangleq \mathrm{E}[\delta\varphi\delta\varphi^*].$$

The solution $\tilde{\varphi}$ such that

$$\tilde{\varphi} = \arg\inf \mathrm{sp} K_{\delta\varphi},$$

is called an *effective solution*.

First of all, we note that the solution of ill-posed problems by the method of least squares (if operator $(L^* K_\varepsilon^{-1} L)^{-1}$ exists) is unbiased, but certainly ineffective: mean error are infinite. In fact,

$$\begin{aligned} \mathrm{E}[\delta\varphi\delta\varphi^*] &= \mathrm{E}[(R\varepsilon)(R\varepsilon)^*] = R\mathrm{E}[\varepsilon\varepsilon^*]R^* = RK_\varepsilon R^* \\ &= (L^* K_\varepsilon^{-1} L)^{-1} L^* K_\varepsilon^{-1} K_\varepsilon K_\varepsilon^{-1} L (L^* K_\varepsilon^{-1} L)^{-1} \\ &= (L^* K_\varepsilon^{-1} L), \quad \mathrm{sp} K_{\delta\varphi} = \infty, \end{aligned}$$

by virtue unboundedness of the Fisher inverse operator in ill-posed problems.

We construct an operator R, which gives an effective solution. Such operator has the minimum errors on the average by all possible realization of errors ε with the correlator K_ε and with the functions with correlator K_φ. We write the error correlator for an arbitrary "decision" operator R:

$$K_{\delta\varphi} = \mathrm{E}bb^* + \mathrm{E}nn^*,$$

$$n = R\varepsilon,$$

$$K_{\delta\varphi} = (RL - I)K_\varphi(RL - I)^* + RK_\varepsilon R^*.$$

The condition of the algorithm R optimum, which gives the solution with the minimum error variance at any space point, will be following:

$$\tilde{R} = \arg\inf \mathrm{sp} K_{\delta\varphi}. \qquad (9.33)$$

We introduce the Hilbert space for operators with a bounded trace (the Hilbert–Schmidt space of operators). In this case the scalar product is determined as

$$(A, B)_H = (A, HB) = \mathrm{sp}(A^* HB),$$

where $H > 0$. Then the problem (9.33) is reduced to the minimization of the quadratic form

$$\begin{aligned}\tilde{R} &= \arg\inf[||(RL - I)^*||^2_{K_\varphi} + ||R^*||^2_{K_\varepsilon}] \\ &= \arg\inf[(L^*R^* - I, K_\varphi(L^*R^* - I)) + (R^*, K_\varepsilon R^*)],\end{aligned}$$

therefore the Euler equation has the form

$$(LK_\varphi L^* + K_\varepsilon)R^* = LK_\varphi.$$

The optimal decision operator $\tilde{R}$ is described by the expression

$$\tilde{R} = K_\varphi L^*(LK_\varphi L^* + K_\varepsilon)^{-1} \qquad (9.34)$$

and gives the following correlator of the solution errors

$$K_{\delta\varphi} = K_\varphi - K_\varphi L^*(LK_\varphi L^* + K_\varepsilon)^{-1}LK_\varphi.$$

In this case the error variance at the point x is minimal in comparison with the solution errors variance, obtained with the help of the linear operator which is distinct from $\tilde{R}$.

We note that the optimal decision operator for $\tilde{R}$ (see expression (9.34)) is the *operator of the linear regression* (Albert, 1972), which is written in the form $\tilde{R} = K_{\varphi u}K_u^{-1}$. The accuracy of the field φ recovery is determined by the correlator

$$K_{\delta\varphi} = K_\varphi - K_{\varphi u}K_u^{-1}K_{u\varphi}.$$

The deviation of the decision rule (ΔR) from the optimal one $\tilde{R}$ leads to increasing of the correlator of the solution errors on the positive operator

$$\Delta K_{\delta\varphi} = \Delta R(LK_\varphi L^* + K_\varepsilon)\Delta R.$$

For the solution of the particular problems it is necessary to introduce the quantitative information measure, as far as only in this case it is possible to carry out the quantitative comparison of the different experiments, set up a mathematical problem for the design of experiment and to give the sense to the statement about the sufficiency or insufficiency of data for the solution of the inverse problem $L\varphi + \varepsilon = u$.

The Shannon functional (9.18), being the measure of the uncertainty, is expressed by means of the probability density p_φ:

$$H_\varphi = -(p(\varphi), \ln p(\varphi)).$$

Starting from the representation of the joint probability density:

$$p(\varphi, u) = p(\varphi)p(u|\varphi) = p(u)p(\varphi|u),$$

it is possible to write the additive property by the following way:

$$H_{u\varphi} = H_\varphi + H_{u|\varphi} = H_u + H_{\varphi|u}, \tag{9.35}$$

where $H_{u|\varphi}$ and $H_{\varphi|u}$ are conditional entropies. As information measure about the field φ, contained in data u, is used the difference a priori (H^{apr}) and a posteriori (H^{apost}) entropies (analogously, ΔS is the difference of the Boltzmann's entropies by transfer of the system from one statement to other):

$$I_{\varphi u} \triangleq H^{apr} - H^{apost} = H_\varphi - H_{\varphi|u},$$

or if it is used the additivity property (9.35),

$$I_{\varphi u} = H_\varphi + H_u - H_{u\varphi} = H_u - H_{u|\varphi}.$$

We show that the information $I_{\varphi u}$ is always nonnegative, i.e. $H_\varphi \geq H_{\varphi|u}$. We prove an auxiliary inequality

$$-\int p(\varphi) \ln \frac{p(\varphi)}{q(\varphi)} d\varphi \leq 0, \tag{9.36}$$

where $q(\varphi) : \int q(\varphi)d\varphi = 1$. Using Jensen's (Rao, 1972) inequality $\mathrm{E}f(\varphi) \leq f(\mathrm{E}\varphi)$, which is valid for any convex functions f, it is possible to write:

$$\begin{aligned} -\int p(\varphi) \ln \frac{p(\varphi)}{q(\varphi)} d\varphi &= \mathrm{E} \ln \frac{q(\varphi)}{p(\varphi)} \leq \ln \mathrm{E} \frac{q(\varphi)}{p(\varphi)} \\ &= \ln \int p(\varphi) \frac{q(\varphi)}{p(\varphi)} d\varphi = \ln 1 = 0. \end{aligned}$$

Replacing $p(\varphi)$ on $p(\varphi|u)$ and $q(\varphi)$ on $p(\varphi)$ in the inequality (9.36), we obtain

$$-\int p(\varphi|u) \ln \frac{p(\varphi|u)}{p(\varphi)} d\varphi \leq 0,$$

i. e.

$$-\int p(\varphi|u) \ln p(\varphi|u) d\varphi \leq -\int p(\varphi|u) \ln p(\varphi) d\varphi.$$

Averaging the inequality with respect to u, we write:

$$-\int\int p(\varphi, u) \ln p(\varphi|u) d\varphi du \leq -\int p(\varphi) \ln p(\varphi) d\varphi,$$

that proves our statement: $H_\varphi \geq H_{\varphi|u}$.

We note, that the inequality (9.36) was a basis for the introduction of the Kullback discriminant measure (Kullback, 1978) in the space of distributions:

$$\rho(p_1, p_2) = \int (p_1 - p_2) \ln \frac{p_1}{p_2} d\varphi = \int p_1 \ln \frac{p_1}{p_2} d\varphi - \int p_2 \ln \frac{p_1}{p_2} d\varphi \geq 0,$$

where $\rho(p_1, p_2) = 0$ only by $p_1 = p_2$.

We remind that a priori an a posteriori distribution densities do not connect with the method of the inverse problem solution and in this sense the amount of information is an objective measure of the decreasing of the system uncertainty as a result of the experiment.

We shall find an explicit form of the information about the field φ, containing in data $u = L\varphi + \varepsilon$ for $\varphi \in N(0, K_\varphi)$ and $\varepsilon \in N(0, K_\varepsilon)$, using the relation (9.21) (the Shannon's entropy for normal distributed function):

$$\begin{aligned} I_{\varphi u} &= H^{apr} - H^{apost} = H_\varphi - H_{\varphi|u} \\ &= -(p_\varphi, \ln\, p_\varphi) + [p_{\varphi|u}, \ln\, p(\varphi|u)] = \frac{1}{2} \ln \mathrm{Det} K_\varphi - \frac{1}{2} \ln\, \mathrm{Det} K_{\varphi|u}, \end{aligned}$$

where K_φ is a priori correlator;

$$K_{\varphi u} \equiv K_{\delta\varphi} = K_\varphi - K_\varphi L^* (L K_\varphi L^* + K_\varepsilon)^{-1} L K_\varphi$$

is a posteriori correlator.

We note, the sense of the amount of information is determined by the difference of the Hartley's measures of information, if we interpret the determinant of the correlator as a number of independent statements in the functional space, which is proportional to the value of the volume $V = [\prod_\alpha \lambda_\alpha]^{1/2}$. Using the equality

$$\mathrm{Det} K_\varphi (\mathrm{Det} K_{\delta\varphi})^{-1} = \mathrm{Det} K_\varphi K_{\delta\varphi}^{-1},$$

it is possible to represent the amount of information as

$$I_{\varphi u} = \frac{1}{2} \ln \mathrm{Det}(I + K_\varphi F),$$

where F is the Fisher information operator. In the proper basis of the operator $K_\varphi F$ (Kozlov information operator (Ermakov and Zhigljavsky, 1987; Turchin *et al.*, 1971)) the amount of information can be presented by the following way

$$I_{\varphi u} = \frac{1}{2} \ln \prod_\alpha (1 + \lambda_\alpha) = \frac{1}{2} \mathrm{sp} \ln (1 + K_\varphi F).$$

We clarify the physical meaning of eigenvalues of the Kozlov information operator $K_\varphi F \varphi_\alpha = \lambda_\alpha \varphi_\alpha$. In this connection we consider the Rayleigh's functional

$$J_1(u) = \frac{(u, L K_\varphi L^* u)}{(u, u)},$$

that has the sense of the variance of the useful signal $s = L\varphi$, where stationary functional values $J_1(u)$, connected with the eigenfunctions of the correlator, $L K_\varphi L^*$, coincide with the correlator eigenvalues:

$$J_2(u) = \frac{(u, K_\varepsilon u)}{(u, u)},$$

which have the sense of the noise component n variance of the model $u = s + n$. The ratio $J = J_1/J_2$ has the physical meaning of the signal/noise ratio:

$$J = \frac{J_1}{J_2} \Rightarrow \frac{s}{n}.$$

This clear physical meaning is observed in the case, if the stationary values of the functional J are reached on the functions, which are simultaneously eigenfunctions both for the operator $LK_\varphi L^*$ and for K_ε (conforming basis).

The functional J can be written in the form

$$J = \frac{(u, LK_\varphi L^* u)}{(u, K_\varepsilon u)},$$

or introducing the change of variables $u = K^{-1/2}\tilde{u}$. The extremal properties of the functional can be analyzed on the set of functions $\tilde{u}$:

$$J = \frac{(\tilde{u}, K_\varepsilon^{-1/2} LK_\varphi L^* K_\varepsilon^{-1/2}\tilde{u})}{(\tilde{u}, \tilde{u})}.$$

The stationary points are determined by equations on eigenvalues:

$$K_\varepsilon^{-1/2} LK_\varphi L^* K_\varepsilon^{-1/2}\tilde{u}_\alpha = \lambda_\alpha \tilde{u}_\alpha,$$

which are identical to the equation

$$K_\varphi F \varphi_\alpha = \lambda_\alpha \varphi_\alpha,$$

where φ_α and $\tilde{u}_\alpha$ are connected by the relation $\tilde{u}_\alpha = K^{-1/2} L\varphi_\alpha$.

Thus the eigenvalues of the operator $K_\varphi F$ have the sense of the ratio signal/noise for different components of the solution in the canonical basis, and the Shannon's information to make it clear, what components of the solution $\varphi = \sum_\alpha \varphi_\alpha(\varphi_\alpha, \varphi)$ are supplied by information:

$$I_{\varphi u} = \frac{1}{2}\sum_\alpha \ln(1 + \lambda_\alpha).$$

9.4 Elements of the Mathematical Design of Experiment

Introduced in Sec. 9.3 information measures are a natural basis for the problem setting of the mathematical design of the experiment. The possibility of the mathematical design is connected with the choice of conditions (σ) from some set Σ. For example, the choice of the measurement channel, choice of the time and frequency intervals for measurements, the choice of spread of receivers in the remote sounding problems. The general mathematical problem definition of the design of experiment (searching of σ) is reduced to:

$$\tilde{\sigma} = \arg \inf_{\sigma \in \Sigma} \Phi(\sigma) = \arg \sup_{\sigma \in \Sigma} \Phi^{-1}(\sigma).$$

It is advisable to choice the functional Φ as convex, where it should be bounded below (above).

The statement of the problem, connected with the solution of the inverse problem $L\varphi + \varepsilon = u$, is called a *regression model*, if the function φ is represented a priori in the finite-dimensional basis. In this case it is looked for the best unbiased estimate, which, as shown above, can be found by the method of least squares. The standard representation of the regression equation

$$u_{ij} = \sum_\alpha \varphi_\alpha \mathbf{f}_\alpha(x_i) + \varepsilon_{ij},$$

(by the lack of a priori information about φ) where $\sum\limits_\alpha \varphi_\alpha \mathbf{f}_\alpha(x_i)$ is the regression function. From the expression

$$L(x_i)\varphi = L\sum_\alpha \psi_\alpha(\psi_\alpha, \varphi) = \sum_\alpha (\psi_\alpha, \varphi)L\psi_\alpha$$

follows the correspondence $\varphi_\alpha = (\psi_\alpha, \varphi)$ and $\mathbf{f}_\alpha(x_i) = L_\alpha(x_i)\psi_\alpha$ ($\{\varphi_\alpha\}$ are unknown parameters; x_i are controlled variables, for example, coordinates of the registration $i = 1 \div n$, index j changes from 1 until r_i, $\sum r_i = N$). The *standardized design of experiment* $\sigma(N)$ is called a totality of values:

$$\sigma = \begin{Bmatrix} x_1 & x_2 & \dots & x_n \\ p_1 & p_2 & \dots & p_n \end{Bmatrix},$$

where $\sum p_i = 1$, $p_i = r_i/N$. In this case as a functional argument is used the Fisher's information matrix:

$$\mathbf{F}(\sigma) = \sum_i p_i \mathbf{f}^T(x_i) K_\varepsilon^{-1}(x_i)\mathbf{f}(x_i),$$

$$\mathbf{f}^T(x_i) = ||f_1(x_i), f_2(x_i) \dots f_\alpha(x_i) \dots ||,$$

$$f_\alpha(x_i) = L(x_i)\psi_\alpha.$$

The *continuous design of experiment* is called the function $p(x)$, such that

$$p(x) \geq 0, \qquad \int_X p(x)dx = 1.$$

As optimum criteria it are used the following (Penenko, 1981; Ermakov and Zhigljavsky, 1987; Fedorov, 1972):

(1) $\tilde{\sigma} = \arg\inf \mathrm{Det}F^{-1}(\sigma) = \arg\sup \mathrm{Det}F(\sigma)$,
where F is the Fisher's operator, depending on the design as on a parametric function, and then the optimal design $\hat{\sigma}$ is called *D-optimal.* At that, it is minimized the volume of the concentration ellipsoid of the estimate obtained by the least squares method. The problem can be empty even in the case of the finite basis for the completely continuous L-operator.

(2) $\tilde{\sigma} = \arg\inf \lambda_{\max}(F(\sigma)) = \arg\sup \lambda_{\min}(F(\sigma))$
($\lambda_{\min}, \lambda_{\max}$ are the minimum and maximum eigenvalues of the Fisher's operator $F(\sigma)$). This criterion leads to the *E-optimal* design of experiment, in this case it is minimized the error of the least informative linear combination of the parameters $\boldsymbol{\varphi}$:

$$F\boldsymbol{\varphi} = \lambda_{\min}\boldsymbol{\varphi}.$$

(3) $\tilde{\sigma} = \arg\inf_{\sigma\in\Sigma} \sup_{x\in X}(\mathbf{f}^T(x), F^{-1}(\sigma)\mathbf{f}(x))$.
This expression is a criterion, which minimizes the maximum value of the variance of the regression function estimate. The corresponding design is called *G-optimal.*

(4) $\tilde{\sigma} = \arg\inf_{\sigma} \int\limits_X \mathbf{f}^T(x)F^{-1}(\sigma)\mathbf{f}(x)dx$.
This criterion minimizes the average on X of the variance value of the regression function estimate.

(5) $\tilde{\sigma} = \arg\inf AF^{-1}(\sigma)$.
This criterion minimizes the risk value which is given by the matrix A by the generalized squared loss:

$$\mathrm{E}(\boldsymbol{\varphi} - \hat{\boldsymbol{\varphi}}, A(\boldsymbol{\varphi} - \hat{\boldsymbol{\varphi}})).$$

This criterion is called *L-optimal.*

The design of experiment problems have the evident practical significance. At that it is involved to the optimum criteria: outlay, resources, qualitative parameters, etc.

In each particular case, the problem of the criterion construction may be very complicated. The averaging measures can possess an essential uncertainty. As a rule, it is impossible to construct an unique criterion. The design of experiment should satisfy to the totality of the criteria, i.e. the problem of the design of experiment becomes multicriterion. By introducing a vector criterion $\{\Phi_\alpha(\sigma)\}$ the plans can be partially ordered. One says that σ_1 dominates over σ_2 in respect with the set $\{\Phi_\alpha(\sigma)\}$, if $\{\Phi_\alpha(\sigma_1)\} \leq \{\Phi_\alpha(\sigma_2)\}$ for all values α, in this case at least for one of all the strong inequality is fulfilled. Any plan is called *Pareto optimal,* if it belongs to Pareto set (with criterion Φ_α), and the elements of this set $\Sigma_0 \subseteq \Sigma$ do not have the dominating conditions σ in the set Σ.

The specific character of the problem statement of the design of experiment, which is connected with the inverse problems of the mathematical physics, is the obligatory inclusion of a priori information, in particular, the use of conforming basis for the solution regularization.

Chapter 10

Construction and interpretation of tomographic functionals

The recovery of the coefficients, that are functions of space coordinates and describe local properties of a medium from known characteristics of the sounding signal field, makes up the subject matter of inverse problems in the mathematical physics (Belishev, 1990; Blagoveshchenskii, 2005, 1978, 1966; Romanov, 1974, 1987; Faddeev, 1976; Newton, 1981). In setting up corresponding inverse problems, it is usual to use the fields that are deterministically defined on the space-time continuum as initial data. A real experiment can be reduced to the traditional mathematical setting, provided that the record is considered as continuous in time and the so-called "reduction to an ideal device", which is widely known the ill-posed problem (Pytev, 1990; Tikhonov and Arsenin, 1977), is first performed and then followed by the space-time interpolation. Although such a procedure causes a loss of the information, it is still applied in certain cases (Yanovskaya and Porokhova, 2004).

To formulate a mathematical model and to set up an inverse problem that is adequate to a real physical experiment (to remote sensing, in particular), it is necessary to take into account basic factors determining the model construction. The recovery of fields of the required parameters of a medium from values of a set of functionals of measurements is a natural mathematical model of interpreting physical experiment. In interpretation problems, the conjugate space of linear functionals arises that provided a mathematical model of linear devices recording the sounding signal. As it will be shown in what follows, the conjugate space enables us to interpret the structure of functionals that directly act on the fields of the parameters of interest in clear physical terms. These functionals are called tomographic functionals.

10.1 Construction of the Model of Measurements

The first step in solving of the inverse problem of remote sensing is the construction of the model of the relationship between measurement data and unknown parameters of the medium. As a rule, the digital records provided by spatially located receivers are used as initial data. The unknown parameters determining the properties of the medium are the elements ($\theta(x)$) of functional spaces (Θ): $\theta(x) \in \Theta$; for instance, the fields of magnetic and gravitational anomalies, the fields of conductivity ($\sigma(x)$), the fields of elastic Lame parameters ($\lambda(x)$ and$\mu(x)$), of the density ($\rho(x)$), etc. The measurement space is the space of functionals $\{h_n\}$ over the fields of sounding signals ($\varphi \in \Phi$); the model of measurements

$$u_n : \ u_n = H_n(\varphi) \triangleq \langle h_n | \varphi \rangle ,$$

where $n = 1 \div N$ is a number of digital samples of the receiving system and $\{h_n\} \in \Phi^*$.

A *tomographic experiment* is determined by a mapping $\Theta(R^3) \longrightarrow R^N$ of the functional space into the measurement space. Here, experimental data include some noise ε and functionals of the known parameter fields, i.e.,

$$u_n = \mathcal{P}_n(\theta) + \varepsilon_n. \tag{10.1}$$

Let the propagation process be described by a linear operator L_θ so that

$$L_\theta \boldsymbol{\varphi} = \mathbf{s}, \tag{10.2}$$

where $\boldsymbol{\varphi}$ is the field of the sounding signal; $\mathbf{s}$ is the source field, and

$$L_\theta \ : L_\theta(\alpha\varphi + \beta\psi) = \alpha L_\theta \varphi + \beta L_\theta \psi.$$

The operator L_θ determines the properties of the medium θ. Mathematically, the problem of interpretation of a tomographic experiment reduces to the recovery of the operator L_θ from the measurement data. To find the solution, it is necessary to reconstruct the functionals $\mathcal{P}_n \in \Theta^*$, from the relations $H_n(\varphi(\theta)) = \mathcal{P}_n(\theta)$ and the propagation law (10.2).

Assume that the field $\boldsymbol{\varphi}$ is generated by a group of sources and it is recorded by a receiver with a directional pattern Ω and a fixed orientation of its principal lobe $\mathbf{e}$. Assuming the size of the receiver to be small as compared with the typical sizes of the problem (the wavelength of the sounding signal and the characteristic scale of the inhomogeneity), we consider the receiver as localized at a point. For example, the complete set of experimental data in the seismic case will be obtained if we take into account that the experiment involves J group of sources,$3 \times K$ (K is the number of receiver points) traces from each group, and L processed samples of digital records of the seismogram. This yields $N = 3\,K\,J\,L$ samples. It should be noted that it is not the field $\boldsymbol{\varphi}$ that is directly recorded, but the result of its transformation by the apparatus function H of the recording channel, which involves time and amplitude quantization. The most general model of the transformation of the sounding signal by the recording channel is given by the linear convolution operator. We introduce a through numbering of samples ($n = 1 \div N$). Then individual sample of the digital record can be represented in the form

$$u_n = H_n\, L_\theta^{-1}\, \mathbf{s} + \varepsilon_n ,$$

$$H : \ H_n\varphi = \iiint dx d\tau d\Omega h_n(\mathbf{e}_n, \mathbf{e}; t_n - \tau, x_n - x)\varphi(x, \mathbf{e}, \tau), \tag{10.3}$$

$$\mathbf{e} : \mathbf{e} \in R^3,$$

Here, all the unknown properties of the medium are included into the operator L_θ^{-1}, and the experimental data are given by the result of the convolution of the field $\boldsymbol{\varphi}$ with the apparatus function h. The experimental value coincides with the projection of the field $\boldsymbol{\varphi}(x,t)$ onto the direction $\mathbf{n}$ only in the idealized problem formulation (where $h_n(t_n - t) = \delta(t - t_n)$, which corresponds to an infinite spectral band of the receiver, never occurring in practice, and where the directional pattern satisfies the condition $h(\mathbf{e}_n\,, \mathbf{e}) = h(\mathbf{e}_n^T\,, \mathbf{e})$), and only if the random error ε is absent. As a realistic model of the determinate part of a particular measurement one may take the functional $h_n = h_n\,(\boldsymbol{\varphi})$, which can be considered as continuous, for physical reasons, and as linear, for technical specifications. Due to nonlinearity of the functional $\mathcal{P}_n\,(\theta)$ from (10.1) (even if an explicit expression for its action on the field θ is available and if $\varepsilon_n \longrightarrow 0$), the solution is necessarily of the interpretational form. As a rule, a linearization of the functional $\mathcal{P}_n$ is the basic component of an each iteration step. Let the medium be described by the field $\Theta_0 = \Theta_0\,(x)$. Then the measurement model (10.1) takes the form

$$u_n = \mathcal{P}_n\,(\Theta_0) + \frac{\delta}{\delta\theta}\Big|_{\Theta_0} \mathcal{P}_n\,(\delta\theta) + \tilde{\varepsilon}_n\,,$$

where $\tilde{\varepsilon}_n$ includes both the random noise ε_n, and the noise caused by the linearization and related to the determinate part of the model.

For the field Θ_0 the propagation equation $L_0\,\boldsymbol{\varphi} = \mathbf{s}$ is satisfied. We assume that the unknown field θ is close to the field Θ_0, i.e. $\theta = \Theta_0 + \delta\theta$, $\delta\theta \ll \Theta_0$. Sometimes the solution φ_0 can be obtained in the analytic form by using an approximate method, for instance, by the ray method. The solution for the medium with the parameter field θ is given by the equality

$$\boldsymbol{\varphi} = \boldsymbol{\varphi}_0 + L_0^{-1}\,\delta L_\theta\,\boldsymbol{\varphi} \tag{10.4}$$

($\delta L_\theta = L_0 - L_\theta$ is the perturbation operator). The equality (10.4) is a consequence of the operator identity

$$L_\theta^{-1} \equiv L_0^{-1} + L_0^{-1}\,(L_0 - L_\theta\,)L_\theta^{-1}$$

for the representation $\boldsymbol{\varphi} = L_\theta^{-1}\,\mathbf{s}$.

We note that the above expression (10.4) can be obtained provided that the field $\boldsymbol{\varphi}_0$ satisfies the homogeneous equation $L_0\,\boldsymbol{\varphi}_0 = 0$. Writing the operator as $L = L_0 - \delta L_\theta$ and the solution $\boldsymbol{\varphi}$ as $\boldsymbol{\varphi} = \boldsymbol{\varphi}_0 + \delta\boldsymbol{\varphi}$, we obtain the equation

$$L_0\,\delta\boldsymbol{\varphi} = \delta L_\theta \boldsymbol{\varphi}\,,$$

i.e., the correction of the perturbed field is described by the same equation as the field $\boldsymbol{\varphi}_0$ in the reference medium, but now with the source $\mathbf{s} = \delta L_\theta \boldsymbol{\varphi}$ including, along with the field φ_0, also the correction $\delta\varphi$.

Taking into account (10.4), the general model (10.3) can be written in the form

$$u_n = H_n\left[\varphi_0 + L_0^{-1}\,\delta L_\theta\,\varphi\right] + \varepsilon_n\,. \tag{10.5}$$

Here the medium properties are reflected both in δL_θ and in the factor φ. Due to dependence of φ upon $\delta\theta$, the determinate part of the measurement model proves to be nonlinear with respect to $\delta\theta$. If $\delta\theta$ is small enough, i.e., if the condition

$$\frac{||H_n\, L_0^{-1}\, \delta L_\theta\, (\boldsymbol{\varphi} - \boldsymbol{\varphi}_0\,)||^2}{E(\varepsilon_n^2\,)} \ll 1 \tag{10.6}$$

is satisfied (E is the expectation operator), then in (10.5) φ can be replaced by $\boldsymbol{\varphi}_0$. Physically, it is the validity of the inequality (10.6) that determines the adequacy of the model $u_i\,(\boldsymbol{\varphi})$ to the real measurement condition, i.e., the model error resulting from replacing $\boldsymbol{\varphi}$ by $\boldsymbol{\varphi}_0$ in (10.5) is much smaller than the measurement error.

We analyze the norm of the linearization error by using the inequality

$$||HL_0^{-1}\, \delta L_\theta\, (\varphi - \varphi_0)|| \le ||HL_0^{-1}\delta L_\theta\, ||\, ||\varphi - \varphi_0\, ||\, .$$

The norm of the difference between the fields φ and φ_0 is bounded by physical considerations (since physical fields do not possess infinite energy), i.e., $||\varphi - \varphi_0\, || \le c < \infty$. The operator $HL_0^{-1}\, \delta L_\theta$ is compact, since so is the operator H (the integral convolution operator), that determines the space-time quantization. As $\delta\theta \longrightarrow 0$, we have $||HL_0^{-1}\delta L_\theta\, || \longrightarrow 0$, and condition (10.6) is trivially satisfied. Taking into account (10.6), we rewrite model (10.5) in the modified form

$$u_n = H_n\, [\varphi_0 + L_0^{-1}\, \delta L_\theta\, \varphi_0\,] + \tilde{\varepsilon}_n\, . \tag{10.7}$$

The errors, including the linearization errors, are suppressed by the action of the operator H_n. Introducing the bilinear form

$$\langle \xi|\eta\rangle_{V,T,\Omega} = \int\limits_\Omega \int\limits_V \xi(\mathbf{e}, x, t) * \eta(\mathbf{e}, x, t)\, dx\, d\Omega,$$

we rewrite (10.7) as follows:

$$u_n = \big\langle h_n|\varphi_0\, \big\rangle_{V,T,\Omega} + \big\langle h_n\, |L_0^{-1}\, \delta L_\theta\, \varphi_0 \big\rangle_{V,T,\Omega} + \tilde{\varepsilon}_n\, ,$$

($*$ is the time convolution sign, V is the sounded domain, T is the time interval of measurements).

Reducing the experimental data by the known value

$$u_n^0 = \mathcal{P}_n\, (\Theta_0\,) \equiv \big\langle h_n\, |\varphi_0\, \big\rangle_{V,T,\Omega}\, ,$$

then yields

$$\tilde{u}_n = \big\langle h_n\, |L_0^{-1}\, \delta L_\theta\, \varphi_0\, \big\rangle_{V,T,\Omega} + \tilde{\varepsilon}_n\, , \tag{10.8}$$

where $\tilde{u}_n = u_n - u_n^0$.

10.2 Tomographic Functional

From the perturbation operator δL_θ we single out the monotone function $\nu(\delta\theta)$ with respect to wish the perturbation operator is linear. Taking into account that in many tomographic problems the operator δL_θ is near-local (for example, it is a differential operator), we rewrite (10.8) in the form

$$\begin{aligned} \tilde{u}_n &= \left\langle (L_0^{-1})^* h_n \left| \delta L_\theta \varphi_0 \right\rangle_{V,T,\Omega} + \tilde{\varepsilon}_n \right. \\ &= \left\langle \left\langle G_0^* h_n \left| \frac{\delta}{\delta\nu} \delta L_\theta \right| G_0 \mathbf{s} \right\rangle_{T,\Omega} |\nu(\delta\theta)\right\rangle_V + \tilde{\varepsilon}_n, \end{aligned} \tag{10.9}$$

$$\text{where} \quad G_0 = L_0^{-1}; \quad \frac{\delta}{\delta\nu}\delta L_\theta : \frac{\delta u}{\delta\nu} = \left\langle G_0^* h_n \left| \frac{\delta}{\delta\nu}\delta L_\theta \right| \varphi_0 \right\rangle_{T,\Omega}.$$

The integral kernel of the functional with respect to $\nu(\delta\theta)$ will be called the *tomographic functional*:

$$p_n^\nu = \left\langle \varphi_{\text{out}} \left| s^\nu \right| \varphi_{\text{in}} \right\rangle_{T,\Omega}, \tag{10.10}$$

where $\boldsymbol{\varphi}_{\text{in}} = \boldsymbol{\varphi}_0$ is the incoming field $\boldsymbol{\varphi}_0$: $L_0 \boldsymbol{\varphi}_0 = \mathbf{s}$, in the known reference medium Θ_0; $\boldsymbol{\varphi}_{\text{out}} : L_0^* \boldsymbol{\varphi}_{\text{out}} = \mathbf{h}_n$ is the reverted outgoing field "generated" by the receiver; $S^\nu = (\delta/\delta\nu)\delta L_\theta$ is the operator of the interaction of the fields $\boldsymbol{\varphi}_{\text{in}}$ and φ_{out}.

Taking into account (10.10), we represent the model (10.9) in the form

$$\tilde{u} = P\nu + \tilde{\varepsilon},$$

$$\tilde{u} = \left\|\tilde{u}_1, \ldots, \tilde{u}_n, \ldots, \tilde{u}_N\right\|^T, \tilde{\varepsilon} = \left\|\tilde{\varepsilon}_1, \ldots, \tilde{\varepsilon}_n, \ldots, \tilde{\varepsilon}_N\right\|^T,$$

$$P = \left\| \begin{matrix} \langle p_{11}| & \ldots & \langle p_{1m}| & \ldots & \langle p_{1M}| \\ & \ldots & & \ldots & \\ \langle p_{n1}| & \ldots & \langle p_{nm}| & \ldots & \langle p_{nM}| \\ & \ldots & & \ldots & \\ \langle p_{N1}| & \ldots & \langle p_{Nm}| & \ldots & \langle p_{NM}| \end{matrix} \right\|, \quad \nu = \left\| \begin{matrix} |\nu_1\rangle \\ \ldots \\ |\nu_m\rangle \\ \ldots \\ |\nu_M\rangle \end{matrix} \right\|.$$

The tomographic functional determines the influence of all elements of the spatial region upon the n-th sampling of experimental data. It should be noted that in traditional ray tomography the tomographic functional is singular and is localized along the ray connecting the source and the receiver, and its weight to along this ray is constant. In the diffraction tomography, however, even if the ray description is applicable to the fields $\boldsymbol{\varphi}_{\text{in}}$ and φ_{out}, every element of the volume of the region under investigation is linked to the receiver and the source by two ray trajectories. Here, to every element of the spatial region its own weight, determined by the interaction of the fields $\boldsymbol{\varphi}_{\text{in}}, \boldsymbol{\varphi}_{\text{out}}$) is attributed. It should be noted that mathematical methods of the computer tomography are based on methods for solving problems of the integral geometry, in which data (projections) are given by integrals of parametric functions on manifolds of lower dimensions (rays, two-dimensional surfaces). By contrast, in diffraction tomography, the supports of tomographic functionals belong to R^3. We note that the main content of a tomographic experiment is relating to the overlapping of supports of tomographic functionals, i.e., the information on one and the same element of the volume is contained in the whole set of measurements. The measurements are related to the changes in the localization of the group of sources (φ_{in} in (10.10)), to the location and orientation of the receiver, and to the sampling (φ_{out} in (10.10)) of the dynamic fields.

A priori data can be represented either in probabilistic or in determinate form, for instance, the type of the spatial symmetry can be specified. The determinate form of the representation of a priori information makes it possible to reduce the tomographic functional to that in a space of lower dimension. Whenever the medium is a priori assumed to be stratified homogeneous, the support of the tomographic functional is one dimensional, whereas the corresponding tomographic functional is the Radon projection of the generalized tomographic functional onto the vertical direction. In the case of the spherical symmetry, the parameter of the kernel of the tomographic functional is the radial coordinate.

10.3 Examples of Construction and Interpretation of Tomographic Functionals

Now we consider some examples of the constructing and interpreting of tomographic functionals.

10.3.1 *Scalar wave equation*

The operators L_0 and L_θ are of the following form:

$$L_0 = -\Delta + c_0^{-2}(x)\frac{\partial^2}{\partial t^2},$$
$$L_\theta = -\Delta + c^{-2}(x)\frac{\partial^2}{\partial t^2},$$

$$\Theta_0 = c_0(x), \quad \theta = c(x), \quad x \in R^3, \quad \nu = \nu(x) = \frac{1}{c_0^2}\left(1 - \frac{c_0^2}{c^2}\right),$$

$$S^\nu = \frac{\partial^2}{\partial t^2}, \qquad p = \left\langle \varphi_{\text{out}} \left| \frac{\partial^2}{\partial t^2} \right| \varphi_{\text{in}} \right\rangle_T.$$

In this case the conjugate Green's operator

$$G^* : G^*(x, x'; t - t') = G(x', x; -(t - t'))$$

determines wave propagation from the receiver in the reverse time. The support of the tomographic functional in a homogeneous reference medium is concentrated in a paraboloidal layer if the incoming wave is plane, and in an ellipsoidal layer if the incoming wave is spherical (assuming a point-type receiver). The section of the support of the tomographic functional passing through the symmetry axis is shown in Fig. 10.1(a). Here the paraboloidal layer is formed by kinematically equivalent points (c_0=const). Fig. 10.1(b) sketches the configuration of the support of the tomographic functional under the assumptions that the source (s) is a point one and the receiver (r) has the apparatus function $h(t) = \delta(t)$. The space-time representations of the field φ_{in} and φ_{out}, are shown in Fig. 10.1(b), and the support of the tomographic functional is indicated in Fig. 10.1(c).

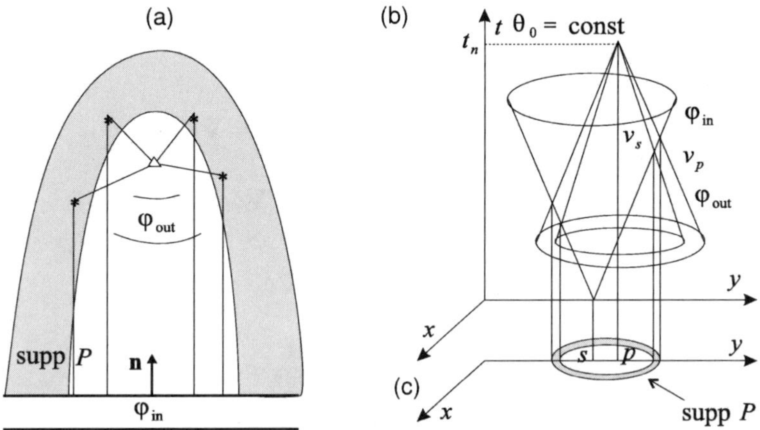

Fig. 10.1 Space-time representation of the fields and the support of the tomographic functional.

10.3.2 *The Lame equation in an isotropic infinite medium*

The operators of the wave field propagation in the reference (L_0) and perturbed (L_θ) media are written as follows

$$\begin{aligned} L_0\,\boldsymbol{\varphi} = \rho_0\,\frac{\partial^2 \boldsymbol{\varphi}}{\partial t^2} - [(\lambda_0 + \mu_0)\boldsymbol{\nabla\nabla}\cdot\boldsymbol{\varphi} + \mu_0\,\Delta\boldsymbol{\varphi} + \boldsymbol{\nabla}\lambda_0\boldsymbol{\nabla}\cdot\boldsymbol{\varphi} \\ + \boldsymbol{\nabla}\mu_0\,\times\boldsymbol{\nabla}\times\boldsymbol{\varphi} + 2(\boldsymbol{\nabla}\mu_0\cdot\boldsymbol{\nabla})\boldsymbol{\varphi}]\,, \end{aligned}$$

$$\begin{aligned} L_\theta\,\boldsymbol{\varphi} = \rho\,\frac{\partial^2 \boldsymbol{\varphi}}{\partial t^2} - [(\lambda + \mu)\boldsymbol{\nabla\nabla}\cdot\boldsymbol{\varphi} + \mu\,\Delta\boldsymbol{\varphi} + \boldsymbol{\nabla}\lambda\boldsymbol{\nabla}\cdot\boldsymbol{\varphi} \\ + \boldsymbol{\nabla}\mu\,\times\boldsymbol{\nabla}\times\boldsymbol{\varphi} + 2(\boldsymbol{\nabla}\mu\cdot\boldsymbol{\nabla})\boldsymbol{\varphi}]\,. \end{aligned} \tag{10.11}$$

Here

$$\Theta_0 = \begin{matrix} |\lambda_0\,(x)\rangle \\ |\mu_0\,(x)\rangle \\ |\rho_0\,(x)\rangle \end{matrix}\,, \qquad \Theta = \begin{matrix} |\lambda(x)\rangle \\ |\mu(x)\rangle \\ |\rho(x)\rangle \end{matrix}\,,$$

$$\nu(\delta\theta) = \delta\theta\,, \qquad \lambda(x) = \lambda_0\,(x) + \delta\lambda(x)\,,$$

$$\mu(x) = \mu_0(x) + \delta\mu(x),$$

$$\rho(x) = \rho_0(x) + \delta\rho(x).$$

In this case, the structure of the operator P can be described by the relation

$$\langle p^\nu| = ||\langle P^\lambda|\langle P^\mu|\langle P^\rho|\,||.$$

The perturbation operator $\delta L = L_0 - L_\theta$ from (10.4) is given by the sum

$$\delta L = \delta L_\lambda + \delta L_\mu + \delta L_\rho,$$

where

$$\delta L_\lambda : \delta L_\lambda \boldsymbol{\varphi} = \delta\lambda \boldsymbol{\nabla}\boldsymbol{\nabla}\cdot\boldsymbol{\varphi} + \boldsymbol{\nabla}\delta\lambda\boldsymbol{\nabla}\cdot\boldsymbol{\varphi} = \boldsymbol{\nabla}(\delta\lambda\boldsymbol{\nabla}\cdot\boldsymbol{\varphi}), \tag{10.12}$$

$$\begin{aligned}\delta L_\mu \ : \ \delta L_\mu \boldsymbol{\varphi} &= \delta\mu\boldsymbol{\nabla}\boldsymbol{\nabla}\cdot\boldsymbol{\varphi} + \delta\mu\Delta\boldsymbol{\varphi} + \boldsymbol{\nabla}\delta\mu\times\boldsymbol{\nabla}\times\boldsymbol{\varphi} \\ &+ 2(\boldsymbol{\nabla}\delta\mu\cdot\boldsymbol{\nabla})\boldsymbol{\varphi} = 2\delta\mu\Delta\boldsymbol{\varphi} + \delta\mu\times\boldsymbol{\nabla}\times\boldsymbol{\nabla}\boldsymbol{\varphi} \\ &+ \boldsymbol{\nabla}\delta\mu\times\boldsymbol{\nabla}\times\boldsymbol{\varphi} + 2(\boldsymbol{\nabla}\delta\mu\cdot\boldsymbol{\nabla})\boldsymbol{\varphi},\end{aligned}$$

$$\delta L_\rho : \delta L_\rho \boldsymbol{\varphi} = -\delta\rho\frac{\partial^2}{\partial t^2}\boldsymbol{\varphi}. \tag{10.13}$$

In the coordinate form with the unit vector $\mathbf{e}_i$, the expression $(\boldsymbol{\nabla}\delta\mu\cdot\boldsymbol{\nabla})\boldsymbol{\varphi}$ can be represented as the sum

$$\sum_i(\boldsymbol{\nabla}\delta\mu\cdot\boldsymbol{\nabla}\varphi_i)\mathbf{e}_i.$$

Using the identity

$$\boldsymbol{\nabla}\cdot(\eta\boldsymbol{\nabla}\xi) = \eta\Delta\xi + \boldsymbol{\nabla}\eta\cdot\boldsymbol{\nabla}\xi,$$

we transform the expression

$$2\delta\mu\Delta\boldsymbol{\varphi} + 2(\boldsymbol{\nabla}\delta\mu\cdot\boldsymbol{\nabla})\boldsymbol{\varphi}$$

to the form

$$2\sum_i\boldsymbol{\nabla}\cdot(\delta\mu\boldsymbol{\nabla}\varphi_i)\mathbf{e}_i.$$

In view of the identity

$$\boldsymbol{\nabla}\times(\xi\mathbf{f}) = \boldsymbol{\nabla}\xi\times\mathbf{f} + \xi\boldsymbol{\nabla}\times\mathbf{f},$$

we obtain

$$\delta\mu\boldsymbol{\nabla}\times\boldsymbol{\nabla}\times\boldsymbol{\varphi} + \boldsymbol{\nabla}\delta\mu\times\boldsymbol{\nabla}\times\boldsymbol{\varphi} = \boldsymbol{\nabla}\times\delta\mu\boldsymbol{\nabla}\times\boldsymbol{\varphi}.$$

Finally, the action of the operator δL_μ is represented in the form

$$\delta L_\mu = \boldsymbol{\nabla}\times(\delta\mu\boldsymbol{\nabla}\times\boldsymbol{\varphi}) + 2\boldsymbol{\nabla}\cdot(\delta\mu\boldsymbol{\nabla}\boldsymbol{\varphi}).$$

We recall that the values of the tomographic functionals (10.10) allow for the representation

$$\langle p^{\theta}|\delta\theta\rangle_V = \int\limits_V \boldsymbol{\varphi}_{\text{out}} \otimes \delta L_{\theta}\, \boldsymbol{\varphi}_{\text{in}}\, dx.$$

For the variations of the arbitrary parameter field $\delta\theta$ the symbol $\otimes$ denotes the convolution over time and summation over all the indices of spatial coordinates of vector and tensor expressions involved. The symbol $*$ is reserved for ordinary convolution over time. For example,

$$\psi_{ik} \otimes \varphi_{lm} \stackrel{\Delta}{=} \psi_{ik} * \delta_{il}\, \delta_{km}\, \varphi_{lm}\,.$$

The determinate part $\bar{u}_n = \tilde{u}_n - \tilde{\varepsilon}_n$ of the model (10.9) can be represented as the sum

$$\bar{u}_n = \int\limits_V \boldsymbol{\varphi}_{\text{out}} \otimes \delta L_{\lambda}\, \boldsymbol{\varphi}_{\text{in}}\, dx + \int\limits_V \boldsymbol{\varphi}_{out} \otimes \delta L_{\mu} \boldsymbol{\varphi}_{\text{in}}\, dx + \int\limits_V \boldsymbol{\varphi}_{\text{out}} \otimes \delta L_{\rho} \boldsymbol{\varphi}_{\text{in}}\, dx. \quad (10.14)$$

Taking into account expression (10.12) for δL_{λ}, we single out the divergent part of the first integrand from (10.14)

$$\begin{aligned}\boldsymbol{\varphi}_{\text{out}} \otimes \delta L_{\lambda}\, \boldsymbol{\varphi}_{\text{in}} &= \boldsymbol{\varphi}_{out} \otimes \boldsymbol{\nabla}(\delta\lambda \boldsymbol{\nabla} \cdot \boldsymbol{\varphi}_{\text{in}}) \\ &\equiv \boldsymbol{\nabla} \cdot (\boldsymbol{\varphi}_{\ out} \otimes \delta\lambda \boldsymbol{\nabla} \cdot \boldsymbol{\varphi}_{\text{in}}) - \delta\lambda \boldsymbol{\nabla} \cdot \boldsymbol{\varphi}_{\ out} \otimes \boldsymbol{\nabla} \boldsymbol{\varphi}_{\text{in}}\end{aligned}$$

and apply the Gauss–Ostrogradskii theorem, which yields

$$\int\limits_V \boldsymbol{\nabla} \cdot (\boldsymbol{\varphi}_{\text{out}} \otimes \delta\lambda \boldsymbol{\nabla} \cdot \boldsymbol{\varphi}_{\text{in}})dx = \int\limits_{\partial V} d\mathbf{s} \cdot (\boldsymbol{\varphi}_{\text{out}} \otimes \delta\lambda \boldsymbol{\nabla} \cdot \boldsymbol{\varphi}_{\text{in}}).$$

Then we choose a large enough integration volume, which enables us to set $\delta\lambda|_{\partial V} \equiv 0$. Then the integral over the surface vanishes, and the first term in (10.14) can be rewritten as follows

$$\begin{aligned}\int\limits_V \boldsymbol{\varphi}_{\text{out}} \otimes \delta L_{\lambda}\, \boldsymbol{\varphi}_{\text{in}}\, dx &= \int\limits_V \delta\lambda \boldsymbol{\nabla} \cdot \boldsymbol{\varphi}_{\text{out}} \otimes \boldsymbol{\nabla} \cdot \boldsymbol{\varphi}_{\text{in}}\, dx \\ &= \int\limits_V p^{\lambda}(x)\delta\lambda(x)dx = \langle\langle \boldsymbol{\varphi}_{\text{out}}|S^{\lambda}|\boldsymbol{\varphi}_{\text{in}}\rangle_T|\delta\lambda\rangle_V\,,\end{aligned}$$

$$S^{\lambda}\, :\, p^{\lambda} = \langle \boldsymbol{\varphi}_{\text{out}}|S^{\lambda}|\boldsymbol{\varphi}_{\text{in}}\rangle = -\boldsymbol{\nabla} \cdot \boldsymbol{\varphi}_{\text{out}} \otimes \boldsymbol{\nabla} \cdot \boldsymbol{\varphi}_{\text{in}}, \quad (10.15)$$

i.e., the action of the operator S^{λ}, describing the interaction of the fields $\boldsymbol{\varphi}_{\text{in}}$ and $\boldsymbol{\varphi}_{\text{out}}$, is reduced to the symmetric transformation of the fields $\boldsymbol{\varphi}_{\text{in}}$ and $\boldsymbol{\varphi}_{\text{out}}$ to the fields of their divergences $\boldsymbol{\nabla} \cdot \boldsymbol{\varphi}_{\text{in}}$ and $\boldsymbol{\nabla} \cdot \varphi_{\text{out}}$. To determine the action of the operator S^{μ}, we take into account the identity

$$\boldsymbol{\nabla} \cdot (\xi \mathbf{f}) = \boldsymbol{\nabla}\xi \cdot \mathbf{f} + \xi \boldsymbol{\nabla} \cdot \mathbf{f},$$

and expression (10.13) for δL_μ. In this way we obtain the relation

$$\int_V \varphi_{out} \otimes \nabla \cdot (\delta\mu \nabla \varphi_{\text{in}})dx$$

$$= \int_V \nabla \cdot (\varphi_{out} \otimes \delta\mu \nabla \varphi_{\text{in}})dx - \int_V \nabla \varphi_{out} \otimes \delta\mu \nabla \varphi_{\text{in}})dx\,. \quad (10.16)$$

In view of the condition $\delta\mu|_{\partial V} = 0$, the application of the Gauss–Ostrogradskii theorem yields that the first integral on the right hand side of the equation (10.16) vanishes while the second integral involves $\delta\mu$ as a factor.

Similarly, the interaction of φ_{out} with the second term in (10.16) yields

$$\int_V \varphi_{out} \otimes \nabla \times (\delta\mu \nabla \times \varphi_{\text{in}})\, dx$$

$$= \int_V \nabla \cdot \int_T dt'\, \varphi_{\text{out}}(t-t') \times \delta\mu \nabla \times \varphi_{\text{in}}(t')dx$$

$$+ \int_V (\delta\mu \nabla \times \varphi_{out} \otimes \nabla \times \varphi_{\text{in}})dx\,. \quad (10.17)$$

Here, the first integral on the right-hand side has the divergent form and vanishes because $\delta\mu|_{\partial V} = 0$. Finally, by using equations (10.16) and (10.17), we obtain

$$\int_V \varphi_{\text{out}} \otimes \delta L_\mu \varphi_{\text{in}}\, dx$$

$$= \int_V dx\, [\nabla \times \varphi_{\text{out}} \otimes \nabla \times \varphi_{\text{in}} - 2\nabla\varphi_{\text{out}} \otimes \nabla\varphi_{\text{in}}]\delta\mu$$

$$= \int_V dx p^\mu (x)\delta\mu(x) = \langle\langle \varphi_{\text{out}} | S^\mu | \varphi_{\text{in}} \rangle_T | \delta\mu \rangle_V\,,$$

i.e.,

$$S^\mu \; : \; p^\mu = \langle \varphi_{\text{out}} | S^\mu | \varphi_{\text{in}} \rangle = \nabla \times \varphi_{\text{out}} \otimes \nabla \times \varphi_{\text{in}} - 2\nabla\varphi_{\text{out}} \otimes \nabla\varphi_{\text{in}}\,. \quad (10.18)$$

In view (10.13), the action of the operator S^ρ is determined by the expression

$$\int_V \varphi_{\text{out}} \otimes \delta L_\rho \varphi_{\text{in}}\, dx = -\int_V \varphi_{\text{out}} \otimes \frac{\partial^2}{\partial t^2} \varphi_{\text{in}}\, dx\,, \quad (10.19)$$

which can be rewritten in the form that is explicitly symmetric with respect to the fields (φ_{out} and φ_{in}). Indeed, performing the integration by parts in (10.19)

$$\varphi_{\text{out}} \otimes \frac{\partial^2}{\partial t^2} \varphi_{\text{in}} = \int_{-\infty}^{\infty} dt \varphi_{\text{out}}\, (T-t) \frac{\partial^2}{\partial t^2} \varphi_{\text{in}}\, (t)$$

$$= \varphi_{\text{out}} \frac{\partial}{\partial t} \varphi_{\text{in}} \Big|_{-\infty}^{\infty} - \int \frac{\partial}{\partial t} \varphi_{\text{out}} \frac{\partial}{\partial t} \varphi_{\text{in}}\, dt$$

and taking into account that conditions $\varphi_{\text{out}}|_{+\infty} = 0$, $\partial/\partial t\, \varphi_{\text{in}}|_{-\infty} = 0$, we obtain

$$S^{\rho} : p^{\rho} = \langle \varphi_{\text{out}} | S^{\rho} | \varphi_{\text{in}} \rangle = \frac{\partial}{\partial t}\varphi_{\text{out}} \otimes \frac{\partial}{\partial t}\varphi_{\text{in}} \,. \tag{10.20}$$

Thus, formulas (10.15), (10.18) and (10.20) determine the interaction operators S^{λ}, S^{μ} and S^{ρ} for the fields φ_{in} and φ_{out}. Physical interpretation of the tomographic functionals is that their integral kernels are spatial functions of the influence of variations of the fields of the required medium parameters on the sampling experimental values of the wave field of the sounding signal. The support of the tomographic functional is determined by the region of interaction of the fields φ_{in} and φ_{out}. This region is certainly bounded on account of physical conditions of the tomographic experiment (a finite temporal interval of measurements and a finite velocity of the field φ_{out} propagation) .

The norm of the tomographic functional is determined by the amplitude of the influence function related to the interaction operator. In Fig. 10.2, the diagrams of the scattering of the plane wave by the elementary perturbations of the shear module μ and the mass density ρ are represented (the perturbation of the Lame parameter λ has an isotropic diagram). The diagrams characterize the interaction of the fields φ_{in} and φ_{out}, if their sources are placed far from the scattering object (near-field region). The field is generated by a "source" with the time dependence determined by the apparatus function of the seismic recording channel. Since the frequency band of the apparatus function is limited, the interaction of the fields φ_{in} and φ_{out}, which is always a convolution in the time domain (and a product in the frequency domain), involves only the overlapping spectral components of the fields φ_{in} and φ_{out}.

The analysis of tomographic functionals provides the possibility of mathematically designing a tomographic experiment by monitoring of the parameters involved in tomographic functional (the localization of the source and receivers of the fields φ_{in} and φ_{out} in space, the shape of the sounding signal φ_{in}, and the choice of the time interval for recording the field φ_{out}). In this case, the character of the interaction is determined only by the influence of the medium parameters upon the physics of propagation of the sounding signal. For example, if the field is pure rotational (the transversal wave in an isotropic reference medium), then the tomographic functional p^{λ} becomes zero. Accordingly, such a tomographic experiment obviously provides no information on the field of the parameter λ. A physically informative tomographic experiment is characterized not only by the norms of tomographic functionals, but also by the domain of the overlapping of their support.

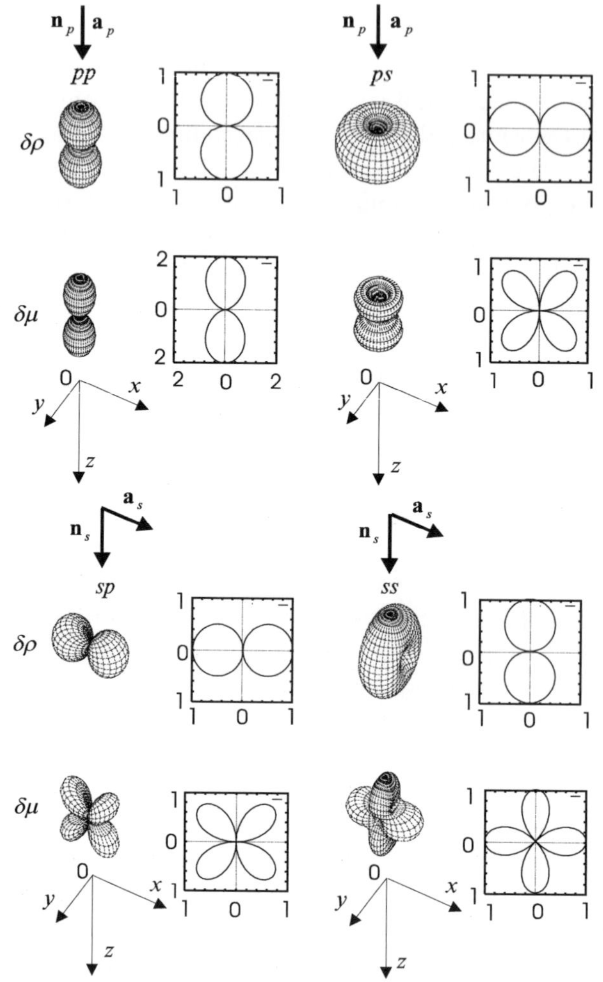

Fig. 10.2 The diagrams of the scattering of shear and longitudinal wave by the elementary perturbations. To the right of the spatial diagrams their sections in the plane (x, z) are represented; $\mathbf{n}_q$ and $\mathbf{a}_q$ are respectively the normal to the wave front and the unit amplitude vector of the incident q-wave $(q = p, s)$.

Consider the case where the reference medium may be treated as homogeneous. Let the variations of the elastic parameters in a homogeneous isotropic medium be such that $|\nabla\lambda|$, $|\nabla\mu|$ and $\delta\rho$ are small enough. We consider two cases where the field $\varphi_{\rm in}$ is either a part of the pure longitudinal wave or the pure transversal wave. In the first case, we introduce the function

$$\nu_p\,(v_p^2\,) = \rho_0 \left(\frac{v_p^2}{v_{p_0}^2} - 1\right),$$

where, in addition, $\rho_0\,\delta(v_p^2\,) = \delta(\lambda + 2\mu)$. Writing the action of the perturbation operator upon the field $\boldsymbol{\varphi}_{\rm in} \equiv \boldsymbol{\varphi}_p$, where $\boldsymbol{\varphi}_p\,:\boldsymbol{\nabla}\times\boldsymbol{\varphi}_p\,\equiv 0$, we obtain

$$\delta L\boldsymbol{\varphi}_p\;= \delta\lambda\boldsymbol{\nabla}\boldsymbol{\nabla}\cdot\boldsymbol{\varphi}_p + 2\delta\mu\Delta\boldsymbol{\varphi}_p = (\delta\lambda + 2\delta\mu)\Delta\boldsymbol{\varphi}_p\;= \delta(\lambda + 2\mu)\Delta\boldsymbol{\varphi}_p\,. \qquad (10.21)$$

Making an observation that at the space points where the sources are absent the equation

$$\left(\rho_0\frac{\partial^2}{\partial t^2} - (\lambda_0 + 2\mu_0\,)\Delta\right)\boldsymbol{\varphi}_p\;= 0\,,$$

holds true, we replace $\Delta\boldsymbol{\varphi}_p$ with $(\rho_0/\lambda_0 + 2\mu_0)\,\partial^2\,\boldsymbol{\varphi}_{\rm p}/\partial t^2$ in (10.21), which yields

$$\delta L\boldsymbol{\varphi}_{\rm p} = \delta\nu_p\,\frac{\partial^2}{\partial t^2}\,\boldsymbol{\varphi}_{\rm p}\,.$$

The corresponding tomographic functional and the interaction operator are of the form

$$S^{\nu_{\rm p}}\;: p^{\nu_{\rm p}}\;= \langle\boldsymbol{\varphi}_{\rm in}|S^{\nu_{\rm p}}|\boldsymbol{\varphi}_{\rm out}\rangle = -\frac{\partial}{\partial t}\boldsymbol{\varphi}_{\rm in}\otimes\frac{\partial}{\partial t}\boldsymbol{\varphi}_{\rm out}. \qquad (10.22)$$

In the second case, we introduce the function

$$\nu_{\rm p}\,(v_{\rm s}^2\,) = \rho_0\left(\frac{v_{\rm s}^2}{v_{\rm s_0}^2} - 1\right).$$

Then

$$\rho\delta(v_{\rm s}^2\,) = \delta\mu\,.$$

The action of the perturbation operator upon the field $\boldsymbol{\varphi}_{\rm in}\;\equiv\boldsymbol{\varphi}_{\rm s}$, where $\boldsymbol{\varphi}_{\rm s}\;:\boldsymbol{\nabla}\cdot\boldsymbol{\varphi}_{\rm s}\equiv 0$, is expressed as

$$\delta L\boldsymbol{\varphi}_{\rm s}\;= \delta\mu\Delta\boldsymbol{\varphi}_{\rm s}\,.$$

By using the equation

$$\left(\rho_0\,\frac{\partial^2}{\partial t^2} - \mu\Delta\right)\boldsymbol{\varphi}_{\rm s}\;= 0$$

for the solenoidal field $\boldsymbol{\varphi}_{\rm s}$, we obtain

$$\delta L\boldsymbol{\varphi}_{\rm s}\;= \delta\nu_{\rm s}\,\frac{\partial^2}{\partial t^2}\,\boldsymbol{\varphi}_{\rm s}\,.$$

The tomographic functional for ν_s is of the form

$$S^{\nu_s} : p^{\nu_s} = \langle \varphi_{\text{in}} | S^{\nu_s} | \varphi_{\text{out}} \rangle = -\frac{\partial}{\partial t}\varphi_{\text{in}} \otimes \frac{\partial}{\partial t}\varphi_{\text{out}} . \tag{10.23}$$

Despite the fact that the operators S^{ν_p} and S^{ν_s} given by (10.22) and (10.23), respectively, and determining the interaction of the fields φ_{in} and φ_{out}, are identical, the supports of the corresponding tomographic functionals are different, which stems from the fact that the velocities of propagation of the compressional (v_p) and shear (v_s) waves are different. To illustrate the configuration of the support of the tomographic functional, we consider the plane incoming wave. Fig. 10.3(a) shows the characteristic cones corresponding to the field of the point source φ_{out}, that is generated by the receiver at the moment t_n and propagates in the reverse time. The outer conical surface corresponds to the propagation with the compressional velocity, and the inner one corresponds to the propagation with the shear velocity. The field is concentrated between these conical surfaces in the three-dimensional space R^3 (for simplicity, we assume that the "source" is simulated by the δ-function in time, and, therefore, in Fig. 10.3(a) there is no action of the convolution over time). The normal to the front of the plane wave is oriented in the direction opposite to that of the vector $\mathbf{e}_Z$. In Fig. 10.3(a), the field φ_{in} represents the plane compressional wave with finite signal length. The support of the tomographic functional (Fig. 10.3(a)) is limited in the space by two surfaces. The outer one is the paraboloid of revolution that is the space projection of the section of the outer characteristic cone (v_p) by the plane front of the incident P-wave. The inner surface is the ellipsoid of revolution that is the space projection of the section of the inner characteristic cone (v_s) by the plane front of the incident P-wave. The symmetry axes of both surfaces limiting the support of the tomographic functional coincide with the vector $\mathbf{e}_Z$.

In Fig. 10.3(c), the incident field φ_{in} is the plane S-wave. In this case, the outer conical surface limiting the support of the functional (Fig. 10.3(d)) is a hyperboloid of revolution, whereas the inner one is a paraboloid of revolution. The symmetry axis is $\mathbf{e}_Z$ as in the previous case.

10.3.3 *The transport equation of the stationary sounding signal*

The transport equation arises in studying the spatial structure of the atmosphere of the Earth and other planets, in studying plasma objects in a wide range of physical experiments where the measurement data are quadratic functions of wave fields, and in describing passages of particle flows through a medium.

Let the field of the sounding signal be

$$\varphi(x, \mathbf{n}) , \quad \mathbf{n} \in \Omega = \{\mathbf{n} : \ \mathbf{n} \in R^3 , |\mathbf{n}| = 1\} .$$

The transport equations for the reference (L_0) and perturbed (L_θ) media in the

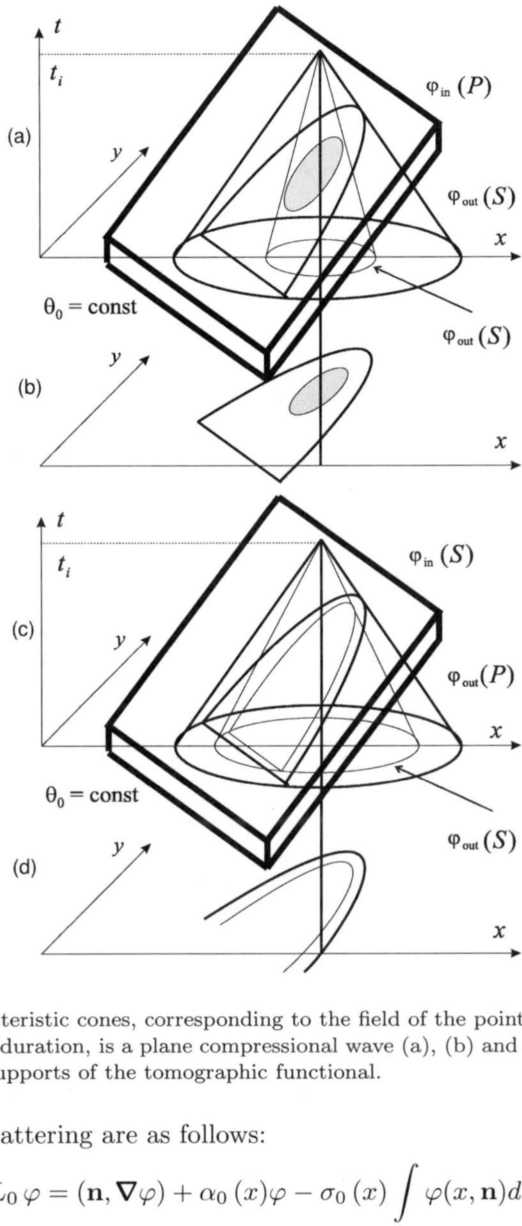

Fig. 10.3 The characteristic cones, corresponding to the field of the point source $\boldsymbol{\varphi}_{\text{out}}$. The field $\boldsymbol{\varphi}_{\text{in}}$, with finite signal duration, is a plane compressional wave (a), (b) and a plane shear wave (c), (d); (b), (d) are the supports of the tomographic functional.

case of isotropic scattering are as follows:

$$L_0\,\varphi = (\mathbf{n}, \boldsymbol{\nabla}\varphi) + \alpha_0\,(x)\varphi - \sigma_0\,(x) \int\limits_{\Omega} \varphi(x, \mathbf{n})d\mathbf{n}, \tag{10.24}$$

$$L_\theta\,\varphi = (\mathbf{n}, \boldsymbol{\nabla}\varphi) + \alpha(x)\varphi - \sigma(x) \int\limits_{\Omega} \varphi(x, \mathbf{n})d\mathbf{n} \tag{10.25}$$

where α is the absorption coefficient, β is scattering coefficient, $\Theta_0 = ||\alpha_0\,(x), \sigma_0\,(x)||$, $\nu(\delta\theta) = \delta\theta)$. The interaction operator has two components: $S^\nu = [S^\alpha\,, S^\sigma\,]$, and the tomographic functional is represented as $\langle p^\nu \,| = [\langle p^\alpha|, \langle p^\sigma\,|]$. The interaction operators S^α and S^σ are easily found from the following representation of the perturbation operator for equations (10.24) and (10.25):

$$\delta L = \delta L_\alpha + \delta L_\sigma \,,$$

$$\delta L_\alpha \,:\, \delta L_\alpha \, \varphi = \delta\alpha\varphi \,, \tag{10.26}$$

$$\delta L_\sigma \,:\, \delta L_\sigma \, \varphi = \delta\sigma \int\limits_\Omega \varphi(x, \mathbf{n}') \, d\mathbf{n}' \,. \tag{10.27}$$

The tomographic functionals are written in the form

$$\left\langle p^\alpha \,|\delta\alpha\right\rangle_\Omega = \int\limits_\Omega \varphi_{\text{out}} \, \delta L_\alpha \, \varphi_{\text{in}} \, d\mathbf{n} \,,$$

$$\left\langle p^\sigma \,|\delta\sigma\right\rangle_\Omega = \int\limits_\Omega \varphi_{\text{out}} \, \delta L_\sigma \, \varphi_{\text{in}} \, d\mathbf{n} \,.$$

Substituting δL_α in the form (10.26) and δL_σ in the form (10.27), we obtain

$$S^\alpha : \left\langle p^\alpha \,(x)\right| = \left\langle \varphi_{\text{out}} \,|S^\alpha \,|\varphi_{\text{in}}\right\rangle_\Omega = \left\langle \varphi_{\text{out}}|S^\alpha|\varphi_{\text{in}}\right\rangle_\Omega \,, \tag{10.28}$$

$$S^\sigma : \left\langle p^\sigma \,\right| = \left\langle \varphi_{out} \,|S^\sigma \,|\varphi_{\text{in}}\right\rangle_\Omega = \left\langle \varphi_{\text{out}}\right| \int\limits_\Omega \varphi_{\text{in}} \,(x, \mathbf{n}') \, d\mathbf{n}' \Big\rangle_\Omega \,. \tag{10.29}$$

Examining formulas (10.28) and (10.29), we see that, in the isotropic case, the action of the tomogarphic functional corresponding to the absorption coefficient α, reduces to the integration of the product of the field φ_{in} and φ_{out} over the full solid angle, whereas the action of the tomographic functional of the scattering coefficient is determined by the product of the integrals of the fields φ_{in} and φ_{out} over the full solid angle. We note that the field φ_{in}, occurring in the expressions for the tomographic functional is the solution of the transport equation (10.24) in the reference medium. The field φ_{out} satisfies the conjugate equation

$$-(\mathbf{n}, \boldsymbol{\nabla}\varphi) + \alpha_0 \,(x) - \sigma_0 \,(x) \int\limits_\Omega \varphi(x, \mathbf{n}') \, d\mathbf{n}' = h \,,$$

where h is the field generated by the source.

10.3.4 *The diffusion equation*

In the geoelectrics it is common to consider the quasistationary fields **E** and **H**. In the case of the absence of charges, these fields satisfy the homogeneous diffusion equation

$$\Delta\varphi - \sigma \frac{\partial}{\partial t} \varphi = 0 \,,$$

where φ coincides with either $\mathbf{E}$ or $\mathbf{H}$. The evolutionary operators in the reference (L_0) and perturbed (L_θ) media are as follows

$$L_0 = \Delta - \sigma_0(x)\frac{\partial}{\partial t},$$
$$L_\theta = \Delta - \sigma(x)\frac{\partial}{\partial t},$$

where $\sigma(x) = \sigma_0(x) + \delta\sigma(x)$. The perturbation operator δL_σ can be represented as $\delta L_\sigma = \delta\sigma\partial/\partial t$. Then the tomographic functional takes the form

$$p^\sigma = \boldsymbol{\varphi}_{\text{in}} \otimes \frac{\partial}{\partial t}\boldsymbol{\varphi}_{\text{out}}\,. \tag{10.30}$$

We note that $\boldsymbol{\varphi}_{\text{in}}$ and $\boldsymbol{\varphi}_{\text{out}}$ occurring in (10.30) satisfy the equation

$$L_0\boldsymbol{\varphi}_{\text{out}} = 0\,, \qquad L_0^*\boldsymbol{\varphi}_{\text{in}} = \mathbf{h}\,,$$

where

$$L_0^* = \Delta + \sigma_0(x)\frac{\partial}{\partial t}\,.$$

10.4 Ray Tomographic Functional in the Dynamic and Kinematic Interpretation of the Remote Sounding Data

The interpretation of the data of remote sounding can be carried out by two ways. As the basic data for the tomography experiment can be used the phase characteristic of the sounding signal (in this case we deal with the ray tomography) or the full field of the sounding signal (it corresponds to the diffraction tomography). In the latter case the wave fields structure carries an information about space domain which is limited by kinematic reason, instead of that in the ray tomography, the information about medium properties is located in the vicinity of the bent ray. The kinematic restriction in the diffraction tomography appears because of a finite time function of the source, if before the switching-on of the source, the medium has been in rest. It is possible to develop the tomographic interpretation using the stationary wave fields. It must take into account both kinematic and dynamic restrictions, moreover, the interference principle of the wave field shaping allows us to construct its spectral amplitude as a superposition of the fields having the ray structure.

Let us show forming of wave amplitude spectrum on the example of acoustic field excited by a point source located inside the uniform layer with thickness h and satisfied the homogeneous boundary conditions on the upper and lower bounds:

$$(\nabla^2 + k^2)\varphi = -\delta(z - z_0)\frac{\delta(\rho)}{2\pi\rho}\,, \tag{10.31}$$

$$\varphi\big|_{z=0} = 0\,, \qquad \varphi\big|_{z=h} = 0\,.$$

Let us consider the solution (10.31) without the consideration of the boundary conditions, i.e. $\varphi_0(x) = e^{ikl}/(4\pi l)$ (l — the ray segment), which joins the source point $(0,z_0)$ and the observation point $x = (\rho, z)$). The rays are reflected by the upper bound $z = 0$ and the reflection is determined by the boundary condition $\varphi|_{z=0} = 0$. In this case the reflection coefficient is equal -1. The reflected ray can be represented as the ray generated by some virtual source with a mirror position relatively to the plane $z = 0$ in the point $z = -z_0$. Then the reflected wave can be described by the formula $e^{ikl}/(4\pi l)$, where l is the ray segment between virtual source and observation point. In the case of lower bound, the reflection coefficient is equal $+1$, and the virtual source is located in the point $2h - z_0$. Multiple reflection of the waves can be described using a set of rays, when each ray has an own virtual source located in the point $2h_n + z_0$, where $n = 0, \pm1, \pm2, \ldots$. The wave field is the sum of the incident wave and multiple reflected waves. Each reflected wave is the spherical wave, generated by appropriate virtual source (mirror-image method):

$$\varphi(x) = \frac{1}{4\pi} \sum_{n=-\infty}^{\infty} (-1)^n \left[\frac{e^{ikl_\uparrow}}{l_\uparrow} - \frac{e^{ikl_\downarrow}}{l_\downarrow} \right], \tag{10.32}$$

where

$$l_\uparrow = [\rho^2 + ((z - z_0) + 2hn)^2]^{1/2}$$

is the ray, which goes from the source toward the upper bound;

$$l_\downarrow = [\rho^2 + ((z + z_0) + 2hn)^2]^{1/2}$$

is the ray, which goes from the source toward the upper bound.

Let us show as simplest ray expansion (10.32) is connected with more complicated approximation of the wave field vertically inhomogeneous medium. Taking into account the axial symmetry of the problem, we write the wave equation for the Hankel image of the field $\tilde{\varphi}(ka, z)$:

$$\tilde{\varphi}(ka, z) = 2\pi \int_0^\infty J_0(ka\rho)\varphi(\rho, z)\rho d\rho,$$

where $J_0(ak\rho)$ is the Bessel function of zero order;

$$2\pi \int_0^\infty J_0(ka\rho) \frac{\partial}{\partial\rho}\left(\rho \frac{\partial\varphi}{\partial\rho}\right) d\rho + \frac{\partial^2 \varphi(ka, z)}{\partial z^2} + k^2 n^2(z)\tilde{\varphi}(ka, z) = \delta(z - z_0); \tag{10.33}$$

$$\tilde{\varphi}(ka, 0) = 0; \qquad \partial\tilde{\varphi}/\partial z - (ka, h) = 0;$$

$n(z)$ is the refraction index. Taking into account that

$$2\pi \int_0^\infty J_0(ka\rho) \frac{\partial}{\partial\rho}\left(\rho \frac{\partial\varphi}{\partial\rho}\right) d\rho = -k^2 a^2 \tilde{\varphi}(ka, z)$$

and we obtain another form for the equation (10.33):

$$\frac{\partial^2 \tilde{\varphi}(ka, z)}{\partial z^2} + k^2 [n^2(z) - a^2]\tilde{\varphi}(ka, z) = -\delta(z - z_0) \tag{10.34}$$

with the same boundary condition. In the case of the solution of the equation (10.34), which satisfies the first boundary condition, we introduce the notation $\tilde{\varphi}_1(ka, z_>)$. And for the solution which satisfies the second boundary condition we introduce the notation $\tilde{\varphi}_2(ka, z_<)$, then we can write general solution in the form

$$\tilde{\varphi}(ka, z) = \tilde{\varphi}_1(ka, z_<)\tilde{\varphi}_2(ka, z_>)/W(ka), \tag{10.35}$$

where

$$W = \begin{pmatrix} \tilde{\varphi}_1 & \tilde{\varphi}_2 \\ \frac{\partial \tilde{\varphi}_1}{\partial z} & \frac{\partial \tilde{\varphi}_2}{\partial z} \end{pmatrix}$$

is the Wronskian, the symbols $z_>$, $z_<$ mean, that $z > z_0$ and $z < z_0$ correspondingly.

We represent $\tilde{\varphi}_1(ka, z_<)$ and $\tilde{\varphi}_2(ka, z_>)$ using up-going and down-going wave fields

$$\tilde{\varphi}_1(ka, z_<) = U(ka, z_<) + r_1(ka, z_<)D(ka, z_<),$$
$$\tilde{\varphi}_2(ka, z_>) = D(ka, z_>) + r_2(ka, z_>)U(ka, z_>).$$

Where U is the solution of the homogeneous equation for up-going waves in the vicinity of z_0; D is down-going wave. At that the normalization of U and D is chosen as follows

$$W[U, D] = -2ik,$$

$$r_1 = \frac{U(ka, 0)}{D(ka, 0)}, \qquad r_2 = \frac{\partial_z D(ka, h)}{\partial_z U(ka, h)}$$

are the reflection coefficients at upper and lower bounds.

Taking into account that

$$W(\tilde{\varphi}_1, \tilde{\varphi}_2) = (1 - r_1 r_2)W(U, D) = 2ik(1 - r_1 r_2)$$

and using the Hankel image $\tilde{\varphi}(ka, z)$ (10.35) we can write the final representation of the wave field

$$\varphi(\rho, z) = \frac{ik}{4\pi}\int_0^\infty J_0(ka\rho)[U(ka, z_<) + r_1 D(ka, z_<)]$$
$$\times [D(ka, z_>) + r_2 U(ka, z_>)](1 - r_1 r_2)^{-1} a\, da.$$

Factoring of the binomial $(1 - r_1 r_2)^{-1}$, we obtain the multiple scattering representation (by analogy with the ray representation)

$$\varphi(\rho, z) = \sum_{n=0}^{\infty} \frac{ik}{4\pi}\int_0^\infty J_0(ka, \rho)[(r_1 r_2)^n U(ka, z_<)D(ka, z_>)$$
$$+ (r_1 r_2)^{n+1} U(ka, z_>)D(ka, z_<) + r_1^{n+1} r_2^n D(ka, z_<)$$
$$\times D(ka, z_>) + r_1^n r_2^{n+1} U(ka, z_>)U(ka, z_<)]a\, da. \tag{10.36}$$

Here the first addend describes the wave which come out from the source in upward direction, then after n reflections for each bound it propagates as down-going wave;

the second addend describes similar wave with $(m+1)$ reflections for each bound; the third and the fourth addends are interpreted as the wave, which goes out from the source in definite direction, then after n reflections from one bound and $n+1$ reflections from other one it propagates in the opposite direction. In the considered cases as reflection we understand the refraction phenomenon also, appropriate turning point is described by the equality $n^2(z) = a^2$ (see Eq. (10.34)).

The wave field (10.36) is generated by the point source, located in the point $\rho = 0$ and $z = z_0$, i.e. the solution is the Green function. The ray interpretation of the Green function, considered above, is based on the ray expansion of the wave field:

$$G(x, x_0) = \sum_j g_j(x, x_0) \exp[i\omega\tau_j(x, x_0)],$$

where g_j is the amplitude of the wave field for j-th ray; τ_j is the eikonal.

10.5 Construction of Incident and Reversed Wave Fields in Layered Reference Medium

The tomographic functional (10.10) includes the wave fields φ_{in} and φ_{out}, which are the solutions of the direct problems. In the case of the arbitrary reference medium such solutions can be connected with the great mathematical and computing difficulty. As usual, the solution of the direct problem can be simplified if the reference medium has some type of the symmetry. The most symmetric is, obviously, uniform infinite medium allows an analytic representation of the wave fields φ_{in} and φ_{out}. The next class of the medium is a class of piecewise-layered media. Such media have a wide spread occurrence owing to shell structure of the Earth and near-Earth space. Under the local description the spherical symmetry layers can be approximated by plane parallel layers. Let us describe the algorithms for calculation of the fields φ_{in} and φ_{out} in the case of the layered medium (Brekhovskikh, 1960; Petrashen *et al.*, 1985; Kennett, 1983).

The problem of propagation of elastic, acoustic and electromagnetic waves with zero-initial conditions can be reduced to the solution of the differential matrix equation of the first order (Ursin, 1983; Ursin and Berteussen, 1986):

$$\frac{\partial}{\partial x_3}\tilde{\varphi} = \beta\tilde{L}\tilde{\varphi} = \beta \begin{Vmatrix} 0 & \tilde{L}_1 \\ \tilde{L}_2 & 0 \end{Vmatrix} \cdot \begin{Vmatrix} \tilde{\varphi}_1 \\ \tilde{\varphi}_2 \end{Vmatrix}. \tag{10.37}$$

So, if the equation $L\varphi = 0$ describes propagation of elastic waves in an isotropic medium with the parameters ρ, λ, μ, then the Fourier transform by lateral coordinates (x_1, x_2) and the temporal frequency ω:

$$\varphi(\omega, p_1, p_2, x_3) = \iiint\limits_{-\infty}^{\infty} dt\, dx_1\, dx_2\, \varphi(t, x_1, x_2, x_3) \times \exp[i\omega t - ip_1x_1 - ip_2x_2]. \tag{10.38}$$

We obtain following representation for $\tilde{\varphi}$:

$$\tilde{\varphi}_1 = \begin{Vmatrix} \partial_t \tilde{\varphi}_{(3)} \\ \tilde{\sigma}_{(1)} \\ \tilde{\sigma}_{(2)} \end{Vmatrix}, \qquad \tilde{\varphi}_2 = \begin{Vmatrix} \tilde{\sigma}_{(3)} \\ \partial_t \tilde{\varphi}_{(1)} \\ \partial_t \tilde{\varphi}_{(2)} \end{Vmatrix},$$

where $\partial_t \varphi_{(i)}$ is the velocity of the particles in the direction of i-th ort; $\sigma_i = \sigma_{3i}$ is the component of the stress tensor; the tilde denotes the Fourier transform.

In this case

$$\tilde{L}_1 = \begin{Vmatrix} \frac{1}{\lambda+2\mu} & \frac{\lambda S_1}{\lambda+2\mu} & \frac{\lambda S_2}{\lambda+2\mu} \\ \frac{\lambda S_1}{\lambda+2\mu} & a & -\frac{\mu(3\lambda+2\mu)S_1 S_2}{\lambda+2\mu} \\ \frac{\lambda S_2}{\lambda+2\mu} & -\frac{\mu(3\lambda+2\mu)S_1 S_2}{\lambda+2\mu} & b \end{Vmatrix},$$

where $a = \rho - \frac{4\mu(\lambda+\mu)S_1^2}{\lambda+2\mu} - \mu S_2^2$; $b = \rho - \mu S_2^1 - \frac{4\mu(\lambda+\mu)S_2^2}{\lambda+2\mu}$;

$$\tilde{L}_2 = \begin{Vmatrix} \rho & S_1 & S_2 \\ S_1 & \mu^{-1} & 0 \\ S_2 & 0 & \mu^{-1} \end{Vmatrix}.$$

Here $S_i = p_i/\omega$ $(i = 1; 2)$ are the components of the slowness vector; $\beta = -i\omega$.

The equation (10.37) describes the plane P–SV-wave with the normal vector $\mathbf{n} = \mathbf{e}_1$, for the initial condition $\tilde{\sigma}_2 = \partial_t \tilde{\varphi}_2 = 0$, and the equation (10.37) describes SH-wave, for the initial condition $S_1 = 0$, $\tilde{\sigma}_1 = \tilde{\sigma}_3 = \partial_t \tilde{\varphi}_1 = \partial_t \tilde{\varphi}_3 = 0$. Finally, if a normal vector to the wave front is $\mathbf{e}_3$ and initial conditions are $\tilde{\sigma}_1 = \tilde{\sigma}_2 = \partial_t \tilde{\varphi}_1 = \partial_t \tilde{\varphi}_2 = 0$, then the equation (10.37) describes P plane wave.

Let us consider an electromagnetic case. The Maxwell equations for an isotropic medium reads as

$$\nabla \times \mathbf{E} = -\mu \frac{\partial \mathbf{H}}{\partial t}, \quad \nabla \times \mathbf{H} = \sigma \mathbf{E} + \varepsilon \frac{\partial \mathbf{H}}{\partial t}.$$

For the Fourier transform, by analogy with the relation (10.38), we obtain the following representation $\tilde{L}_1$, $\tilde{L}_2$, $\tilde{\varphi}_1$ and $\tilde{\varphi}_2$ from the equation (10.37):

$$\beta = -i\omega, \qquad \tilde{\varphi}_1 = \begin{Vmatrix} E_1 \\ E_2 \end{Vmatrix}, \qquad \tilde{\varphi}_2 = \begin{Vmatrix} -H_2 \\ H_1 \end{Vmatrix},$$

$$\tilde{\mathbf{L}}_1 = \begin{Vmatrix} \mu - \frac{S_1^2}{\gamma} & -\frac{S_1 S_2}{\gamma} \\ -\frac{S_1 S_2}{\gamma} & \mu - \frac{S_2^2}{\gamma} \end{Vmatrix}, \quad \tilde{\mathbf{L}}_2 = \begin{Vmatrix} \gamma - \frac{S_2^2}{\mu} & \frac{S_1 S_2}{\mu} \\ \frac{S_1 S_2}{\mu} & \gamma - \frac{S_1^2}{\mu} \end{Vmatrix}.$$

Where $S_i = p_i/\omega$, $i = 1; 2$ and $\gamma = \sigma/(\omega i) + \varepsilon$.

According to the boundary conditions the tangential components $\mathbf{E}$ and $\mathbf{H}$ must be continuous at the crossing of the boundary. In the case of horizontally layered medium it leads to continuity the vectors $\tilde{\boldsymbol{\varphi}}_1$ and $\tilde{\boldsymbol{\varphi}}_2$. The components of the vector $\tilde{\boldsymbol{\varphi}}$ provide the following representation of the energy flow in x_3 direction

$$R_3 = -\frac{1}{4}(\tilde{\varphi}_1^* \tilde{\varphi}_2 + \tilde{\varphi}_2^* \tilde{\varphi}_1) = -\frac{1}{4}\tilde{\varphi}^* M \tilde{\varphi},$$

$$\mathbf{M} = \begin{Vmatrix} 0 & \mathbf{I} \\ \mathbf{I} & 0 \end{Vmatrix}$$

($\mathbf{I}$ is identity matrix with dimension 2×2).

In the case of the acoustic medium, the vectors $\tilde{\boldsymbol{\varphi}}_1$, $\tilde{\boldsymbol{\varphi}}_2$ and matrices $\tilde{\mathbf{L}}_1$, $\tilde{\mathbf{L}}_2$ (from the equation (10.37)) are represented as

$$\beta = -i\omega\,, \qquad \tilde{\varphi}_1 = ||\partial_t\, \tilde{\varphi}_3||\,, \qquad \tilde{\varphi}_2 = ||\tilde{\sigma}||\,,$$

$$\tilde{L}_1 = \left\| \frac{1}{\lambda} - \frac{S^2}{\rho} \right\|, \quad \tilde{L}_2 = ||\rho||\,, \quad S^2 = \frac{S_1^2 + S_2^2}{\omega^2}\,.$$

The solution of the evolutionary equation (10.37) in the general form can be represented with the help of the transition matrix $P(x_3, x_3^0)$, which satisfies the equation

$$\frac{\partial}{\partial x_3} P(x_3, x_3^0) = \beta \tilde{L}(x_3)\, P(x_3, x_3^0) \tag{10.39}$$

and initial conditions $P(x_3^0, x_3^0) = I$.

Taking into account that the equation (10.39) can be represented in the equivalent form:

$$P(x_3, x_3^0) = I + \beta \int\limits_{x_3^0}^{x_3} dx_3'\, L(x_3')\, P(x_3', x_3^0)\,,$$

we can write the transition matrix in an iterative form:

$$\mathbf{P} = I + \int\limits_{x_3^0}^{x_3} \tilde{L}(x_3')\, dx_3' + \int\limits_{x_3^0}^{x_3} dx_3'\, \tilde{L}(x_3') \int\limits_{x_3^0}^{x_3'} dx_3''\, \tilde{L}(x_3'')$$

$$+ \int\limits_{x_3^0}^{x_3} dx_3'\, \tilde{L}(x_3') \int\limits_{x_3^0}^{x_3'} dx_3''\, \tilde{L}(x_3'') \int\limits_{x_3^0}^{x_3''} dx_3'''\, \tilde{L}(x_3''') + \ldots$$

Let note, that the transition matrix for the adjoint equation has an opposite sign of β.

Chapter 11

Tomography methods of recovering the image of medium

The interpretation of the reconstruction tomography as a problem of recovering the image of an object from its ray projections has gradually transformed to its interpretation as a problem of integral geometry, i.e. as the problem of recovering a function $\theta(x)$ from the values of a given function

$$u(x) = \int\limits_{\Sigma(x)} \theta(x')\, w(x,x')d\sigma\,,$$

where $x \in R^n$; $\Sigma(x)$ is a set of k-dimensional manifolds in R^k, $k < n$; $w(x,x')$ is a weight function; $d\sigma$ is an element of a measure on $\Sigma(x)$.

As we have seen in Chap. 10, where the notion of the tomographic functional was introduced, any problem of recovering distributed parameters of a system (in particular, an arbitrary problem of remote sensing) can be considered as a problem of the reconstruction tomography. Integrals of parametric functions defined on manifolds of the dimension coinciding with that of the definition domain are treated as the initial data of the corresponding tomographic problem. These manifolds are fixed in the linear approximation; in the general case, they are implicitly transformed at each iteration.

11.1 Elements of Linear Tomography

Mathematically, classical tomography is based on *the Radon transform* (Helgason, 1999; Herman, 1980):

$$u(p,\mathbf{n}) = \hat{R}\varphi = \int\limits_{X} \varphi(\mathbf{x})\ delta(p - \mathbf{x}\cdot\mathbf{n})d\mathbf{x} \triangleq \langle\delta(p-\mathbf{x}\cdot\mathbf{n})|\varphi(\mathbf{x})\rangle,$$

where $u(p,\mathbf{n})$ is the Radon transform of a function $\varphi(\mathbf{x})$, defined in an N-dimensional space; $\delta(p - \mathbf{x}\cdot\mathbf{n})$ is the singular kernel of the Radon transform, which cuts out an $(N-1)$-dimensional hyperplane (a straight line in the two-dimensional case), that is defined by the normal vector $\mathbf{n}$ and the aiming-distance parameter p, i.e., the distance from the origin of coordinates to the hyperplane:

$$\mathbf{x}\cdot\mathbf{n} = p.$$

The Radon transform arose in the classical tomography as a mathematical model for the description multiforeshortening radiographies by X-rays of biological "soft" objects, on which the scattering of X-rays may be assumed small. The transport equation for a point (x_0), monoenergetic (E_0), collimation (ν_0) source of X-rays radiation is a reduced transport equation of the form

$$(\boldsymbol{\nu}\cdot\nabla)I(x)+\varphi(x)I(x)=c\delta(x-x_0)\delta(\nu-\nu_0)\delta(E-E_0). \tag{11.1}$$

Here, $I(x)$ is the beam of primary X-ray photons; $\boldsymbol{\nu}$ is the direction of collimation of the source; $\varphi(x,E)$ is a linear coefficient; E is the energy of photons. It is assumed that the vectors $\boldsymbol{\nu}$ are coplanar, i.e. the problem considered is two-dimensional. The solution of (11.1) can be written, upon performing the change of variables $\mathbf{x}\longrightarrow(p,\mathbf{n})$, as

$$I(p,\mathbf{n},E_0)=c_0\exp\{-\int\limits_{\mathcal{L}(p,\mathbf{n})}\varphi(x)\,dx\}\,, \tag{11.2}$$

where $\mathcal{L}(p,\mathbf{n})$ is the line which radiation propagates. Taking the logarithm of the relation (11.2), we obtain the Radon transform for the two-dimensional problem

$$-\ln\frac{I(p,\mathbf{n},E_0)}{c}\triangleq u(p,\mathbf{n})=\int\limits_{\mathcal{L}}\varphi(x)\,dx=\int\limits_{X}\varphi(x)\delta(p-\mathbf{x}\cdot\mathbf{n})\,dx\,.$$

Here, in the rightmost expression, the integration is carried out over the entire domain of the definition of $\varphi(x)$.

The model of a planar cross section of an object arises in describing plasma sounding by a planar laser ray ("light knife"). After its transition through the volume of plasma under investigation, the planar beam is focused on a photodiode. As it is not difficult to see, in both cases tomographic experiments are based on the adequate use of the ray description of the propagation of the sounding signal (the wavelength is small in comparison with the characteristic size of an inhomogeneity with a strongly stretched forward scattering indicatrix).

Before passing to the basic properties of the direct Radon transform, we recall some properties of linear continuous functionals, i.e., of generalized functions

$$l(\varphi)\triangleq(l,\varphi)=(l(x),\varphi(x)),$$

where $x\in R^n$.

11.1.1 *Change of variables*

Let $x=Ly+x_0$ be a nonsingular mapping, then

$$[l(Ly+x_0),\varphi(y)]=|L|^{-1}\,[l(x),\varphi(L^{-1}\,(x-x_0))]\,,$$

whence for the delta-function we obtain

$$\langle\delta(Ly+x_0)|\varphi(y)\rangle\triangleq\langle\delta(x)||L|^{-1}\,\varphi(L^{-1}\,(x-x_0))\rangle\,.$$

In particular,

(a) $\langle \delta(x - x_0) | \varphi \rangle = \varphi(x_0)$;
(b) $\langle \delta(ay) | \varphi(y) \rangle = \langle \delta(x) | \, |a|^{-n} \varphi(a^{-1} x) \rangle$;
(c) if $L:\; L^* L = I$ and $x_0 = 0$, then

$$\langle \delta(Ly) | \varphi(y) \rangle = \langle \delta(x) | \varphi(L^* x) \rangle;$$

(d) $$\langle \delta(\mathcal{L}(x)) \varphi(x) \rangle = \sum_k \langle \delta(y) | | \, L'|_{x_k} |^{-1} \varphi(L'|_{x_k}^{-1} y + x_k) \rangle = \sum_k | \, L'|_{x_k} |^{-1} \varphi(x_k),$$

where $x_k \;:\; L(x_k) = 0$.

For example:

$$\delta(x^2 - c^2 t^2) = \frac{1}{2ct} \left[\delta(x - ct) + \delta(x + ct) \right];$$

$$\delta(\sin x) = \sum_{k=-\infty}^{\infty} \delta(x - k\pi);$$

$$\delta(x - x_0) = \left[\prod_i |h_i| \right]^{-1} \prod_i \delta(q_i - q_i^0).$$

Here, $q_i = q_i(x)$, and $h_i = \sum_k (\partial x_k / \partial q_i)^2$, which corresponds to the passage from Cartesian coordinates to curvilinear orthogonal coordinates $q_i(x)$.

11.1.2 *Differentiation of generalized function*

Let $l \in C^n$ and let $|\alpha| \leq n$. Then

$$(D^\alpha l, \varphi) = (-1)^\alpha (l, D^\alpha \varphi).$$

In particular,

$$\langle D^\alpha \delta | \varphi \rangle \triangleq (-1)^{|\alpha|} D^\alpha \varphi(0).$$

Below we list the basic properties of the Radon transform (Pikalov and Preobrazhenski, 1983; Helgason, 1999).

1. Its linearity

$$\hat{R}[\alpha_1 \varphi_1 + \alpha_2 \varphi_2] = \alpha_1 \hat{R}[\varphi_1] + \alpha_2 \hat{R}[\varphi_2]$$

follows from the definition.

2. In view of (11.2), the rotation of the object leads to the same rotation of the vector $\mathbf{n}$ in the argument of the Radon transform, i.e.,

$$\hat{R}[\varphi(Ox)] = u(p, On); \tag{11.3}$$

the similarity transformation with coefficient $\alpha \neq 0$ increases the aiming distance by the factor α and decreases the Radon transform by the factor α^{1-n}, i.e.,

$$R[\varphi(\alpha x)] = \alpha^{-n} u\left(p, \frac{\mathbf{n}}{\alpha}\right) = \alpha^{1-n} u(\alpha p, \mathbf{n}).$$

3. The shift of the object by x_0 amounts to the shift of the aiming distance by $\mathbf{n} \cdot \mathbf{x}_0$, i.e.,

$$\hat{R}[\varphi(x - x_0)] = u(p - \mathbf{n} \cdot \mathbf{x}_0, \mathbf{n}) .$$

4. The Radon transform of the derivative of the function φ, which describes the object, is given by

$$\hat{R}[(\mathbf{k} \cdot \boldsymbol{\nabla})\varphi] = \mathbf{k} \cdot \mathbf{n} \frac{\partial u(p, \mathbf{n})}{\partial p},$$

for the mixed second derivative we have

$$\hat{R}[(\mathbf{k} \cdot \boldsymbol{\nabla})(\mathbf{l} \cdot \boldsymbol{\nabla})\varphi] = (\mathbf{k} \cdot \mathbf{n})(\mathbf{l} \cdot \mathbf{n}) \frac{\partial^2 u(p, \mathbf{n})}{\partial p^2},$$

whence it follows that

$$\hat{R}[(\boldsymbol{\nabla} \cdot \boldsymbol{\nabla})\varphi(x)] = |\mathbf{n}|^2 \frac{\partial^2 u}{\partial p^2} = \frac{\partial^2 u(p, n)}{\partial p^2}. \tag{11.4}$$

Finally, for an arbitrary linear differential operator $\hat{L} = L(\partial/\partial x_1, \partial/\partial x_2, \ldots, \partial/\partial x_k)$ with constant coefficients, we have

$$\hat{R}[\hat{L}\varphi(x)] = \hat{L}\left(n_1 \frac{\partial}{\partial p}, \ldots, n_k \frac{\partial}{\partial p}\right) u(p, n) .$$

An example of using property (11.4) is provided by the application of the direct Radon transform to the 3-dimensional wave equation

$$\left(\nabla^2 - \frac{1}{c^2} \frac{\partial^2}{\partial t^2}\right)\varphi(x, t) = 0 ,$$

$$\hat{R}\left[\left(\nabla^2 - \frac{1}{c^2} \frac{\partial^2}{\partial t^2}\right)\varphi\right] = \left(\frac{\partial^2}{\partial p^2} - \frac{1}{c^2} \frac{\partial^2}{\partial t^2}\right)u(p, \mathbf{n}) ,$$

which makes possible to reduce the 3-dimensional wave equation to the corresponding 1-dimensional equation.

5. Differentiation the Radon transform

$$(\mathbf{k} \cdot \boldsymbol{\nabla})\hat{R}[\varphi(x)] = -\frac{\partial}{\partial p}\hat{R}[(\mathbf{k} \cdot \mathbf{x})\varphi(x)]$$

for two vectors $\mathbf{k}$ and $\mathbf{l}$ and the derivatives with respect to components of the unit vector $\mathbf{n}$, we arrive at the equality

$$(\mathbf{k} \cdot \boldsymbol{\nabla})(\mathbf{l} \cdot \boldsymbol{\nabla})u(p, \mathbf{n}) = \frac{\partial^2}{\partial p^2}\hat{R}[(\mathbf{k} \cdot \mathbf{x})(\mathbf{l} \cdot \mathbf{x})\varphi(x)] .$$

6. The Radon transform of the convolution of functions is as follows:

$$\hat{R}\left[\int g(x - x')s(x')dx'\right] = \int s(x')dx' \int g(x - x')\delta(p - \mathbf{n} \cdot \mathbf{x})dx.$$

Performing the change of variables $x - x' = y$, we derive

$$u(p, \mathbf{n}) = \int s(x')\, dx' \int g(y)\delta(p - \mathbf{n}\cdot\mathbf{x}' - \mathbf{n}\cdot\mathbf{y})\, dy$$
$$= \int s(x')\hat{R}[g(p - \mathbf{n}\cdot y, \mathbf{n})]\, dx'$$
$$= \int \hat{R}[g(p - p', \mathbf{n})]\, dp' \int s(x')\delta[p' - \mathbf{n}\cdot\mathbf{x}']\, dx',$$

i.e.

$$\hat{R}[g * s] = \hat{R}[g] * \hat{R}[s].$$

Note that the Radon transform of a multidimensional convolution is equal to the one-dimensional convolution of the Radon transforms with respect to the aiming distance p^*.

7. The relation between the Radon and Fourier transforms.
We write the Fourier transform with respect to the aiming distance of the Radon transform $u(p, \mathbf{n})$:

$$\tilde{u}(p, \mathbf{n}) = (2\pi)^{-1/2} \int e^{-ip\nu}\, dp \int \varphi(x)\, \delta(p - \mathbf{x}\cdot n)\, dx$$
$$= (2\pi)^{-1/2} \int dx\, \varphi(x) \int e^{-ip\nu}\, \delta(p - \mathbf{x}\cdot n)\, dp$$
$$= (2\pi)^{-1/2} \int \varphi(x)\, e^{-i\nu(\mathbf{x}\cdot n)}\, dx,$$

or, in operator form,

$$F_p\left[u(p, \mathbf{n})\right] \triangleq F_p\left[\hat{R}[\varphi(x)]\right] = (2\pi)^{(n-1)/2}\, F_{\mathbf{x}}\left[\varphi(x)\right]. \qquad (11.5)$$

The one-dimensional Fourier of the Radon transform is equivalent to the multidimensional Fourier transform of the function $\varphi(x)$, which characterizes the object. This relation between the Fourier transform of the Radon transform and multidimensional Fourier transform of φ is formulated as the following generalized theorem on the central section: for a fixed $\mathbf{n}$, the one-dimensional Fourier transform of the Radon transform yields the central section of the multidimensional Fourier spectrum along the ray $\mathbf{n}\nu$, passing through the origin of coordinates.

Based on the relation (11.5), the inverse Radon transform is derived in the following way:

$$\varphi(x) = (2\pi)^{(-n+1)/2}\, F_x^{-1}\, F_p\, u(p, \mathbf{n})$$
$$= C_n \int d(\nu\mathbf{n})\, e^{i\nu\mathbf{n}\cdot\mathbf{x}} \int dp\, e^{-i\nu p}\, u(p, \mathbf{n})$$
$$= C_n \int d(\nu\mathbf{n}) \int dp\, e^{i\nu(q-p)}\, u(p, \mathbf{n}),$$

where $q = \mathbf{n}\cdot\mathbf{x}$. Taking into account that $d(\nu\mathbf{n}) = d\mathbf{n}\nu^{n-1}\, d\nu$, we represent $\varphi(x)$ in the form

$$\varphi(x) = C_n \int\limits_{|\mathbf{n}|=1} d\mathbf{n} \int\limits_0^\infty d\nu\, \nu^{n-1} \int dp\, e^{i\nu(q-p)}\, u(p, \mathbf{n}) = C_n \int\limits_{|\mathbf{n}|=1} d\mathbf{n}\, J(\mathbf{n}, q)\,.$$

We extend the limits of integration with respect to ν to $-\infty$ and $+\infty$, either by introducing an even function $J_e(\mathbf{n}, q)$,

$$J_e(\mathbf{n}, q) = \frac{1}{2}[J(\mathbf{n}, q) + J(-\mathbf{n}, -q)]$$

or by using the following explicit representation of the integral $J(\mathbf{n}, q)$:

$$\begin{aligned} J_e(\mathbf{n}, q) &= \frac{1}{2}\int_0^\infty \nu^{n-1}\, d\nu \int_{-\infty}^\infty u(p, \mathbf{n})\, e^{i\nu(q-p)}\, dp \\ &+ \frac{1}{2}\int_0^\infty \nu^{n-1}\, d\nu \int_{-\infty}^\infty u(p, -\mathbf{n})\, e^{-i\nu(q-p)}\, dp \\ &= \frac{1}{2}\int_{-\infty}^\infty |\nu|^{n-1}\, d\nu \int_{-\infty}^\infty u(p, \mathbf{n})\, e^{i\nu(q-p)}\, dp\,. \end{aligned}$$

The cases of even and odd values of n are considered separately.

1. Let n be odd. Integrating $J_o(\mathbf{n}, q)$ by parts, $n-1$ times, we obtain

$$\begin{aligned} J_o(\mathbf{n}, q) &= \frac{1}{2}(i)^{1-n}\int_{-\infty}^\infty d\nu \int_{-\infty}^\infty dp \left(\frac{\partial}{\partial p}\right)^{n-1} u(p, \mathbf{n})\, e^{i\nu(q-p)} \\ &= \frac{1}{2}(i)^{1-n}\int_{-\infty}^\infty \left(\frac{\partial}{\partial p}\right)^{n-1} u(p, \mathbf{n})\, \delta(q-p)\, dp \\ &= \frac{1}{2}(i)^{1-n}\left(\frac{\partial}{\partial p}\right)^{n-1}\Bigg|_{p=q=\mathbf{n}\cdot x} u(p, \mathbf{n})\,. \end{aligned}$$

Thus, for odd values of n, the calculation of the integral $J_o(\mathbf{n}, q)$ is reduced to the computation of the derivative of order $(n-1)$ with respect to the aiming distance p of the Radon transform $u(p, \mathbf{n})$ at the point $p = g = \mathbf{n}\cdot x$. The ultimate formula for Radon inversion in this case is

$$\varphi(x) = C_n \int_{|\mathbf{n}|=1} \left(\frac{\partial}{\partial p}\right)^{n-1}\Bigg|_{p=\mathbf{n}\cdot x} u(p, \mathbf{n})\, d\mathbf{n}\,, \tag{11.6}$$

where $C_n = (2\pi i)^{1-n}/2$.

2. Let n be even. Similarly to the case of add values of n, we integrate $J_e(\mathbf{n}, g)$ by parts $n-1$ times, which yields

$$\begin{aligned} J_e(\mathbf{n}, q) &= \frac{1}{2}(i)^{1-n}\int_{-\infty}^\infty \operatorname{sgn}(\nu) d\nu \int dp e^{-i\nu(p-q)} \left(\frac{\partial}{\partial p}\right)^{n-1} u(p, \mathbf{n}) \\ &= \frac{1}{2}(i)^{1-n}\int_{-\infty}^\infty dp \left(\frac{\partial}{\partial p}\right)^{n-1} u(p, \mathbf{n}) \int_{-\infty}^\infty d\nu \operatorname{sgn}(\nu) e^{-i\nu(p-q)} \\ &= \frac{(i)^{-n}}{2\pi}\int \frac{(\partial/\partial p)^{n-1}\, u(p, \mathbf{n})}{p-q}\, dp\,. \end{aligned}$$

Here we have used the relation

$$F[\mathrm{sgn}(\nu)] = \frac{i}{\pi}\mathcal{P}\left(\frac{1}{p}\right),$$

where $\mathcal{P}$ is the symbol of integration in the sense of Cauchy principal value. Denoting the Hilbert transform by $\mathcal{H}$, i.e., setting

$$\mathcal{H}(u) = -\frac{1}{\pi}\mathcal{P}\left(\frac{1}{p} * u(p)\right),$$

we can write the following representation of the Radon inversion in an even-dimensional space:

$$\begin{aligned}\varphi(x) &= -iC_n \int\limits_{|\mathbf{n}|=1} d\mathbf{n}\left[\mathcal{H}\left\{\left(\frac{\partial}{\partial p}\right)^{n-1} u(p,\mathbf{n})\right\}\right](\mathbf{n}\cdot x, n) \\ &= -\frac{iC_n}{\pi}\int\limits_{|\mathbf{n}|=1} d\mathbf{n}\,\mathcal{P}\int\limits_{-\infty}^{\infty}\frac{(\partial/\partial p)^{n-1}\,u(p,\mathbf{n})}{p-\mathbf{n}\cdot x}\,dp\,. \end{aligned} \tag{11.7}$$

Thus, in the case of an even value of n, in order to recover the function $\varphi(x)$ one needs to compute the derivative of the order $(n-1)$ with respect to the aiming distance p of the Radon transform $u(p,\mathbf{n})$, thereafter perform the Hilbert transform with respect to the aiming distance p, and, finally, carry out integration over the unit sphere.

Analyzing representations (11.6) and (11.7) of the Radon inversion for even- and odd-dimensional spaces, we readily see that the Radon inversion is "local" for odd values of n. In this case, Radon surfaces on hyperplanes passing in the vicinity of the space point x of the recovery are used. In even-dimensional spaces, the Radon inversion is essentially nonlocal, which is explicitly demonstrated by the integral Hilbert transform. The manifestation of the locality or nonlocality of the Radon inversion is caused by a close relation between the Radon inversion and harmonic functions. In particular, this relation manifests itself in the nonlocality of wave fronts in even-dimensional spaces, where a wave front is followed by a diffusion train, which is in contrast to sharply pronounced wave fronts in odd-dimensional spaces.

We want to represent the Radon transform in the operator form. To this end, for an arbitrary function $\psi(p,\mathbf{n})$ (where $p=\mathbf{n}\cdot\mathbf{x}$), such that $\psi(p,\mathbf{n})=\psi(-p,-\mathbf{n})$, we introduce the conjugate Radon operator R^*:

$$R^*\psi = \int\limits_{|\mathbf{n}|=1}\psi(\mathbf{n}\cdot\mathbf{x},\mathbf{n})d\mathbf{n}\,.$$

Then, for odd values of n, the Radon inversion can be represented in the operator form as

$$R^- = R^*\,\Gamma_e\,,$$

where

$$\Gamma_e = C_n \left(\frac{\partial}{\partial p}\right)^{n-1}\Bigg|_{p=\mathbf{n}\cdot x},$$

whereas for even values of n, it can be represented in the form

$$R^- = R^* \Gamma_a, \quad \Gamma_a = -iC_n \mathcal{H}\left(\frac{\partial}{\partial p}\right)^{n-1}.$$

The operators Γ_o and Γ_e are counterparts of the operator $(RR^*)^-$, which occurs in the following representation of the solution of the integral equation $R\varphi = u$ provided by the least squares method:

$$\varphi = R^* (RR^*)^- u.$$

The subscript in the formula for the operator R^- of the Radon inversion is used because Radon inversion is an unstable procedure, and thus a regularization must be used (Pikalov and Preobrazhenski, 1983; Tikhonov *et al.*, 1987).

11.2 Connection of Radon Inversion with Diffraction Tomography

The linearized model of measurements in problems of sounding an object by a signal φ with the equation of propagation $L\varphi = s$ is represented by the relation (10.9):

$$\tilde{u}_n = \langle (L_0^{-1})^* h_n | \delta L_\theta \varphi_0 \rangle + \tilde{\varepsilon}_n \triangleq \langle \varphi_{n\,(out)} | \delta L_\theta | \varphi_{in} \rangle + \tilde{\varepsilon}_n .$$

The idealized analog of this model is obtained by representing the apparatus function (h_n) by a point receiver with the infinite band-pass, i.e., by letting $h_n \Rightarrow \delta(x - x_n)\delta(t - t_n)$ and by neglecting noises: $\tilde{\varepsilon}_n \Rightarrow 0$. Then $\tilde{u}_n$ will be represented as the scattering field at the measurement point x_n, i.e.,

$$\tilde{u}_n = \tilde{\varphi}|_{x_n} \triangleq (\varphi - \varphi_{in})|_{x_n} .$$

We write this model in the form $\tilde{\varphi} = \langle \varphi_{out} | \delta L_\theta | \varphi_{in} \rangle$.

As an example, consider the scalar wave equation $(c^{-2}\, \partial^2/\partial t^2 - \Delta)\varphi = 0$ in the case of a homogeneous reference medium, i.e., in the case $c_0(x) = c_0 = \text{const}$. As it was shown in Sec. 10.3, the tomographic functional p for the scalar wave equation is of the form

$$p : \ \tilde{\varphi} = \langle p | \nu(x) \rangle ,$$

where $\nu(x) = c_0^{-2}(1 - c_0^2/c^2(x))$, and the integral kernel of the tomographic functional is given by the expression

$$p = \langle \varphi_{out} \left| \frac{\partial^2}{\partial t^2} \right| \varphi_{in} \rangle_T .$$

For a point source s in the homogeneous medium considered, the fields φ_{in} and φ_{out} are of the same structure, i.e.,

$$\varphi_{in} = \frac{1}{4\pi|x - x_s|}\,\delta\left(t - t_s - \frac{|x - x_s|}{c_0}\right),$$

$$\varphi_{out} = \frac{1}{4\pi|x - x_r|}\,\delta\left(t_r - t - \frac{|x - x_r|}{c_0}\right)$$

where x_s and t_s specify the location of the source s and the time of its engagement, respectively; x_r and t_r describe the location of the receiver and the current time of recording respectively. In this particular case, the tomographic functional is of the form

$$p_r = \frac{1}{(4\pi)^2}\int\limits_T dt\,\frac{1}{|\Delta x_r|}\delta\left(t_r - t - \frac{|\Delta x_r|}{c_0}\right)\frac{\partial^2}{\partial t^2}\times\frac{1}{|\Delta x_s|}\delta\left(t - t_s - \frac{|\Delta x_s|}{c_0}\right),$$

$$|\Delta x_s| = |x - x_s|, \qquad |\Delta x_r| = |x - x_r|.$$

Expanding $|\Delta x_r|$ and $|\Delta x_s|$ into series and retaining the first terms of the expansions in the denominators and the first two terms in the arguments of the δ-function (which corresponds to the Fraunhofer approximation), we obtain

$$\begin{aligned}
p &= \frac{1}{(4\pi)^2\,|\mathbf{x}_r|\,|\mathbf{x}_s|}\int\limits_T dt\,\delta\left(t_r - t - \frac{|\mathbf{x}_r|}{c_0} - \frac{\mathbf{n}_r\cdot\mathbf{x}}{c_0}\right)\\
&\times\delta''\left[t - t_s - \frac{|\mathbf{x}_s|}{c_0} - \frac{\mathbf{n}_s\cdot\mathbf{x}}{c_0}\right] = \frac{1}{(4\pi)^2\,|\mathbf{x}_r|\cdot|\mathbf{x}_s|}\\
&\times\delta''\left[(t_r - t_s) - \frac{(|\mathbf{x}_s| + |\mathbf{x}_r|)}{c_0} - \frac{(\mathbf{n}_r + \mathbf{n}_s, \mathbf{x})}{c_0}\right]\\
&= \frac{c_0}{(4\pi)^2\,|\mathbf{x}_r|\,|\mathbf{x}_s|}\delta''\left(q_r - \mathbf{n}\cdot\mathbf{x}\right),
\end{aligned}$$

$$q_r = (t_r - t_s)C_0 - (|\mathbf{x}_s| + |\mathbf{x}_r|),$$

$$\mathbf{n}_r = \frac{\mathbf{x}_r}{|\mathbf{x}_r|}, \qquad \mathbf{n}_s = \frac{\mathbf{x}_s}{|\mathbf{x}_s|}, \qquad \mathbf{n} = \mathbf{n}_r + \mathbf{n}_s.$$

Here, the vector $\mathbf{n}$ corresponds to the normal to the plane relative to which the beam φ_{in} with vector $\mathbf{n}$ is mirror-likely reflected in the direction $(-\mathbf{n}_r)$. Finally, the scattering field is related to the function $\nu(x)$ via the Radon transform, i.e.,

$$\begin{aligned}
\tilde{\varphi} = \langle p|\nu(x)\rangle &= \frac{c_0}{(4\pi)^2\,|\mathbf{x}_r|\,|\mathbf{x}_s|}\int \nu(x)\delta''(q_r - \mathbf{n}\cdot\mathbf{x})\,dx\\
&= \text{const}\frac{\partial^2}{\partial p^2}\int \nu(x)\delta(p - \mathbf{n}\cdot\mathbf{x})dx.
\end{aligned}$$

The Radon transform, formally represented as the integral over the points of the plane $\{X : p = \mathbf{n}\cdot\mathbf{x}\}$, provides, for different values of the recording time (t_r), the following approximate representation of $\tilde{\varphi}$:

$$\tilde{\varphi} = \frac{c_0}{(4\pi)^2\,|\mathbf{x}_r|\,|\mathbf{x}_s|}\delta''[(t_r - t_s)c_0 - (|\Delta x_r| + |\Delta x_s|)].$$

Here, for different values of the recording time t_r, it is implied that integration is carried out over the corresponding ellipsoid of revolution with axes of the symmetry passing through the points of the source ($\mathbf{x}_s$) and receiver (x_r) (which are the focuses of the ellipsoids), which leads to the following clear physical interpretation: the integration is performed over sets of kinematically equivalent points in the space, i.e., over level of the constant total phase.

We consider the generalized Radon inversion in the idealized diffraction tomography on the more complicated example presented in (Beylkin, 1985). In scalar wave equation in a reference medium with the velocity of propagation $c_0(x)$

$$\left(\frac{1}{c_0^2(x)}\frac{\partial^2}{\partial t^2} - \Delta\right)\varphi(x,t) = 0$$

can be represented in the form of the Helmholtz equation

$$(k^2 n_0^2(x) + \Delta)\varphi_{in}(x,k) = 0\,,$$

where $k^2 n^2(x) = \omega^2/c_0^2(x)$, ω is a cyclic frequency. Introducing the dimensionless function $\mu(x)$ via the relations

$$\mu(x) = -\frac{\omega^2}{k^2}\nu(x)\,, \quad \nu(x) = \frac{1}{c_0^2(x)}\left(1 - \frac{c_0^2(x)}{c^2(x)}\right),$$

we write the Helmholtz equation in a medium with the velocity $c(x)$ in the following form:

$$[k^2(n_0^2(x) + \mu(x)) + \Delta]\varphi(x,k) = 0\,.$$

The perturbation operator (an operator of multiplication) is of the form $\delta L = k^2\mu(x)$. We assume that the reference medium is such that the fields φ_{in} and φ_{out} are sufficiently accurately approximated by the zero-order terms of their ray expansion. Then the scattering field is represented in the form

$$\begin{aligned}\tilde{\varphi}(k,x_r) &= k^2\langle\varphi_{out}|\mu(x)|\varphi_{in}\rangle = \langle p_r|\mu(x)\rangle \\ &= k^2\int \mathcal{A}_{out}(x,x_r)e^{ik\tau(x,x_r)}\mu(x)\mathcal{A}_{in}(x,x_s)e^{-k\tau(x,x_s)}\,dx\,.\end{aligned}$$

The amplitudes $\mathcal{A}$ satisfy the transport equation, and τ satisfies the eikonal equation, so that

$$(\nabla\tau,\nabla\tau) = n_0^2(x)\,,$$

$$\left[(\nabla\tau,\nabla) + \frac{1}{2}\Delta\tau\right]\mathcal{A} = 0\,.$$

The tomographic functional p is represented by the integral kernel

$$p = k^2\mathcal{A}_{out}(x)\mathcal{A}_{in}\exp\{ik[\tau(x,x_r) + \tau(x,x_s)]\} \triangleq k^2\mathcal{A}(x)\exp(ik\tau)\,.$$

We represent the generalized causal Radon transform associated with the representation of the scattering field $\tilde{\varphi} = \langle p|\mu\rangle$ in the form

$$R\mu = \langle\mathcal{A}(x)\delta(q - \tau(x))|\mu(x)\rangle\,,$$

and set $R \equiv 0$ for $t < 0$.

Note that the scattering field $\tilde{\varphi}$ can be represented up to the factor as the Fourier transform of the generalized Radon transform $R\mu$, i.e.,

$$\tilde{\varphi} = c\, F_q\, R\mu\,.$$

Let the point x_s of the source lies inside a closed region V with the boundary ∂V. We assume that the reference medium and the boundary ∂V are such that each point of the boundary can be connected with the point of the source by a unique ray. In addition, we assume that the rays do not intersect. In this case, the points of the boundary ∂V can be parameterized by the corresponding points of the unit sphere centered at the source point. Then the action of the conjugate operator R^* is determined in the following way:

$$R^*\,\psi = \int\limits_{\partial V} \psi(q, x_r)|_{q=\tau(x_r, x_s, x)}\; w(x, x_r)\, dx_r\,,$$

where $w(x, x_r)$ is a smooth nonnegative weight function, which is chosen in the form

$$w(x, x_r) = D(x, x_r)\mathcal{A}^{-1}(x, x_r)h(x, x_r)\,.$$

Here, $h(x, x_r)$ is a cut-off factor insuring the nonnegativity of $w(x, x_r)$, which implies that the points of the boundary ∂V and of the unit sphere are in one-to-one correspondence;

$$D(x, x_r) = \begin{Vmatrix} \partial^2_{x1}\,\tau & \partial^2_{x2}\,\tau & \partial^2_{x3}\,\tau \\ \partial^2_{x1\,x_r1}\,\tau & \partial^2_{x2\,x_r1}\,\tau & \partial^2_{x3\,x_r1}\,\tau \\ \partial^2_{x1\,x_r2}\,\tau & \partial^2_{x2\,x_r2}\,\tau & \partial^2_{x3\,x_r2}\,\tau \end{Vmatrix}\,,$$

$$\mathcal{A}^{-1}(x, x_r) = [\mathcal{A}_{out}(x, x_r)\,\mathcal{A}_{in}(x)]^{-1}\,.$$

It can be shown (Beylkin, 1985), that $D\,dx_r = D_0\,d\Omega$ (Ω is a solid angle) and that $D_0\,(x, x_r) = n(x)[1 + \cos\theta(x)]$, where ($\cos\theta(x) = \mathbf{e}_{out}\cdot\mathbf{e}_{in}$ and $\mathbf{e} = \mathbf{e}(x)$ is the unit ray vector at the point x).

We write the generalized integral Fourier operator, defined as follows:

$$\Phi\mu(x) = \frac{1}{(4\pi)^3}\int\limits_0^\infty\int\limits_{\partial V}\int\limits_V e^{ik\tau(x,x',x_r)}\,\tilde{a}(x, x', x_r)\mu(x')\,dx'dx_r k^2 dk\,, \tag{11.8}$$

where $\tau(x, x', x_r) = \tau(x', x_r, x_s) - \tau(x, x_r, x_s)$,

$$a(x, x', x_r) = \frac{\mathcal{A}(x', x_r)}{\mathcal{A}(x, x_r)}\,D(x, x_r)\,h(x, x_r)\,.$$

Observing that the generalized Fourier operator (Φ) includes the Fourier operator (F^+), where

$$F^+\;:\;F^+_k\,\varphi = \frac{1}{(2\pi)^{3/2}}\int\limits_0^\infty \tilde{\varphi}\,e^{-ikq}\,dk\,,$$

we can represent the generalized Fourier transform as the following composition of three operators:

$$\Phi = R^* \, F_k^+ \, F_q \, .$$

In order to analyze the meaning of the generalized Fourier operator introduced above, we write the linear term of the phase function:

$$\tau(x, x') \approx \nabla_x \, \tau(x, x_r \, , x_s \,)(x - x') \, ,$$

$$I_{\partial V^0} \mu = \frac{1}{(4\pi)^3} \int\limits_0^\infty \int\limits_{\partial V^0} \int\limits_V e^{ik\nabla_x \, \tau(x, x_r, x_s)(x - x')} \times$$

$$\times \, D(x, x_r) \mu(x') dx' dx_r k^2 dk \, .$$

Introducing the new variable $p = k\nabla_x \, \tau(x, x_r \, , x_s \,)$, we represent the latter integral in the form

$$I_{\partial V^0} \, \mu = \frac{1}{(4\pi)^3} \int\limits_{\Omega^0(x)} \int\limits_V e^{ip(x - x')} \, \mu(x) \, dx' \, dp \, .$$

Here, $\Omega^0 \, (x)$ is the image of the domain of integration under the above change of variable.

Examining formula (11.8), we see that the generalized Fourier operator provides for the recovery of the spectrum of the function $\mu(x)$, which is determined by the spectral domain $\Omega^0 \, (x)$. In its turn, the latter domain depends on the part of the surface of the boundary ∂V^0 over which the scattering field $\tilde{\varphi}$ is integrated, and also on the nonsingularity of the Jacobian of the passage from the coordinates on the surface (coordinates of recording) to the ray coordinates. Therefore, the approximate operator equality

$$I_{\partial V^0} \, \approx R^* \, F_k^+ \, F_q$$

determines the reconstruction operator

$$\mu \approx R^* \, F_k^+ \, \tilde{\varphi} \, .$$

We note that in our opinion, the solution obtained is mathematically elegant. But it does not actually concern to practical interpretation problems. Even under our idealized assumptions on the relationship between the scattering field and the medium structure to be recovered, the linearized model, which is based on the ray representation of the wave fields in the reference medium, has been used. The Jacobian occurring in the generalized conjugate operator R^* certainly fails to be nonsingular if the dependence of the field φ_{in} on time differs from the δ-function. This is easily seen from the generalized Radon transform, which involves the manifold $\delta(q - \tau(x))$ of the dimension coinciding with that of the problem. Each value corresponds to the integration over a three-dimensional domain. In the case of a point source, this domain is the surface of the ellipsoid with focuses at the points x_r and x_s if the temporal dependence is described by the δ-function and a layer of an ellipsoid of revolution if the duration of the field φ_{in} is finite (see Fig. 10.1(b)).

11.3 Construction of Algorithms of Reconstruction Tomography

In order to construct a recovery algorithm, we write the model (10.9) using the notion of a tomographic functional:

$$u_n = \sum_{\mu} \langle p_{\mu\, n} \, | \nu_\mu \, (\delta\theta) \rangle + \varepsilon_n \tag{11.9}$$

Here, u_n is a sampling of the seismogram, $\rho_{\mu n}$ is the tomographic functional, $\nu_\mu(\delta\theta)$ is an unknown function of the variation of parameters (μ is the index of the corresponding field, $\mu = 1 \div M$). The error ε_n includes a random component, which is generated by the fields of unidentified sources and instrument noises, as well as the potential inadequacy of the model. The latter is caused, first, by errors in the representation of the apparatus function; second, by errors in the prior picture of the medium, ignoring dissipative and nonlinear effects in the description of the propagation of sounding signals; third, by errors caused by the linearization of the field φ; fourth, by assumptions on the smoothness of the function $\theta(x)$, and so on. Since, physically the stochastic structure of the noise field can be regarded as the sum of a large number of independent sources, it is natural to assume that the random component is additive and normally distributed (Goltsman, 1971; Troyan, 1984, 1985). The statistical analysis of experimental data is generally confined to the first and second moments of the random component. As it was mentioned in Sec. 9.3, given the first and second moment, the distribution of maximum entropy is normal, which provides yet another reason for the choosing of Gaussian random variables in the constructing of the model.

In constructing of practical algorithms, the most simple and frequently used model (K_ε) is that of uncorrelated equal-accuracy errors, i.e., $\mathbf{K}_\varepsilon = \sigma_\varepsilon^2 \mathbf{I}$, where $\mathbf{I}$ is an identity matrix, and the only parameter σ_ε^2 is the variance of the random component. In the case of uncorrelated measurements with the unequal accuracy, the correlation matrix is of the form

$$\mathbf{K}_\varepsilon = \begin{Vmatrix} \sigma_1^2 & 0 & \ldots 0 \\ 0 & \sigma_2^2 & \ldots 0 \\ . & . & \ldots . \\ 0 & 0 & \ldots \sigma_2^2 \end{Vmatrix} .$$

The simplest mathematical description of the correlated errors of measurements is provided by the Markov's model of a random process, where the correlation matrix is of the form

$$\mathbf{K}_\varepsilon = \begin{Vmatrix} 1 & \gamma & \gamma^2 & \ldots \gamma^N \\ \gamma & 1 & \gamma & \ldots \gamma^{N-1} \\ \gamma^2 & \gamma & 1 & \ldots \gamma^{N-2} \\ . & . & . & \ldots . \\ \gamma^N & \gamma^{N-1} & \gamma^{N-2} & \ldots 1 \end{Vmatrix} .$$

This matrix has the advantage that its inverse can be represented in the explicit form, namely

$$\mathbf{K}_{\varepsilon}^{-1} = \frac{1}{1-\gamma^2} \left\| \begin{matrix} 1 & -\gamma & 0 & \dots 0 \\ -\gamma & 1+\gamma^2 & -\gamma & \dots 0 \\ \cdot & \cdot & \cdot & \dots\dots \\ 0 & 0 & 0 & \dots -\gamma \\ 0 & 0 & 0 & \dots 1 \end{matrix} \right\| .$$

Note that the matrix $\mathbf{K}_{\varepsilon}^{-1}$ is the finite difference counterpart of the symmetric positive defined operator $(-\Delta + \gamma I)$, which is frequently used in the Tikhonov's regularization (Tikhonov and Arsenin, 1977).

In what follows, we will assume that the covariance matrix of the measurement errors is known. However, the simultaneous estimation of the distribution parameters and of the desired parameters of the medium presents no great difficulties, although additional is needed because, in this case, the estimation problem becomes certainly nonlinear (Turchin *et al.*, 1971).

Since the random component of ε is principally unremovable, any method of solving inverse problems related to the interpretation of practical data must necessarily deal with a problem of the statistical estimation. Note that regularization methods make it possible to construct formally stable solutions of tomographic problems, which represent the desired fields $\nu(\delta\theta)$ at any point of the domain under the investigation. The fields representing the medium can be considered as multicomponent one (Tarantola, 1984). However, the solution of such problems in interpreting data of remote sensing is useless because of the insufficiency of information, i.e., practically a priori picture of the fields of parameters of the medium is not refined. The idea of recovering fields from values of a finite number of functionals of measurement has been elaborated in geophysics (see (Backus and Gilbert, 1967, 1968)). In accordance with this approach, we try to replace the recovery of the field of parameters of the medium in the whole space by the recovery of values of a number of linear functionals l of the field $\nu(\delta\theta)$. This is a well-posed problem only in the case where the unknown functional l belongs to the linear hull of the tomographic functionals $\{p_n\,, n = 1 \div N\}$, i.e., to the subspace

$$\overset{\sqcap}{\Phi}{}^{*} = \{l : \exists\alpha : (l - P^*\alpha)\nu(\delta\theta) = 0\,\forall\nu\}\,.$$

This regularization method has evolved spontaneously when solving practical inverse problems, and different modifications of this method are commonly used, although they are not properly justified.

If no a priori information on $\nu(\delta\theta)$ is available, then the stability of the solution obtained by the maximum likelihood method is determined by the Fisher's operator, which is certainly degenerate. The degeneracy of the Fisher's operator means that there is a direction in the space Θ that also determines elements of the functional space, the information on which is unavailable, whence the variance of the corresponding linear functional is unbounded. On the other hand, if the functional

l is the linear combination of the tomographic functionals $\{p_n\}$ or, in other words, if the conjugate equation $P^* \alpha - l = 0$ is solvable, then the variance of its solution is finite.

In practice, one frequently solves inverse problems by representing the solution $\nu(\delta\theta)$ in the form of a finite series:

$$|\nu(\delta\theta)\rangle = \sum_q |\psi_q\rangle\langle\psi_q|\nu(\delta\theta)\rangle \leftrightarrow \sum_q \nu_q |\psi_q\rangle.$$

In this case, it is necessary that estimations of the coefficients ν_q have a finite variance, i.e., the equations $P^*_\alpha - \nu_q = 0$ must be solvable. As the basis of $\{\nu_q\}$ one uses either a number of functions from a complete orthonormalized set (for example, Fourier series, Fourier-Bessel series, Korhunen–Loeve basis and so on; the choice of the basis is usually determined by an intuitive a priori idea of the medium), or a certain incomplete basis (characteristic functions of nonoverlapping space domains, B-splines etc. (Troyan, 1981a, 1983)), or a number of heuristic (incomplete, nonorthogonal) basis functions of the kind of "model bodies" (for example, the field representations in the magnetic and gravimetric prospecting).

The error $\Delta\nu$ of the recovery of the field $\nu(\delta\theta)$ when using the representation in the form of finite series

$$|\nu(\delta\theta)\rangle = \sum_q \nu_q |\psi_q\rangle$$

is determined, first, by errors $\Delta\nu_q$ of the recovery of the coefficients ν_q, i.e., of the values of linear functionals $\langle\psi_\nu|\nu\rangle$, and, second, by the component of the field $\nu(\delta\theta)$ not involved in the above representations, i.e.,

$$\Delta\nu(\delta\theta) = \sum_q \Delta\nu_q |\psi_q\rangle + \tilde{\nu},$$

where $\tilde{\nu} \in \Theta = \left(I - \sum_q |\psi_q\rangle\langle\psi_q|\right)\Theta$.

Note that infinite error related to $\tilde{\theta}$ is usually ignored when analyzing solutions of practical problems, i.e., it is intuitively assumed that the solution sought belongs to the subspace corresponding to the basis chosen.

The problem of the recovery of an arbitrary linear functional can be made a well-posed one if one invokes additional information by introducing either a (nondegenerate) probability measure in the space Θ or deterministic (usually quadratic) constraints. The introduction of a probability measure, for example, of the Gaussian one with positive correlator K_θ, ensures that the variance σ_l^2 of an arbitrary linear functional l becomes finite:

$$\sigma_l^2 = l(K_\theta \Delta l) = l((F + K_\theta^{-1})^{-1} l).$$

Here $F = P^* K_\varepsilon^{-1} P$ is the Fisher's operator and $\Delta l = l - P^* \alpha$. If one introduces a deterministic constraint by specifying a closed domain in the parameter space, for instance, the sphere $||\nu(\delta\theta)||_{L_2} = 1$, then, obviously, the variance of any linear functional will be finite. But such an approach is sometimes senseless because the error can prove to be of the same order as the diameter of a priori chosen domain, although the equation $P^* \alpha = l$ is solvable (Morozov, 1993).

The regularization through the introducing of compacts in the space Θ, for example, of a set of monotone bounded functions in L_2 (Bakushinski and Goncharski, 1994), is performed by applying methods of mathematical programming. In contrast to the case of quadratic constraints, the corresponding algorithms are multistep methods and they can not be represented by closed formulas

It should be noted that all the conclusions mentioned above are implied by the identity $(\widehat{l, \nu}) = l(\hat{\nu})$, which means that the best estimate of a linear functional of the field ν coincides with the value of this functional at the best estimate of the field. All basic formulas and relations can be obtained in the same way as the optimal decision operator, the construction of which was proposed in Sec. 9.3. Nevertheless, in order to emphasize the similarities and difference in the Backus-Gilbert approach and that suggested here, below we outline the algorithm for solving the inverse problem.

The problem is to recover the value of a given liner functional l of an unknown field ν_μ of parameters of the medium from N values u_n of the sum of M linear functional p_μ. For any linear functional $l(\nu)$, we seek a solution in the form of a linear combination of measurements, i.e.,

$$l(\nu) = \sum_n \alpha_n \, u_n = (\alpha, u)\,, \tag{11.10}$$

where $l(\nu) \equiv \langle l|u\rangle$. Using the model of experimental data (11.9), in the vicinity of a point X the linear estimation (11.10) can be represented as follows:

$$\nu_\mu\,(\delta\theta) = \sum_{\mu'} \langle(\alpha_\mu^X\,, p_{\mu'}\,)|\nu_{\mu'}\,(\delta\theta)\rangle_V + (\alpha_\mu^X\,, \varepsilon)\,.$$

Now we will refine the meaning of the term "the field in the vicinity of a point X". As $\nu(\delta\theta_\mu^X\,)$ one can accept, e.g., the mean value in the sphere of the radius ρ centered at the point X, i.e., the value of the linear functional of the unknown field $\nu(\delta\theta_\mu^X\,)$:

$$\nu_\mu\,(\delta\theta^X\,) = \langle l^x\,|\nu_\mu\,\rangle_V\,,$$

$$l^X = \left(\frac{4}{3}\pi\rho^3\right)^{-1} H\,(\rho - |x - X|)\,,$$

$$H(\xi) = \begin{cases} 0 & \text{if} \quad \xi < 0\,, \\ 1 & \text{if} \quad \xi \geq 0\,. \end{cases}$$

Irrespective of the specific choice of the linear procedure, the error η_μ^X of the recovery $\langle l^X\,|\nu_\mu\,\rangle$ can be formally represented as

$$\begin{aligned} \eta_\mu^l &= \langle l|\nu_m\,\rangle - \sum_{\mu'} (\alpha, \langle p_{\mu'}\,)|\nu_{\mu'}\,\rangle) - (\alpha, \varepsilon) \\ &= \langle l - (\alpha, p_\mu\,)|\nu_\mu\,\rangle \sum_{\mu' \neq \mu} \langle(\alpha, p_{\mu'}\,)|\nu_{\mu'}\,\rangle - (\alpha, \varepsilon)\,. \end{aligned} \tag{11.11}$$

The quantitative interpretation of this expression is as follows. The error η_μ depends on the unknown field $\nu_\mu(\delta\theta)$ (the first term in (11.11)) as well as on the other fields $\nu_{\mu'}$ ($\mu' \neq \mu$), and also on the random component ε. The less the norm of the functional $l - (\alpha, p_\mu)$, i.e., the better the approximation of the functional l under the estimation by the linear combination of the tomographic functionals p_μ, the less the component of the error that depends on the field ν_μ. As is readily seen, a reasonable approximation of the functional l, whose support is localized in the vicinity of the point X can be obtained only if the set $\{p_{\mu n}\}$ includes functionals whose supports have nonempty intersections with this vicinity. The error η^l_μ is the sum of linear forms of the unknown ν_μ ($\mu = 1 \div M$) and ε, and each of these forms depends on the coefficient vector α, which specifies a particular recovery procedure.

The vector α should be chosen in such a way that the recovery errors $|\eta_\mu|$ or $(\eta_\mu)^2$ be minimized. In turn, the squared error can be described by a quadratic form, which must involve all of the linear forms occurred in the representation (11.11) of η_μ.

We consider the solution that minimizes the squared recovery error, i.e.,

$$\hat{\alpha} = \operatorname{arginf}\left[(\Lambda - Q^*\alpha)^* K (\Lambda - Q^*\alpha)\right],$$

where $Q = ||P|I||$ and I is the identity operator of the dimension equal to the number of the measurement;

$$(Q^*\alpha)^* = ||\alpha_1 \dots \alpha_n \dots \alpha_N||$$
$$\times \begin{Vmatrix} \langle p_{11}(x)| & \dots & \langle p_{\mu 1}(x)| & \dots & \langle p_{M1}(x)| & 1 & \dots & 0 \\ \dots & \dots & \dots & \dots & \dots & \dots & \dots & 0 \\ \langle p_{1n}(x)| & \dots & \langle p_{\mu n}(x)| & \dots & \langle p_{Mn}(x)| & 0 & \dots & 0 \\ \dots & \dots & \dots & \dots & \dots & \dots & \dots & 0 \\ \langle p_{1N}(x)| & \dots & \langle p_{\mu N}(x)| & \dots & \langle p_{MN}(x)| & 0 & \dots & 1 \end{Vmatrix},$$
$$\Lambda^* = ||\lambda_1^* \dots \lambda_\mu^* \dots \lambda_M^* \dots \lambda_{M+1}^* \dots \lambda_{M+N}^*||,$$

where for $i = 1, \dots, M$, the λ_i^* are equal to $\langle 0|$ (except for $\lambda_\mu^* = \langle l(x)|$), whereas for $i = (M+1), \dots, (M+N)$, the λ_i^* are zero. The operator K must be positive definite.

In this case, the optimal estimation $\hat{\alpha}$, which, by the linearity of the model and the estimation of the functional, also determines the optimal estimation of the functional l itself, is given by the expression

$$\hat{\alpha} = (QKQ^*)^{-1} QK\Lambda. \tag{11.12}$$

Since the estimation error (11.11) depends on unknown fields $\nu_\mu(\delta\theta)$ and on the random variable ε, the choice of a specific operator K in the criterion determines a class of functions $\{\nu_\mu,\ \mu = 1, \dots, M\}$ and the statistical structure of the random component ε, which is equivalent to the introduction of a priori information.

Consider the statistical interpretation of the operator K and the estimation $\hat{\alpha}$. If the set of fields $\{\nu_\mu(\delta\theta)\}$ accepted as admissible solutions is characterized by the Gaussian measure in the corresponding functional space and the random component is normally distributed, then the solution that minimizes the squared error of the recovery $E(\eta^l_\mu)$ (E denotes the mean over the fields $\{\nu_\mu(\delta\theta)\}$ and realizations of ε):

$$K = \begin{Vmatrix} K_{\nu_1\nu_1} & \dots & K_{\nu_1\nu_\mu} & \dots & K_{\nu_1\nu_M} & \dots & K_{\nu_1\varepsilon} \\ \dots & \dots & \dots & \dots & \dots & \dots & \dots \\ K_{\nu_\mu\nu_1} & \dots & K_{\nu_\mu\nu_\mu} & \dots & K_{\nu_\mu\nu_M} & \dots & K_{\nu_\mu\varepsilon} \\ \dots & \dots & \dots & \dots & \dots & \dots & \dots \\ K_{\varepsilon\nu_1} & \dots & K_{\varepsilon\nu_\mu} & \dots & K_{\varepsilon\nu_M} & \dots & K_{\varepsilon\varepsilon} \end{Vmatrix}.$$

Here $K_{\nu_\mu\nu_{\mu'}}$ is an operator of the integral type such that, for any linear functionals ξ and η, the condition

$$\langle\xi|K_{\nu_\mu\,\nu_{\mu'}}|\eta\rangle = \mathrm{E}\langle\xi|\nu_\mu\rangle\langle\nu_{\mu'}|\eta\rangle\,,$$

where

$$K_{\varepsilon\varepsilon} = \mathrm{E}(\varepsilon\varepsilon^T)\,, \quad K_{\varepsilon\nu_\mu} \;:\; \mathrm{E}\varepsilon\langle\nu_\mu|\eta\rangle = K_{\varepsilon\nu_\mu}|\eta\rangle$$

is satisfied

The optimal estimation $\hat{\alpha}$ in (11.12) determines the structure of the recovery algorithm with the minimal error variance. This estimation can be represented in the following form, in which all statistical correlations of the fields ν_μ and $\nu_{\mu'}$ occur in the explicit form (under the assumption that the fields ν_μ and the random component ε are not correlated, we have $K_{\mu\varepsilon} = 0$):

$$\hat{\alpha} = \Bigg[P_\mu\,K_{\nu_\mu\,\nu_\mu}\,P^*_\mu + \sum_{\mu'\neq\mu}(P_{\mu'}\,K_{\nu_{\mu'}\,\nu_\mu}\,P^*_\mu + P_\mu\,K_{\nu_\mu\,\nu_{\mu'}}\,P^*_{\mu'}) + \sum_{\mu'\neq\mu}\sum_{\mu''\neq\mu} P_{\mu'}\,K_{\nu_{\mu'}\,\nu_{\mu''}}\,P^*_{\mu''} + K_{\varepsilon\,\varepsilon}\Bigg]^{-1} \times \Bigg(P_\mu K_{\nu_\mu\,\nu_\mu} + \sum_{\mu'\neq\mu} P_{\mu'}K_{\nu_{\mu'}\,\nu_\mu}\Bigg)|l\rangle\,.$$

In the particular case where there is no statistical correlations either between distinct field $\nu_\mu(\delta\theta)$ and $\nu_{\mu'}(\delta\theta)$ or between, $\nu_\mu(\delta\theta)$ and ε, the optimal estimation (11.12) reduces to the form

$$\hat{\alpha}_M = \Bigg(P_\mu\,K_{\nu_\mu\,\nu_\mu}\,P^*_\mu + \sum_{\mu'\neq\mu} P_{\mu'}\,K_{\nu_{\mu'}\,\nu_{\mu'}}\,P^*_{\mu'} + K_{\varepsilon\,\varepsilon}\Bigg)^{-1} P_\mu\,K_{\nu_\mu\,\nu_\mu}\,|l\rangle\,,$$

where

$$P_\mu\,K_{\nu_\mu\,\nu_\mu}\,P^*_\mu = \begin{Vmatrix} \langle p_{1\mu}|K_{\nu_\mu\,\nu_\mu}|p_{1\mu}\rangle & \dots & \langle p_{1\mu}|K_{\nu_\mu\,\nu_\mu}|p_{N\mu}\rangle \\ \dots & \dots & \dots \\ \langle p_{n\mu}|K_{\nu_\mu\,\nu_\mu}|p_{1\mu}\rangle & \dots & \langle p_{n\mu}|K_{\nu_\mu\,\nu_\mu}|p_{N\mu}\rangle \\ \dots & \dots & \dots \\ \langle p_{N\mu}|K_{\nu_\mu\,\nu_\mu}|p_{1\mu}\rangle & \dots & \langle p_{N\mu}|K_{\nu_\mu\,\nu_\mu}|p_{N\mu}\rangle \end{Vmatrix}$$

is the covariance matrix of the legitimate (with respect to the μ-th parameter field) signal;

$$\left(\sum_{\mu' \neq \mu} P_{\mu'} \, K_{\nu_{\mu'} \, \nu_{\mu'}} \, P^*_{\mu'} + K_{\varepsilon\varepsilon} \right)$$

is the covariance matrix of the effective noise that includes the statistical correlations of the fields $\{\nu_{\mu'}\,,\mu' \neq \mu\}$ as well as of the random component ε. The error variance of the estimation $\hat{\alpha}$ is of the form

$$\mathrm{E}(\eta^l_\mu)^2 = \langle l | K_{\nu_\mu \, \nu_\mu} - K_{\nu_\mu \, \nu_\mu} \, P^*_\mu \Bigg(P_\mu \, K_{\nu_\mu \, \nu_\mu} \, P^*_\mu + \sum_{\mu' \neq \mu} P_{\mu'} \, K_{\nu_{\mu'} \, \nu_{\mu'}} \, P^*_{\mu'} + K_{\varepsilon\varepsilon} \Bigg)^{-1} P_\mu \, K_{\nu_\mu \, \nu_\mu} \, |l\rangle,$$

or, in terms of the Fisher's operator,

$$\mathrm{E}(\eta^l_\mu)^2 = \langle l | F^{-1}_\mu \, |l\rangle \,,$$

$$F_\mu = \left[P^*_\mu \left(\sum_{\mu' \neq \mu} P_{\mu'} \, K_{\nu_{\mu'} \, \nu_{\mu'}} \, P_{\mu'} + K_{\varepsilon\varepsilon} \right)^{-1} P_\mu + K^{-1}_{\nu_\mu \, \nu_\mu} \right] .$$

The latter expression clearly demonstrates that the higher the sensitivity of the data processed to the variations of the μ-th field and the less the influence of the variations of the other fields ($\mu' \neq \mu$) on the experimental data, the higher quality of the estimation of the μ-th field.

The model of experimental data (11.9) formally includes (Troyan, 1982) all the fields of the parameters of the medium that occur in the propagation equation of the sounding signal. Taking into account the observation system under a chosen experimental design, we know a priori that the magnitude of the signal-to-noise ratio is not necessarily large for all of the fields $\nu_\mu\,(\delta\theta)$. In order to construct an adequate model of experimental data, it is necessary to introduce a quantitative measure of the influence of the variations of each of the parameters on the measurement field. To this end, we introduce the notation of the information sensitivity of the observation field (u) with respect to a fixed linear parameter field ($\nu(\delta\theta)$) as the limit of the derivative of the Shannon information $I_\alpha\,(l(\nu)/u)$ on α with respect to $\alpha \longrightarrow 0$ (signal-to-nose ratio), as α tends to zero:

$$S^P_l = \lim_{\alpha \longrightarrow 0} \frac{\partial}{\partial \alpha} I_\alpha \left(\frac{l(\nu)}{u_P} \right),$$

$$I_\alpha \left(\frac{l(\nu)}{u_P} \right) = \frac{1}{2} \ln \frac{\mathrm{E}(\eta^l_{\mathrm{apr}})^2}{\mathrm{E}(\eta^l_\alpha)^2} \,, \qquad \mathrm{E}(\eta^l_{\mathrm{apr}})^2 = \langle l | K_{\nu\nu} \, |l\rangle \,,$$

$$\mathrm{E}(\eta^l_\alpha)^2 = \langle l | (P^* \, K^{-1}_{\varepsilon\varepsilon} \, P + K^{-1}_{\nu\nu})^{-1} \, |l\rangle \,,$$

$$K_{\varepsilon\varepsilon} = \sigma_\varepsilon^2 I\,, \quad K_{\nu\nu} = \sigma_\nu^2 I\,, \quad \alpha = \sigma_\nu^2/\sigma_\varepsilon^2\,.$$

The explicit expression for the derivative with respect to α is as follows:

$$\frac{\partial}{\partial\alpha} I_\alpha\left(\frac{l(\nu)}{u_p}\right) = -\frac{1}{2}\frac{\langle l|l\rangle}{\langle l|I - \alpha P^*(\alpha PP^* + I)^{-1}P|l\rangle} \times\left[-\frac{\langle l|P^*(\alpha PP^* + I)^{-1}P|l\rangle}{\langle l|l\rangle} + \right.$$
$$\left. + \alpha\frac{\langle l|P^*(\alpha PP^* + I)^{-1}PP^*(\alpha PP^* + I)^{-1}P|l\rangle}{\langle l|l\rangle}\right].$$

By passing to the limit as $\alpha \longrightarrow 0$, we obtain

$$S_l^P = \lim_{\alpha\longrightarrow 0} I_\alpha\left(\frac{l(\nu)}{u_P}\right) = \frac{1}{2}\frac{\langle l|P^*P|l\rangle}{\langle l|l\rangle} = \frac{1}{2}\frac{||P|l>||^2_{R^N}}{||\,|l>||^2_{L_2}}\,. \tag{11.13}$$

The area of application of the information sensitivity includes the construction of the model of experimental data and the design of tomographic experiment.

Physically, expression (11.13) means that the greater energy of the legitimate signal that has passed through the medium with a nonhomogeneity, the field of which coincides with the weight function of the estimate functional l the higher the information sensitivity of a given system of tomographic functionals. In the case of the ray tomography, such a qualitative conclusion is obvious if the functional under the estimation is of the type of mean over volume elements. An important property of the measure of information sensitivity introduced above is that it provides a quantitative characteristic for arbitrary tomographic functionals (we recall that the functionals of the diffraction tomography are alternating Sec. 4.3) and arbitrary functionals to be estimated (e.g. the coefficients of the space Fourier transform of the unknown parameter field). This measure of the information sensitivity makes it possible to determine the contribution of each of the fields involved in the model (11.9) and to construct an adequate model, from which all fields not informationally covered by a chosen design of the experiment (the set of tomographic functionals) are excluded.

11.4 Errors of Recovery, Resolving Length and Backus and Gilbert Method

To analyze the errors of recovering, we shall follow to the scheme from Sec. 9.3. Neglecting the influence of all field ν_μ, except one field, we rewrite the representation for the error of the estimate (11.11):

$$\eta^l = \langle l - (\alpha, p)\nu\rangle - (\alpha, \varepsilon) \tag{11.14}$$

then, introducing the notation b (displacement) and e (noise):

$$\eta = b + e,$$

where $b = \mathrm{E}_\nu \eta = \langle l - (\alpha, p)\nu\rangle \equiv \langle l - \tilde{l}|\nu\rangle$; $e = -(\alpha, \varepsilon)$; E_ν denotes the conditional mathematical expectation.

We should note, that the general reason for an ill-posed statement of the estimation problem consists in an absence of a priori information about the solution set $\{\nu(\delta\theta)\}$ (Backus and Gilbert, 1967, 1968). Actually, an unbiased estimate can be obtained if estimated functional l belongs to the linear span of the tomographic functionals (α, p). But the latter condition a fortiori not holds, if, for example, l is singular, but p_n are regular ($n = 1 \div N$). So, the error of the estimate η^l in the general case can be infinite by the virtue of the unboundness of the set $\{\nu(\delta\theta)\}$. Therefore it is impossible to introduce any optimal criterion for the estimation.

Taking into account the mentioned above problem, we will analyze the Backus and Gilbert method (B – G) (Sec. 6.16), which is wide represented in scientific literature (Jensen *et al.*, 2000; Troyan, 1985; Yanovskaya and Porokhova, 2004; Nolet, 1985). Let us briefly explain the idea and the algorithmic structure of (Backus and Gilbert, 1968), in which a presence of the random component in a measurement data is assumed, but a priori information about the set of solutions $\{\nu(\delta\theta)\}$ is supposed to be absent. Using the notations of the formula (11.14), we write the optimality criterion in a sense of Backus and Gilbert criterion. In the work (Ryzhikov and Troyan, 1987) a one-dimensional problem ($x \in R^1$) is considered. The criterion $J_{\text{B-G}}$ contains two components. The first one (*R-criterion "delta-likeness"*): a singular functional $\langle \hat{l}_{\text{B-G}}| = \delta(x - X)$ is estimated. If we denote $(\alpha, p) \equiv (\alpha(X), p(x)) \equiv \tilde{l}$, then

$$R = \langle l_{\text{B-G}} - \tilde{l}|K_{\text{B-G}}|l_{\text{B-G}} - \tilde{l}\rangle,$$

where the integral operator $K^X_{\text{B-G}}$ has a singular kernel of the form

$$k^X_{\text{B-G}}(x', x'') = c(x' - X)(x'' - X)\delta(x' - x''),$$

which contains the estimated point X. We write the value of R relatively to α parameter

$$R = (\alpha, V^*\alpha),$$

where

$$\begin{aligned} V^*\, v_{nm} &= c\int\!\!\int dx'dx'' p_n(x')(x' - X)\delta(x' - x'')(x'' - X)p_m(x'') \\ &= c\int\!\!\int p_n(x)p_m(x)(x - X)^2 dx. \end{aligned}$$

The value of R is connected with the resolution of the method on x coordinate; for this purpose the constant c in R^1 is assumed to be 12, because, if we put to criterion

$$R = 12\int \tilde{l}(x)(x - X)^2 dx$$

an uniform at the interval Δ distribution, instead of the linear combination

$$\tilde{l} = \sum_{n=1}^{N} \alpha_n(X)p_n(x),$$

(which produce "solving kernel"), then R will be numerically equal to a value of the interval Δ, i.e.

$$R = 12 \int \tilde{l}^2(x)(x-X)^2 dx \leftarrow 12 \int\limits_{X-\Delta/2}^{X+\Delta/2} \left[\frac{1}{\Delta}(x-X)^2\right] dx \equiv \Delta.$$

Here Δ is the resolving length on the coordinate x.

An influence of the *random component* ε, cased, for example, by measurement errors, is taken into account by the *second component* of the $J_{\mathrm{B-G}}$ criterion:

$$N(\alpha, K_\varepsilon \alpha)$$

($K_\varepsilon = E(\varepsilon\varepsilon^T)$ is a covariance matrix of the measurement errors).

Finally, the optimality criterion for the estimation of the functional l consists in minimizing on α the sum of the weighted quadratic forms R and N:

$$J_{\mathrm{B-G}} = (1-\gamma)R + \gamma N, \quad \gamma \in [0,1],$$

or

$$J_{\mathrm{B-G}} = (1-\gamma)\langle l - \tilde{l}|K_{\mathrm{B-G}}|l - \tilde{l}\rangle + \gamma(\alpha, K_\varepsilon \alpha)$$

under the condition $\langle \tilde{l}|1\rangle = 1$. The supplementary condition is introduced by analogy with $\langle l_{\mathrm{B-G}}|1\rangle \equiv \int \delta(x-X)dx$. It is possible to see, that without the supplementary condition the criterion produces only an unique zero solution.

It is assumed to minimize the criterion $J_{\mathrm{B-G}}$ under various values of the parameter γ, and the optimality estimate $(\hat{\alpha})$ corresponds to the minimum or the sum R and N, that provides under ideology of the Backus and Gilbert method a compromise between "resolving length" and "error".

We shall compare the optimality criterion $J_{\mathrm{B-G}}$ and optimality criterion introduced by Ryzhikov and Troyan (R–T) (Ryzhikov and Troyan, 1994), which is considered in Sec. 11.3:

$$J_{\mathrm{P-T}} = [(\Lambda - Q^*\alpha)^* K_{\mathrm{P-T}}(\Lambda - Q^*)].$$

This criterion for the considered case of the single one-dimensional estimated field without any a priori link between $\nu(\delta\theta)$ and ε) we write down in the form

$$J_{\mathrm{P-T}} = \langle l_{\mathrm{P-T}} - \tilde{l}|K_{\nu\nu}|l_{\mathrm{P-T}} - \tilde{l}\rangle + (\alpha, K_\varepsilon \alpha).$$

It was mentioned, the first term of $J_{\mathrm{P-T}}$ criterion, which is connected with the displacement b, must include a priori information about $\nu(\delta\theta)$. In the criterion $J_{\mathrm{P-T}}$ this a priori information is introduced by the operator $K_{\mathrm{P-T}}$, which determines a set of feasible solutions $\{\nu(\delta\theta)\}$. A priori information can be introduced in the probability form, for example, by determination of the Gaussian measure on the set $\{\nu(\delta\theta)\}$. In this case the operator $K_{\mathrm{P-T}}$ makes a sense of the a priori covariance operator of the field $\{\nu(\delta\theta)\}$.

The class of the solution $\nu(\delta\theta)$ can be determined also by inequality $\langle\nu|H|\nu\rangle \le C$, where H is the Hermitian positive operator. For example, in the case of

$$x \in R^1, \quad H = \sum_{q=0}^{r} \partial_q(u\partial_q), \quad u < 0,$$

then with accurate within a factor γ, which is a regularization parameter, the operator $K_{\mathrm{P-T}}$ is the Green operator for the operator H, in addition the class $\{\nu(\delta\theta)\}$ is interpreted as a class of solutions of the r-th degree of smoothness.

The first component of the criterion $J_{\mathrm{B-G}}$ is a quadratic form of the displacement. The integral operator $K_{\mathrm{B-G}}$ with a singular kernel brings a priori information about the solution set. This operator implicitly depends on the point X, which location is not determined. So, a priori information in the Backus and Gilbert method is transformed together with a change of the point X, that is inadmissible in the processing of the fixed volume of experimental data. The solving kernel $(\alpha(X), p(x))$ is oscillating one and the interpretation of the first component of $J_{\mathrm{B-G}}$ as a resolving length is incorrect. Probably it is better instead of the functional $l^x = \delta(x - X)$ to use the functional which has a sense of a middle value of a desired field $\nu(\delta\theta)$ in the vicinity of the point X. In this case a diameter of the vicinity is determined "resolving length", and the error of the estimate has a variance $\langle l|F|l\rangle$, which includes the displacement b and the noise e.

The linear constraint $\langle\tilde{l}|1\rangle$, which is used in $J_{\mathrm{B-G}}$, is an artificial one. Its appearance is caused by two reasons: first, the estimated functional is localized in the point X; second, the kernel $K^X_{\mathrm{B-G}}$ is equal to zero in the same point. So, without the linear constraint we can not obtain the proper estimate. It should underscore, that the delta function is not defined by the condition $\langle l^X|1\rangle = 1$, it is defined by the condition $\langle l^X|\psi\rangle = \psi(X)$ The logic of the Backus and Gilbert method leads to the necessity to introduce an infinite number of the (delta-like) linear constraints $\langle\tilde{l}|\psi_r\rangle = \psi(X)$ (where $\{\psi_r\}$ is a complete system of the basis functions).

If the kernel $k^X_{\mathrm{B-G}}$ does not equal to zero in the point X, then the optimality criterion does not need the linear constraints The various shapes of the kernel $k^X_{\mathrm{B-G}}$, are considered in (Backus and Gilbert, 1968), but most of the applications use the kernel of the form

$$k^X_{\mathrm{B-G}} = c(x' - X)(x'' - X)\delta(x' - x'').$$

One can show, that the introducing of the linear constraint in (R–T)-algorithm leads to increasing of the estimate variance. In this case (R–T)-criterion has a form

$$\tilde{J}_{\mathrm{P-T}} = J_{\mathrm{P-T}} + \gamma\langle l - \tilde{l}|\nu_{\mathrm{H}}\rangle$$

(ν_{H} is a model field) and an optimal estimate

$$\tilde{\alpha} = \arg\inf \tilde{J}_{\mathrm{P-T}}$$

is determined as

$$\tilde{\alpha} = K_{uu}^{-1}(u_r + \varkappa u_{\mathrm{H}}),$$

where $u_r = PK_{\nu\nu}|l\rangle$; $u_{\rm H} = P\nu_{\rm H}$; $K_{uu} = PK_{\nu\nu}P^* + K_\varepsilon$;

$$\varkappa = \frac{\langle l^X|\nu_{\rm H}\rangle - u_r^T K_{uu}^{-1} u_{\rm H}}{u_{\rm H}^T K_{uu}^{-1} u_{\rm H}}.$$

Using the expression for $\varkappa$, we, finally, obtain

$$\tilde{\alpha} = \left[I - \frac{1}{u_{\rm H}^T K_{uu}^{-1} u^{\rm H}} K_{uu}^{-1} u_{\rm H} u_{\rm H}^T\right] K_{uu}^{-1} u_r + \langle l^X|\nu_{\rm H}\rangle \frac{K_{uu}^{-1} u_{\rm H}}{u_{\rm H}^T K_{uu}^{-1} u_{\rm H}}.$$

Let write an increasing of the variance of the estimate of the functional l^X after introducing of the linear constraint to the criterion

$$\Delta\eta^2 = (\tilde{\eta} - \eta)^2 = \frac{(\langle l^X|\nu_{\rm H}\rangle - u_{\rm H}^t K_{uu}^{-1} u_r)^2}{u_{\rm H}^T K_{uu}^{-1} u_{\rm H}} \geq 0.$$

Analysis of the $\Delta\eta^2$ shows, that introducing of the supplementary linear constraint $\langle l^X - \tilde{l}|\nu_{\rm H}\rangle = 0$ (like to usage of $\langle \tilde{l}|1\rangle = 1$ in the Backus and Gilbert criterion), leads to loss of the estimation efficiency.

After carrying out the analysis we can conclude, that the correct restoration algorithm for the parameters using seismic data should be based on the quadratic optimality criterion:

$$J_{\rm P-T} = \langle l_{\rm P-T}^X - \tilde{l}|K_{\rm P-T}|l_{\rm P-T}^X - \tilde{l}\rangle,$$

where as the estimated functional $l_{\rm P-T}^X$ we use the mean value of the desired field in the vicinity of the point X; for example, in a rectangular parallelepiped abc with center in the point (X_1, X_2, X_3):

$$l_{\rm P-T}^X = (abc)^{-1} H\left(\frac{a}{2} - |x_1 - X_1|\right) H\left(\frac{b}{2} - |x_2 - X_2|\right) H\left(\frac{c}{2} - |x_3 - X_3|\right),$$

$$H(\xi) = \begin{cases} 1, & \text{if} \quad \xi \geq 0, \\ 0, & \text{if} \quad \xi < 0. \end{cases}$$

The variance of the estimate reads as

$$\mathrm{E}\eta^2 = \langle l_{\rm P-T}^X|F^{-1}|l_{\rm P-T}^X\rangle$$

and it includes both the displacement b and noise e.

The criterion $J_{\rm P-T}$ in the full measure takes into account both a priori information about the elastic field and statistical structure of observation errors. In Fig. 11.1 the integral kernels for the operators of the criterion $K_{\rm B-G}$ *(a)*,

$$K_{\rm B-G}(x', x'') = (x' - X)(x'' - X)\delta(x' - x''),$$

are represented: $K_{\rm P-T}$ for the noncorrelated homogeneous field *(b)* and $K_{\rm P-T}$ for the Markovian homogeneous field *(c)*, which determine an introducing a priori information in (B–G) and (P–T) methods.

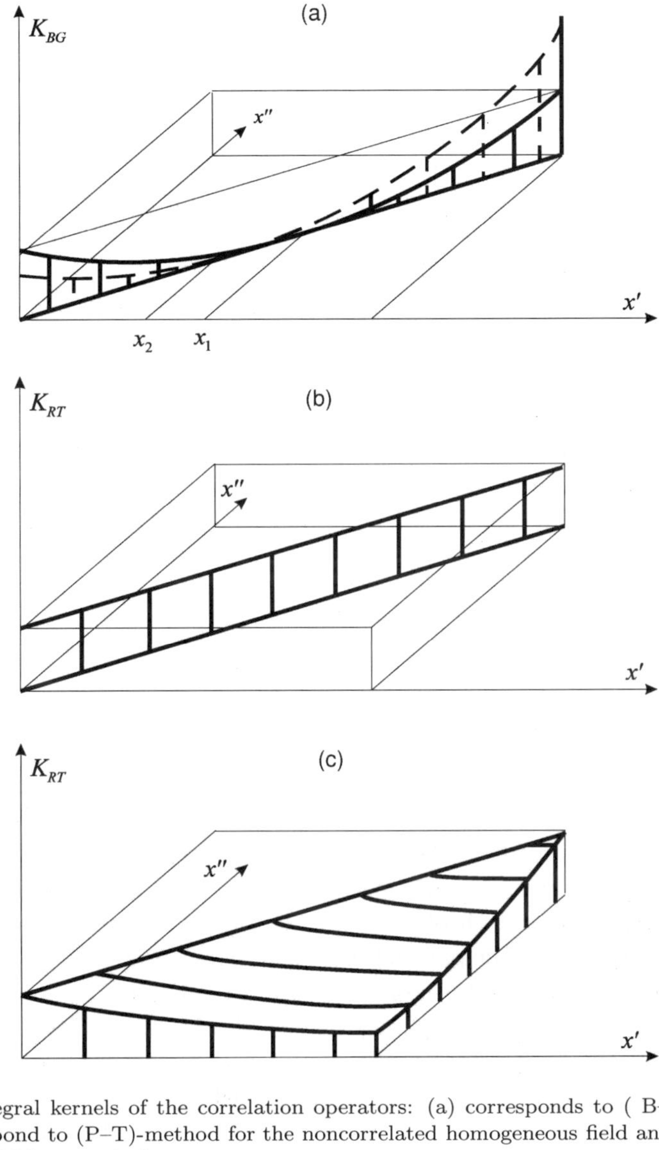

Fig. 11.1 Integral kernels of the correlation operators: (a) corresponds to (B–G)-method, (b) and (c) correspond to (P–T)-method for the noncorrelated homogeneous field and the Markovian homogeneous field respectively.

11.5 Back Projection in Diffraction Tomography

Let us remind the model for the diffraction tomography experiment:

$$u = P\theta + \varepsilon,$$

where the tomographic operator P is

$$P\theta \Rightarrow \int p_t(\xi, x)\theta(x)dx,$$

$$P_t(\xi, x) = \langle G_0^* h_t^{\xi_r} | S_{[x]}^{\theta} | G_0 s^{\xi_s} \rangle \equiv \langle \varphi_{\text{out}\{t\}} | S_{[x]}^{\theta} | \varphi_{\text{in}}^{\xi_s} \rangle,$$

where G_0^* includes the time inversion under the interpretation of the nonstationary sounding signal; the integral kernel of the tomography functional $p(\cdot, x)$ is the weight function connected with the variation of the medium parameters in the space point x and in the temporary point t. This weight function corresponds to the ξ source-receiver pair and it produces by local interaction $S_{[x]}^{\theta}$ of the inverted field $\varphi_{\text{out}\{t\}}^{\xi_r}$ $L_0^* \varphi_{\text{out}} = h_t^{\xi_r}$, which is produced by the source $h_t^{\xi_r}$ located in the point ξ_r and working in the inverse time t, and the incident field $\varphi_{\text{in}}^{\xi_s}$ $L_0 \varphi_{\text{in}} = s^{\xi_s}$.

The inversion of the data u using the regularized algorithm $\hat{R}\,\hat{\theta} = Ru$ is constructed on the basis of the solution of an extremum problem:

$$\hat{R} = \arg\inf(\|RP - I\|_{K_\theta}^2 + \|R\|_{K_\varepsilon}^2),$$

where $\|A\|_B = (\text{sp}ABA^*)^{1/2}$ reads as

$$\hat{R} = K_\theta P^* (P K_\theta P^* + K_\varepsilon)^{-1}, \tag{11.15}$$

and provides a minimum variance of the restoration error, i.e.

$$\hat{R} = \arg\inf_R \text{sp} K_{\delta\theta},$$

$$K_{\delta\theta} = K_\theta - R K_u R^*, \; K_u = P K_\theta P^* + K_\varepsilon.$$

The construction of the optimum operator $\hat{R}$ leads to some problems. The analogous problems appear at the realization of the exact algorithms in the conventional ray tomography. But in the computational tomography these problems are solved in the way of the construction of heuristic algorithms, which provide to use with more or less effectiveness some part of the information contained in the experimental data. One of the routine methods of a such kind is the back projection procedure (Vasilenko and Taratorin, 1986; Kireitov and Pikalov, 1985; Beylkin, 1985). We should note that the back projection is a part of the exact analytical inversion of the Radon transform (Helgason, 1999; Herman, 1980), which as a data for the back projection are used the derivatives of the Hilbert images of the Radon projections.

Let us consider a suboptimal algorithm $\tilde{R}$ of the restoration with the use of the inversion of the measured wave fields and the back projection. To write down the algorithm $\tilde{R}$ in a form

$$\tilde{\theta}(x) = \tilde{R}u(x) = \alpha^{-1} \sum_{\xi} \langle \Phi_{\text{out}}^{\xi} | S_{[x]}^{\theta} | \Phi_{\text{in}}^{\xi} \rangle,$$

where the summation is implemented on the results of processing of the isolated experiments; Φ^{ξ}_{in} is a direct continuation of the wave field, which produces by the s-th set of the sources; Φ^{ξ}_{out} is a backward continuation, measured by the r-th set of receivers; α is a dimension factor, which is proportional to a volume of the processed experimental data. After the parametrization of the experimental record in the form

$$u^{\xi}_{(t)} = \sum_{t_n}\sum_{\xi_s}\sum_{\xi_r} u^{\xi_r\xi_s t_n}\delta(t-t_n)\delta(x_s-\xi_s)\delta(x_r-\xi_r),$$

we rewrite the expression for the estimate $\tilde{\theta}(x)$:

$$\begin{aligned}\tilde{\theta}(x) &= \alpha^{-1}\sum_{t_n}\sum_{\xi_s}\sum_{\xi_r} u^{\xi_r\xi_s t_n}p^{\xi_r\xi_s t_n}\\ &\equiv \alpha^{-1}\sum_{t_n}\sum_{\xi} u^{\xi}(t_n)p_{t_n}(\xi,x) = \alpha^{-1}P^*u.\end{aligned} \tag{11.16}$$

In this case the kernel of the tomography functional $p_{t_n}(\xi, x)$ is produced by the hardware function of the recording channel $h^{\xi_r}_{tn} = \delta(t-t_n)\delta(x_r-\xi_r)$, that corresponds to a wide band recording of the experimental data. Let note, that the last expression for the estimate $\tilde{\theta}(x)$ contains the back projection procedure in the explicit form.

We consider the back projection as a part of the generalized inversion algorithm described in Sec. 11.3. For this purpose we write down the result of the restoration of an arbitrary linear functional $l(\theta) \equiv \langle l|\theta\rangle$ (Ryzhikov and Troyan, 1990):

$$\hat{l}(\theta) = \widehat{\langle l|\theta\rangle} = \langle u(PK_\theta P^* + K_\varepsilon)^{-1}PK_\theta|l\rangle$$

and consider its simplified version for $K_\theta \Rightarrow I$, $\langle l| \Rightarrow \langle l_{[x]}| \equiv$ $\equiv \langle\delta(x-[x])|$. Then the expression for $\hat{l}(\theta)$ reads as

$$\hat{\theta}_{[x]} \equiv \widehat{\langle l_{[x]}|\theta\rangle} = \langle u|(PP^* + K_\varepsilon)^{-1}P|l_{[x]}\rangle.$$

Taking into account, that

$$P|l_{[x]}\rangle \Rightarrow \int P_t(\xi,x)\delta(x-[x])dx = P_t(\xi,[x]),$$

and assuming $(PP^* + K_\varepsilon) \sim \alpha I$, we obtain an analog of an action of the algorithm $\tilde{R}$:

$$\hat{\theta}_{[x]} \sim \alpha^{-1}\sum_{t_n}\sum_{\xi} u^{\xi}(t_n)P_{t_n}(\xi,[x]).$$

The comparison of the formulas (11.15) and (11.16) allows us to see the disadvantages of the generalized inversion (or generalized back projection): it does not taking into account a statistics of the measurement errors, a priori representations about medium parameters are absent, the links between various generalized Radon projections are ignored.

The certain advantage of the back projection consists in its manufacturing, because the algorithm is reduced to the well known procedures of the wave field continuation (Zhdanov *et al.*, 1988; Claerbout, 1985; Kozlov, 1986; Petrashen and Nakhamkin, 1973; Berkhout, 1984.; Stolt and Weglein, 1985).

Let us remind that classical methods of the elastic wave fields continuation (Claerbout, 1976; Petrashen and Nakhamkin, 1973; Timoshin, 1978; Berkhout, 1985) reduce to the procedure $\tilde{u} = G_0^* u$. As it follows from the analysis of the generalized back projection algorithm (11.16), the procedure of the back continuation is one of the component of the procedure of restoration of the image of medium: $\Phi_{\text{out}} = G_0^* u$. From the representation (11.15) it follows, that an application of the continuation as the method of the restoration needs an information about properties of the sounding signal Φ_{in} and the information about the interaction character of the direct and back continuation of the wave fields (S^θ). The algorithm of generalized back projection is virtual to the back continuation under conditions that the incident field Φ_{in} has a plane wave-front and the interaction operator S^θ is a multiplication operator.

Let note, that the procedure of the generalized back projection can be interpreted as a hologram processing. We would remind that the optical holography is found on the registration both the amplitude and the phase of an optic signal at the same time. In optics it implements by recording an interference picture, which appears as a result of the superposition between the monochromatic (laser) reference beam and the wave field scattered by the object. In the optic holography using a monochromatic beam is a single opportunity to fix both the amplitude and phase, but in many sounding tasks recording fields contain both these parameters of the scattering field. The analogy with the holography is found in the explicit form in the generalized back projection procedure:

$$\hat{\theta} \sim \alpha^{-1} \langle G^*(u - u_0) | S^\theta | \varphi_{\text{in}} \rangle,$$

$$u_0 = \varphi_{\text{in}} \big|_{x=\xi_r}.$$

In this case the arbitrary field φ_{in} can be considered as the reference beam, but as in the case of the optical holography, if it does not bring any information. The "holography" image appears as the result of the interaction of the reference field φ_{in} and an illuminated "hologram" $u - u_0$, at that an information transmission to the space satisfies the same laws as the reference field φ_{in}. The interaction operator S^θ is no multiplication operator in general case, and it determines only by the propagation law for the sounding signal.

Let us show that the procedure of the construction of the field φ_{out} in the form $G^* u$ can be represented in the explicit form as an analytical extension of the field observed on some part of the space surface. Let us consider an example of the extension of the scalar wave field.

Using the second Green theorem in the form

$$\int_V [\varphi \nabla^2 \psi - \psi \nabla^2 \varphi] dV = \oint_{\partial V} [\varphi \boldsymbol{\nabla} \psi - \psi \boldsymbol{\nabla} \varphi] \cdot d\boldsymbol{\sigma},$$

where $d\boldsymbol{\sigma} = \mathbf{n}d\sigma$, we can chose ψ such as

$$\psi\left(\nabla^2 - \frac{1}{c^2}\frac{\partial^2}{\partial t^2}\right)\psi = -4\pi\delta(x - x'),$$

where x' lays inside the volume V, bounded by the closed surface ∂V, the field φ produced by the sources, located outside of the volume V. Let us write the Kirchhoff integral (Lavrentev, 1967; Berkhout, 1985, 1984.):

$$\varphi(x') = \frac{1}{4\pi}\int\limits_{\partial V}[\psi\boldsymbol{\nabla}\varphi - \varphi\nabla\psi]\cdot d\boldsymbol{\sigma}. \tag{11.17}$$

The expression (11.17) represents an explicit form of the link of the field parameters in the arbitrary point x' in the volume with the field in the closed surface, i.e. the Kirchhoff integral gives an analytic continuation of the field from the surface to the volume, which is bounded by that surface. This kind of continuation is not always convenient for the practice, because it is necessary to state the normal derivative $(\mathbf{n}\cdot\boldsymbol{\nabla})\varphi$ of the field on the surface ∂V besides the function φ itself. Let φ be a field of the pressure, then, taking into account that $\boldsymbol{\nabla}\varphi = -i\omega\rho\mathbf{v}$ (v), the time Fourier representation of the Kirchhoff integral on time has the form:

$$\varphi(x') = \frac{1}{4\pi}\int\limits_{\partial V}\left[\omega(x)\frac{\partial}{\partial n}\left(\frac{\exp(-ik|x - x'|)}{|x - x'|}\right) + i\omega\rho v_n\frac{\exp(-ik|x - x'|)}{|x - x'|}\right]d\boldsymbol{\sigma}.$$

Let us add the field ψ_0 to the field ψ, where ψ_0 is satisfied to the homogeneous wave equation:

$$\left(\nabla^2 - \frac{1}{c^2}\frac{\partial^2}{\partial t^2}\right)\psi_0 = 0$$

and the boundary condition on ∂V:

$$(\mathbf{n}\cdot\boldsymbol{\nabla}|_{\partial V}\psi' \triangleq \mathbf{n}\cdot\boldsymbol{\nabla}|_{\partial V})(\psi + \psi_0),$$

i.e.

$$\left.\frac{\partial\psi^0}{\partial n}\right|_{\partial V} = -(\mathbf{n}\cdot\mathbf{e})\frac{1 + ik|x - x'|}{|x - x'|62}\exp(-ik|x - x'|),$$

$$\mathbf{e} = (\mathbf{x}' - \mathbf{x})|x - x'|^{-1}.$$

The boundary condition can be satisfied, if it takes a part of the plane $x_3 = x_3^0$ as a boundary, then ψ^0 is a field of a point source with a mirror location relatively to the point x'. For such field ψ' the Kirchhoff integral takes the form

$$\varphi(x') = \frac{i\omega\rho}{2\pi}\int(\mathbf{n}\cdot\mathbf{v}(x))\frac{\exp(-ik|x - x'|)}{|x - x'|}d\boldsymbol{\rho}, \tag{11.18}$$

where $\boldsymbol{\rho} = (x_1, x_2)$. Integral on the part of the sphere which is closed the boundary ∂V, is assumed to be zero, because it is supposed that the radius of the sphere is big enough, and a continuable wave field φ, induced by a source located outside the volume V, switched at the moment $t = 0$, has an interest for us in the restricted time interval $(0, T)$ only. The integral (11.18) describes the continuation of the normal component of the velocity $\mathbf{v}(x)$, which is stated on a part of the plane $x_3 = x_3^0$, to the pressure $\varphi(x')$ in the direction $(-\mathbf{n})$, where $\mathbf{n}$ in the Green formula is external normal vector. This integral representation is known as the *Rayleigh integral.*

We can obtain the second Rayleigh integral by choosing the field ψ^0 in the accordance with the constraint

$$\psi'\big|_{\partial V} \triangleq (\psi + \psi^0)\big|_{\partial V} = 0.$$

Now we can rewrite the Kirchhoff integral in the form

$$\varphi(x') = \frac{1}{2\pi} \int (\mathbf{n} \cdot e(x))\varphi\big|_{x_3 = x_3^0}(x) \frac{1 + ik|x - x'|}{|x - x'|} \times \exp(-ik|x - x'|) d\rho, \quad (11.19)$$

where $\boldsymbol{\rho} = (x_1, x_2)$. The expression (11.19) describes the direct continuation of the pressure field from plane $x_3 = x_3^0$ to $(-\mathbf{n})$ direction. This expression is also known as the *Rayleigh integral.*

The formula (11.19) can be obtained using an evolutionary representation for the wave equation, with a parameter of the evolution $x_3 \equiv z$. For the Fourier components of the wave field $\tilde{\varphi}(k_x, k_y, z, \omega)$ the wave equation transforms to the Helmholtz equation:

$$\frac{\partial^2}{\partial z^2}\tilde{\varphi} + k_z^2 \tilde{\varphi} = 0,$$

where $k_z^2 = -k_x^2 - k_y^2 + \omega^2/c^2$.

Representing the approximate solution $\tilde{\varphi}$ in the form of non-interacting fields $\tilde{\varphi}_\uparrow$ and $\tilde{\varphi}_\downarrow$, that satisfy the equations

$$\left(\frac{\partial}{\partial z} - ik_z\right)\tilde{\varphi}_\uparrow = 0, \quad \left(\frac{\partial}{\partial z} + ik_z\right)\tilde{\varphi}_\downarrow = 0,$$

it is possible to write down the solution in the form

$$\tilde{\varphi}_{\{\uparrow\downarrow\}}(z_2) = \tilde{\varphi}_{\{\uparrow\downarrow\}}(z_1 \exp\left\{\pm i \int_{z_1}^{z_2} k_z dz\right\}. \quad (11.20)$$

The similar expression can be obtained for the case of the WKB approximation (see Sec. 5.3), where the eikonal equation with $\varphi = \varphi_0 \exp[i\tau(z)]$ $(\partial\tau(z)/\partial z)^2 = k_z^2$ under the assumption of the weak inhomogeneity of the medium on the coordinates x, y, i.e. a quasi-layered model of the medium is used. In this case the eikonal equation is disintegrated on two equations:

$$\frac{\partial \tau}{\partial z} = k_z, \quad \frac{\partial \tau}{\partial z} = -k_z.$$

At small values of Δz the expression (11.20) can be approximately written as

$$\tilde{\varphi}(z + \Delta z) \approx \tilde{\varphi}(z) \exp\{\pm ik_z \cdot \Delta z\},$$

where $\exp(\pm ik_z \cdot \Delta z)$ is a continuation operator in the frequency domain. Let us note, that sign + corresponds to an opposite orientations of the direction of wave propagation and direction of the continuation, that corresponds to the continuation in inverse time, but sign – leads coincidence of these directions:

$$\tilde{H}_- = \exp(ik_z \cdot \Delta z), \quad \tilde{H}_+ = \exp(-ik_z \cdot \Delta z),$$

here the operators of the direct and inverse continuation are connected of the relation $\tilde{H}_- = \tilde{H}_+^*$.

To recognize a link between the second Rayleigh integral and the operator of the direct continuation, let us consider a field of the point source, located in the origin of coordinates:

$$\tilde{\varphi}_0\big|_{z=0} = \frac{1}{2\pi} \int\int_{-\infty}^{\infty} \frac{\exp(-ik\rho)}{\rho} \exp(-i(k_x x + k_y y)) d\boldsymbol{\rho} = \frac{1}{ik_z},$$

$$\rho = \sqrt{x^2 + y^2}.$$

This field is continued from the plane $z = 0$ to some plane $\Delta z \neq 0$. The field after the continuation should has a form $\exp[ikr]/r$. After the direct continuation of the field $\varphi_0\big|_{z=0}$, we obtain the equality

$$\frac{e^{-ikr}}{r} = \int_{-\infty}^{\infty} \frac{1}{ik_z} \exp\{-ik_z \Delta z\} \exp\{i(k_x x + k_y y)\} dk_x dk_y,$$

where $r = (x^2 + y^2 + z^2)^{1/2}$. The differentiation both parts of this equality by z, we obtain

$$-\frac{\partial}{\partial z} \frac{e^{-ikr}}{r} \int \exp\{-ik_z \Delta z\} \exp\{i(k_x x + k_y y)\} dk_x dk_y \triangleq 2\pi F_{k_x}^{-1} F_{k_y}^{-1} \tilde{H}_+.$$

So far, the space-frequency representation of the operator $\tilde{H}_+$ can be written down as follows

$$H_+ = \frac{\mathbf{n} \cdot e}{2\pi} \frac{(1 + ikr)}{r^2} \exp(-ikr) = \frac{\Delta z}{2\pi} \frac{(1 + ikr)}{r^3} \exp(-ikr).$$

Thus, we have shown that continuation of the wave field, produced of the evolutionary equation with the evolution parameter z, is identically to continuation of the wave field representing by the second Rayleigh integral.

Once again we shall point out, that the construction of the field φ_{out} is connected with the conjugate Green function, which corresponds in the space-time domain to the propagation of the signal in the reverse time, but in the space-frequency domain it corresponds to the complex conjugate Green function.

Let us show the natural growth of the idea of the back projection while the solving of the nonlinear problem

$$u = \mathcal{P}(\theta) + \varepsilon,$$

where$\mathcal{P}(\theta)$ is a generalized nonlinear tomographic operator.

As it was shown in Sec. 9.1, the Newton's method of the solution of the extremum problem

$$\tilde{\theta} = \arg\inf\{\|u - \mathcal{P}(\theta)\|^2_{K_\varepsilon^{-1}} + \|\theta - \theta_0\|^2_{K_\theta^{-1}}\}$$

reduces to an iterative process

$$\theta_{n+1} = \theta_n - [P_n^* K_\varepsilon^{-1} P_n + K_\theta^{-1}]^{-1} \times [P_n^* K_\varepsilon^{-1}(u_n - u) + K_\theta^{-1}(\theta_n - \theta_0)].$$

The first step of this process gives an exact solution if the expression

$$u = \mathcal{P}(\theta_0) + \left.\frac{\delta}{\delta\theta}\right|_{\theta_0} \mathcal{P}\delta\theta + \varepsilon$$

approximates an initial model with the satisfactory accuracy.

Each step of the gradient method applied to the solution of the extremum problem includes in the explicit form the back projection operator

$$\theta_{n+1} = \theta_n - \alpha[P_n^* K_\varepsilon^{-1}(u_n - u) + K_\theta^{-1}(\theta_n - \theta_0)],$$

therefore, if the linearized model is enough satisfied, then the direction of the gradient descent is determined by the expression

$$\delta\theta = \tilde{\alpha} P^* K_\varepsilon^{-1}(u - \mathcal{P}(\theta_0)).$$

We can also change the initial model by some equivalent one:

$$K_\varepsilon^{-1/2} u = K_\varepsilon^{-1/2}\mathcal{P}(\theta) + K_\varepsilon^{-1/2}\varepsilon$$

or $\tilde{u} = \tilde{\mathcal{P}}(\theta) + \tilde{\varepsilon}$, and to obtain an explicit formula for the back projection:

$$\delta\tilde{\theta} = \tilde{\alpha}\tilde{P}^*(\tilde{u} - \tilde{\mathcal{P}}(\theta_0)).$$

We note, that the Newton's procedure at the solution of a nonlinear problem can be represented by some variant of the generalized gradient method, if to propose that the quantity of information about θ, which contains in U, is small enough, i.e. the Fisher operator F_n does not contribute to a priori information: $\|F_n\| \ll \|K_\theta^{-1}\|$. In this case an inverse derivative of the initial regularized functional is equal to K_θ, and the first step produces the back projection with the consequent filtering:

$$\delta\tilde{\theta} = \tilde{\alpha}\tilde{K}_\theta\tilde{P}^*(\tilde{u} - \tilde{\mathcal{P}}(\theta_0)).$$

And the filtration is entirely determined by a priori information and corresponds to the smoothing with a kernel of the operator $\tilde{K}_0$. If the correlator $\tilde{K}_\theta$ describes the homogeneous fields θ the filtration is reduced to the convolution of the image, received by the back projection, because the kernel $k_\theta(x, x')$ of the correlator K_θ is such, that $k_\theta(x, x') = k_\theta(|x - x'|)$ and $\delta\tilde{\theta} = k_\theta * \delta\tilde{\theta}$.

Moreover, just the back projection with the subsequent filtering are a part of the formal algorithm of the gradient descent using in (Ryzhikov and Troyan, 1990; Tarantola, 1984; Nolet *et al.*, 1986), where the models of the seismotomography are written under the assumption that the physical equipments have an infinite bandwidth. The algorithm represented in these works, can be interpreted as the next consequence: the propagation of the seismic signal, which is produced by a source with the direct time φ_{in}; the back projection of the field of the discrepancy from the observation surface φ_{out}; the interaction of the fields φ_{in} and φ_{out} with the consequent filtering. The similar algorithm in (Tarantola, 1984) is represented as follows

$$\theta^{(n+1)} = \theta^{(n)} - \alpha^{(n)} K_\varphi (S^\theta \varphi_{\text{in}}^{(n)})^T \varphi_{\text{out}}^{(n)},$$

where $\varphi_{\text{in}}^{(n)}\ L_{\theta_n} \varphi_{\text{in}}^{(n)} = s$; s is the source function;

$$\varphi_{\text{out}}^{(n)}\ L_{\theta_n}^* \varphi_{\text{out}}^{(n)} = \delta s^{(n)},$$

where $\delta s^{(n)} = K_\varepsilon^{-1} \varphi_{\text{in}}^{(n)} - \varphi_{obs}$; φ_{obs} observed field. It is assumed that the field φ_{obs} coincides with the field of the seismic signal in the observation points up to an additive value ε.

11.6 Regularization Problems in 3-D Ray Tomography

The traditional tomographic interpretation of remote sensing data is based on the ray representation of the propagation of a signal. In this case, the path of a ray is assumed to be known in the reference medium (see (Bukhgeim, 2000)). The use of the analytic Radon inversion leads to rough errors in the case of an obvious incompleteness of data (Alekseev *et al.*, 1985). The methods of the algebraic reconstruction use either a priori block decomposition of a medium (Alekseev *et al.*, 1983) or a finite basis of space functions (Marchuk, 1982).

As a consequence of the inevitable observation errors, it is necessary to apply regularization methods even in the processing the data of "complete foreshortening" tomography. In actual the tomography experiment, the errors related to a small foreshortening of observations are added to these errors. From the mathematical viewpoint, the solution of an inverse problem is related to the inversion of a completely continuous operator and the action on the function that does not belong certainly to the image of the operator, i.e., the inversion problem is essentially incorrect.

Here we consider the potentialities of the regularization method in the solving 3-D ray tomography problem and give a physical interpretation of algorithms suggested. Here, we take into account that it is advisable to perform the regularization for equations represented in functional spaces. Only in this case one can guarantee the fact that algebraic analogs of these equations will be regularized independently of the dimension of the respective Euclidean space.

As a model of experimental data u in ray tomography (in the seismic case, the travel time or amplitudes of seismic waves), we use an operator equation of the first kind:

$$P\nu + \varepsilon = u, \tag{11.21}$$

where ν is a field associated with physical parameters of a medium such as the velocity of propagation of an elastic wave or the slack coefficient involving attenuation and scattering; P is an integral operator with the singular kernel $\delta(\mathcal{L}_{\{i=1,\ldots,n\}}(x))\varphi_0(x)$, ε is the residual of the model and experimental data that is assumed to be random variable normally distributed with the zero mathematical expectation and positive-defined covariance matrix $\mathbf{K}_\varepsilon = \mathrm{E}\varepsilon\varepsilon^T$ and that includes the measurement and linearization errors.

The model of experimental data in the form (11.21) is introduced under the assumption that $\nu(x)$ represents small smooth deviations relative to μ_0 (the field of parameters in the reference medium), i.e., $\nu(x) = (\mu - \mu_0)/\mu_0$ (ν vanish nowhere), where μ is the field of absolute values of physical parameters that assumed to be smooth with respect to the characteristic scale of a problem, in our case, with respect to the characteristic wavelength of a sounding signal. It should be noted that, first, the smoothness is a necessary condition for the ray interpretation of the propagation of an elastic wave and, second, the smallness of $\mu(x)$ makes possible to apply the linearization:

$$\mathcal{P}(\mu) = \mathcal{P}(\mu_0) + \mathcal{P}'\big|_{\mu_0}(\mu - \mu_0) = \mathcal{P}(\mu_0) + P\nu,$$

$$P = \mathcal{P}'\big|_{\mu_0}\mu_0.$$

In the general case, one may consider the linearization as the first step of the iterative solution of a nonlinear problem, but in the processing of the data of the ray seismic tomography, subsequent iterations, as a rule, are not provided with the sufficient information.

Precisely the physical requirements to the application of the model (11.21) determine the regularization methods for the solution of the inverse problem: find

$$\hat{\nu} = \arg\inf \|P\nu - u\|^2_{K_\varepsilon^{-1}}$$

under some conditions of the smoothness and boundedness of the function ν. The formalization of these conditions reduces to the well-known Tikhonov regularization (Tikhonov and Arsenin, 1977) (considered in Sec. 9.2), which, with the use of the regularization parameter α, leads to the solution of the extremum problem

$$\hat{\nu} = \arg\inf \|P\nu - u\|^2_{K_\varepsilon^{-1}} + \alpha(\nu, (I - \Delta)\nu), \tag{11.22}$$

where

$$\|f\|^2_{K_\varepsilon^{-1}} = \sum_i \sum_j [k_\varepsilon^{-1}]_{ij} f_i f_j;$$

I is the identity operator; Δ is the Laplace operator; $(\cdot,\cdot)$ denotes the scalar product in the Hilbert space:

$$(\nu, (I - \Delta)\nu) \Rightarrow (\nu, \nu) + (\nabla\nu, \nabla\nu)$$

under the condition of vanishing ν or $\nabla\nu$ on the boundaries of the integration domain.

One can write the solution of the problem (11.22) in the form

$$\hat{\nu} = D^{-1}P^*(PD^{-1}P^* + K_\varepsilon)^{-1}u,$$

where P^* is the adjoint operator in the sense of Lagrange to the operator P and $D^{-1} = \alpha^{-1}(I - \Delta)^{-1}$ is the integral operator with the kernel equal to the Green's function of the operator $D \triangleq \alpha(I - \Delta)$:

$$D^{-1}(\mu(x)) = \int dx' \frac{1}{\alpha} \frac{\exp(-|x - x'|)}{|x - x'|} \mu(x') \equiv \frac{1}{\alpha} \frac{\exp(-|x|)}{|x|} * \mu(x), \tag{11.23}$$

where $\alpha^{-1}\exp(-|x|)/|x|$ is the Yukawa potential, which is well known in the quantum mechanics and the field theory.

A "simplified" Tikhonov regularization corresponds to the solution of the extremum problem in the class of smooth functions ν, i.e.,

$$\hat{\nu} = \arg\inf[\|P\nu - u\|^2_{K_\varepsilon^{-1}} + \alpha\|\nabla\nu\|^2],$$

and yields an operator D_1^{-1} such that

$$D_1^{-1}(\mu(x)) = \int dx' \frac{1}{\alpha} \frac{1}{|x - x'|} \mu(x') = \frac{1}{\alpha} \frac{1}{|x|} * \mu(x), \tag{11.24}$$

where $\alpha^{-1}|x|^{-1}$ is the Coulomb potential.

Consider the composition of the operators $PD^{-1}P^*$ and $D^{-1}P^*$. Taking into account the action of the operator P, which corresponds to the integration of the space function $\nu(x)$ over ray parts, and the action of the operators D^{-1} and D_1^1, we see that the matrix element $(PD^{-1}P^*)_{ij}$ is obtained as a result of the convolution of the δ-function that are localized on the paths of the i-th and j-th rays with respective potentials:

$$\begin{aligned}(PD_1^{-1}P^*)_{ij} &= \frac{1}{\alpha} \int\int \mu_0(x')\delta(\mathcal{L}_i(x'))\delta(\mathcal{L}_j(x'')) \frac{\mu_0(x'')}{|x' - x''|} dx'dx'' \\ &= \mu_0\delta(\mathcal{L}_i(x)) * \frac{1}{\alpha} \frac{1}{|x|} * \mu_0\delta(\mathcal{L}_j(x)),\end{aligned} \tag{11.25}$$

$$\begin{aligned}(PD^{-1}P^*)_{ij} &= \frac{1}{\alpha} \int\int \delta(SL_i(x'))\delta(|SL_j(x'')) \frac{\exp(|x' - x''|)}{|x' - x''|} dx'dx'' \\ &= \delta(\mathcal{L}_i(x)) * \frac{\exp(-|x|)}{\alpha|x|} * \delta(SL_j(x)).\end{aligned} \tag{11.26}$$

The physical meaning of the matrix element $(PD_1^{-1}P^*)_{ij}$ in the formula (11.25) is that it represents the energy of the Coulomb interaction of two charged threads with charge densities $(4\pi/\alpha)^{1/2}\mu_0$ that are localized along the i-th and j-th rays; the matrix element in the formula (11.26) describes the energy of the i-th and j-th threads that are interacted by means of the Yukawa potential.

The result of the action of the operator $D^{-1}P^*$ is a set of m fields (in accordance with the number of rays) created at each point of the space by each of the m rays with Coulomb or Yukawa field.

In the statistical interpretation of the algorithm (11.22), the estimate of the field corresponds to the optimal statistical extrapolation or regression

$$\hat{\nu} = K_{\nu u} K_{uu}^{-1} u,$$

where $K_{\nu u} = D^{-1}P^*$ and $K_{uu} = PD^{-1}P^* + K\varepsilon$, i.e., it corresponds to a priori ideas of the field ν as a random homogeneous and isotropic field with correlation functions coinciding in (11.23) with the Yukawa potential and in (11.24) with the Coulomb potential.

If we use the concept of the local regularization(see (Ryzhikov and Troyan, 1986a)), it is not difficult to write a regularization algorithm that adequately includes more complete a priori information. Such a priori knowledge may be an idea of the field ν as a random homogeneous and isotropic one. This may be related to geophysical ideas of the space distribution of an inhomogeneous medium: so, for example, in the case of the stratified model of the reference medium the situation may occur where the orientation of inhomogeneities coincides with the spatial orientation of the layers and one can indicate the typical sizes of inhomogeneities, i.e., the sizes of the axes of a priori ellipsoid concentration of the random field ν. If a priori information on the primary localization of nonhomogeneities is available, then the method of the local regularization makes it possible to interpret the field ν as a random isotropic inhomogeneous one and to include this information adequately into the computational procedure. Moreover, a correspondence should be obeyed between the a priori ideas of the characteristic sizes of inhomogeneities and the extend of the correlation and the accepted ray model of propagation of a sounding signal: the least radius should be greater than or equal to the characteristic wavelength. Thus, physical requirements on the ray interpretation of the seismic tomography necessarily leads to the application of the Tikhonov's regularization of the first or higher order when solving these problems.

The application of the local regularization is based on the fact that in the regularizing functional $(\nu, (I_\Delta)\nu)$, instead of the elliptic operator Δ an operator $K_{ij}\partial_i\partial_j$ is used with a positive defined matrix K, which in terms of correlation functions corresponds to the change $|x| \to (x, K^{-1}x)^{1/2}$. It should be noted that the introduction of the local regularization leaves in the force all proofs of Tikhonov's regularization theorems, because the norms of the respective Sobolev space are equivalent.

In practical implementation of the algorithms, the matrix $\mathbf{K}^{-1}$ is given by six parameters: the sizes of three axes of the correlation and three Euler angles determined the orientation of the ellipsoid relative to the coordinate system associated with the observation scheme. The matrix elements $(K^{-1})_{ij}$ are computed with the help of the similarity transformation induced by the matrix of rotation from the proper coordinate system of the ellipsoid to the coordinate system of the observation scheme.

11.7 Construction of Green Function for Some Type of Sounding Signals

Let us consider a weak sounding signal, for example, a small seismic displacement, then the process of the signal propagation inside the seismic medium can be represented in a form of the linear operator equation

$$\hat{L}_\theta \boldsymbol{\varphi} = \mathbf{s},$$

where $\hat{L}_\theta$ is a linear tensor operator, which depends on the field of the medium parameters; $\boldsymbol{\varphi}$ is a vector field of the sounding signal; $\mathbf{s}$ is a vector function of the source, which is assumed to be known. One of the main elements of the inverse problem solution of the remote sounding is the problem of the recovery of the field $\varphi(\mathbf{x}, t)$ in the reference medium with a known field $\theta_0(x)$ with a given source function $\mathbf{s}$:

$$\boldsymbol{\varphi} = \hat{L}^{-1}\,\mathbf{s}, \qquad \text{or} \quad \boldsymbol{\varphi} = \hat{G}\mathbf{s},$$

where $\hat{L}\hat{G} = \hat{I}$; $\hat{G}$ is the Green operator.

Let us explain some examples of the Green function construction for the sounding signals of the various physical nature.

11.7.1 *Green function for the wave equation*

Let us to write the wave equation in the form

$$\left(\frac{\partial^2}{\partial t^2} + H\right)\varphi = 0, \tag{11.27}$$

where H is an operator that acts to the space variables. To find the Green function of (11.27), let's write the solution of (11.27) in the form of infinite series over eigenfunctions of the operator H:

$$\varphi(x,t) = \sum_{n=0}^{\infty} A_n(t)\varphi_n(x)\,, \tag{11.28}$$

where $\varphi_n(x)$ is such that $H\varphi_n = \lambda_n\varphi_n$ and $(\varphi_n, \varphi_m) = \delta_{nm}$. The substitution of the notation (11.28) into the wave equation (11.27), we obtain

$$\sum_n (\ddot{A}_n + A_n\lambda_n)\varphi_n = 0. \tag{11.29}$$

Scalar multiplying the equality (11.29) on $\varphi_{n'}$, we obtain for each n the equation for the coefficients $A_n(t)$:

$$\ddot{A}_n(t) + A_n(t)\lambda_n = 0\,,$$

with the next solution

$$A_n(t) = a_n\, e^{i\sqrt{\lambda_n}\,t} + b_n\, e^{-i\sqrt{\lambda_n}\,t}\,. \tag{11.30}$$

We note, that $A_n(0) = a_n + b_n$, and the derivative in $(\cdot)$ 0 $\dot{A}_n(0) = i\sqrt{\lambda}(a_n - b_n)$. For the initial conditions

$$\dot{\varphi}(x,t)|_{t=0} \triangleq \varphi_0 \tag{11.31}$$

$$\text{and} \quad \dot{\varphi}(x,t) \triangleq \dot{\varphi}_0\,, \tag{11.32}$$

it is possible to obtain the system of equations for the coefficients a_n, b_n:

$$\varphi_0 = \sum_n (a_n + b_n)\varphi_n\,,$$
$$\dot{\varphi}_0 = i\sum_n \sqrt{\lambda_n}(a_n - b_n)\varphi_n\,. \tag{11.33}$$

By scalar multiplying (11.33) on φ_n, we obtain

$$(\varphi_n, \varphi_0) = a_n + b_n\,,$$
$$(\varphi_n, \dot{\varphi}_0) = i\sqrt{\lambda_n}(a_n - b_n)\,.$$

Each of the coefficients has the next representation:

$$a_n = \frac{1}{2}\left[(\varphi_n, \varphi_0) - \frac{i}{\sqrt{\lambda_n}}(\varphi_n, \dot{\varphi}_0)\right], \tag{11.34}$$

$$b_n = \frac{1}{2}\left[(\varphi_n, \varphi_0) + \frac{i}{\sqrt{\lambda_n}}(\varphi_n, \dot{\varphi}_0)\right]. \tag{11.35}$$

Because in the case of the wave equation the initial condition are represented by two functions (11.31), (11.32), it is possible for each function to construct the Green function and to represent the general solution as follows

$$\varphi = G_1\,\varphi_0 + G_2\,\dot{\varphi}_0\,.$$

Substituting the representations (11.34), (11.35) to the right hand side of equality (11.30), and the coefficients $A_n(t)$ we substitute into (11.28), we obtain

$$\varphi(x,t) = \sum_n \cos\sqrt{\lambda_n}\,t\varphi_n(x)(\varphi_n, \varphi_0) + \sum_n \frac{\sin\sqrt{\lambda_n}\,t}{\sqrt{\lambda_n}}\varphi_n(x)(\varphi_n, \dot{\varphi}_0). \tag{11.36}$$

We note, that the representation of the operator H using eigenfunctions can be written down as

$$H = \sum_n \lambda_n\,\varphi_n\,\varphi_n^*,$$

hence

$$f(H) = \sum_n f(\lambda_n)\varphi_n\,\varphi_n^*\,,$$

we can write the Green functions G_1 and G_2 in the operator expression:

$$\hat{G}_1 = \cos\sqrt{H}t\,, \qquad \hat{G}_2 = \frac{\sin\sqrt{H}\,t}{\sqrt{H}}\,.$$

Using (11.36) we can obtain a coordinate form of the Green function

$$g_1(x,t;x',t') = \sum_n \cos\sqrt{\lambda_n}(t-t')\varphi_n(x)\varphi_n^*(x'),$$

$$g_2(x,t;x',t') = \sum_n \frac{\sin\sqrt{\lambda_n}(t-t')}{\sqrt{\lambda_n}}\varphi_n(x)\varphi_n^*(x').$$

If we have the system of eigenfunctions for the operator H, then we can write the particular representation for the Green functions of the wave equation.

Let us consider, for example, the *wave equation without the source* (*homogeneous wave equation*) for the case of the uniform space:

$$\Delta\varphi - \frac{1}{c^2}\frac{\partial^2}{\partial t^2}\varphi = 0.$$

The eigenfunctions of the operator, which in this case is equal to $(-\Delta)$ $(x \in R^3)$, are determined as

$$\varphi_k = \frac{1}{(2\pi)^{3/2}}e^{ikx}, \qquad x \in R^3.$$

Here the normalized coefficient corresponds to the condition $(\varphi_k, \varphi_{k'}) = \delta(k-k')$. Thus, the coordinate form of $\hat{G}_2$ reads as

$$\begin{aligned} g_2(x,t;x',t') &= \frac{1}{(2\pi)^3}\int_{-\infty}^{\infty}\frac{\sin c|k|(t-t')}{c|k|}e^{ik(x-x')}d^3k \\ &= \frac{1}{4\pi c|x-x'|}\{\delta(|x-x'|+c(t-t')) - \delta(|x-x'|-c(t-t'))\}. \end{aligned}$$

From this representation it is followed that under condition $t > t'$ just only the second part of the Green function "works", i.e.

$$g_2(x,t;x',t') = \frac{1}{4\pi c}\left\{-\frac{\delta(|x-x'|-c(t-t'))}{|x-x'|}\right\} \triangleq G_{(-)} \qquad (11.37)$$

for $t > t'$, that corresponds the signal propagation in the direct time. In the case of the inverse time $t < t'$ the Green function is equal to

$$g_2(x,t;x',t') = \frac{1}{4\pi c}\left[-\frac{\delta(|x-x'|+c(t-t'))}{|x-x'|}\right] \triangleq G_{(+)}$$

for $t < t'$.

11.7.2 *Green function for "Poisson equation"*

Let us write the Poisson equation in the form

$$H\varphi = s,$$

where $H > 0$, $s = s(x)$ is the source function. The Green function in the operator form can be written as $G = H^{-1}$, and in the coordinate form it reads as

$$G(x,x') = \sum_{\lambda_n \neq 0} \lambda_n^{-1}\varphi_n(x)\varphi_n^*(x'). \qquad (11.38)$$

Here λ_n are eigenvalues of the operator H. In the expression (11.38) the summation is implemented only if $\lambda_n \neq 0$, that is equivalent to the representation of the solution in the form

$$\varphi = H^{-1}\, s + k\varphi_0\,,$$

where φ_0 is an eigenfunction of the operator H that corresponds to $\lambda_n = 0$, that leads to the equality

$$H\varphi = s = H\varphi + kH\varphi_0$$

for an arbitrary k. If the operator H is equal to $-\Delta$ ($x \in R^3$) (we consider the Poisson equation in the infinite space), then the Green function in the coordinate form can be written as

$$\begin{aligned} G(x, x') &= \frac{1}{(2\pi)^3} \int \frac{\exp[ik(x - x')]}{k^2}\, d^3k \\ &= \frac{1}{2\pi^2} \frac{1}{|x - x'|} \int_0^\infty \frac{\sin k|x - x'|}{k} dk = \frac{1}{4\pi} \frac{1}{|x - x'|}\,. \end{aligned} \tag{11.39}$$

The solution of the *Poisson equation* can be written down as follows

$$\varphi(x) = \frac{1}{4\pi} \int \frac{s(x')\, dx'}{|x - x'|} + k\varphi_0\,(x),$$

where φ_0: $\Delta\varphi_0 = 0$; k is an arbitrary number.

11.7.3 *Green function for Lame equation in uniform isotropic infinite medium*

The vector Lame equation $L\boldsymbol{\varphi} = \mathbf{s}$ has an unique solution, which can be represented with the help of the Green tensor $\hat{G}$ such that its integral kernel is the tensor function $\hat{g}(x, t; x', t')$ which satisfies the equation

$$L\hat{g} = \delta(x - x')\, \delta(t - t')I.$$

In the infinite uniform space the retarded Green function $g_-\,(x, x'; t, t')$ satisfies the condition $g_- \equiv 0$ where $t < t'$. To construct the fundamental solution of the Lame equation in the explicit form, we use, taking into account the Helmhotz theorem, the representation of the vector field by scalar and vector potentials (4.20):

$$\begin{aligned} \boldsymbol{\varphi} &= -\boldsymbol{\nabla}\Phi + \boldsymbol{\nabla} \times \mathbf{A}\,, \\ \mathbf{s} &= -\boldsymbol{\nabla}\Phi_s + \boldsymbol{\nabla} \times \mathbf{A}_s\,, \end{aligned}$$

which satisfy the scalar and vector wave equation respectively

$$\ddot{\Phi} - v_p^2\, \Delta\Phi = \frac{\Phi_s}{\rho}\,, \tag{11.40}$$

$$\ddot{\mathbf{A}} - v_s^2\, \Delta\mathbf{A} = \frac{\mathbf{A}_s}{\rho}\,. \tag{11.41}$$

The representation of the vector $\mathbf{s}$ as a sum of the vectors of potential and solenoidal fields allow us to introduce the vector field $\mathbf{H}$ such as

$$-\nabla^2 \mathbf{H} = \mathbf{s} = -\boldsymbol{\nabla}\Phi_s + \boldsymbol{\nabla} \times \mathbf{A}_s \,.$$

Taking into account the identity law $-\nabla^2\mathbf{H} = -\boldsymbol{\nabla}(\boldsymbol{\nabla}\cdot\mathbf{H}) + \boldsymbol{\nabla}\times(\boldsymbol{\nabla}\times\mathbf{H})$, it is possible to find that

$$\Phi_s = \boldsymbol{\nabla}\cdot\mathbf{H}\,, \tag{11.42}$$

$$\mathbf{A}_s = \boldsymbol{\nabla}\times\mathbf{H}\,. \tag{11.43}$$

Let us write a solution for the vector Poisson equation with the help of the solution for the scalar Poisson equation for the case of the vector point source, located in the origin of coordinates and oriented in the direction of the first ort of the coordinates, i.e. $\mathbf{s}(\mathbf{x},t) = \mathbf{e}_1\, s(t)\delta(x)$. The solution of the vector Poisson equation $\nabla^2\mathbf{H} = -\mathbf{s}$, using Green function for the scalar wave equation (11.39) reads as

$$H(\mathbf{x},t) = \frac{1}{4\pi}\int\limits_V dx'\,\frac{s(t)\delta(x')}{|x-x'|} = \frac{1}{4\pi}\mathbf{e}_1\,\frac{s(t)}{|\mathbf{x}|}\,. \tag{11.44}$$

Using the relations for the scalar (11.42), vector (11.43) potential and the representation for $\mathbf{H}$ in the form (11.44), we obtain

$$\Phi_s = \boldsymbol{\nabla}\cdot\left(\frac{1}{4\pi}\mathbf{e}_1\,\frac{s(t)}{|\mathbf{x}|}\right) = \frac{s(t)}{4\pi}\partial_1\,|\mathbf{x}|^{-1}\,, \tag{11.45}$$

$$\mathbf{A}_s = \boldsymbol{\nabla}\times\left(\frac{1}{4\pi}\mathbf{e}_1\,\frac{s(t)}{|\mathbf{x}|}\right) = \frac{s(t)}{4\pi}\left\|\begin{matrix} 0 \\ \partial_3\,|\mathbf{x}|^{-1} \\ \partial_2\,|\mathbf{x}|^{-1}\end{matrix}\right\|. \tag{11.46}$$

Substituting the source functions in the forms (11.45) and (11.46), we rewrite the wave equations (11.40) and (11.41):

$$\ddot{\Phi} - v_{\mathrm{p}}^2\,\Delta\Phi = \frac{s(t)}{4\pi\rho}\,\partial_1\,|\mathbf{x}|^{-1}\,, \tag{11.47}$$

$$\ddot{\mathbf{A}} - v_{\mathrm{s}}^2\,\Delta\mathbf{A} = \frac{s(t)}{4\pi\rho}\left\|\begin{matrix} 0 \\ \partial_3\,|\mathbf{x}|^{-1} \\ -\partial_2\,|\mathbf{x}|^{-1}\end{matrix}\right\|. \tag{11.48}$$

We write retarded Green functions for the wave equation (11.47), (11.48), using the Green function, obtained in the case of $L = (\Delta - \frac{1}{c^2}\,\partial_t^2\,)$ (see (11.37)):

$$\begin{aligned} g_\Phi\,(x,x';t-t') &= \frac{1}{4\pi v_{\mathrm{p}}^3}\,\frac{\delta(|x-x'| - v_{\mathrm{p}}\,(t-t'))}{|x-x'|} \\ &= \frac{1}{4\pi v_{\mathrm{p}}^2}\,\frac{\delta((t-t') - |x-x'|/v_{\mathrm{p}}\,)}{|x-x'|}\,, \end{aligned} \tag{11.49}$$

$$g_A\,(x,x';t-t') = I\frac{1}{4\pi v_{\rm s}^2}\,\frac{\delta((t-t')-|x-x'|/v_{\rm s}\,)}{|x-x'|}\,. \tag{11.50}$$

To represent the solutions of the wave equations (11.47) and (11.48) in the form

$$\Phi(\mathbf{x},t) = \int dx'\,dt'\,g_\Phi\,(\mathbf{x},\mathbf{x}';t-t')\left[\frac{s(t')}{4\pi\rho}\,\partial_1\,|\mathbf{x}'|^{-1}\right], \tag{11.51}$$

$$\mathbf{A}(\mathbf{x},t) = \int dx'\,dt'\,g_{\mathbf{A}}\,(x,x';t-t')\frac{s(t')}{4\pi\rho}\left\|\begin{matrix}0\\ \partial_3\,|\mathbf{x}|^{-1}\\ -\partial_2\,|\mathbf{x}|^{-1}\end{matrix}\right\|. \tag{11.52}$$

Then, using the expressions for the Green function (11.49) and (11.50) the solutions (11.51), (11.52) we write down as follows

$$\Phi(\mathbf{x},t) = \frac{1}{(4\pi)^2\,\rho v_{\rm p}^2}\int dx'\,dt'\,\frac{\delta((t-t')-|x-x'|/v_{\rm p}\,)}{|x-x'|}s(t')\,\partial_1\,|\mathbf{x}'|^{-1}, \tag{11.53}$$

$$\mathbf{A}(\mathbf{x},t) = \frac{1}{(4\pi)^2\rho v_{\rm s}^2}\int dx'dt'\frac{\delta((t-t')-|x-x'|/v_{\rm s})}{|x-x'|}$$
$$\times\, s(t')\left\|\begin{matrix}0\\ \partial_3|\mathbf{x}'|^{-1}\\ -\partial_2|\mathbf{x}'|^{-1}\end{matrix}\right\|. \tag{11.54}$$

After the integration over t' the solution (11.53) and after the change of variables $|\mathbf{x}-\mathbf{x}'|/v_{\rm p} = \tau$, we obtain the representation for the scalar potential

$$\Phi(\mathbf{x},t) = \frac{1}{(4\pi)^2\,\rho v_{\rm p}^2}\int\limits_0^\infty d\tau\,\frac{s(t-\tau)}{\tau}\oint\limits_{|x-x'|=v_{\rm p}\,\tau}\partial_1\,|\mathbf{x}'|^{-1}d\sigma\,. \tag{11.55}$$

To bear in mind the calculation the surface integral from the right hand side of (11.55), we consider the following integral

$$\partial_i\oint\limits_{|x-x'|=r}\frac{d\sigma}{|\mathbf{x}'|}, \tag{11.56}$$

where $r = v_{\rm p}\,\tau$, $d\sigma = 2\pi r^2\sin\theta\,d\theta$, $\mathbf{x}' = \mathbf{x}-\mathbf{r}$,

$$(\mathbf{x}',\mathbf{x}') = (\mathbf{x},\mathbf{x}) - 2(\mathbf{x},\mathbf{r}) + (\mathbf{r},\mathbf{r})$$

(Fig. 11.2) or $x'^2 = x^2 - 2xr\cos\theta + r^2$. To write down an exact differential of the last expression and to obtain

$$x'\,dx' = xr\sin\theta\,d\theta\,,$$

and from it follows

$$d\sigma = \frac{2\pi r|\mathbf{x}'|\,dx'}{|\mathbf{x}'|}.$$

To calculate the integral (11.56):

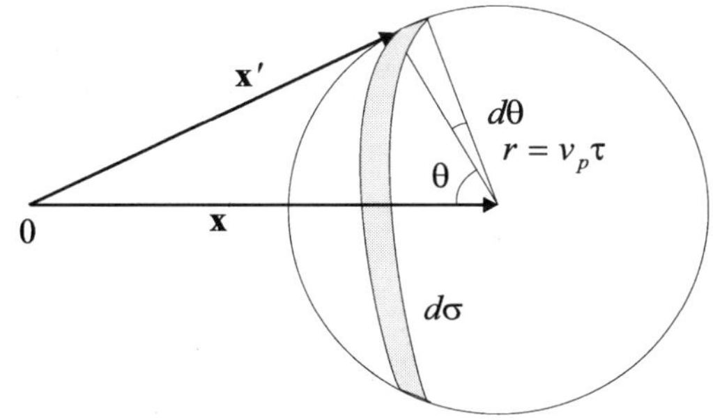

Fig. 11.2 Illustration of the calculation of the integral from right hand side of the equality (11.55).

$$\partial_i \oint\limits_{|\mathbf{x}-\mathbf{x}'|=\mathbf{r}} \frac{d\sigma}{|\mathbf{x}'|} = 2\pi r \partial_i |\mathbf{x}|^{-1} \int\limits_{|\mathbf{x}|-r}^{|\mathbf{x}|+r} dx' = \begin{cases} 4\pi r^2 \partial_i |\mathbf{x}|, & \text{if } \; r < |\mathbf{x}|, \\ 0, & \text{if } \; r > |\mathbf{x}|. \end{cases} \tag{11.57}$$

Using representation (11.57) for the surface integral, we rewrite the expression (11.55) for the potential $\Phi(\mathbf{x}, t)$:

$$\Phi(\mathbf{x}, t) = \frac{1}{4\pi\rho} (\partial_1 |\mathbf{x}|^{-1}) \int\limits_0^{|\mathbf{x}|/v_\mathrm{p}} \tau s(t-\tau) d\tau. \tag{11.58}$$

Taking into account the representation (11.57) and after change of variable $|\mathbf{x} - \mathbf{x}'|/v_\mathrm{p} = \tau$ in the solution (11.54) we can calculate the vector potential:

$$\mathbf{A} = \frac{1}{4\pi\rho} \int\limits_0^{|\mathbf{x}|/v_\mathrm{s}} \tau s(t-\tau) d\tau \begin{Vmatrix} 0 \\ \partial_3 |\mathbf{x}|^{-1} \\ -\partial_2 |\mathbf{x}|^{-1} \end{Vmatrix}. \tag{11.59}$$

Using obtained solutions for scalar and vector potentials (11.58) and (11.59) we restore the potential and solenoidal components of the fields $\boldsymbol{\varphi}_\mathrm{p}$ and $\boldsymbol{\varphi}_\mathrm{s}$:

$$\boldsymbol{\varphi}_\mathrm{p} = -\boldsymbol{\nabla}\Phi = -\frac{1}{4\pi\rho} \begin{Vmatrix} \partial_1^2 |\mathbf{x}|^{-1} \\ \partial_2 \partial_1 |\mathbf{x}|^{-1} \\ \partial_3 \partial_1 |\mathbf{x}|^{-1} \end{Vmatrix} \int\limits_0^{|\mathbf{x}|/v_\mathrm{p}} d\tau s(t-\tau)\tau$$

$$- \frac{|\mathbf{x}| s(t - |\mathbf{x}|/v_\mathrm{p})}{4\pi \rho v_\mathrm{p}^2} \begin{Vmatrix} \partial_1 |\mathbf{x}|^{-1} \partial_1 |\mathbf{x}| \\ \partial_1 |\mathbf{x}|^{-1} \partial_2 |\mathbf{x}| \\ \partial_1 |\mathbf{x}|^{-1} \partial_3 |\mathbf{x}| \end{Vmatrix},$$

$$\boldsymbol{\varphi}_\mathrm{s} = \boldsymbol{\nabla} \times \mathbf{A} = \frac{1}{4\pi\rho} \begin{Vmatrix} -\partial_2^2 |\mathbf{x}|^{-1} - \partial_3^2 |\mathbf{x}|^{-1} \\ \partial_1 \partial_2 |\mathbf{x}|^{-1} \\ \partial_1 \partial_3 |\mathbf{x}|^{-1} \end{Vmatrix} \int\limits_0^{|\mathbf{x}|/v_\mathrm{s}} \tau s(t-\tau) d\tau$$

$$+ \frac{|\mathbf{x}| s(t - |\mathbf{x}|/v_\mathrm{s})}{4\pi \rho v_\mathrm{s}^2} \begin{Vmatrix} -\partial_2 |\mathbf{x}|^{-1} \partial_2 |\mathbf{x}| - \partial_3 |\mathbf{x}|^{-1} \partial_3 |\mathbf{x}| \\ \partial_2 |\mathbf{x}|^{-1} \partial_1 |\mathbf{x}| \\ \partial_3 |\mathbf{x}|^{-1} \partial_1 |\mathbf{x}| \end{Vmatrix}.$$

Introducing the notations $\gamma_i = \partial_i |\mathbf{x}|$ and under the assumption that the source direction is $\mathbf{e}_j$, i.e. after change 1 on j, we obtain, after some mathematical manipulations,

$$\varphi = \varphi_{\mathrm{p}} + \varphi_{\mathrm{s}} = \frac{\hat{P}}{4\pi\, \rho v_{\mathrm{p}}^2 |\mathbf{x}|} \mathbf{s}\left(t - \frac{|\mathbf{x}|}{v_{\mathrm{p}}}\right) + (\hat{I} - \hat{P})$$

$$\times \frac{1}{4\pi\, \rho v_{\mathrm{s}}^2 |\mathbf{x}|} \mathbf{s}\left(t - \frac{|\mathbf{x}|}{v_{\mathrm{s}}}\right) + (3\hat{P} - \hat{I}) \frac{1}{4\pi\, \rho |\mathbf{x}|^3} \int\limits_{|\mathbf{x}|/v_{\mathrm{p}}}^{|\mathbf{x}|/v_{\mathrm{s}}} \tau \mathbf{s}(t - \tau) d\tau.$$

The retarded Green function after the integration over τ can be represented in the form

$$g(x, x'; t - t') = \frac{1}{4\pi\rho} \left\{ \frac{\hat{P}}{v_{\mathrm{p}}^2} \frac{\delta(t - t' - r/v_{\mathrm{p}})}{r} + \frac{(\hat{I} - \hat{P})}{v_{\mathrm{s}}^2} \frac{\delta(t - t' - r/v_{\mathrm{s}})}{r} \right.$$
$$\left. + (3\hat{P} - \hat{I}) \frac{(t - t')[\theta(t - t' - r/v_{\mathrm{p}}) - \theta(t - t' - r/v_{\mathrm{s}})]}{r^3} \right\},$$

where $\theta(t)$ is the rump function,

$$\theta(t) = \begin{cases} 1, & \text{if } t \geq 0, \\ 0, & \text{if } t < 0; \end{cases}$$

$$\mathbf{r} = \mathbf{x} - x', \quad r = |\mathbf{x} - \mathbf{x}'|; \; e_r = \frac{\mathbf{r}}{|\mathbf{r}|}, \quad P = (\mathbf{e}_r, \mathbf{e}_r^T).$$

11.7.4 *Green function for diffusion equation*

Let us to write a "diffusion equation" in the form

$$\left(\frac{\partial}{\partial t} + H\right) \varphi = 0, \tag{11.60}$$

where $H > 0$. The operator H acts on space variables only. To find the Green function of the equation (11.60) we represent its solution as expansion on eigenfunctions of the operator H:

$$\varphi(x, t) = \sum_{n=0}^{\infty} A_n(t) \varphi_n(x). \tag{11.61}$$

Substituting the expansion (11.61) to the equation (11.60), we obtain the equation for the coefficients $A_n(t)$:

$$\sum_{n=0}^{\infty} (\dot{A}_n + A_n \lambda_n) \varphi_n = 0,$$

i.e.

$$A_n(t) = A_n(0) e^{-\lambda_n t}. \tag{11.62}$$

In the expansion (11.61) the coefficient $A_n(t)$ is equal to projection of the function $\varphi(x,t)$ on the ort of the proper basis of the operator H φ_n, i.e.

$$A_n(t) = (\varphi_n, \varphi).$$

Denoting $\varphi(x, t = 0)$ using φ_0, we write down:

$$A_n(0) = (\varphi_n, \varphi_0). \tag{11.63}$$

Substituting the expression (11.63) into the representation (11.62) and, further, substituting the representation (11.62) into the expansion (11.61), we find a link between $\varphi(x,t)$ and $\varphi(x, t = 0)$:

$$\varphi(x,t) = \sum_{n=0}^{\infty} e^{-\lambda_n t} \varphi_n(x)(\varphi_n, \varphi_0). \tag{11.64}$$

From the formula (11.64) we can see, that the Green function in the operator form can be written as $\hat{G} = e^{-\hat{H}t}$, in the coordinate representation we have

$$G(x,x';t,t') = \sum_{n=0}^{\infty} e^{-\lambda_n (t-t')} \varphi_n(x)\varphi_n^*(x').$$

Let the operator is $H = -a^2\nabla^2$ ($x \in R^3$), then the Green function for the *diffusion equation* ($t > t'$) reads as

$$G(x,t;x',t') = \frac{1}{a(2\sqrt{\pi(t-t')})^3} \exp\left\{-\frac{(x-x')^2}{4a^2(t-t')}\right\}.$$

11.7.5 *Green function for operator equation of the second genus*

Let us write the equation under the consideration in the form

$$\varphi + H\varphi = s, \tag{11.65}$$

where $H > 0$. If we represent the equation (11.65) in the form $(I + H)\varphi = s$, then we write the coordinate representation of the Green operator $G = (I + H)^{-1}$ as follows

$$G(x,x') = \sum_{n=0}^{\infty} (1+\lambda_n)^{-1} \varphi_n(x)\varphi_n^*(x').$$

As an example to find the Green function of the equation

$$(-\nabla^2 + m)\varphi = s.$$

The coordinate representation of the Green function can be written as follows

$$G(x,x') = \frac{1}{8\pi^3}\int d^3k\, \frac{\exp[ik(x-x')]}{k^2+m}$$
$$= \frac{1}{4\pi^2\, i|x-x'|}\int_{-\infty}^{\infty} \frac{\exp[ik|x-x'|]}{k^2+m}\, k\, dk = \frac{1}{4\pi}\,\frac{\exp[-m|x-x'|/2]}{|x-x'|}.$$

This expression is known as *Yukawa's potential.*

11.8 Examples of the Recovery of the Local Inhomogeneity Parameters by the Diffraction Tomography Method

Let us consider the results of the numerical simulation on the recovery of the parameters of local inhomogeneities by the diffraction tomography method. The resolution of the diffraction tomography is demonstrated by the recovery results of a bulk of inhomogeneities with sizes much smaller of the wavelength, which are placed in the 3-D uniform reference medium (Troyan and Ryzhikov, 1994, 1995). The results of the recovery of the parameters of local inhomogeneities of a complex shape and the sizes comparable with the wavelength (Kiselev and Troyan, 1998) allows us to estimate the errors sequent from the Born approximation, which is used here to linearize the inverse problem solution.

11.8.1 *An estimation of the resolution*

Let us consider the numerical experiments (Troyan and Ryzhikov, 1994, 1995), for the of the resolution of the diffraction tomography method under the simply ordered array of three-component seismometers. The reference medium is assumed to be infinite and uniform. The incident wave field is a plane p-wave with a definite time shape and a given normal vector to the wave front. The resolution is tested on a regular set of the point diffractors. The synthetic seismograms (direct problem solution) are solved under the Born approximation. The geometry of such experiments is represented at Fig. 11.3. All results are normalized on a dominant wavelength λ of the incident p-wave in the reference medium (v_{p_0} =4000 m/s, v_{s_0} =2000 m/s). The velocity of the p-wave inside diffractors is equal to 3300 m/s. The recovery of the set of diffractors is implemented by the processing of the several experiments with the various normal to the wave front and the various signal shape. The recovery consists in the estimation of the parameter $w = (v_{p_0}/v_p - 1)$. In the experiment the following normal vectors are used: $\mathbf{n}_0$, $\mathbf{n}_1$ and $\mathbf{n}_2$. The next equalities are valid $(\mathbf{n}_0, \mathbf{e}_z) = 1$, $(\mathbf{n}_1, \mathbf{e}_x) = \sqrt{2}/2$, $(\mathbf{n}_2, \mathbf{e}_x) = \sqrt{2}/2$, where $\mathbf{e}_z$ and $\mathbf{e}_x$ are unit vectors, directed along z and x axis respectively. For the simulation of the real seismic data the white noise with the zero mathematical expectation and the variance σ is added.

The resolution of the diffraction tomography method is investigated depending on the distance between receivers, the type of the incident wave, noise level.

At Fig. 11.4 the results of the recovery for distances between receivers $d = 0,5$ λ, $d = 3$ λ and $d = 5$ λ are represented. The noise level is 5 %. Two incident plane waves are used for the calculation of the synthetic seismogram. The best recovery is obtained for the case of $d = 3$ λ , Fig. 11.4(b). The better recovery is observed in the case of two plane incident waves in comparison with one plane wave. The applied algorithm can be used for the case of the noise level 20 %.

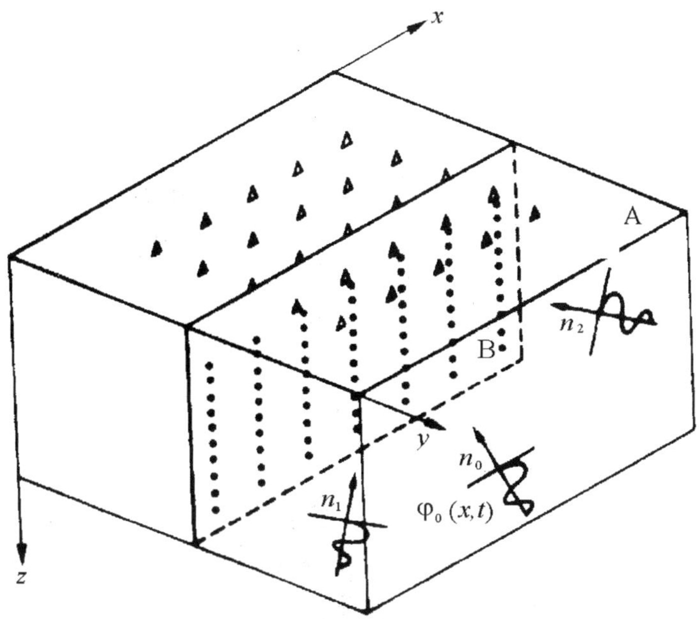

Fig. 11.3 The geometry of the numerical experiment: A is the plane ($z = 0$) contained the geophones, B is the plane with diffractors (y = const), intervals between diffractors along z axis are equal to 1.5 λ and intervals between diffractors along x axis are equal to 1.0 λ).

11.8.2 *An estimation of the recovery accuracy of inhomogeneities parameters*

Let us consider the results of the numerical simulation on the estimation of the parameters of local elastic inhomogeneities and inhomogeneities of an electrical conductivity by the diffraction tomography method. In the cases of both types of inhomogeneities, the direct problem on the elastic wave propagation (the solution of the Lame equation (4.17), (10.11)) and the direct problem on the electromagnetic wave propagation (the solution of the Maxwell equation (4.61)–(4.64)) is solved by the finite difference method. It allows us to take into account the diffraction phenomenon on the target object with the complex geometry and estimate the accuracy of the diffraction tomography method depending on the contrast of the target object relatively to the reference medium.

Let us consider the restoration of the elastic parameters. Using notion of the tomography functional (see Secs. 10.2, 10.3), the vector of the difference field (which is the field, scattered by the desired inhomogeneity — see Sec. 10.1) can be written as follows

$$\begin{aligned}\tilde{\mathbf{u}}(\mathbf{x}_s, \mathbf{x}_r, t) = \int_S [&\delta\lambda(\mathbf{x})\mathbf{p}^\lambda(\mathbf{x}, \mathbf{x}_s, \mathbf{x}_r, t) + \delta\mu(\mathbf{x})\mathbf{p}^\mu(\mathbf{x}, \mathbf{x}_s, \mathbf{x}_r, t) \\ &+ \delta\rho(\mathbf{x})\mathbf{p}^\rho(\mathbf{x}, \mathbf{x}_s, \mathbf{x}_r, t)]d\mathbf{x}, \end{aligned} \tag{11.66}$$

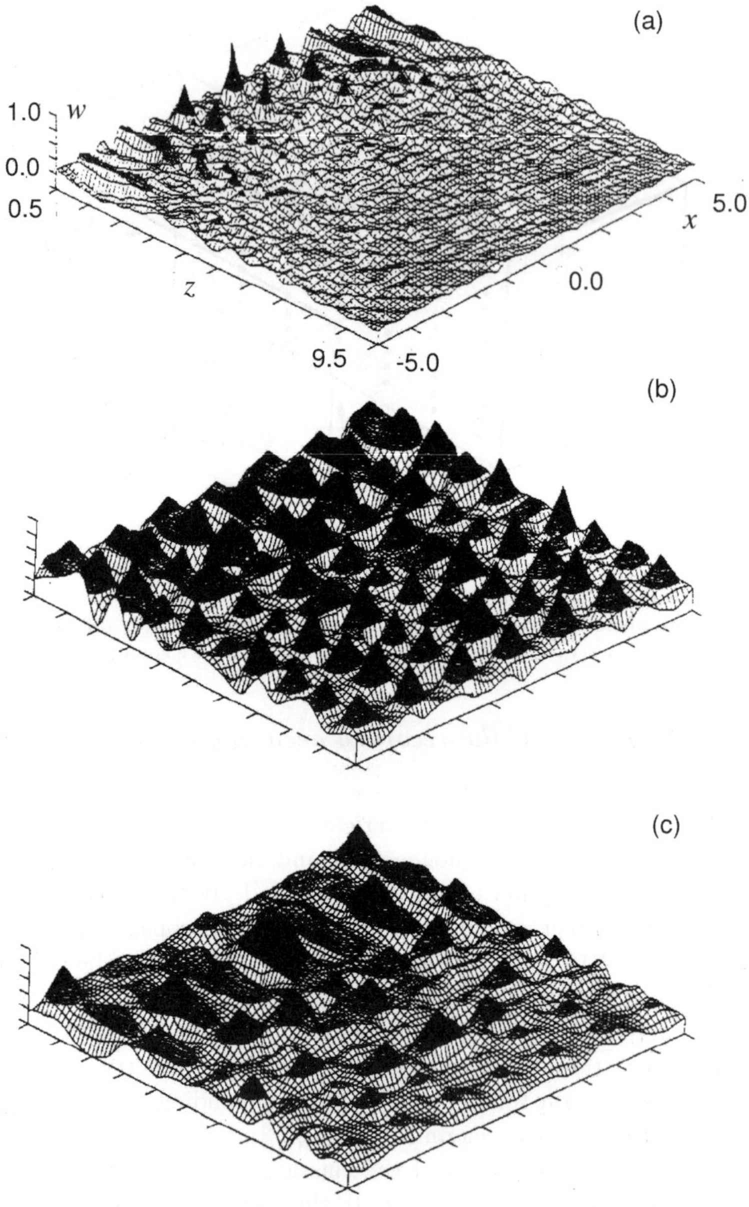

Fig. 11.4 The results of the recovery for two incident plane waves with the normals $\mathbf{n}_1$ and $\mathbf{n}_2$ to the wave front ($w = v_{p0}^2/v_p^2 - 1$): (a) — $d = 0.5\ \lambda$; (b) — $d = 3\ \lambda$; (c) — $d = 5\ \lambda$.

where $\mathbf{x}_s$, and $\mathbf{x}_r$ are the source and receiver positions respectively; $\mathbf{p}^\lambda$, $\mathbf{p}^\mu$ and $\mathbf{p}^\rho$ are the vectors of the tomographic functionals, which corresponds to the next perturbations of the medium $\delta\lambda$, $\delta\mu$ and $\delta\rho$ respectively. Under assumption of the linear relation between perturbations of parameters of the elastic medium ($\delta\lambda$, $\delta\mu$, $\delta\rho$):

$$\delta\lambda(\mathbf{x}) = c_\lambda \delta\mu(\mathbf{x}), \quad \delta\rho(\mathbf{x}) = c_\rho \delta\mu(\mathbf{x}), \tag{11.67}$$

the integral equation (11.66) can be written as follows

$$\begin{aligned}\tilde{\mathbf{u}}(\mathbf{x}_s, \mathbf{x}_r, t) \approx \int\limits_S & \left[c_\lambda \mathbf{p}^\lambda (\mathbf{x}, \mathbf{x}_s, \mathbf{x}_r, t) + \mathbf{p}^\mu (\mathbf{x}, \mathbf{x}_s, \mathbf{x}_r, t)\right. \\ & \left. + \, c_\rho \mathbf{p}^\rho (\mathbf{x}, \mathbf{x}_s, \mathbf{x}_r, t)\right] \delta\mu(\mathbf{x}) d\mathbf{x}.\end{aligned} \tag{11.68}$$

After digitization of the equation (11.68) the system of linear equations relatively to the vector $\mathbf{d}_\mu$ of the desired values $\delta\mu(\mathbf{x})$ can be written as

$$P\mathbf{d}_\mu = \mathbf{d}_u, \tag{11.69}$$

where $\mathbf{d}_u$ are the samples of the field scattered by the inhomogeneity. The solution of the system of linear equations (11.69) we obtain by the minimizing of the sum of squares of the deviations between left hand side and right hand side of (11.69), that, after introducing regularizing terms, leads the system of linear equations

$$[P'P + \alpha_1(B'_x B_x + B'_z B_z) + \alpha_2 C'C + \alpha_3 D'D]\mathbf{d}_\mu = P'\mathbf{d}_u. \tag{11.70}$$

In equation (11.70) α_1, α_2, α_3 are the regularizing coefficients; matrices B_x and B_z are finite differences images of second partial derivatives with respect to x and z respectively; C and D are penalty matrices for non-zero values of the desired values ($\mathbf{d}_\mu$) in the boundary point and the 'near' boundary point of the recovery region S respectively.

Minimizing of the sum of squares of the deviations between the left hand side and the right hand side of (11.69) permits to find the coefficients c_λ and c_ρ. In this case the system of equations (11.70) is solved iteratively, and the minimum of such deviation (connected with the values of c_λ, c_ρ and $\delta\mu(\mathbf{x})$) can be found, for example, by the gradient method.

The parameters of inhomogeneity can be determined using the parametric representation of inhomogeneity. In this case for the minimization of the sum of squares of the deviations between left hand side and right hand side of (11.69) the gradient method can be used also.

The accuracy of the direct problem solution is estimated on the intrinsic convergence for the following local inhomogeneity

$$\delta\mu(\mathbf{x}) = \begin{cases} 0.25\mu_m\left[1 + \cos\left(\pi\frac{x-\hat{x}}{\Delta_x}\right)\right]\left[1 + \cos\left(\pi\frac{z-\hat{z}}{\Delta_z}\right)\right], \\ \quad \text{if} \quad |x - \hat{x}| < \Delta_x \quad \text{and} \quad |z - \hat{z}| < \Delta_z, \\ 0, \text{ if} \quad |x - \hat{x}| > \Delta_x \quad \text{or} \quad |z - \hat{z}| > \Delta_z, \end{cases} \tag{11.71}$$

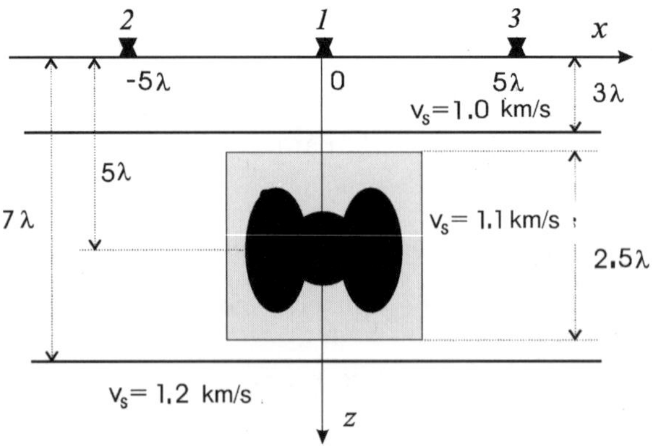

Fig. 11.5 Seismic model and observation scheme: *1–3* are the source and receiver location; S is the recovery region; $\lambda \equiv \lambda_s = 0.02$ km.

where $(\hat{x}, \hat{z})$ is the location of the maximum deviation μ_m the value $\mu(\mathbf{x})$ from the value $\mu_0(\mathbf{x})$ in the reference medium, Δ_x and Δ_z determine inhomogeneity sizes in horizontal ($\mathbf{e}_1$) and vertical directions ($\mathbf{e}_3$) respectively. The solution of the direct problem is implemented with a net for which in the case of $\Delta_x = \Delta_z \approx \lambda_s = v_s T_0$ decreasing of the distance between nodes on two times leads to the deviation between signal amplitudes approximately of 2–5 %.

The recovered inhomogeneities are described by the formula (11.71). The recovery region S and the desired inhomogeneity are represented in Fig. 11.5. The inhomogeneity with a complex geometry is obtained by the superposition of the simple shapes (11.71). So, the model, which is marked in Fig. 11.5 by more dark color, is constructed using three shapes (11.71) with the next parameters:

$$\begin{array}{lll} \hat{x}_1 = \hat{x}_0, & \hat{z}_1 = \hat{z}_0 & (\Delta_x = \Delta_z = 0.45\lambda_s), \\ \hat{x}_2 = \hat{x}_0 + 0.5\lambda_s, & \hat{z}_2 = \hat{z}_0 & (\Delta_x = 0.5\lambda_s,\ \Delta_z = \lambda_s), \\ \hat{x}_3 = \hat{x}_0 - 0.5\lambda_s, & \hat{z}_3 = \hat{z}_0 & (\Delta_x = 0.5\lambda_s,\ \Delta_z = \lambda_s), \\ \hat{x}_0 = 0, & \hat{z}_0 = 5.0\lambda_s, & \mu_{m\,1} = \mu_{m\,2} = \mu_{m\,3} = 1.2. \end{array} \tag{11.72}$$

An asymmetric inhomogeneity described by the parameters

$$\begin{array}{lll} \hat{x}_1 = \hat{x}_0 - 0.25\lambda_s, & \hat{z}_1 = \hat{z}_0 + 0.6\lambda_s & (\Delta_x = \Delta_z = 0.5\lambda_s), \\ \hat{x}_2 = \hat{x}_0 + 0.25\lambda_s, & \hat{z}_2 = \hat{z}_0 + 0.1\lambda_s & (\Delta_x = \Delta_z = 0.5\lambda_s), \\ \hat{x}_3 = \hat{x}_0 + 0.25\lambda_s, & \hat{z}_3 = \hat{z}_0 - 0.5\lambda_s & (\Delta_x = \Delta_z = 0.45\lambda_s), \\ \hat{x}_0 = 0, & \hat{z}_0 = 5.0\lambda_s & \mu_{m\,1} = \mu_{m\,2} = \mu_{m\,3} = 1.2 \end{array} \tag{11.73}$$

is considered also. The inhomogeneity is located in the uniform space ($v_s = 1.1$ km/s), which contacts with the half-space ($v_s = 1.2$ km/s) and with the uniform layer ($v_s = 1.0$ km/s), which is bounded by the free surface ($z = 0$). The maximum velocity contrast of the inhomogeneity is 20 % ($\delta v_{p\,\max} = 0.37$ km/s, $\delta v_{s\,\max} = 0.21$ km/s). The observation points and the source points (*1–3* Fig. 11.5) are placed under the free surface ($z = 0$) with the depth ~ 3 m.

The parameters of the inhomogeneities (11.72), (11.73) are recovered by the solution of the system of linear equations (11.70) and with the help of the optimizing methods together with the parametric representation of inhomogeneities using more wade function set than one given by the formula (11.71). In the last case the desired parameters are found by minimizing of the sum of squared differences of the left hand side and right hand side of the equation (11.69). The elementary test function

$$\delta\mu(\mathbf{x}) = \begin{cases} 0, & r > \Delta + \Delta_1, \\ 0.5\mu_m \left[1 + \cos\left(\pi \frac{r-\Delta_1}{\Delta}\right)\right], & \Delta_1 < r < \Delta + \Delta_1, \\ \mu_m, & r < \Delta_1, \end{cases}$$

$$r = \sqrt{(x-\hat{x})^2 + (z-\hat{z})^2 h_z^2 + (x-\hat{x})(z-\hat{z}) h_{xz}^2} \qquad (11.74)$$

is determined by the seven parameters: the location of its center $\hat{x}$, $\hat{z}$; the values Δ and Δ_1, which determine an ellipse with the constant value of $\delta\mu = \mu_m$ and the region of a smooth change of $\delta\mu$ from the value μ_m to zero value; the maximum value of the shear module μ_m; multipliers h_x and h_{xz}. The restoration of the parameters is implemented using one, two or three functions of the form (11.74). For finding from 9 to 23 desired values (including c_λ and c_ρ from the integral equation (11.68)) a gradient method is used. The convergence of the gradient method to the values close to true values occurs, when the deviation of the initial values is not greater than 50 %, and the deviation for $\hat{x}$ and $\hat{z}$ is 0.25–0.3 λ_s.

Before consideration of the results of the numerical simulation, we make a few remarks. The system of the linear equations (11.70) is solved at nonzero values of the regularizing coefficients α_1 and α_2. As it follows from the numerical simulation the smoothness condition, for recovered parameters of the medium, assigned as a restriction on the value of the second derivative, leads to the better result than the application of the first derivative. The regularization using a penalty function on the nonzero value of the desired parameters in the boundary points of the recovery region S is a very natural procedure for the recovered region with a limited size. Similar regularization ($\alpha_2 \neq 0$) leads to decrement by 10–20 % of the error of the parameters (for example, the velocity of the p-wave) recovery. The penalty function connected with the regularizing coefficient $\alpha_3 \neq 0$ can be used as a way for introducing a priori information about the location and sizes of the desired inhomogeneity. A choice of the regularization parameter α_1 does not involve difficulties, because in the great size of its changing (1–3 orders), after the transition from the bad conditionality of the system of equation (11.70), the recovered function is a very stable and tends to more smooth one with increasing of α_1. The parameter α_2 is chosen so that the recovered values in the boundary of the recovered region S (see Fig. 11.5) are much smaller then their maximum value.

The number of nodes in the recovery region S is equal to 25×25. The represented below examples of the parameters recovery are obtained using two (x- and z-) component of the wave field, "observed" in the points *1–3* (see Fig. 11.5). The wave field is excited at the same points (9 source-receiver pairs). Let's not, that recovery of the inhomogeneity parameters using only z-component of the wave field leads to increasing of the recovery error by 10–20 %.

Let us consider the results of recovery of the inhomogeneities (11.72) and (11.73) (which are represented on Fig. 11.6(a) and 11.7(a) respectively) with the maximum velocity perturbation of p-wave $v_{p\ \max} = 0.37$ km/s when the velocity of p-wave in the reference medium is equal to $v_p = 1.9$ km/s. In Fig. 11.6(b) (11.7(b)) are

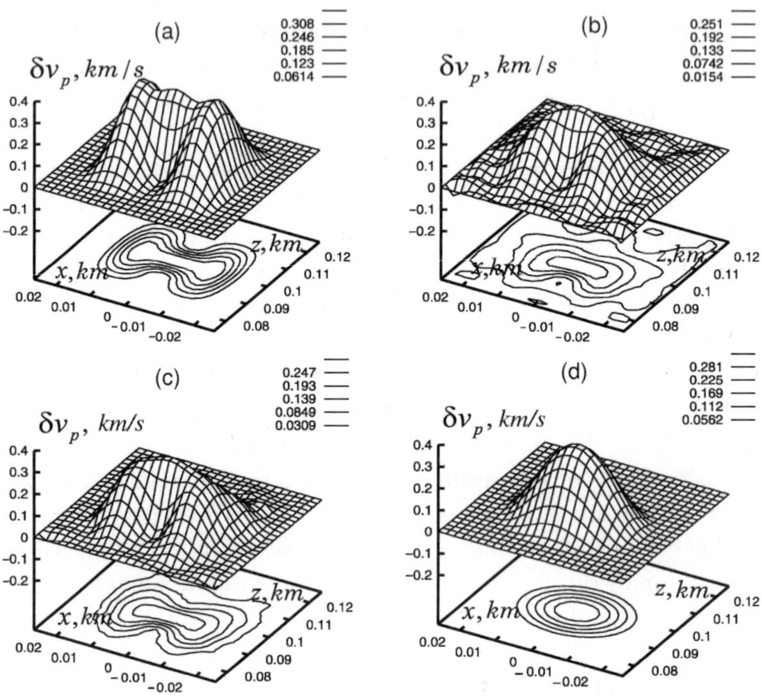

Fig. 11.6 The recovery of v_p for the inhomogeneity (11.72):
(a) is the model; (b) – (d) are the results of recovery: (b), (c) the recovery by the solution of the system of equations (11.70) with $\alpha_3 = 0$ and $\alpha_3 \neq 0$ respectively; (d) the recovery with the use of the parametric representation of the inhomogeneity.

represented the results of recovery (the maximum value $v_{p\ \max} = 0.31$ ($v_{p\ \max} = 0.29$)), which are obtained by the solution of the system of linear equations (11.70) together with the calculation of c_λ and c_ρ (from the equation (11.68)) at $\alpha_3 = 0$. The values c_λ and c_ρ corresponds to the minimum value of the sum of squared differences of the left hand side and the right hand side of the equation (11.69), which is finding by gradient method. The effect of the introducing of penalty functions ($\alpha_3 \neq 0$, equation (11.70)), is demonstrated on Fig. 11.6(c) ($v_{p\ \max} = 0.30$). As a rule, four or five iteration are enough for calculation of c_λ and c_ρ with error not greater than 1 %, if initial values ($c_{\lambda 0}$ and $c_{\rho 0}$), differ from the true values by 50 % at least.

In the considered cases the errors of recovery c_λ and c_ρ are not greater of 10 %

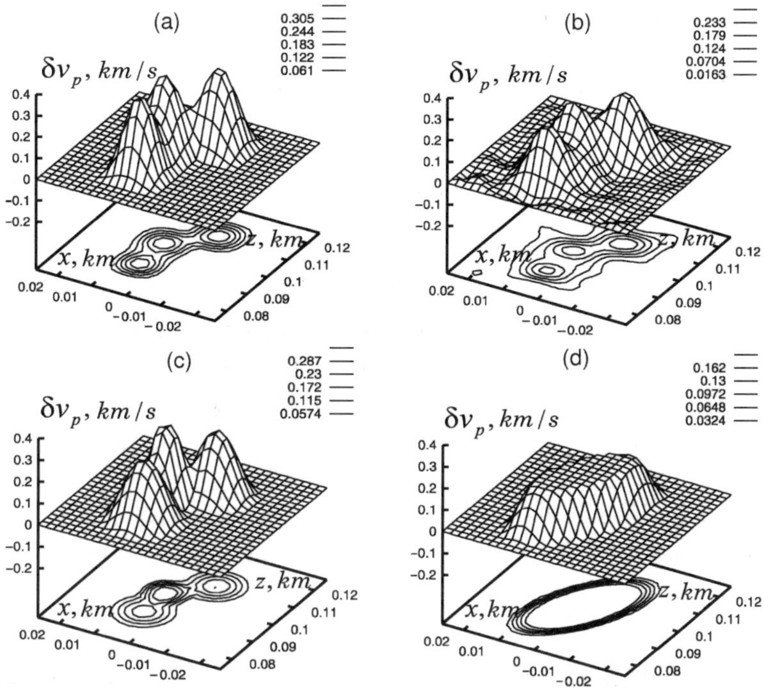

Fig. 11.7 Recovery of v_p for inhomogeneity (11.73): (a) is the model of inhomogeneity; (b)–(d) are the results of recovery: (b) recovery by the solution of the system of equations (11.70) with $\alpha_3 = 0$; (c) and (d) recovery with the use of the parametric representation of inhomogeneity with the help of three and one function (11.74) correspondingly.

and 90 % respectively. Let's note, that c_ρ always is overstated. Such considerable error of the $\delta\rho$ recovery can be explained so that the part of the wave field scattered by the perturbation of $\delta\rho$, "at the average" has a small contribution to the full scattering field.

Similar errors are obtained for the values c_λ and c_ρ in the case of the parametric representation of inhomogeneities using the formula (11.74). The results of the recovery of inhomogeneity (11.72) (see Fig. 11.6(a)) with the help of the one function of (11.74) are represented in Fig. 11.6(d) ($v_{p\ \max} = 0.337$ km/s). The recovered inhomogeneity of the form (11.73) is represented in Fig. 11.7(c) and 11.7(d) correspondingly for the cases of three and one function (11.74).

The calculation of c_λ allows as to recover $\gamma = v_s/v_p$. In the considered model is assumed $\lambda = \mu$, and recovery error of γ is about 1 %.

From the results of numerical simulation, it follows that in the case of the velocity contrast 40–80 % of the target inhomogeneity relatively to the reference medium, the recovery error can be 40–50 %. In the cases of the symmetric target objects the satisfactory result can be obtained using three or even two source-receiver pairs.

Let us consider the results of the numerical simulation of the recovery of perturbation of an electrical conductivity using a nonstationary wave field. The algorithm of the recovery in this case is very close to considered above algorithm. The expression for the tomographic functional $\mathbf{p}^{\sigma}$, which is connected with a scattering by the perturbation of the electrical conductivity σ, has a form of the tomographic functional, which is connected with the perturbation of the mass density (see an expression (10.19)). The tomographic functional $\mathbf{p}^{\sigma}$ is expressed through the electric intensities $\mathbf{E}_{\text{out}}$ and $\mathbf{E}_{\text{in}}$:

$$\mathbf{p}^{\sigma} = \mathbf{E}_{\text{out}} \otimes \frac{\partial}{\partial t} \mathbf{E}_{\text{in}}.$$

Let us note, that the tomographic functional $\mathbf{p}^{\varepsilon}$, which is connected with the perturbation of the electric permittivity ε accurate within the name of variables coincide the mentioned above tomographic functional for the mass density:

$$\mathbf{p}^{\varepsilon} = \mathbf{E}_{\text{out}} \otimes \frac{\partial^2}{\partial t^2} \mathbf{E}_{\text{in}}.$$

With the help of the tomographic functional, the integral equation For the desired perturbation of the electric permittivity $\delta\sigma$ can be written as

$$\delta \mathbf{E} = \int\limits_S \delta\sigma \mathbf{p}^{\sigma} d\mathbf{x}, \tag{11.75}$$

where $\delta\mathbf{E}$ is the difference field. In Fig. 11.8 it is represented the result of recovery of the inhomogeneity with a simple shape. The inhomogeneity size is equal approximately the wavelength in the reference medium with ε =10. The inhomogeneity

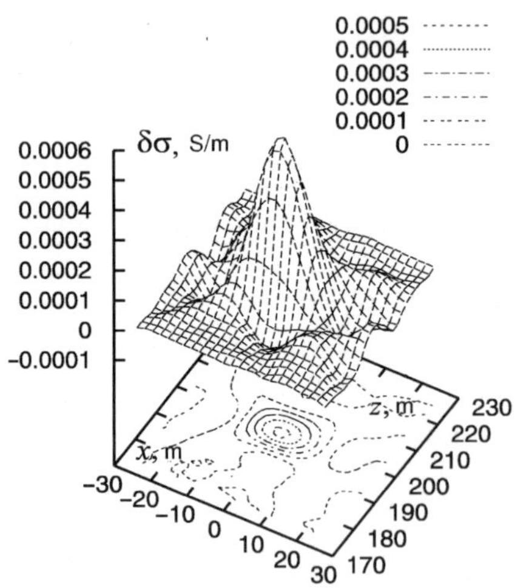

Fig. 11.8 The recovery of the electrical conductivity.

is determined by the function (11.71) with a maximum quantity of the electrical conductivity 10^{-3} S/m. Sounding is implemented by the signal with apparent frequency $5 \cdot 10^6$ Hz. The observation scheme is similar the considered above scheme of the elastic case. The distance between the inhomogeneity and the observation line is equal to 10 λ (λ=20 m). The system of linear equations (11.70) is solved at zero values of the regularizing coefficients α_2 and α_3.

As in the case of the elastic medium, the recovery error is approximately equal to the value of the maximum contrast of inhomogeneity relatively to the reference medium.

Chapter 12

Methods of transforms and analysis of the geophysical data

12.1 Fourier Transform

12.1.1 *Fourier series*

If the integral

$$\int_{-T/2}^{T/2} |f(\tau)| d\tau \tag{12.1}$$

exists, the real function $f(t)$ in the interval $-T/2 < t < T/2$ generates the Fourier series, which is infinite trigonometric series

$$\frac{1}{2}a_0 + \sum_{n=1}^{\infty}(a_n \cos n\omega_0 t + b_n \sin n\omega_0 t) \equiv \sum_{n=-\infty}^{+\infty} c_n \exp\{in\omega_0 t\}, \tag{12.2}$$

where $\omega_0 = 2\pi/T$,

$$a_n = \frac{2}{T}\int_{-T/2}^{T/2} f(\tau)\cos n\omega_0\tau d\tau, \quad b_n = \frac{2}{T}\int_{-T/2}^{T/2} f(\tau)\sin n\omega_0\tau d\tau,$$

$$c_n = \bar{c}_n = \frac{1}{2}(a_n - ib_n) = \frac{1}{T}\int_{-T/2}^{T/2} f(\tau)\exp\{n\omega_0\tau\}d\tau. \tag{12.3}$$

If the Fourier series is created for the interval $(a, a+T)$, then the integration in (12.3) should be implemented between a and $a+T$.

If the integral

$$\int_{-T/2}^{T/2} [f(\tau)]^2 d\tau,$$

exists, then the mean square error

$$\frac{1}{T}\int_{-T/2}^{T/2} [f(\tau) - P_n(\tau)]^2 d\tau,$$

where

$$P_n(t) = \frac{1}{2}\alpha_0 + \sum_{n=1}^{n}(\alpha_n \cos n\omega_0 t + \beta_n \sin n\omega_0 t)$$

is an arbitrary trigonometric polynomial, which for every n has a minimum value, if it is a partial sum

$$s_n(t) = \frac{1}{2}a_0 + \sum_{n=1}^{n}(a_n \cos n\omega_0 t + b_n \sin n\omega_0 t)$$

of the Fourier series of the function $f(t)$. The coefficients a_n and b_n are calculated by the formula (12.3).

If the trigonometric series (12.2) converges uniformly to $f(t)$ inside interval $(-T/2, T/2)$, then its coefficients are the Fourier coefficients (12.3) of the function $f(t)$ with the necessity.

The real coefficients a_n, b_n and complex coefficients c_n are connected by the formulas

$$\begin{aligned} a_n &= c_n + c_{-n}, \quad b_n = i(c_n - c_{-n}), \\ c_n &= \frac{1}{2}(a_n - ib_n), \quad c_n = \frac{1}{2}(a_n + ib_n) \ \ (n = 0, 1, 2 \ldots), \end{aligned} \tag{12.4}$$

where $b_0 = 0$. Fourier series (12.3) for the even function or the odd function $f(t)$ reduces to the sine Fourier series or cosine Fourier series correspondingly.

The necessary conditions for the expansion of the function $f(t)$ in the Fourier series consist in the absolute convergence and the constrained variation of $f(t)$ (12.1) (a finite interval contains a finite number of extremum points). Last condition is always applied in practice.

12.1.2 *Fourier integral*

Let us show transformation of the Fourier series to Fourier integral in the case of $T \to \infty$. Substituting the expression (12.3) to the right hand side of equalities (12.4), we obtain

$$f(t) = \sum_{n=-\infty}^{\infty} \left[\int_{-T/2}^{T/2} f(t) \exp\{-in\omega t\} dt \right] \exp\{in\omega_0 t\}(1/T).$$

If T tends to the infinity, then $1/T$ tends to zero, hence

$$1/T \to d\nu_0 = d\omega_0/2\pi.$$

The difference between neighboring harmonics $n\omega_0$ and $(n+1)\omega_0$ becomes infinitely small, i.e. $n\omega_0$ transforms to continuous variable ω. So, the discrete spectrum $n\omega_0$, which is connected with the Fourier series, transforms to the continuous spectrum.

The subscripts in ν and ω can be omitted, because after proceeding to limit it does not has any value. The sum

$$f(t) = \sum_{n=-\infty}^{\infty} a_n \exp\{in\omega_0 t\}$$

transforms to the integral

$$f(t) = \int_{-\infty}^{\infty} \left[\int_{-\infty}^{\infty} f(t) \exp\{i\omega t\} dt \right] \exp\{i\omega t\} d\omega / 2\pi. \qquad (12.5)$$

The inner integral from the right hand side of the equality (12.5) is called the *Fourier transform* of the function $f(t)$:

$$\tilde{f}(\omega) = \int_{-\infty}^{\infty} f(t) \exp\{-i\omega t\} dt, \qquad (12.6)$$

and outer integral from the right hand side of the equality (12.5) is called the *inverse Fourier transform* of the function $\tilde{f}(\omega)$:

$$f(t) = \frac{1}{2\pi} \int_{-\infty}^{\infty} f(\omega) \exp\{i\omega t\} d\omega. \qquad (12.7)$$

There are other kinds of the Fourier and inverse Fourier transforms, for example, with multiplier $\sqrt{1/2\pi}$ in both Fourier transform and inverse Fourier transform.

The sufficient condition for the existence of the Fourier transform of the function $f(t)$ consists in a boundedness of the integral

$$\int_{-\infty}^{\infty} |f(t)| dt.$$

If the function $f(t)$ is real one, then inverse Fourier transform can be written down as

$$f(t) = \frac{1}{2\pi} \int_{-\infty}^{\infty} R(\omega) \cos \omega t d\omega - \frac{1}{2\pi} \int_{-\infty}^{\infty} X(\omega) \sin \omega t d\omega, \qquad (12.8)$$

where

$$R(\omega) = \int_{-\infty}^{\infty} f(y) \cos \omega y dy \qquad (12.9)$$

is cosine transform of $f(t)$,

$$-X(\omega) = \int_{-\infty}^{\infty} f(y) \sin \omega y dy \qquad (12.10)$$

is sine transform of $f(t)$. In addition the following formula is valid

$$\tilde{f}(\omega) = R(\omega) + iX(\omega) = A(\omega)\exp\{i\gamma(\omega)\},$$

where

$$A(\omega) = \left[R^2(\omega) + X^2(\omega)\right]^{1/2} \tag{12.11}$$

is a *amplitude spectrum*,

$$\gamma(\omega) = \operatorname{arctg}\left[X(\omega)/R(\omega)\right] \tag{12.12}$$

is a *phase spectrum.*

In the case of multivariable function $f(t_1, t_2, \ldots, t_n)$, if the following integral exists

$$\int_{-\infty}^{\infty}\int_{-\infty}^{\infty}\cdots\int_{-\infty}^{\infty} |f(\tau_1, \tau_2, \ldots, \tau_n)| d\tau_1 d\tau_2 \ldots d\tau_n,$$

then the direct

$$C(\omega_1, \ldots, \omega_n) = \frac{1}{(2\pi)^{1/2}} \int_{-\infty}^{\infty}\int_{-\infty}^{\infty}\cdots\int_{-\infty}^{\infty} f(\tau_1, \tau_2, \ldots, \tau_n) \times \exp\left[-i\sum_{j=1}^{n} \omega_j \tau_j\right] d\tau_1 d\tau_2 \ldots d\tau_n$$

and inverse

$$f(t_1, \ldots, t_n) = \frac{1}{(2\pi)^{1/2}} \int_{-\infty}^{\infty}\int_{-\infty}^{\infty}\cdots\int_{-\infty}^{\infty} C(\omega_1, \omega_2, \ldots, \omega_n) \times \exp\left[-i\sum_{j=1}^{n} \omega_j t_j\right] d\omega_1 d\omega_2 \ldots d\omega_n.$$

Fourier transforms are determined.

12.2 Laplace Transform

The Laplace transform is closely connected with the Fourier transform. If we do not use generalized functions, then many functions, for example, $\sin at$ and $\cos at$ functions, do not have the Fourier image, because the direct Fourier integral is divergent one. But if we multiply appropriate function by $\exp(-\sigma|t|)$, where σ is a real positive value, which satisfies the condition $\lim_{t\to\pm\infty}\{\exp(-\sigma|t|)f(t)\} = 0$, then the function $\{\exp(-\sigma|t|)f(t)\}$ has the Fourier image. Such transform is called the

Laplace transform of $f(t)$. If $f(t) = 0$ at $t < 0$, then we obtain the *one-side Laplace transform*

$$F(s) = \int\limits_{0}^{\infty} f(t) \exp(-st) dt, \tag{12.13}$$

where $s = \sigma + i\omega$, in addition the following condition $\lim\limits_{t \to \pm\infty} \{\exp(-s|t|) f(t)\} = 0$ must be satisfied. The *inverse Laplace transform*, can be written as

$$f(t) = \frac{1}{2\pi i} \int\limits_{\sigma - i\infty}^{\sigma + i\infty} F(s) \exp(st) ds, \tag{12.14}$$

where the path of integration is a straight line, which is parallel to the imaginary axis and placed in the right half-plane of the complex plane.

In many cases Laplace transform is more preferable than Fourier transform, for example, in the cases when the function $F(s)$ from the transform of (12.14) has the pole particularities in the imaginary axis.

12.3 Z-Transform

Z-transform is a special kind of the transform, which is useful for calculations with the functions represented by their discrete sequences.

Let us consider the sequence of numbers b_k, which is a discrete representation of the continuous function $f(t)$ at given time points $t_k = k\Delta$ $(k = \ldots, -1, 0, 1, \ldots)$, where Δ is a sampling interval. Let us write the discrete image $f_d(t)$ of the function $f(t)$ as

$$f_d(t) = \sum_{k=-\infty}^{\infty} b_k \delta(t - k\Delta), \tag{12.15}$$

where $\delta(t)$ is delta function. The Fourier transform of $f_d(t)$ can be written as the following sum

$$B(\omega) = \int\limits_{-\infty}^{\infty} f_d(t) \exp(-i\omega t) dt = \sum_{k=-\infty}^{\infty} b_k \exp(-ik\omega\Delta). \tag{12.16}$$

Introducing the notation $Z = \exp(-i\omega\Delta)$ we can write $B(\omega)$ (12.16) as polynomial

$$B(Z) = \sum_{k=-\infty}^{\infty} b_k Z^k, \tag{12.17}$$

where $B(Z)$ is Z-transform of $f_d(t)$[1].

[1]Once Z is determined as $Z = \exp(i\omega\Delta)$, then a polynomial with reversed sign of Z is obtained in comparison with the sign of the polynomials which are considered above.

12.4 Radon Transform for Seismogram Processing

The direct Radon transform reads as

$$s(p,\tau) = \int_{-\infty}^{\infty} u(x, px+\tau)dx$$

(Fig. 12.1). The inverse Radon transform is determined as

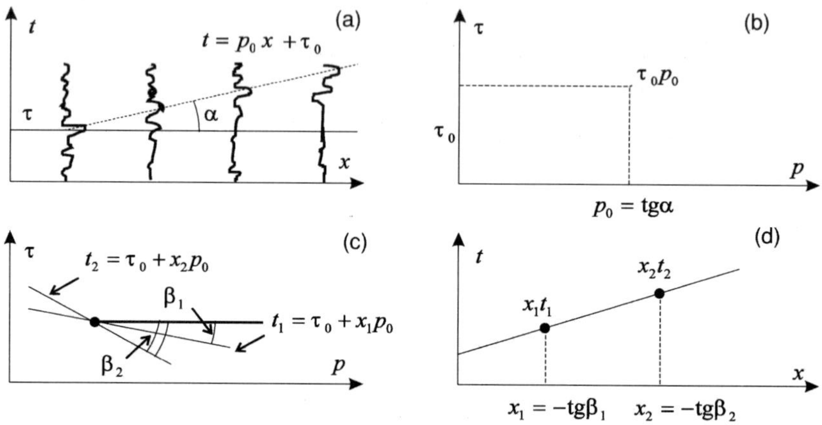

Fig. 12.1 The direct and inverse Radon transform. The straight line in the space-time plane (a) and the point in the plane angle coefficient – reflection time (b). The point in the plane angle coefficient – reflection time (c) and the line in space-time plane (d).

$$u(x,t) = -\int_{-\infty}^{\infty} H\frac{d}{d\tau}s(p,\tau)dp = -PHD[s(p,\tau)],$$

or

$$u(x,t) = -\int_{-\infty}^{\infty} \left[-\frac{1}{\pi\tau} * \frac{\partial}{\partial\tau}s(p,\tau)\right] dp.$$

Let us write the two-dimensional Fourier transform of the seismogram (Fig. 12.2):

$$u(k_x,\omega) = \iint_{-\infty}^{\infty} u(x,t)\exp[i(\omega t - k_x x)]dxdt.$$

Using the formula $k_x = p\omega$, we can write the spectrum $u(k_x,\omega)$ in the form

$$u(p\omega,\omega) = \iint_{-\infty}^{\infty} u(x,t)\exp[i(\omega t - p\omega x)]dxdt,$$

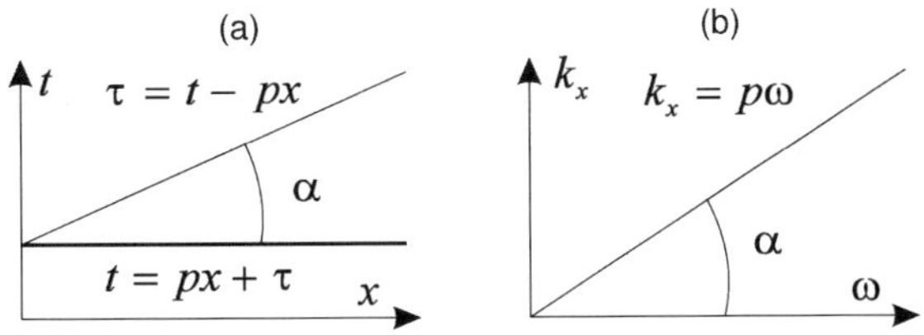

Fig. 12.2 The line in the space-time plane (a) and in the wave number – frequency plane (b).

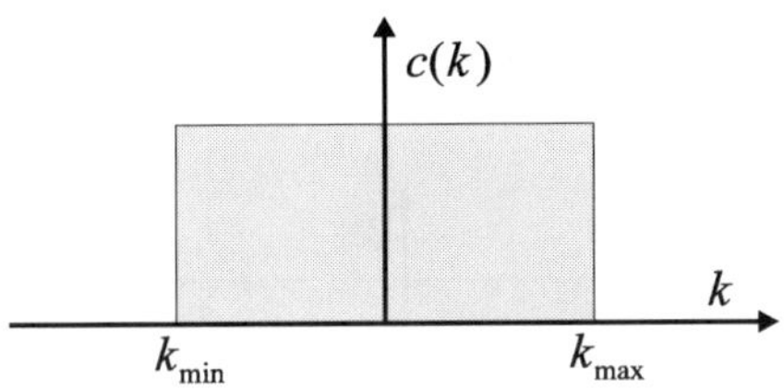

Fig. 12.3 The window function for the integration in k-domain.

or, by introducing variable $t = \tau + px$, we obtain the following expression:

$$\begin{aligned} u(p\omega, \omega) &= \int_{-\infty}^{\infty} e^{i\omega\tau} \Big[\int_{-\infty}^{\infty} u(x, \tau + px) dx \Big] d\tau \\ &= \int_{-\infty}^{\infty} e^{i\omega\tau} s(p, \tau) d\tau = FR[u(x,t)], \end{aligned}$$

where $s(p,\tau) = \int_{-\infty}^{\omega} u(x, \tau + px)dx$ is the Radon transform of the seismogram u. Thus, the two-dimensional Fourier transform can be represented as the Radon transform and the one-dimensional Fourier transform.

The filtration in the domain wavenumber–frequency $(k - \omega)$ reduces to the integration with the limits k_{min} and k_{max} (Fig. 12.3):

$$u(x) = \frac{1}{2\pi} \int_{k_{\min}}^{k_{\max}} u(k) \exp(ikx) dk,$$

where $k = |\omega| p$, or

$$u(x) = \frac{|\omega|}{2\pi} \int_{p_{\min}}^{p_{\max}} u(p) \exp(i\omega px) dp \quad (dk = |\omega| dp).$$

The complete filtration of the seismogram in (ω, k) plane can be written down as following (Fig. 12.4)

$$u(x,t) = \frac{1}{4\pi^2} \int\limits_{\omega_{\min}}^{\omega_{\max}} \int\limits_{p_{\min}}^{p_{\max}} u(\omega,p)|\omega| \exp[-i\omega(t-px)]dpd\omega$$

$$= \frac{1}{2\pi} \int\limits_{p_{\min}}^{p_{\max}} dp\left[\frac{1}{2\pi} \int\limits_{\omega_{\min}}^{\omega_{\max}} \exp[-i\omega(t-px)] \times FR[u(x,t)]|\omega|d\omega\right]$$

$$= \frac{1}{2\pi} \int\limits_{p_{\min}}^{p_{\max}} R[u(x,t)] * (F^{-1}[|\omega|])dp.$$

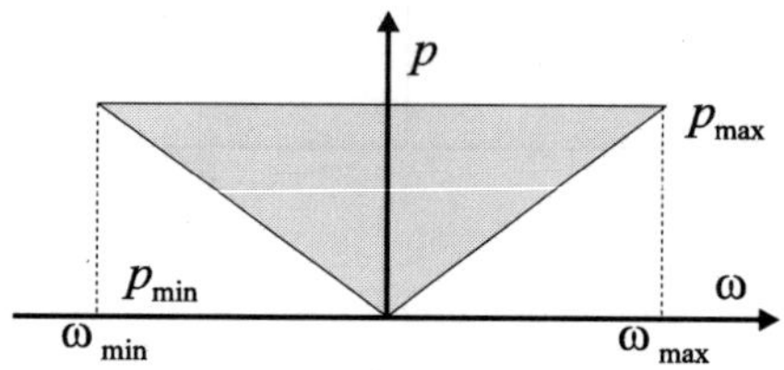

Fig. 12.4 The domain of integration in (ω, k) plane.

12.5 Gilbert Transform and Analytical Signal

Let the function $S(t)$ is a bounded function with the norm L^p, i.e.

$$S(t) \in L^p(-\infty,\infty), \qquad \int\limits_{-\infty}^{\infty} |S(t)|^p dt < \infty, \quad p \geq 1.$$

In this case it is possible to define the direct

$$Q(t) = -\frac{1}{\pi}\text{V.P.} \int\limits_{-\infty}^{\infty} \frac{S(\tau)}{t-\tau} d\tau = -S(t) * \frac{1}{\pi t}$$

and inverse Gilbert transform

$$S(t) = \frac{1}{\pi}\text{ V.P.} \int\limits_{-\infty}^{\infty} \frac{Q(\tau)}{t-\tau} d\tau,$$

where V.P. denotes the Cauchy principal value of the integral at the point $t = \tau$.

Let us calculate the direct Fourier transform of the Gilbert transform $Q(t)$:

$$Q(\omega) = \int\limits_{-\infty}^{\infty} Q(t)\exp(i\omega t)dt = -\frac{1}{\pi} \int\limits_{-\infty}^{\infty} S(\tau) \int\limits_{-\infty}^{\infty} \frac{\exp(i\omega t)}{\tau - t} dt d\tau.$$

After change of variable $\tau - t = u$ we obtain the simple expression for the spectrum:

$$Q(\omega) = -\frac{1}{\pi}\int_{-\infty}^{\infty} S(\tau)\exp(i\omega\tau)\int_{-\infty}^{\infty}\frac{\exp(i\omega u)}{u}du d\tau$$
$$= -S(\omega)\frac{1}{\pi}\int_{-\infty}^{\infty}\frac{\exp(i\omega u)}{u}du$$
$$= -S(\omega)\frac{2i}{\pi}\int_{0}^{\infty}\frac{\sin\omega u}{u}du = -iS(\omega)\,\mathrm{sign}\,\omega,$$

because (Fig. 12.5)

$$\int_{0}^{\infty}\frac{\sin\omega u}{u}du = \frac{\pi}{2}\,\mathrm{sign}\,\omega.$$

We should note, that $|Q(\omega)| = |S(\omega)|$, arg $Q(\omega) =$ arg $S(\omega) \pm \pi/2$.

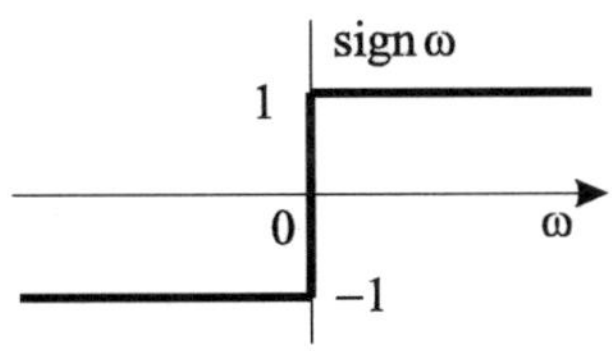

Fig. 12.5 The graphic representation of the function sign ω.

The function $Q(\omega)$ is a *quadrature filter*, which changes a phase of the signal by $\pm\pi/2$, but does not change its amplitude.

Let $Z(t)$ is a complex (temporal) function or an analytical signal:

$$Z(t) = S(t) + iQ(t), \quad \text{or} \quad Z(t) = a(t)\exp(i\varphi(t)),$$

where

$$S(t) = \mathrm{ReZ(t)} = \mathrm{a(t)}\cos\varphi(\mathrm{t}),$$
$$Q(t) = \mathrm{ImZ(t)} = \mathrm{a(t)}\sin\varphi(\mathrm{t}).$$

The amplitude and phase function can be represented as following

$$a(t) = (S^2(t) + Q^2(t))^{1/2} \quad \text{and} \quad \varphi(t) = \arctan(Q(t)/S(t)).$$

The amplitude and phase function are used for the interpretation of seismic data.

12.6 Cepstral Transform

The convolution can be realized as a multiplication of the spectra of the initial signals, which are obtained with the help of the direct Fourier transform (from the time domain to the spectrum domain). The cepstral transform (from the time domain to the cepstral domain) allows to substitute the convolution of the signals for the summation of their cepstral images. Besides, in some cases a separation of the signals is better to implement in the cepstral domain than in the spectral domain, that can leads to more effective filtration.

The inverse Fourier transform of the logarithm of the frequency spectrum yields a *cepstrum*

$$\hat{f}(\zeta) = 1/2\pi \int_{-\infty}^{\infty} \ln[\tilde{f}(\omega)] \exp(i\omega\zeta) d\omega. \tag{12.18}$$

The transform from the time domain to the cepstral domain can be implemented per four steps:

$$\begin{aligned}
&1)\ \tilde{f}(\omega) = \textstyle\int_{-\infty}^{\infty} f(t) \exp(-i\omega t) dt, \\
&2)\ \tilde{f}(\omega) = |\tilde{f}(\omega)| \exp[i\varphi(\omega)], \\
&3)\ \ln \tilde{f}(\omega) = \ln |\tilde{f}(\omega)| + i\varphi(\omega), \\
&4)\ \hat{f}(\zeta) = 1/2\pi \int_{-\infty}^{\infty} [\ln |\tilde{f}(\omega)| + i\varphi(\omega)] \exp(i\omega\zeta) d\omega.
\end{aligned} \tag{12.19}$$

To return in the time domain we should implement the inversion of the transforms (12.19):

$$\begin{aligned}
&1)\ \overline{f}(\omega) = \textstyle\int_{-\infty}^{\infty} \hat{f}(\zeta) \exp(-i\omega\zeta) d\zeta, \\
&2)\ \tilde{f}(\omega) = \exp[\overline{f}(\omega)], \\
&3)\ f(t) = 1/2\pi \int_{-\infty}^{\infty} \tilde{f}(\omega)| \exp(i\omega t) d\omega.
\end{aligned} \tag{12.20}$$

The variable ζ is called the *cepstral frequency*. By analogy with a filtering in the time and frequency domains, the filtering in the cepstral domain is called the *cepstral filtering*.

Let's note that during passing to the cepstral domain an important step consists in the phase determination $\varphi(\omega)$. It is due to the periodic property of the tangent, which has the period $n\pi$. Therefore the phase value $\varphi(\omega)$ can be determined accurate within $n\pi$. If the phase $\varphi(\omega)$ is assumed a continuous value, then at their calculation we should use, for example, a criterion of the smallness of the absolute value of the phase function.

12.7 Bispectral Analysis

In Sec. 1.10.5 it was introduced the correlation function for a real random function. Let us consider an autocovariance function of the second order[2] $b(\tau_1, \tau_2)$ for the real random function $\xi(t)$:

$$b(\tau_1, \tau_2) = \int\limits_{-\infty}^{\infty} \xi(t)\xi(t + \tau_1)\xi(t + \tau_2)dt. \tag{12.21}$$

After Fourier transform of $b(\tau_1, \tau_2)$, we obtain the two-dimensional spectrum

$$B(\omega_1, \omega_2) = \int\limits_{-\infty}^{\infty}\int\limits_{-\infty}^{\infty} b(\tau_1, \tau_2) \exp\{-i(\omega_1\tau_1 + \omega_2\tau_2)\}d\tau_1 d\tau_2 \tag{12.22}$$

of the autocorrelation function, at that the next inequality is valid

$$B(\omega_1, \omega_2) = f(\omega_1)f(\omega_2)f^*(\omega_1 + \omega_2),$$

where $f(\omega)$ is the Fourier transform of $\xi(t)$, symbol $*$ denotes a complex conjugation. Let us note, that the function $b(\tau_1, \tau_2)$ from the formula (12.21) is a symmetric one:

$$\begin{aligned} b(\tau_1, \tau_2) &= b(\tau_2, \tau_1) = b(-\tau_2, \tau_1 - \tau_2) \\ &= b(\tau_1 - \tau_2, -\tau_2) = b(-\tau_1, \tau_2, -\tau_1) = b(\tau_2 - \tau_1, -\tau_1), \end{aligned}$$

from that follows the next equalities for the spectral function $B(\omega_1, \omega_2)$:

$$\begin{aligned} B(\omega_1, \omega_2) &= B(\omega_2, \omega_1) = B(\omega_1, -\omega_1 - \omega_2) \\ &= B(-\omega_1 - \omega_2, -\omega_1) = B(\omega_2, -\omega_1 - \omega_2) = B(-\omega_1 - \omega_2, -\omega_2). \end{aligned}$$

Since we consider here a real process $\xi(t)$, the next equality is valid also

$$B(\omega_1, \omega_2) = B^*(-\omega_1, -\omega_2).$$

Let us consider the relation between the phase function of a linear filter and the phase characteristic of the bispectral function of the output signal. Let us the process x_t is represented as the discrete convolution of the filter w_s with the "input" process g_t

$$x_t = \sum_{s=0}^{\infty} w_s g_{t-s}.$$

If the process g_t is the stationary and uncorrelated one, then the second order autocovariance of the process x_t can be written as

$$b(\tau_1, \tau_2) = \gamma_g^3 \sum_{s=0}^{\infty} w_s w_{s+\tau_1} w_{s+\tau_2}, \tag{12.23}$$

[2]Here we do not consider the living conditions of the autocovariance function of the second order. In action, the random function have a limited length, therefore the existence of similar function does not raise doubts.

where $\gamma_g^3 = M[g_t^3]$, $w_s = 0$, $s < 0$. We represent spectrum $B_x(\omega_1, \omega_2)$ of the autocovariance function $b(\tau_1, \tau_2)$ from the relation (12.23) as multiplication of the filter $W(\omega)$ spectral functions:

$$B_x(\omega_1, \omega_2) = \gamma_g^3 \sum_{\tau_1=-\infty}^{\infty} \sum_{\tau_2=-\infty}^{\infty} \sum_{s=-\infty}^{\infty} s_s w_{s+\tau_1} w_{s+\tau_2}$$
$$\times \exp[-i(\omega_1\tau_1 + \omega_2\tau_2)] = \gamma_g^3 W(\omega_1) W(\omega_2) W(-(\omega_1 + \omega_2))$$
$$= \gamma_g^3 W(\omega_1) W(\omega_2) W^*(\omega_1 + \omega_2),$$

where

$$W(\omega) = \sum_{s=-\infty}^{\infty} w_s \exp(-i\omega s), \quad \begin{cases} -\pi \le \omega_1 \le \pi, \\ -\pi \le \omega_2 \le \pi, \\ -\pi \le \omega_1 + \omega_2 \le \pi. \end{cases}$$

We note, that the spectrum of the second order autocovariance function of x_t accurate within multiplier coincides with spectrum of the second order autocovariance function of the filter w_s:

$$B_x(\omega_1, \omega_2) = \gamma_g^3 B_\omega(\omega_1, \omega_2).$$

If we denote the phase of the spectra of the second order covariance function of the filter and the filter phase function as $\psi(\omega_1, \omega_2)$ and $\varphi(\omega)$ respectively

$$B_\omega(\omega_1, \omega_2) = G_\omega(\omega_1, \omega_2) \exp(i\psi(\omega_1, \omega_2)),$$
$$W(\omega) = H(\omega) \exp(i\varphi(\omega),$$

then we can write the equality

$$\psi(\omega_1, \omega_2) = \varphi(\omega_1) + \varphi(\omega_2) - \varphi(\omega_1 + \omega_2). \tag{12.24}$$

So, the calculation of the spectrum phase $\psi(\omega_1, \omega_2)$ of the second order autocovariance function of x_t, it is possible to find a spectrum phase of the filter $\varphi(\omega)$.

12.8 Kalman Filtering

The recurrence procedure, which has been considered in Sec. 6.11, can be assumed as a basis of the construction of the algorithm of the *dynamic (Kalman) filtration.* A shape of the seismic signals can be changed during the propagation. Such changes can be taken into account using the following model:

$$u_i = \boldsymbol{\varphi}_i \boldsymbol{\theta}_i + \varepsilon_i, \quad \varepsilon_i \in N(0, \sigma_{\varepsilon_i}^2),$$
$$\boldsymbol{\theta}_i = F_{ii-1} \boldsymbol{\theta}_{i-1} + \mathbf{G}\delta_{i-1}, \quad \delta_i \in N(0, \sigma_{\delta_i}^2),$$

where $\boldsymbol{\theta}_i$ is an amplitude vector for i-th sample:

$$\boldsymbol{\theta_i} = \begin{bmatrix} \theta_i^{(1)} \\ \vdots \\ \theta_i^{(M)} \end{bmatrix}, \quad \boldsymbol{\theta}_{i-1} = \begin{bmatrix} \theta_{i-1}^{(1)} \\ \vdots \\ \theta_{i-1}^{(M)} \end{bmatrix}, \quad \mathbf{G} = \begin{bmatrix} 1 \\ 0 \\ \vdots \\ 0 \end{bmatrix}.$$

The vector components $\boldsymbol{\theta}_i$ and $\boldsymbol{\theta}_{i-1}$ are satisfied the next relations

$$\begin{bmatrix} \theta_i^{(2)} = \theta_{i-1}^{(1)} \\ \dots \\ \theta_i^{(M)} = \theta_{i-1}^{(M-1)} \end{bmatrix}.$$

The row vector $\boldsymbol{\varphi}_i = [\varphi_i^{(1)}, \dots, \varphi_i^{(M)}]$ contains the signal shapes for i-th sample. Matrix

$$F_{ii-1} = \begin{bmatrix} r_{i-1}(1) & r_{i-1}(2) & \dots & r_{i-1}(M-1) & r_{i-1}(M) \\ 1 & 0 & \dots & 0 & 0 \\ 0 & 1 & \dots & 0 & 0 \\ \dots & \dots & \dots & \dots & \dots \\ 0 & 0 & \dots & 1 & 0 \end{bmatrix},$$

is a transition matrix from $\boldsymbol{\theta}_{i-1}$ to $\boldsymbol{\theta}_i$, $r_{i-1}(\mu)$ are the elements of a priori autocorrelation of the components of $\boldsymbol{\theta}_{i-1}$. The illustration of the Kalman filtering procedure is represented at Fig. 12.6.

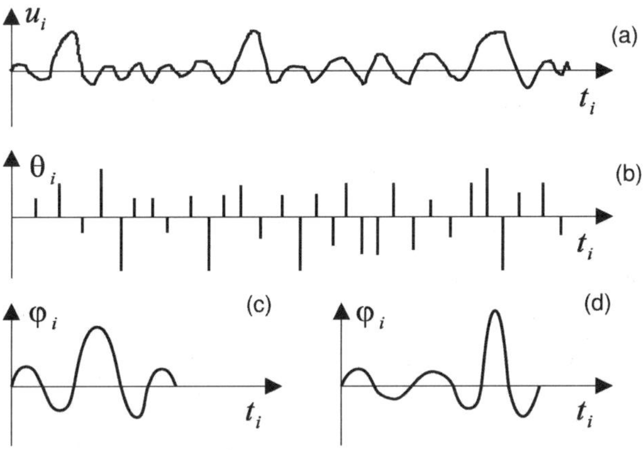

Fig. 12.6 The illustration of the Kalman filtering: (a) is an initial seismic trace; (b) is an impulse seismic trace; (c), (d) are the signal shapes for different parts of the trace.

The estimate of $\boldsymbol{\theta}_i$ can be obtained with the help of the quadratic criterion using the minimizing procedure

$$l(\boldsymbol{\theta_i}) = \frac{-1}{2\sigma_{\varepsilon i}^2}(u_i - \boldsymbol{\varphi}_i\boldsymbol{\theta}_i)^2 - \frac{1}{2}(\boldsymbol{\theta}_i - \hat{\boldsymbol{\theta}}_i{}')^T R_{\theta_i'}^{-1}(\boldsymbol{\theta}_i - \hat{\boldsymbol{\theta}}_i{}'),$$

where $\hat{\boldsymbol{\theta}}_i' = F_{ii-1}\boldsymbol{\theta}_{ii-1}$,

$$R_{\boldsymbol{\theta}_i'}^{-1} = \langle \boldsymbol{\theta}_i \boldsymbol{\theta}_i^T \rangle = F_{ii-1} R_{\theta_{i-1}} F_{ii-1}^T + \mathbf{G}\sigma_{\delta_{i-1}}^2 \mathbf{G}^T.$$

The estimate $\boldsymbol{\theta}_i$ has the form

$$\tilde{\boldsymbol{\theta}}_i = \left(\frac{\boldsymbol{\varphi}_i^T \boldsymbol{\varphi_i}}{\sigma_{\varepsilon i}^2} + R_{\theta_i'}^{-1}\right)^{-1} \left(\frac{\boldsymbol{\varphi}_i^T u_i}{\sigma_{\varepsilon i}^2} + R_{\theta_i'}^{-1}\hat{\boldsymbol{\theta}}_i{}'\right).$$

Using the identities (6.39), we can write a recurrence formula for the parameters estimation together with the appropriate covariance matrix

$$\tilde{\boldsymbol{\theta}}_i = \hat{\boldsymbol{\theta}}_i{}' + R_{\theta_i'}\boldsymbol{\varphi}_i^T(\boldsymbol{\varphi}_i R_{\theta_i'}\boldsymbol{\varphi}_i^T + \sigma_{\varepsilon i}^2)^{-1}(u_i - \boldsymbol{\varphi}_i\hat{\boldsymbol{\theta}}_i{}'),$$
$$R_{\theta_i} = R_{\theta_i'} - R_{\theta_i'}\boldsymbol{\varphi}_i^T(\boldsymbol{\varphi}_i R_{\theta_i'}\boldsymbol{\varphi}_i^T + \sigma_{\varepsilon i}^2)^{-1}\boldsymbol{\varphi}_i R_{\theta_i'}.$$

The recurrence formula needs an initial vector $\boldsymbol{\theta}_0$ and an initial matrix R_{θ_0}. Without a priori information usually the next initial data

$$\boldsymbol{\theta}_0 = \begin{bmatrix} 0 \\ \vdots \\ 0 \end{bmatrix}, \quad R_{\theta_0} = \begin{bmatrix} 1 & 0 & \dots & 0 \\ 0 & 1 & \dots & 0 \\ \dots & \dots & \dots & \dots \\ 0 & 0 & \dots & 1 \end{bmatrix}$$

are used.

12.9 Multifactor Analysis and Processing of Time Series

We consider a time series

$$\{x_i\}_{i=1}^N \Rightarrow x_i = f((i-1)\Delta t), \quad i = 1, 2, \dots, N.$$

The function $f(t)$ might be random.

12.9.0.1 *I. Evolvent of 1-D time series to multidimensional one*

We chose some number $M < N$, named *Maximum Number of Principal Components (MNPC)*, and present the first M values of the time series x_i as a first row of the matrix X. As elements of the second row we take values from x_2 to x_{M+1}. The last row with the number $k = N - M + 1$ will be last M of the time series $x_k, x_{k+1}, \dots, x_N$:

$$X = (x_{ij})_{i,j=1}^{k,M} = \begin{bmatrix} x_1 & x_2 & x_3 & \dots & x_M \\ x_2 & x_3 & x_4 & \dots & x_{M+1} \\ \vdots & \vdots & \vdots & \ddots & \vdots \\ x_k & x_{k+1} & x_{k+2} & \dots & x_N \end{bmatrix}. \tag{12.25}$$

This matrix, with elements $x_{ij} = x_{i+j-1}$, can be considered as M-dimensional sampling with k sample size.

12.9.0.2 *II. Analysis of principal components: Singular decomposition of the sample correlation matrix*

1. The calculation of the mean value and the standard deviation along the columns of the matrix X:

$$\bar{x}_j = \frac{1}{k}\sum_{i=1}^{k} x_{i+j-1}, \quad s_j = \sqrt{\frac{1}{k}\sum_{i=1}^{k}(x_{i+j-1} - \bar{x}_j)^2}, \tag{12.26}$$

where $\bar{x}_j$ has meaning of the moving average, s_j has meaning of the standard deviation with the rectangular window with the width k.

2. We introduce matrix X^*, that is centered along

$$x^*_{ij} = \frac{(x_{ij} - \bar{x}_j)}{s_j}, \quad i = 1, \dots, k; \quad j = 1, \dots, M. \tag{12.27}$$

3. Calculation of the sample correlation matrix

$$R = \frac{1}{k} X^{*T} X^* \tag{12.28}$$

with elements

$$r_{ij} = \frac{1}{k} \sum_{l=1}^{k} \frac{1}{s_i s_j} (x_{i+l-1} - \bar{x}_i)(x_{j+l-1} - \bar{x}_j).$$

4. Calculation of eigenvalues and eigenvectors of the matrix R (*Singular Value Decomposition – SVD*):

$$R = P \Lambda P^T, \tag{12.29}$$

where

$$\Lambda = \begin{bmatrix} \lambda_1 & 0 & \dots & 0 \\ 0 & \lambda_2 & \dots & 0 \\ \vdots & \vdots & \ddots & \vdots \\ 0 & 0 & \dots & \lambda_M \end{bmatrix}$$

is a diagonal matrix of eigenvalues and

$$P = (p_1, p_2, \dots, p_M) = \begin{bmatrix} p_{11} & p_{21} & \dots & p_{M1} \\ p_{12} & p_{22} & \dots & p_{M2} \\ \vdots & \vdots & \ddots & \vdots \\ p_{1M} & p_{2M} & \dots & p_{MM} \end{bmatrix}$$

is an orthogonal matrix of the eigenvectors of the matrix R.

In this case the following relations are valid:

$$P^T = P^{-1}, \quad P^T P = P P^T = I_M,$$

$$\Lambda = P^T R P, \quad \sum_{i=1}^{M} \lambda_i = M, \quad \prod_{i=1}^{M} \lambda_i = \det R.$$

Interpretation of the matrices P and Λ. The matrix P might be considered as a transfer function from the matrix X^* to the *Principal Components*

$$X^* P = Y = (y_1, y_2, \dots, y_M).$$

If we consider random time series x_i, then the eigenvalues of the matrix R are sampling variances and square roots of its are sampling standards. The sampling standards are proportionally to the semiaxes of the ellipsoid of the concentration, described by matrix R.

If the considered time series $\{x_i\}$ is stationary, in the sense that any part of time series with the length $s = \min(M, k)$ carries out all basic information about the structure of the process $\{x_i\}$, it is possible to average the matrix R along diagonals and to obtain the autocorrelation function $r(\tau)$:

$$r(\tau) = \frac{1}{M-\tau} \sum_{j=1}^{M-\tau} r_{j,j+\tau}, \quad \tau = 1, \ldots, M. \tag{12.30}$$

12.9.0.3 *III. Selection of the principal components*

As it was defined above the principal components is defined as

$$Y = X^* P = (y_1, y_2, \ldots, y_M).$$

We normalize the principal components on the sampling standards (by $\lambda_M \neq 0$)

$$Y^* = Y\Lambda^{-1/2} = (y_1^*, y_2^*, \ldots, y_M^*).$$

In this case the principal components are orthonormalized:

$$Y^{*T} Y^* = I_M,$$

i.e. we obtain the decomposition of the initial M-dimensional process on orthogonal components.

The transform

$$y_j = X^* p_j$$

is very close to the linear transform of the initial time series by means of the discrete convolution operator

$$\begin{aligned} y_j(l) &= \sum_{q=1}^{M} X_{lq}^* p_{jq} = \sum_{q=1}^{M} (x_{l+q-1} - \bar{x}_q) \frac{p_{jq}}{s_q} \\ &= \sum_{q=1}^{M} x_{l+q-1} \frac{p_{jq}}{s_q} - \sum_{q=1}^{M} \bar{x}_q \frac{p_{jq}}{s_q}. \end{aligned}$$

Thus, the procedure of the multifactor analysis creates a set of the linear filters. In this case the eigenvectors are the transition functions of linear filters.

The visual analysis of the eigenvectors and principal components, obtained after linear filtering can give a lot of interesting information about the structure of the process.

It is very useful the visual information, which gives a two-dimensional diagram (on the x-axis – p_i or y_i, on the y-axis p_j or y_i).

Example 12.1. If the two-dimensional diagram for p_i and p_{i+1} is closed to a circle, it means that we can interpret these eigenvectors as a pair of sin–cos in the Fourier transform. Let us consider the functions

$$\begin{cases} x(t) = a(t)\cos(\omega(t)t + \psi(t)), \\ y(t) = a(t)\sin(\omega(t)t + \psi(t)), \end{cases}$$

assuming $a(t)$, $\omega(t)$ and $\psi(t)$ slowly varying functions

$$\begin{cases} x(t) = r(t)\cos(\alpha(t)t), \\ y(t) = r(t)\sin(\alpha(t)t), \end{cases}$$

where

$$r(t) = \sqrt{x^2(t) + y^2(t)} = a(t),$$
$$\alpha(t) = \arctan\frac{y(t)}{x(t)} = \omega(t)t + \psi(t).$$

We choose two sequences with the time delay τ and find a difference of the polar angles:

$$\begin{aligned}\Delta\alpha &= \alpha(t+\tau) - \alpha(t) = \omega(t+\tau)(t+\tau) \\ &+ \psi(t+\tau) - \omega(t)t - \psi(t) \\ &= (\omega(t+\tau) - \omega(t))t + \psi(t+\tau) - \psi(t) \\ &= (\omega'(t)\tau + o(\tau))t + (\omega(t) \\ &+ \omega'(t)\tau + o(\tau))\tau + \psi'(t)\tau + o(\tau).\end{aligned}$$

If $\omega(t)$ and $\psi(t)$ are slowly varied functions such that $\omega'(t)$ and $\psi'(t)$ have an order $O(\tau^2)$, we can write

$$\Delta\alpha(t) = \omega(t)\tau = \frac{2\pi\tau}{T(t)},$$
$$\omega(t) = \frac{\Delta\alpha(t)}{\tau} \quad \text{and} \quad T(t) \approx \frac{2\pi\tau}{\Delta\alpha(t)}.$$

Thus, calculating the difference of the polar angles, we can estimate an instantaneous frequency $\omega(t)$ and an instantaneous period $T(\tau)$ of a "harmonic oscillation" corresponding to chosen pair of eigenvectors.

It should be noted, that it is impossible to estimate the frequency more than

$$\omega_N = \frac{\pi}{\tau}.$$

12.9.0.4 *IV. Recovery of 1-D time series*

Taking into account the orthogonal property of the matrix P we have

$$X^* = YP^T = (y_1, y_2, \ldots, y_M) \cdot \begin{pmatrix} p_1^T \\ p_2^T \\ \vdots \\ p_M^T \end{pmatrix}$$

$$= \sum_{l=1}^{M} y_l P_l^T = \sum_{l=1}^{M} X_l^*.$$

After the canceling of the normalization and decentering of the matrix X^*

$$X = \bar{x}\, I_k^T + X^*S = X_0^* + \sum_{l=1}^{M} X_l^* S = \sum_{l=0}^{M} X_l^* S$$

we obtain an initial matrix as a sum of $M + 1$ matrix.

To obtain an initial time series we implement averaging over near main diagonals. To determine the mean operator $\mathcal{A}$ as

$$x = \mathcal{A}(X) = \sum_{l=0}^{M} \mathcal{A}(X_l^* S), \tag{12.31}$$

we obtain the representation for the initial time series as a sum of $M+1$ time series.

We should note, that the interactive property of the considered algorithm is its natural peculiarity.

12.9.0.5 Selection of parameters

The main control parameter is the MNPC – M (M should be $< N/2$).
The choice of the length M depends on the current problem:

(1) *The search of a latent periodicity.*
First of all, we calculate eigenvalues by the maximum M and estimate a number of the eigenvalues l, which satisfy the inequality $\lambda_i > 0$. On the next step we carry out our analysis using $M = l$.
(2) *Smoothing of the time series.*
In this case we can interpret an action of the considered algorithm as a filtering. The reconstruction of some principal components is reduced to the filtering of the time series by the transition function which is equal to the eigenvector. The more M, the more narrow will be the band of the filter.
(3) *Recovery of the periodicity with the known period.*
M should be equal to the period T and N should be multiply to the period T.

It should be noted that the method is very stable to the variation of the length M.

Intermediate results for interpretation.

(1) Eigenvalues of the correlation matrix of the M-dimensional presentation of the time series.
(2) Eigenvectors of the correlation matrix. They take up the orthogonal system.
(3) Principal components of the M-dimensional presentation. They also form the orthogonal system.
(4) Reconstructed time series on the different sets of the principal components is operator of the transfer from components to the M-dimensional matrix X and the operator $\mathcal{A}$ for the transition from the M-dimensional matrix X to the 1-dimensional time series $\{x_i\}_{x=1,\dots,N}$.

It should be pointed out two extreme cases.

- $M \ll N$. We can interpret the eigenvectors as transition functions of the linear filters, and the principal components appear as the action of these filters.

- $M \simeq N/2$. The method can be interpreted as an approximated method with respect to an initial time series. For example, to search of the harmonic components.

12.10 Wiener Filter

The convolution of two vectors $\mathbf{h}$ and $\mathbf{x}$ (see Sec. 6.4) in discrete form is determined by the formula

$$y_k = \sum_{j=0}^{L_h} h_j x_{k-j}, \quad k = 0, \ldots, L_y \quad (L_y = L_h + L_x - 1), \tag{12.32}$$

where L_h is a length of the filter $\mathbf{h}$, L_x is a length of the filtered signal $\mathbf{x}$. The discrete convolution can be represented in a matrix form as following

$$\begin{bmatrix} y_0 \\ y_1 \\ y_2 \\ \vdots \\ y_{L_y} \end{bmatrix} = \begin{bmatrix} x_0 & 0 & 0 & \ldots & 0 \\ x_1 & x_0 & 0 & \ldots & 0 \\ x_2 & x_1 & x_0 & \ldots & 0 \\ \vdots & \vdots & \vdots & \ddots & \vdots \\ 0 & 0 & 0 & \ldots & x_{L_x} \end{bmatrix} \begin{bmatrix} h_0 \\ h_1 \\ h_2 \\ \vdots \\ h_{L_h} \end{bmatrix}, \tag{12.33}$$

or

$$\mathbf{y} = X\mathbf{h}. \tag{12.34}$$

If the vector $\mathbf{y}$ and the matrix X are known, then it is possible to find the vector $\mathbf{h}$, using, for example, the least squares method (see Sec. 6.4), which leads to the system of linear equations

$$X^T X\mathbf{h} = X^T\mathbf{y}. \tag{12.35}$$

The system (12.35) after introducing regularization can be written down as

$$(X^T X + \alpha I)\mathbf{h} = X^T\mathbf{y}, \tag{12.36}$$

where I is an identity matrix, α is a regularization coefficient.

If only one element of the vector $\mathbf{y}$ is equal to 1, and others elements are equal to zero, then the filter $\mathbf{h}$ is called the *Wiener filter*. In this case the vector $\mathbf{x}$ is transformed to $\mathbf{y} = [\ldots, 1, \ldots]$ by the filter $\mathbf{h}$.

Let us consider a filtering of the signal $x(t)$, which is a continuous function of t. Let us find the filter $h(t)$, which minimizes the sum of squares deviations E between some "required" signal $\hat{y}(t)$ and "actual" signal $y(t)$:

$$E = \lim_{T\to\infty}(1/2T)\int_{-T}^{T}[y(t)-\hat{y}(t)]^2 dt$$

$$= \lim_{T\to\infty}(1/2T)\int_{-T}^{T}\left\{\left[\int_{-\infty}^{\infty}h(\tau)x(t-\tau)d\tau\right]-\hat{y}(t)\right\}^2 dt$$

$$= \lim_{T\to\infty}(1/2T)\left\{\int_{-T}^{T}\left[\int_{-\infty}^{\infty}h(\tau)x(t-\tau)d\tau\int_{-\infty}^{\infty}h(\sigma)x(t-\sigma)d\sigma\right]dt\right.$$

$$\left.-2\int_{-T}^{T}\int_{-\infty}^{\infty}[h(\tau)x(t-\tau)d\tau]\hat{y}(t)dt+\int_{-T}^{T}\hat{y}^2(t)dt\right\}. \quad (12.37)$$

After the interchange of integrations we represent E using autocorrelation functions of the input signal R_{xx}, "required" signal $R_{\hat{y}\hat{y}}$ and their cross-correlation function $R_{x\hat{y}}$:

$$E = \int_{-\infty}^{\infty}h(\tau)d\tau\int_{-\infty}^{\infty}h(\sigma)d\sigma\left[\lim_{T\to\infty}(1/2T)\int_{-T}^{T}x(t-\tau)x(t-\sigma)dt\right]$$

$$-2\int_{-\infty}^{\infty}h(\tau)d\tau\left[\lim_{T\to\infty}(1/2T)\int_{-T}^{T}\hat{y}(t)x(t-\tau)dt\right]+\lim_{T\to\infty}(1/2T)\int_{-T}^{T}\hat{y}^2(t)dt$$

$$= \int_{-\infty}^{\infty}h(\tau)d\tau\int_{-\infty}^{\infty}h(\sigma)R_{xx}(\tau-\sigma)d\sigma-2\int_{-\infty}^{\infty}h(\tau)R_{x\hat{y}}(\tau)d\tau+R_{\hat{y}\hat{y}}(0).$$

$$(12.38)$$

Using the methods of the calculus of variations to the expression (12.38), it is possible to obtain an integral equation for the desired function $h(t)$

$$\int_{-\infty}^{\infty}h(t)R_{xx}(\tau-t)dt = R_{x\hat{y}}(\tau), \quad \tau\geq 0. \quad (12.39)$$

This equation is called the *Wiener–Hopf equation.* The solution of the equation (12.39) is considered in (Lee, 1960).

12.11 Predictive-Error Filter and Maximum Entropy Principle

Let us consider the filter $\mathbf{h} = \{h_1,\ldots,h_n\}$, which predicts the values of the causal series $\mathbf{x} = \{x_1,\ldots,x_m\}$ $(m > n)$ for one step on time with the help of the past and the current values of $\mathbf{x}$:

$$x_j = h_1 x_{j-1} + f_2 x_{j-2} + \cdots + h_j x_0 = \sum_{k=1}^{n} h_k x_{j-k},$$
$$j = 1, \ldots, m; \quad x_{j-k} = 0 \text{ if } k > j. \tag{12.40}$$

It is supposed usually, that predicted value x_0 is equal to zero. The prediction error (e_j) of x_j can be written down as

$$e_j = \sum_{k=1}^{n} h_k x_{j-k} - x_j = \sum_{k=0}^{n} h_k x_{j-k}, \tag{12.41}$$

where $h_0 = -1$. The filter $-1, h_1, h_2, \ldots, h_n$ is called the *predictive error filter of the order $(n+1)$ for the prediction on one step.*

In addition to considered above the predictive error filter, there are other filters which are used for different transforms of the time series (see, for example, (Box and Jenkins, 1976)).

With the help of the least squares method (see Secs. 6.4, 12.10) we can write the system of linear equations (see also (12.35)) to find the filter $\mathbf{h}$:

$$\sum_{k=1}^{n} h_k R_{xx}(r-k) = R_{xx}(r), \quad r = 1, 2, \ldots, n, \tag{12.42}$$

where $R_{xx}(r)$ is an autocorrelation function of the input signal.

Sum of squares of the values e_j (12.41) can be represented as

$$E = \sum_{j=0}^{m} e_j^2 = \sum_{j=0}^{m} \left(\sum_{k=1}^{n} h_k x_{j-k} - x_j \right)^2$$
$$= \sum_{j=0}^{m} \left[\left(\sum_{k=1}^{n} h_k x_{j-k} \sum_{l=1}^{n} h_l x_{j-l} \right) - 2x_j \left(\sum_{k=1}^{n} h_k x_{j-k} \right) + x_j^2 \right].$$

After the change of the order of summation and taking into account equation an (12.42), we can represent E with the help of the autocorrelation function R_{xx}:

$$E = \sum_k h_k \left[\sum_l h_l \left(\sum_j x_{j-k} x_{j-l} \right) \right]$$
$$-2 \sum_k h_k \left(\sum_j x_j x_{j-k} \right) + R_{xx}(0)$$
$$= \sum_k h_k \left[\sum_l h_l R_{xx}(k-l) \right] - 2 \sum_k h_k R_{xx}(-k) + R_{xx}(0)$$
$$= \sum_k h_k R_{xx}(k) - 2 \sum_k h_k R_{xx}(-k) + R_{xx}(0).$$

Taking into account that $h_0 = -1$, we obtain

$$E = -\sum_k h_k R_{xx}(k). \tag{12.43}$$

from the relations (12.42) and (12.43) it is possible to obtain the following matrix equation:

$$\begin{Vmatrix} R_{xx}(0) & R_{xx}(-1) & \dots & R_{xx}(-n) \\ R_{xx}(1) & R_{xx}(0) & \dots & R_{xx}(1-n) \\ \dots & \dots & \dots & \dots \\ R_{xx}(n) & R_{xx}(n-1) & \dots & R_{xx}(0) \end{Vmatrix} \begin{Vmatrix} h_0 \\ h_1 \\ \vdots \\ h_n \end{Vmatrix} = \begin{Vmatrix} -E \\ 0 \\ \vdots \\ 0 \end{Vmatrix}. \tag{12.44}$$

The concept of the information and entropy, considered in Sec. 1.9, plays the important role in the probability theory and its applications. In the information theory the entropy characterizes a measure of the uncertainty of time series. The amount of information which can be extracted, is increased with the growth of the entropy. Completely predictable signal does not carry any information. Absolutely unpredictable signal, for example, white noise, carries maximal information potentially. At a filtration by the *maximum entropy method* aspire so to transform an input signal that it is unpredictable as much as possible with the preservation of its autocorrelation function. The purpose of a such transformation can be to extract of the signals on a background of the ordered noise.

In Sec. 1.9 the entropy H for the system of events with an unequal probability was introduced (1.54). It is possible to extend this notion to the cases of stationary time series of the finite length n (H_n) and stationary time series of the infinite length $n \to \infty$ (H) (see, for example, (Stratonovich, 1975)). At that a notion of the specific entropy

$$H = \lim_{n\to\infty} [H_n/(n+1)] \tag{12.45}$$

is introduced. In (Smylie *et al.*, 1973) it is shown that the entropy of the time series with infinite length

$$H = \frac{1}{2}\ln E_\infty = \frac{1}{2}\ln(\omega_N/\pi) + \frac{1}{4\omega_N}\int_{-\omega_N}^{\omega_N} \ln[R_\infty(\omega)]d\omega \tag{12.46}$$

is connected with an energy of the prediction error E_∞ (12.43) (at $n = \infty$)

$$E_\infty = \frac{\omega_N}{\pi}\exp\left[\frac{1}{2\omega_N}\int_{-\omega_N}^{\omega_N} \ln[R_\infty(\omega)]d\omega\right], \tag{12.47}$$

and the spectral function $R_\infty(\omega)$ of the input signal $\mathbf{x}$

$$R_\infty(\omega) = R_\infty(z) = \sum_{k=-\infty}^{\infty} R_{xx}(k)z^k. \tag{12.48}$$

In the formulas (12.47), (12.46) the value ω_N is the *Nyquist frequency*.

In practice the autocorrelation function $R_{xx}(i)$ has a finite length. One can suppose that $R_{xx}(i) = 0$, $|k| > n$ outside the measuring span. But such values of the autocorrelation function can be chosen so, that H is stable with respect to the value $R_{xx}(i)$ at $|k| > n$ (Burg, 1972). Using the formulas (12.46), (12.48), we obtain the system of equations:

$$\frac{\partial H}{\partial R_{xx}(k)} = 0 = \int\limits_{-\omega_N}^{\omega_N} \frac{z^k}{R_\infty(z)} d\omega, \quad |k| > n. \tag{12.49}$$

We will written down the final result without mathematical manipulations:

$$R_\infty(z) = E[h(z)h(z^{-1})]^{-1}, \tag{12.50}$$

where E and $h(z)$ are the sum of predictive error squares (12.43) and z-transform of the predictive error filter (12.44). The solution of the system of equations (12.44) is considered in (Anderson, 1974; Smylie *et al.*, 1973).

Consider a recurrent procedure for the solution of the system of equations (12.44) to find the predictive error filter $1, -h_1, \ldots, -h_n, \ldots, -h_N$ $(n = 0, 1, 2, \ldots, N)$ in the case of the time series $x_1, \ldots, x_m, \ldots, x_M$ $(m = 1, 2, \ldots, M)$ of a finite length. We define the autocorrelation function taking into account a number of the terms in considered time series:

$$R_m = \frac{1}{M-m} \sum_{k=1}^{M-m} x_k x_{k+m} \quad (m = 0, 1, 2, \ldots, M-1).$$

By analogy the sum of the squared differences of the predictive error filter E is normalized.

Let's note that the following identities

$$R_0 \cdot 1 = E_0, \quad \text{where} \quad E_0 = \frac{1}{M} \sum_{k=1}^{M} x_k^2$$

in initial step $n = 0$ are satisfied. At the first step $(n = 1)$ the coefficient h_{nn} of the predictive error filter $1, -h_{11}$ (where the first number of the subscript is the step number) is found by minimizing the sum of the prediction error

$$E_1 = \frac{1}{2(M-1)} \sum_{j=1}^{M-1} (e_j^2 + \tilde{e}_j^2),$$

which calculated by averaging of the predictions in the direct

$$e_j = x_{j+1} - h_{11} x_j$$

and inverse

$$\tilde{e}_j = x_j - h_{11} x_{j+1}$$

directions,

$$E_1 = \frac{1}{2(M-1)} \sum_{j=1}^{M-1} \left[(x_{j+1} - h_{11} x_j)^2 + (x_j - h_{11} x_{j+1})^2 \right]. \tag{12.51}$$

Considering E_1 as a function of h_{11}, we find h_{11} as the solution of the equation

$$dE_1/dh_{11} = 0$$
$$= \frac{1}{2(N-1)} \sum_{j=1}^{M-1} 2[(x_{j+1} - d_{11}x_j)(-x_j) + (x_i - d_{11}x_{j+1})(-x_{j+1})]$$
$$= \frac{1}{2(N-1)} \sum_{j=1}^{M-1} [-(x_j x_{j+1} - x_{j+1}x_j) + h_{11}(x_{j+1^2} + x_j^2)].$$

Thus,

$$h_{11} = 2 \sum_{j=1}^{M-1} x_j x_{j+1} / \sum_{j=1}^{M-1} (x_j^2 + x_{j+1}^2).$$

Using expression (12.51), we find E_1. Let's note, that R_1 it is possible to find from the first equation of the system of equations

$$\begin{Vmatrix} R_0 & R_1 \\ R_1 & R_0 \end{Vmatrix} \begin{Vmatrix} 1 \\ -h_{11} \end{Vmatrix} = \begin{Vmatrix} E_1 \\ 0 \end{Vmatrix}.$$

Generally the average total error for the direct and inverse filters can be written down as follows:

$$E_n = \frac{1}{2(M-n)} \sum_{j=1}^{M-n} \left[\left(x_j - \sum_{k=1}^{n} h_{nk} x_{j+k} \right)^2 + \left(x_{j+n} - \sum_{k=1}^{n} h_{nk} x_{j+nkk} \right)^2 \right].$$

The average error also can be written down as

$$E_n = \frac{1}{2(M-n)} \sum_{j=1}^{M-n} [(b_{nj} - h_{nn}b'_{nj})^2 + (b'_{nj} - h_{nn}b_{nj})^2],$$

where

$$b_{1j} = x_j, \quad b'_{1j} = x_{1+j},$$
$$b_{nj} = \sum_{k=0}^{n} h_{n-1,k} x_{j+k} = \sum_{k=0}^{n} h_{n-1,n-k} x_{j+m-k},$$
$$b'_{nj} = \sum_{k=0}^{n} h_{n-1,k} x_{j+m-k} = \sum_{k=0}^{n} h_{n-1,n-k} x_{j+k},$$

at that for the coefficients b and b' and the coefficients of the filter h the next recurrent relations

$$b_{nj} = b_{n-1,j} - h_{n-1,n-1} b'_{n-1,j+1},$$
$$b'_{nj} = b_{n-1,j+1} - h_{n-1,n-1} b'_{n-1,j+1},$$
$$h_{nk} = h_{n-1,k} - h_{n,m-k},, \quad k = 1, 2, \ldots, n-1,$$
$$h_{n0} = -1, \quad h_{nk} = 0 \quad (k > n)$$

are valid. By solution of the equation for h_{nn} — $dE_n/dh_{nn} = 0$, we obtain

$$h_{nn} = 2\sum_{j=1}^{M-n} b_{nj}b'_{nj} / \sum_{j=1}^{M-n} (b_{nj} + b'^{2}_{nj}). \tag{12.52}$$

Let's show that the value E_n also satisfies to a recurrent relation. Let's write the system of equations determining the predictive error filter for the case of $n = 3$:

$$\begin{Vmatrix} R_0 & R_1 & R_2 & R_3 \\ R_1 & R_0 & R_1 & R_2 \\ R_2 & R_1 & R_0 & R_1 \\ R_3 & R_2 & R_1 & R_0 \end{Vmatrix} \begin{Vmatrix} 1 \\ -h_{n1} \\ -h_{n2} \\ -h_{n3} \end{Vmatrix} = \begin{Vmatrix} E_n \\ 0 \\ 0 \\ 0 \end{Vmatrix}. \tag{12.53}$$

Substituting to the system of equation (12.53) the recurrent relation for h_{nj}, we obtain

$$\begin{Vmatrix} R_0 & R_1 & R_2 & R_3 \\ R_1 & R_0 & R_1 & R_2 \\ R_2 & R_1 & R_0 & R_1 \\ R_3 & R_2 & R_1 & R_0 \end{Vmatrix} \left\{ \begin{Vmatrix} 1 \\ -h_{n-1,1} \\ -h_{n-1,2} \\ 0 \end{Vmatrix} - h_{n3} \begin{Vmatrix} 1 \\ -h_{n-1,1} \\ -h_{n-1,2} \\ 0 \end{Vmatrix} \right\}$$

$$= \begin{Vmatrix} E_{n-1} \\ 0 \\ 0 \\ h_{n3}E_{n-1} \end{Vmatrix} - h_{n3} \begin{Vmatrix} h_{n3}E_{n-1} \\ 0 \\ 0 \\ E_{n-1} \end{Vmatrix} = \begin{Vmatrix} E_n \\ 0 \\ 0 \\ 0 \end{Vmatrix}.$$

Generally the recurrent formula for E_n can be written down as follows:

$$E_n = E_{n-1}(1 - h_{nn}^2).$$

Let's note, that from equality (12.52) it follows $|h_{nn}| < 1$, therefore the inequalities $0 < E_n < E_{n-1}$ are valid.

Appendix A

Computer exercises

The computer exercises represented in the Appendix, make a basis of the computing practical work which go with lectures "Statistical methods of processing of the geophysical data" and "Geophysical inverse problems", given to master students and to post-graduate students in the Earth Physics Department of the Physical Faculty of Saint-Petersburg State University within last ten years. The computer exercises will allow the students to understand the essence and opportunities of the considered methods of analysis and processing of the geophysical information more deeply.

Basis of the computer exercises is the set of programs[3], written in MATLAB language. The programs contain a brief description of the input and output data, demonstration example and recommendations for change input data.[4] The computer exercises require the initial rudiments about MATLAB, such which allow to change the input data in programs written in MATLAB language.

A.1 Statistical Methods

A.1.1 *Examples of numerical simulation of random values*

In Matlab package there are procedures for simulation of the random values. The next script produces the *samples* belonging to uniform (at the interval $(0, 1)$) and normal ($N(0, 1)$) distributions.

```
% Script r0a.m
% Examples of random values
% indd=1 : uniform distribution at interval (0,1)
% indd=2 : Gaussian distribution (N(0,1))
clear
% INPUT
```

[3]The programs are placed in http://geo.phys.spbu.ru/~vtroyan.

[4]The programs together with demonstration examples was prepared in Geophysical Institute of Lausanne University. Authors consider as the pleasant debt to express gratitude to professor R.Olivier for creation of excellent conditions for work and a friendly climate.

```
indd=1;
m=25;        % sample size
% END INPUT
n=1:m;        % n=1,2,...,m
if indd==1
   s=rand(1,m);        % s : row vector
else
   s=randn(length(n),1);        % s : column vector; length(n)=m
   s=s'        % s : row vector
end
rect=[0.1 0.1 .8 0.5];        % size of figure
axes('position',rect)
stem(n,s);        % graphic representation
```

A.1.1.1 *Exercises*

(1) With the use of script `r0a.m` to implement a graphic representation of the samples belonging to the *uniform* and *normal* distributions.
(2) By using the command line to implement graphic presentation of the samplings belonging to the *uniform* and *normal* distributions with the use of `plot` command.

A.1.2 *Construction of histogram*

The histogram characterizes the probability properties of a random variable. The next script (`p0hist.m`) constructs the histogram for a sampling belonging to the normal distribution.

```
%Script p0hist.m
% Construction of histogram
% nr : sample size
clear
% INPUT
nr=500;        % sample size
del=0.5;        % input of bin for xcoordinate
% END INPUT
x=-5:del:5; % construction of the bins for x-coordinate
d=randn(1,nr);                 % d : a batch of numbers
              %(row vector)
% hist(d,x);
[n,x]=hist(d,x);
bar(x,n./(nr.*del)); % graphic representation
              % of the normalized histogram
```

```
hold on
d2=d_gauss(x,0,1); % normal probability density
              % (row vector)
plot(x,d2,'g') % graphic representation of                % the normal density function
```

A.1.2.1 *Exercises*

(1) To compare the histograms obtained using the different sample sizes (for example, 100 – 500).
(2) To compare the histograms obtained for a different number of the bins on the x axis (for example, 4 – 20). Do not change the interval of the histogram determination on the x axis.

A.1.3 *Description of a random variable*

A few probability distributions for the discrete random variables and the density functions for the continuous random variables are represented below.

A.1.3.1 *Beta distribution*

Density function for the beta distribution reads as

$$f_\xi(x|\alpha,\beta) = \begin{cases} \frac{\Gamma(\alpha+\beta)}{\Gamma(\alpha)\Gamma(\beta)} x^{\alpha-1}(1-x)^{\beta-1}, & \text{if } 0 < x < 1, \\ 0, & \text{if } x \le 0 \text{ or } x \ge 1, \end{cases} \tag{A.1}$$

where $\alpha > 0$, $\beta > 0$;

$$M\xi = \frac{\alpha}{(\alpha+\beta)}, \quad D\xi = \frac{\alpha\beta}{(\alpha+\beta)^2(\alpha+\beta+1)}.$$

Graphic presentation of (A.1) is realized by the script `d0beta.m`. The function `d=d_beta(x,al,bt)` returns a row vector `d` with values of the density function for the beta distribution.

A.1.3.2 *Binomial distribution*

Binomial distribution function is determined by the formula (1.38). Graphic presentation of the distribution function (1.38) is realized in the script `d0bin.m`. The function `d=d_bin(n,p)` returns row vector `d` — the probabilities of the binomial distribution.

The script `d0bin.m` is represented below.

```
% Script d0bin.m
% Binomial distribution
% call function d=d_bin(n,p)
% n : n=1,2,...; - the maximum value of x
%        x=0,1,...,n
```

```
% p : 0 ¡ p ¡ 1
% d : row vector, contains the distribution function
%
clear
% INPUT
n=50;
p=0.7;
% END INPUT
d=d_bin(n,p);
x=0:n;
% graphic representation
rect=[0.1 0.1 .8 0.5];        % size of the figure
axes('position',rect)
stem(x,d);
comp_prob=sum(d)      % computation of the total probability
math_exp= n .* p       % mathematical expectation
stand_dev=sqrt(n .* p .* (1-p))     % standard deviation
m1=sum(x.*d)        % mathematical expectation
s1=sqrt(sum((x-m1).^2.*d))       % standard deviation
pause
% comparison with normal distribution
hold on
dn=d_gauss(x,math_exp,stand_dev);
plot(x,dn,'g');
```

The function `d=d_bin(n,p)` is represented below.

```
function d=d_bin(n,p);
% binomial distribution d=d_bin(n,p)
% n : n=1,2,...; - the maximum value of x
%         x=0,1,...,n
% p : 0 ¡ p ¡ 1
%
n1=prod(1:n);
d=zeros(1,n+1);       % determination of the row vector d
for m=0:n
    if m==0
        m1=1;
    else
        m1=prod(1:m);
    end
    nm=n-m;
    if nm==0
```

```
        nm1=1;
    else
        nm1=prod(1:nm);
    end
    d(m+1)=(n1/(m1*nm1))*p^m * (1-p)^nm;
end
```

A.1.3.3 *Cauchy distribution*

The density function of the Cauchy distribution is given by the formula (1.51). Graphic presentation of the Cauchy density function is realized by the script `d0cauch.m`. The function `d=d_cauch(x)` return the row vector `d`, which contains the values of the Cauchy density function.

A.1.3.4 χ^2-*distribution*

Graphic presentation of the density function for χ^2-distribution (1.45) is realized in the script `d0chisq.m`. The function `d=d_chisq(x,n)` returns the row vector `d`, which contains the values of the density function.

The script `d0chisq.m` is represented below.

```
% Script d0chisq.m
% Calculation of the density function for χ2 distribution
% call d=d_chisq(x,n)
% n : a number of degree of freedom
% x : an argument of the density function, a ¡ x ¡ b
% d : returns the values of the density
        % function
clear
% INPUT
n=50
% to determine an interval on x axis for computing the density
        % function
a=0;        % (a ¡ x ¡ b)
b=100.0;
nx=500;         % sample size
% END INPUT
del=(b-a)/(nx-1);        % to determine the values of argument
x=a:del:b;
d=d_chisq(x,n);        % row vector
plot(x,d);        % graphic presentation
comp_prob=trapz(x,d)        % computing of the total probability
math_exp=n        % mathematical expectation
stand_dev=sqrt(2 .* n)        % standard deviation
```

```
m1=trapz(x,x.*d)      % mathematical expectation
s1=sqrt(trapz(x,(x-m1).^2.*d))      % standard deviation
%
pause
% comparison with the normal distribution
hold on
dn=d_gauss(x,math_exp,stand_dev);
plot(x,dn,'g');     % graphic presentation
                    % of the normal distribution
```

The function `d=d_chisq(x,n)` is represented below.

```
function d=d_chisq(x,n)
% χ2-distribution d=d_chisq(x,n)
% n : a number of the degree of freedom
% x : a row vector contains the values of argument (0 ¡ x ¡ inf)
% d : a row vector contains the values of the density function
%
d=((x./2).^((n./2)-1).*exp(-x ./ 2))/(2.*gamma(n./2));
```

A.1.3.5 *Exponential distribution*

Graphic presentation of the exponential density function (1.49) is realized by the script `d0expon.m`. The function `d=d_expon(x,l)` returns row vector `d`, Which contains the values of the exponential density function.

A.1.3.6 *Fisher distribution*

Graphic presentation of the Fisher density function (1.48) is realized by the script `d0fish.m`. The function `d=d_fish(x,n1,n2)` returns row vector `d`, Which contains the values of the Fisher density function.

A.1.3.7 *Univariate normal (Gaussian) distribution*

Graphic presentation of the normal density function (1.41) is realized by the script `d0gauss.m`. The function `d=d_gauss(x,xm,s)` returns row vector `d`, Which contains the values of the normal density function.

A.1.3.8 *Geometrical distribution*

Graphic presentation of the geometrical density function (1.40) is realized by the script `d0geom.m`. The function `d=d_geom(n,p)` returns row vector `d`, Which contains the values of the normal density function.

A.1.3.9 *Laplace distribution*

Graphic presentation of the Laplace density function (1.50) is realized by the script `d0lapl.m`. The function `d=d_lapl(x,l)` returns row vector `d`, Which contains the values of the Laplace density function.

A.1.3.10 *Logarithmic normal distribution*

Graphic presentation of the *logarithmic normal* density function (1.52) is realized by the script `d0lnorm.m`. Function `d=d_lnorm(x,xm,s)` returns row vector `d`, which contains the values of the *logarithmic normal* density function.

A.1.3.11 *Poisson distribution*

Graphic presentation of the *Poisson* probability function (1.39) is realized by the script `d0poiss.m`. Function `d=d_poiss(n,a)` returns row vector `d`, which contains the values of the *Poisson* probability function.

A.1.3.12 *Student distribution*

Graphic presentation of the *Student* density function (1.46) is realized by the script `d0stu.m`. Function `d=d_stu(x,n)` returns row vector `d`, which contains the values of the *Student* density function.

A.1.3.13 *Uniform distribution*

Graphic presentation of the *uniform* density function (1.44) is realized by the script `d0unif.m`. Function `d=d_unif(x,al,bt)` returns row vector `d`, which contains the values of the *uniform* density function.

A.1.3.14 *Two-dimensional normal distribution*

Graphic presentation of the *two-dimensional normal* density function (see Sec. 1.8.2) is realized by the script `d0gauss2.m`. Function `d=d_gauss2.m` returns matrix `d`, which contains the values of the *two-dimensional normal* density function.

A.1.3.15 *Exercises*

(1) To compare the binomial distribution with the normal density function for the following values of the parameters n ($n = 1, 5, 20$) and p ($p = 0.1, 0.5, 0.9$).
(2) To compare the Poisson distribution with the normal density function for the following values of the parameter a ($a = 0.1, 0.5, 1.0, 5.0$).
(3) To compare a density function of the χ^2-distribution with the normal density function for the following values of the parameter n ($n = 2, 5, 10, 20$).
(4) To compare a density function of the Student distribution with the normal density function for the following values of the parameter n ($n = 2, 5, 10, 20$).

(5) To implement a graphic presentation of the two-dimensional normal density function for different mean values m_1 and m_2 ($m_1 = 0, 1, 10$; $m_2 = 0, 1, 10$), standard deviations σ_1 and σ_2 ($\sigma_1 = 0, 1, 10$; $\sigma_2 = 0, 1, 10$), correlation coefficient ρ ($\rho = -0,95; 0; 0,95$).

A.1.4 *Computer simulation of random values*

There are the algorithms for simulation of the random values (see, for example, (Ermakov and Mikhailov, 1982)). The scripts `rnd0ald.m` and `rnd0alc.m` return the random values which belong respectively to discrete and continuous distributions considered in Sec. A.1.3.

For these both cases a generator for the uniform random distribution on interval [0; 1] with the following density function

$$f(x) = \begin{cases} 1, \ 0 \leq x \leq 1, \\ 0, \ x < 0 \text{ if } x > 0. \end{cases} \tag{A.2}$$

is used.

For the discrete case the general method for the simulation of random variable ξ with the probability function $p_m = P(\xi = x_m)$ is based on the relation

$$P\left(\sum_{n=0}^{m-1} p_n \leq \alpha < \sum_{n=0}^{m} p_n\right) = p_m, \tag{A.3}$$

where a value α belongs to the uniform distribution (A.2).

For the continuous case the formula for simulation of random variable using cumulative distribution function $F(x)$ can be written as

$$\xi = F^{-1}(\alpha), \tag{A.4}$$

where $F^{-1}(\alpha)$ is the inverse (cumulative) distribution function, α belongs to the uniform distribution from (A.2) and a value ξ belongs to the distribution with the cumulative distribution function $F(x)$ (in (Ermakov and Mikhailov, 1982) the various algorithms of the simulation of random values are considered). Because of the strongly monotone continuous function $F(x)$, we can write the next equalities

$$P(F^{-1}(\alpha) < x) = P(\alpha < F(x)) = F(x). \tag{A.5}$$

The scripts `rnd0gau.m` and `rnd0uni.m` return random values belonging to Gaussian and uniform distributions correspondingly. Script `rnd0n2.m` returns a sampling belonging to the 2-D Gaussian distribution.

A.1.4.1 *Kolmogorov criterion*

Script `gc0kolm.m` allows to realize the Kolmogorov criterion of belonging of the sample cumulative distribution function to one or other type of the continuous distribution. The Kolmogorov criterion gives probability of the random event, that a difference between cumulative distribution function $F_\xi(x)$ and its sample image $F_N^*(x)$ is smaller than given quantity $\gamma/\sqrt{N}$. In explicit form this criterion reads as

$$\lim_{N\to\infty} P\left(\sup_{-\infty<x<\infty} |F_\xi(x) - F_N^*(x)| < \gamma/\sqrt{N}\right)$$
$$= \sum_{k=-\infty}^{\infty} (-1)^k \exp(-2k^2\gamma^2), \quad \text{(A.6)}$$

where N is sample size. Criterion is realized for continuous distributions from Sec. A.1.3.

A.1.4.2 *Exercises*

(1) Using scripts `rnd0alc.m` and `rnd0ald.m` from Sec. A.1.4 to compare mean values and standard deviation with their population values for different sample sizes (f.e. 10 – 500) for the cases of chi-square and Student distributions.
(2) Using script `rnd0n2.m` to investigate an influence of the sample size (f.e. 100 – 5000) on the sample mean and sample covariance matrix (correlation and covariance coefficients).
(3) With the use of script `gc0kolm.m` to calculate the probability for the Kolmogorov criterion for the different types of the continuous distributions (Gaussian, χ^2, Student). To pay the main attention to a behaviour of the sample (cumulative) distribution function in dependence of sample size (f.e. 10 – 1000).

A.1.5 *Confidence intervals*

Script `g0conf.m` returns the confidence interval for the sample mean of the random value belonging to the Gaussian distribution (see Sec. 1.8.14).

A.1.5.1 *Exercises*

(1) Using script `g0conf.m` to find the confidence intervals for the sample mean of the random value belonging to the Gaussian distribution for different values of the confidence probability β (f.e. β=0.2, 0.5, 0.9) and sample size (10 – 100).

A.1.6 *Time series*

A.1.6.1 *Correlated and uncorrelated time series*

Script `p0covrp.m` returns uncorrelated time series and Markov chain (see Sec. 1.10), creates sample covariance matrix.

A.1.6.2 *Autocorrelation and crosscorrelation functions*

Script `p0corf.m` returns autocorrelation function or crosscorrelation function computed for correlated and uncorrelated processes.

A.1.6.3 *Exercises*

(1) Using script `p0covrp.m` to compare the covariance and correlation matrices for correlated and uncorrelated time series. To use a different length of the time series (f.e. 100, 1000, 10000).
(2) Using script `p0corf.m` to compare autocorrelation functions and crosscorrelation functions for correlating and uncorrelated time series.

A.2 Transforms

A.2.1 *Fourier transform*

The MATLAB functions `fft` and `ifft` are realized the discrete Fourier transform and inverse discrete Fourier transform respectively. These transforms are determined by the next relations (see also Sec. 12.1)

$$y_k = \sum_{r=0}^{N-1} x_r \exp\left[-i\frac{2\pi kr}{N}\right], \quad k = 0, \ldots, N-1 \tag{A.7}$$

(Fourier transform),

$$x_r = \frac{1}{N}\sum_{k=0}^{N-1} y_k \exp\left[i\frac{2\pi kr}{N}\right], \quad r = 0, \ldots, N-1 \tag{A.8}$$

(inverse Fourier transform).

A.2.2 *Signals and their spectral characteristics*

Script `p0wlet.m` creates a simple seismic trace: wavelet and Gaussian noise. The spectral characteristics, autocorrelation function, power spectrum of the trace are calculated (see Sec. 1.10). The signals created by function `p_wlet.m`, that allows to use the simple signals in time domain.

(1) Discrete δ-function:

$$x_r = \begin{cases} 1, \text{ if } r = j, \\ 0, \text{ if } r \neq j. \end{cases} \tag{A.9}$$

(2) Like δ-function signal:

$$x_r = \exp\left[-\pi(r - t_p)/t_c)^2\right], \quad r = 0, \ldots, N-1, \tag{A.10}$$

where t_c determines a width of the signal, t_p determines the shift of the signal relatively to the zero sample $r = 0$.
(3) Ricker wavelet:

$$x_r = \exp\left[-(\pi * (r - (t_c + t_p))/t_c)^2\right] \times \left(1 - 2\left(\pi(r - (t_c + t_p))/t_c\right)^2\right),$$
$$r = 0, \ldots, N-1, \tag{A.11}$$

where t_c determines a width of the signal, t_p determines the shift of the signal relatively to the zero sample $r = 0$. An example of the graphic output of the script `p0wlet.m` is represented at Fig. A.1.

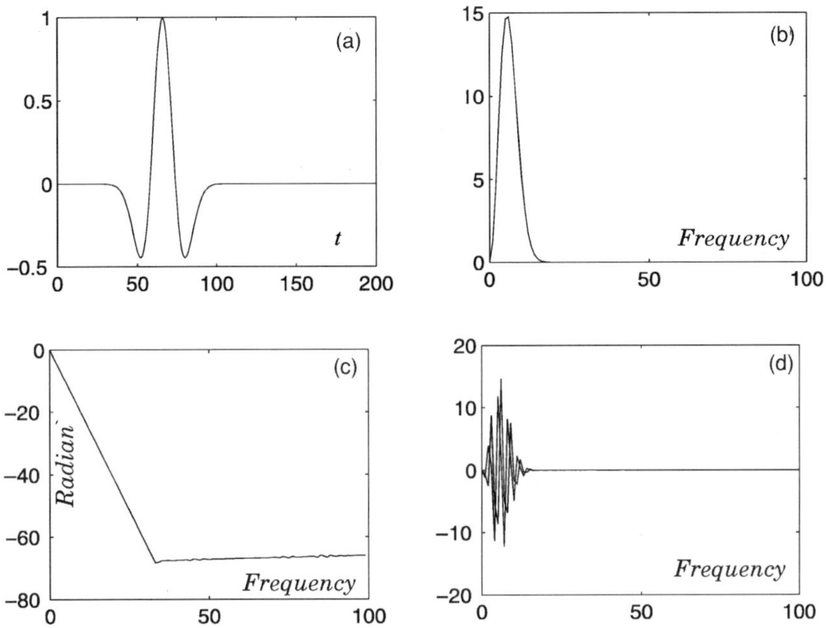

Fig. A.1 An example of the graphic output of`p0wlet.m`. Ricker wavelet (a), its spectrum module (b), spectrum phase (c), real part and imaginary part of the spectrum (d). The abscissa axes contain the samples (see formulas (A.7), (A.8)).

A.2.2.1 *Exercises*

(1) Using script `p0wlet.m` to calculate the spectra of signals (Ricker wavelet, δ-function) for different values of t_p and different signal-to noise ratio.

A.2.3 *Multifactor analysis*

Script `p0cater1.m` is realized the multifactor analysis (see Sec. 12.9).

Let us consider an example of the application of the script `p0cater1.m` to processing of the simple trace "signal with noise" (see Fig.A.2). The input data are:

np=500; % length of the trace (sample)
jm=20; % maximum number of principal
% components
unzer=2; % number of principal components
% (nonzero eigenvalues)

```
                    % for recovery
op=3;                    % sinal shape
tp=100;                   % the time shift of the wavelet
xm=0;                    % expectation of the noise
s=0.05;                   % st deviation of the noise
```

The covariance matrix and its eigenvalues (and eigenvectors) are presented at Fig. A.3 and Fig. A.4 correspondingly. The diagrams of the pairs of the principal components and a complete set of the principal components are presented at Fig. A.5 and Fig. A.6 correspondingly (see Example from 12.9). The result of recovery of the record using a different number (1 – 4) of the principal components are presented at Fig. A.7.

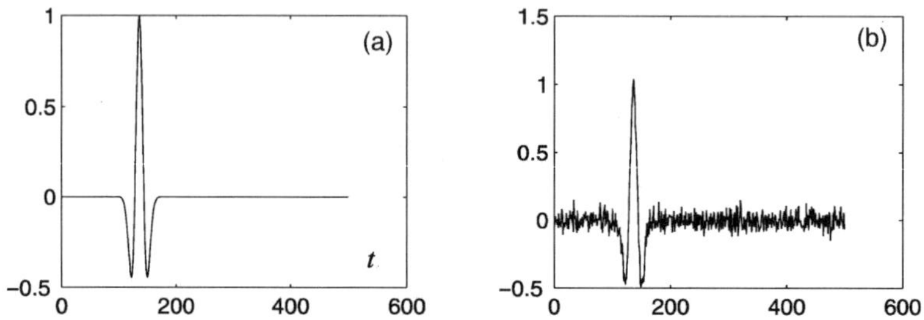

Fig. A.2 Initial record without noise (a) and with additive uncorrelated noise (b).

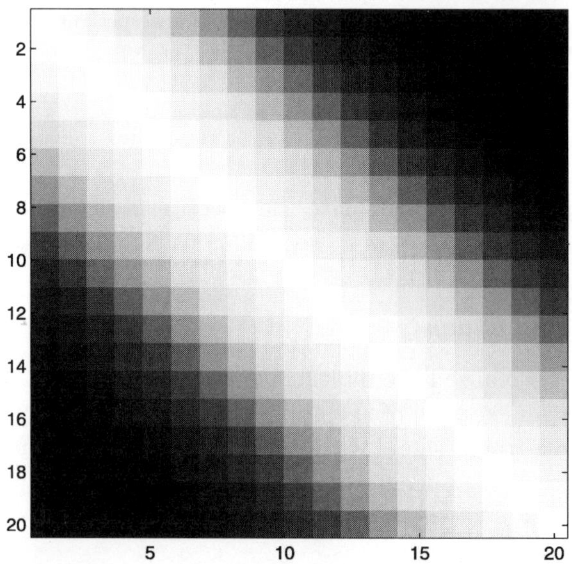

Fig. A.3 Correlation matrix $M = 20$.

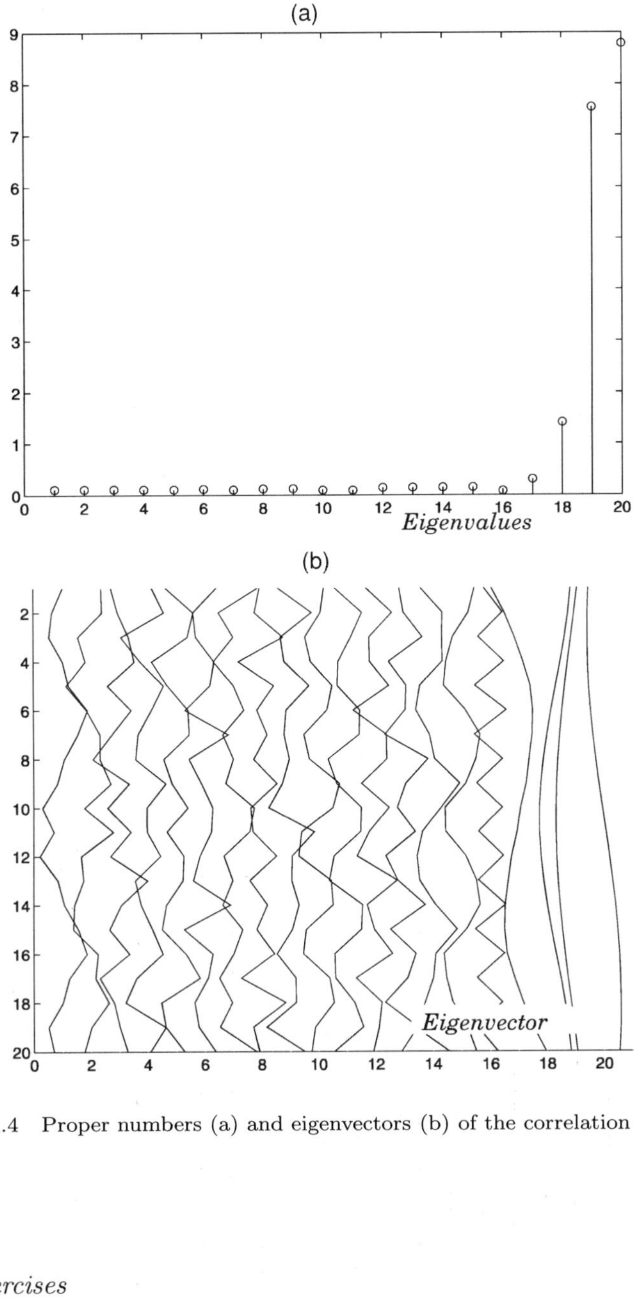

Fig. A.4 Proper numbers (a) and eigenvectors (b) of the correlation matrix.

A.2.3.1 *Exercises*

(1) Using script `p0cater1.m` to implement processing of the trace (wavelet with Gaussian noise) for the cases of different values of the maximum principal components (5 – 20), different number of the principal components using for the recovery (2 – 10), different values of signal-to-noise ratio.

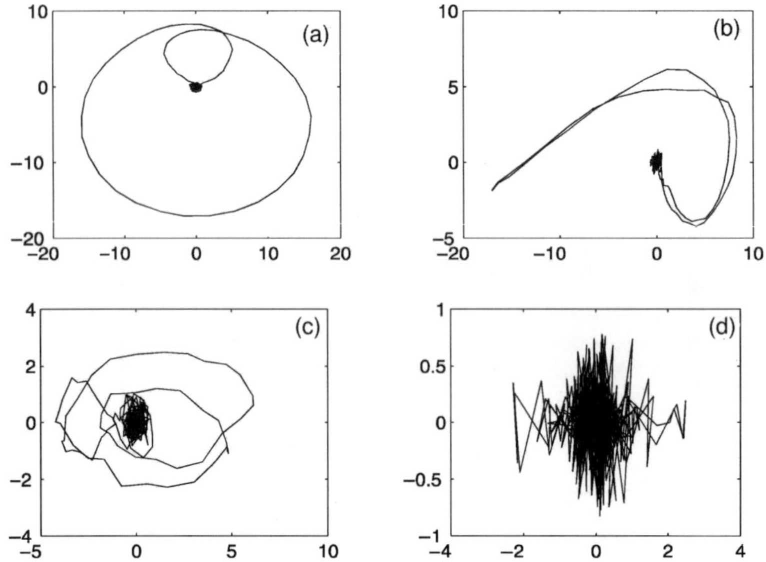

Fig. A.5 The scatterplots of pairs of the principal components: M and $M-1$ (a); $M-1$ and $M-2$ (b); $M-2$ and $M-3$ (c); $M-3$ and $M-4$ (d). The maximum number of principal components is 20.

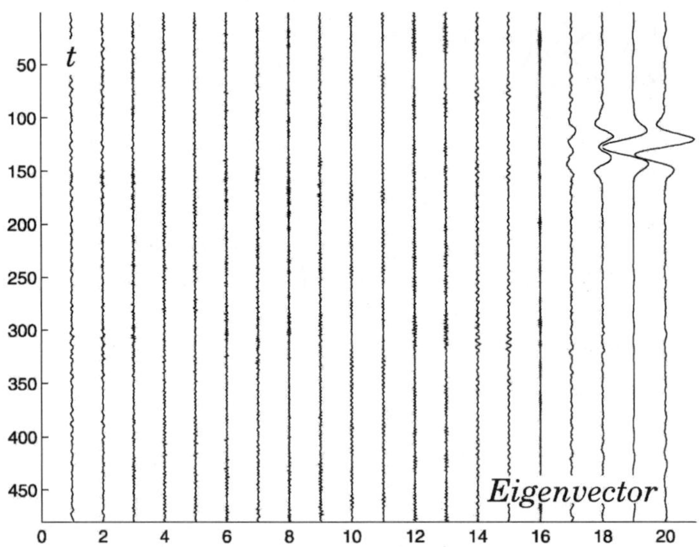

Fig. A.6 The complete set of principal components ($M = 20$).

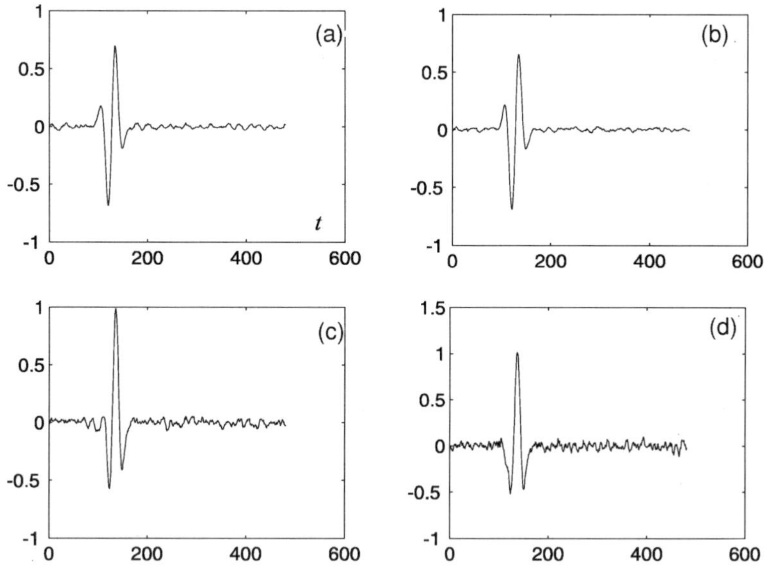

Fig. A.7 Recovery of the record from Fig. A.2(b). Recovery using the first principal component (a), two (b), three (c) and four (d) principal components (see Fig. A.6).

A.2.4 *Cepstral transform*

Script `p0cep1.m` is realized the cepstral transform (see Sec. 12.6).

The cepstral transform in the script `p0cep1.m` is realized as follows:

```
d=fft(trace);                 % Fourier transform
              % of the vector "trace"
dabs=abs(d);                  % computing of the absolute value
              % of the spectrum
dang=unwrap(angle(d));               % calculation of the phase
dcep=fft(log(dabs)+i*dang);        % Fourier transform
              % of the logarithm of spectrum (cepstrum)
```

The inverse cepstrum transform is realized as following:

```
dback=ifft(dcep);              % inverse Fourier
dexp=exp(dback);               % transform
dback0=ifft(dexp);              % inverse Fourier
              % transform
```

In Fig. A.8 and Fig. A.9 two signals and their cepstral characteristics are represented respectively. The result of the convolution of the two signals is compared with the inverse cepstral transform of the sum of their cepstrums (see Fig. A.10).

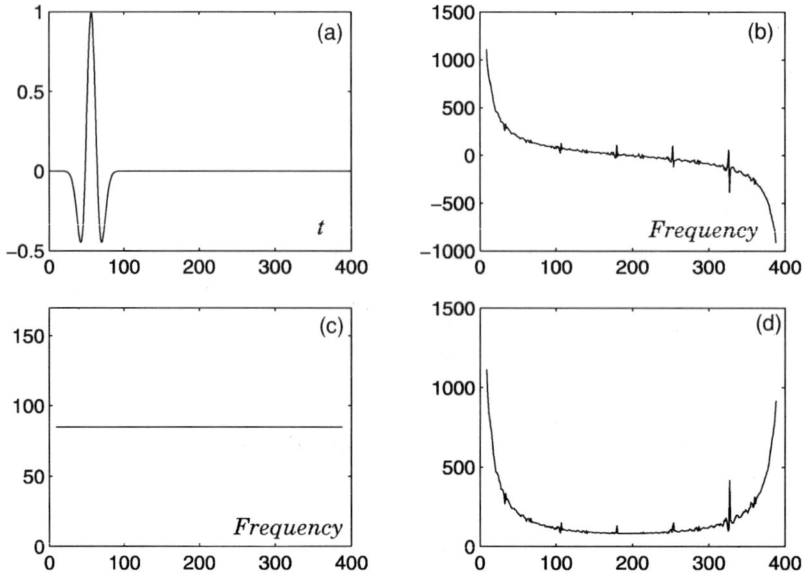

Fig. A.8 The first signal (a) and its cepstral characteristics (b) – (d). The Ricker wavelet (a), real (b), imaginary (c) parts and the module of its cepstrum (d).

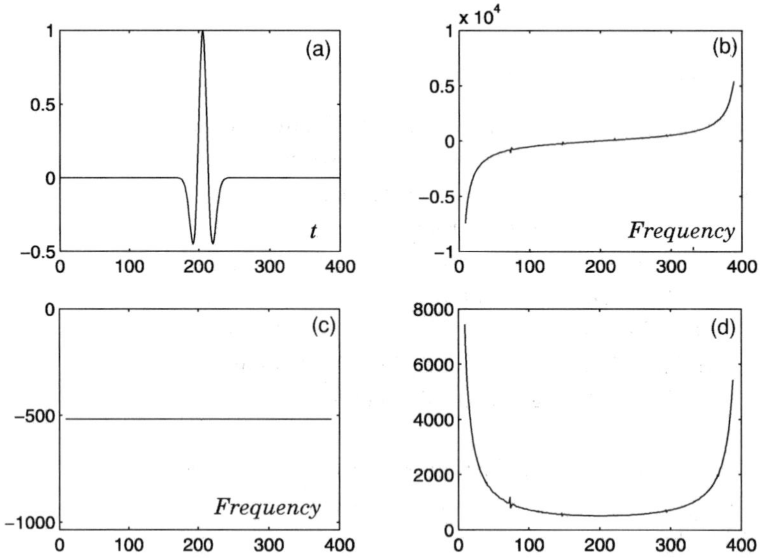

Fig. A.9 The second signal and its cepstrum characteristics (see Fig. A.8).

A.2.4.1 *Exercises*

Using script `p0cep1.m` to compare the convolution in time domain with summation in cepstral domain.

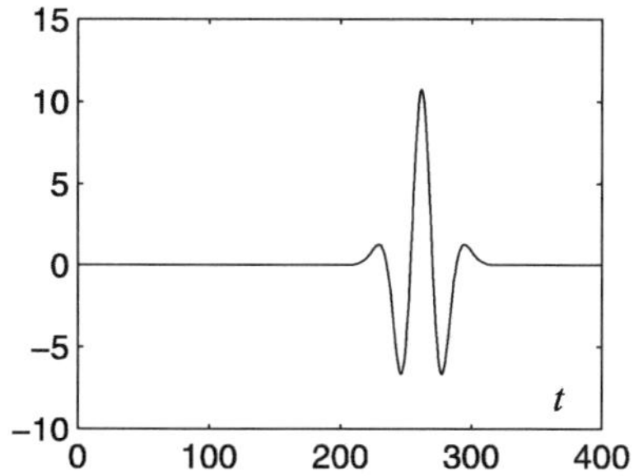

Fig. A.10 The convolution of the signals from Figs. A.8(a), A.9(a) and inverse cepstrum transform of the sum of their cepstrums.

A.3 Direct and Inverse Problem Solution

A.3.1 *Computer simulation of gravitational attraction*

A.3.1.1 *Gravitational attraction of the sphere*

Script `f0sph1.m` is realized the gravitational attraction of the uniform sphere.

Gravitational potential V of the uniform sphere is given by

$$V = \gamma \frac{4}{3} \frac{\pi a^3 \rho}{r}, \tag{A.12}$$

where γ is the gravitational constant; a is a radius of the sphere; ρ is a mass density; r is a distance between a center of the sphere and an observation point.

Thus the gravitational attraction reads as

$$\boldsymbol{g} = -\gamma \frac{4}{3} \frac{\pi a^3 \rho}{r^2} \boldsymbol{n}, \tag{A.13}$$

where $\boldsymbol{n}$ is a unit vector, directed from a center of the sphere to an observation point.

A.3.1.2 *Gravitational attraction of the uniform cylinder*

Script `f0cyl1.m` is realized the gravitational attraction of the uniform cylinder.

Gravitational potential of the uniform cylinder V is given by the formula

$$V = 2\pi a^2 \rho \gamma \log \frac{1}{r}, \tag{A.14}$$

where γ is the gravitational constant; a is a radius of the cylinder; ρ is a mass density; r is a perpendicular distance between a center of the cylinder and an observation point.

Thus the gravitational attraction reads as

$$\boldsymbol{g} = -\frac{2\pi a^2 \rho \gamma}{r} \boldsymbol{n}, \tag{A.15}$$

where $\boldsymbol{n}$ is a unit vector, directed from a center of the cylinder to an observation point.

A.3.2 *Computer simulation of magnetic field*

A.3.2.1 *Magnetic induction of a dipole*

Script `f0dip1.m` is realized the magnetic induction of the dipole.

Magnetic potential V of the magnetic dipole is given by

$$V = C_m \frac{(\boldsymbol{m} \cdot \boldsymbol{r})}{r^3}, \tag{A.16}$$

where C_m is equal $\mu_0/4\pi$ in SI units; $\boldsymbol{m}$ is a dipole moment; r is a distance between the dipole and the observation point. Thus the magnetic induction reads as

$$\boldsymbol{B} = C_m \frac{m}{r^3}[3(\boldsymbol{n}_m \cdot \boldsymbol{n}_r)\boldsymbol{n}_r - \boldsymbol{n}_m], \tag{A.17}$$

where $m = |\boldsymbol{m}|$, $\boldsymbol{n}_m = \boldsymbol{m}/|\boldsymbol{m}|$, $\boldsymbol{n}_r = \boldsymbol{r}/|\boldsymbol{r}|$.

A.3.2.2 *Magnetic induction of the horizontal cylinder*

Script `f0mcyl1.m` is realized the magnetic induction of the magnetic cylinder.

Magnetic potential of the cylinder is given by

$$V = 2C_m \frac{(\boldsymbol{m} \cdot \boldsymbol{n}_r)}{r}, \tag{A.18}$$

where C_m is equal $\mu_0/4\pi$ in SI units; $\boldsymbol{m}$ is the dipole moment per unit length; $r = |\boldsymbol{r}|$ is a distance between the center of the cylinder and the observation point; $\boldsymbol{n}_r = \boldsymbol{r}/|\boldsymbol{r}|$.

Thus the magnetic induction reads as

$$\boldsymbol{B} = \frac{2C_m m}{r^2}[2(\boldsymbol{n}_m \cdot \boldsymbol{n}_r)\boldsymbol{n}_r - \boldsymbol{n}_m], \tag{A.19}$$

where $m = |\boldsymbol{m}|$, $\boldsymbol{n}_m = \boldsymbol{m}/|\boldsymbol{m}|$.

A.3.3 *Computer simulation of the seismic field*

In the script `fd0mn1.m` the numerical simulation of the propagation of the elastic waves in the piecewise half-space (1-D case) is realized by the finite difference method. The component of displacement φ is satisfied the partial equation

$$(\lambda + 2\mu)\frac{\partial^2 \varphi}{\partial z^2} + \frac{\partial(\lambda + 2\mu)}{\partial z}\frac{\partial \varphi}{\partial z} + f = \rho \frac{\partial^2 \varphi}{\partial t^2}, \tag{A.20}$$

the boundary condition at the free surface ($z = 0$)

$$\left.\frac{\partial \varphi}{\partial z}\right|_{z=0} = 0 \tag{A.21}$$

and at welded interfaces ($z = z_i$)

$$\varphi|_{z=z_i-0} = \varphi|_{z=z_i+0}, \quad (\lambda + 2\mu)\left.\frac{\partial \varphi}{\partial z}\right|_{z=z_i-0} = (\lambda + 2\mu)\left.\frac{\partial \varphi}{\partial z}\right|_{z=z_i+0}, \tag{A.22}$$

where ρ is a mass density, λ and μ are Lame parameters. The wave field is excited by the point source

$$f = \delta(z)\hat{f}(t), \tag{A.23}$$

located in the point $z = 0$.

The piecewise homogeneous medium can be perturbed by a smooth inhomogeneity of $E = \lambda + 2\mu$ parameter, which is given by the formula

$$\delta E(z) = \begin{cases} \delta E = 0, & \text{if } |z - \tilde{z}| > \Delta, \\ \delta E = E_m 0.5[1 + \cos(\pi(z - \tilde{z})/\Delta)], & \text{if } |z - \tilde{z}| < \Delta, \end{cases} \tag{A.24}$$

or the smooth inhomogeneity of ρ parameter

$$\delta\rho(z) = \begin{cases} \delta\rho = 0, & \text{if } |z - \tilde{z}| > \Delta, \\ \delta\rho = \rho_m 0.5[1 + \cos(\pi(z - \tilde{z})/\Delta)], & \text{if } |z - \tilde{z}| < \Delta, \end{cases} \tag{A.25}$$

where $\tilde{z}$ is the location of inhomogeneity, Δ is its half-size, E_m and ρ_m are the maximum values of perturbation of E and ρ respectively.

A.3.3.1 *Exercises*

To calculate the seismic wave field for the piecewise homogeneous half-space:

- the source location is $z_s = 0.003$ km;
- the receiver locations are $z_r = 0.003, 0.1, 0.15, 0.2, 0.25, 0.3, 0.35, 0.4, 0.45, 0.5$ km;
- location of the interfaces z=0.3, 0.5, 0.7, 0.9 km;
- Lame parameters $E = \lambda + 2\mu = 1$;
- mass density $\rho = 1.0, 1.2, 0.8, 1.2, 1.0$;
- time sample dt =0.001 s;
- as the source time dependence the Ricker wavelet of 25 Hz frequency is used.

In Fig. A.11 the model, wavelet and the first trace are represented. The seismogram is represented Fig. A.12 to implement a qualitative analysis of the wave field.

A.3.4 *Deconvolution by the Wiener filter*

Script `p0decon2.m` gives an example of deconvolution using the Wiener filter (see Sec. 12.10)[5].

Let consider processing of the seismogram (see Sec. A.3.3, Fig. A.12) with the additional noise by the script `p0decon2.m`. We will try to decrease the signal-to-noise merit and to bring a shape of the signal to the δ-function. In Fig. A.13 a fragment of the seismogram from Fig. A.12 with addition of the Gaussian noise is represented ($N(0, \sigma A_m)$. Here σ=0.1, A_m is a maximum value of the seismic signal on the seismogram). In Fig. A.14 the fragment of the seismogram from Fig. A.13 after the deconvolution is represented.

[5]The Wiener filter can be considered as an example of the application of the method of least squares Sec. 6.4.

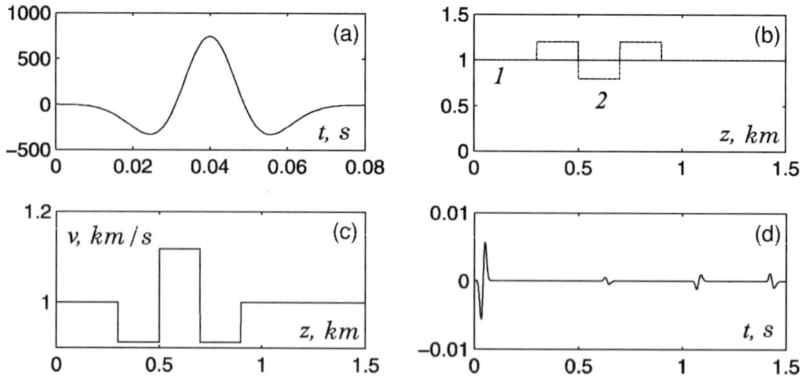

Fig. A.11 Graphic presentation of the model and the first trace. Time dependence in the source (a); *1* — ρ, *2* — $\lambda + 2\mu$ (b); longitudinal wave velocity (c); the wave field in the first observation point (d).

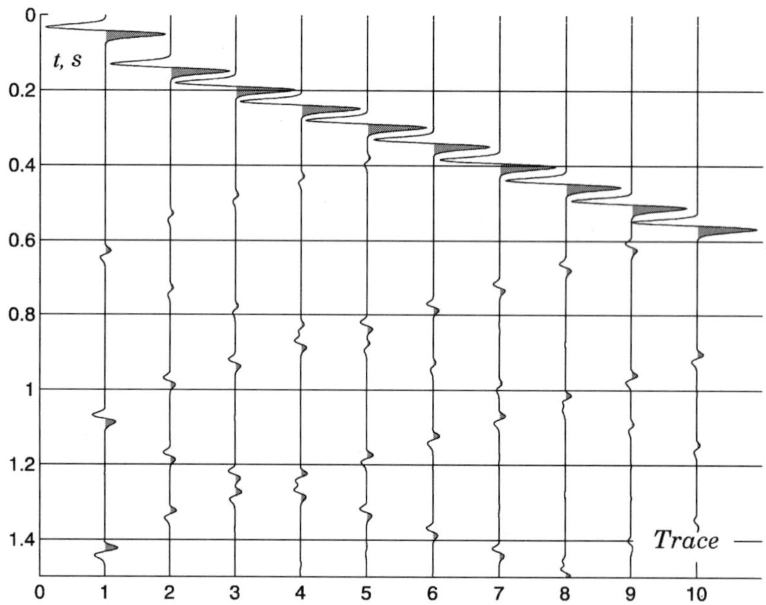

Fig. A.12 The seismogram.

A.3.4.1 *Exercise*

To implement the deconvolution of the seismic field:

- without the noise (`sig1=sig2=0`);
- for different values of the regularizing coefficient α (see formula (12.36)) (at `eps1=2.0, 1.0, 0.5`);

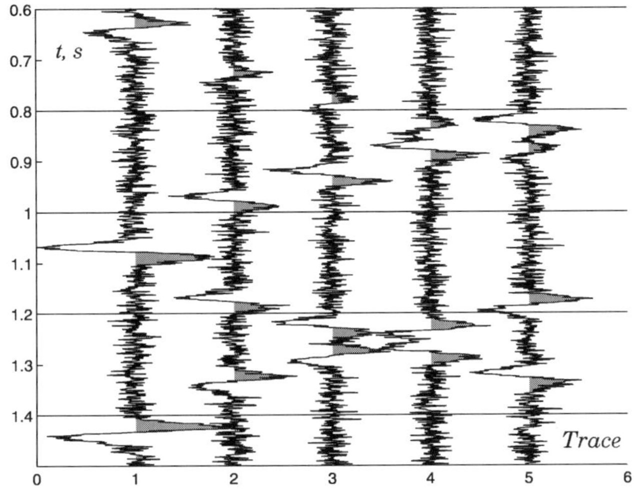

Fig. A.13 The fragment of a seismogram from Fig. A.12 with the addition of the uncorrelated Gaussian noise.

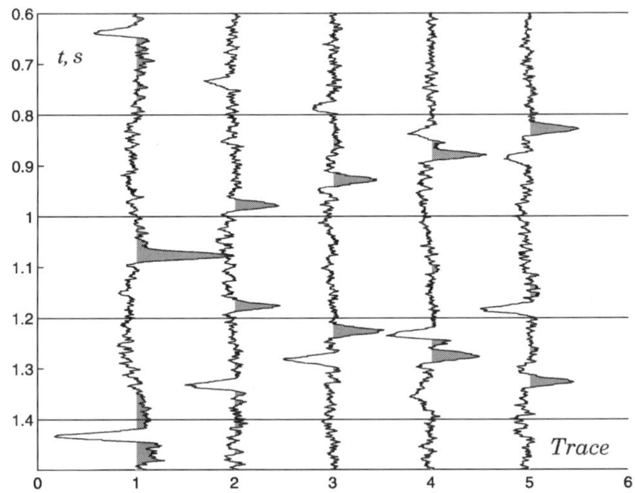

Fig. A.14 The deconvolution of the seismogram from Fig. A.13.

- for different values of the noise (`sig1=sig2=0.5, 0.1`).

A.3.5 *Quantitative interpretation*

Let us consider an application of the Newton–Le Cam method (see Secs. 3.2, 6.2, 6.3) to the point estimation of the parameters $\boldsymbol{\theta}$ of the geophysical objects (script

`pOmg1.m`). As an examples, the magnetic bodies with vertical magnetization are considered:
thread of poles

$$f_k(\boldsymbol{\theta}) = \frac{2Mh}{h^2 + (x_k - \xi)^2} \quad \boldsymbol{\theta} = \{M, h, \xi\}; \tag{A.26}$$

thread of dipoles

$$f_k(\boldsymbol{\theta}) = \frac{2M[h^2 - (x_k - \xi^2)^2]}{[h^2 + (x_k - \xi)^2]^2}, \quad \boldsymbol{\theta} = \{M, h, \xi\}; \tag{A.27}$$

point magnetic pole

$$f_k(\boldsymbol{\theta}) = \frac{M(2h^2 - x_k^2)}{(h^2 + x_k^2)^{5/2}}, \quad \boldsymbol{\theta} = \{M, h\}; \tag{A.28}$$

point magnetic dipole

$$f_k(\boldsymbol{\theta}) = \frac{Mh}{(h^2 + x_k^2)^{3/2}}, \quad \boldsymbol{\theta} = \{M, h\}; \tag{A.29}$$

where M is a magnetic moment, h is a depth, ξ is a coordinate of the function $f_k(\boldsymbol{\theta})$ maximum.

Applying the maximum likelihood method to the additive model of the data

$$u_k = f_k(\boldsymbol{\theta}) + \varepsilon_k \quad (\boldsymbol{u} = \boldsymbol{f}(\boldsymbol{\theta}) + \boldsymbol{\varepsilon}), \tag{A.30}$$

with uncorrelated Gaussian noise ε_k ($N(0, \sigma)$), we can get the next procedure for estimation of $\boldsymbol{\theta} = (\theta_1, \ldots, \theta_S)$. At the initial point to choose an initial value $\boldsymbol{\theta}_0$ of $\boldsymbol{\theta}$. After that to solve a system of the linear equations (see Sec. 6.3)

$$\tilde{C}\Delta\boldsymbol{\theta}^{(1)} = \boldsymbol{d}, \tag{A.31}$$

where

$$d_s = \frac{1}{\sigma^2}\left[(\boldsymbol{u} - \boldsymbol{f}(\boldsymbol{\theta}))^T \frac{\partial \boldsymbol{f}(\boldsymbol{\theta})}{\partial \theta_s}\right]\Bigg|_{\boldsymbol{\theta}=\boldsymbol{\theta}^{(0)}},$$

$$\tilde{c}_{ss'} = \frac{1}{\sigma^2} \frac{\partial \boldsymbol{f}^T(\boldsymbol{\theta})}{\partial \theta_s}\Bigg|_{\boldsymbol{\theta}=\boldsymbol{\theta}^{(0)}} \frac{\partial \boldsymbol{f}(\boldsymbol{\theta})}{\partial \theta_{s'}}\Bigg|_{\boldsymbol{\theta}=\boldsymbol{\theta}^{(0)}},$$

$$s = 1, \ldots, S$$

and to calculate the first $\hat{\boldsymbol{\theta}}^{(1)}$ approximation of $\boldsymbol{\theta}$

$$\hat{\boldsymbol{\theta}}^{(1)} = \boldsymbol{\theta}^{(0)} + \Delta\boldsymbol{\theta}^{(1)}. \tag{A.32}$$

The iterative process is finished if the threshold condition is satisfied, for example,

$$|\Delta\hat{\theta}_s^{(n)} / \hat{\theta}_s^{(n)}| \approx 10^{-2} \div 10^{-3}. \tag{A.33}$$

An example of the similar numerical simulation for the field given by the formula (A.26) is represented in Fig. A.15 for the next input data:

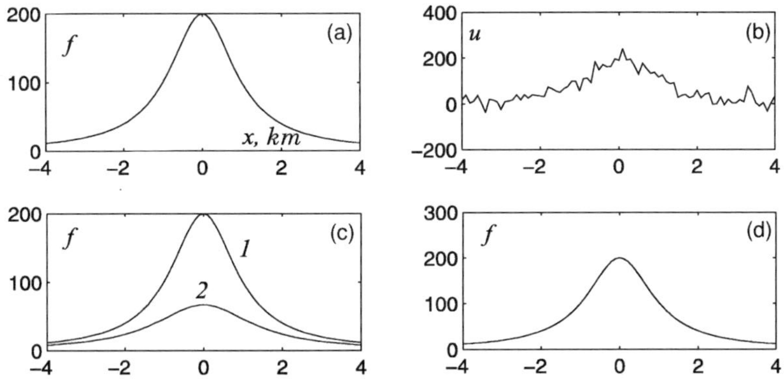

Fig. A.15 Qualitative interpretation. The model field without noise (a) and with noise (b), the model field (*1*) and the field (*2*) computed for the initial parameters (c), the model field and the field computed for the estimated parameters (d).

```
x=-4:0.1:4;   % observation points
m=1;   % intensity
h=1;   % depth
xi=0.0;   % location of the object
xm=0.0;   % expectation of the noise
s=0.1;   % standard deviation of the noise
m0=0.5;   % initial value of m
h0=1.5;   % initial value of h
xi=0.;   % initial value of xi
eps_m=1.0e-5;   % threshold condition for m
eps_h=1.0e-5;   % threshold condition for h
```

The values of M and h are estimated. As the result of numerical simulation, the Fisher matrix and its inverse matrix are calculated.

A.3.5.1 *Exercises*

Using the script `p0mg1.m` to implement the next exercises for the models (A.26)–(A.29) (to pay attention to the Fisher matrix and its inverse matrix):

(1) To investigate restoration of M and h for the different values of the standard deviation of the Gaussian noise (for example, σ =0.1, 0.2, 0.3).
(2) To investigate the convergence of the iterative process in dependence on initial values of the parameters M and h.
(3) To implement the numerical simulation for the different sample sizes (for example, 10, 20, 50, 100).

A.3.6 *Qualitative interpretation*

Using script `p0mg2.m` to implement the qualitative interpretation (see Sec. 3.4, signal and noise (H_0) or noise (H_1)) with the help of a posteriori probability ratio (see Secs. 7.1, 7.2) for the signal models from (A.26)–(A.29). For the case of the uncorrelated Gaussian noise this criterion reads as

$$\alpha = \ln \frac{P(1)}{P(0)} + \frac{1}{2\sigma^2} \left[\boldsymbol{u}^T \boldsymbol{u} - (\boldsymbol{u} - \boldsymbol{f}_1)^T (\boldsymbol{u} - \boldsymbol{f}_1) \right] . \tag{A.34}$$

If $\alpha \geq 0$, then hypothesis H_1 ("noise"). If $\alpha < 0$, then hypothesis H_0 ("signal and noise").

We assume that a model function $f_1(\boldsymbol{\theta})$ differs from the real function $f_1(\boldsymbol{\theta}+\Delta\boldsymbol{\theta})$. Such difference is given by the deviation $\Delta\boldsymbol{\theta}$ of the "true" parameters $\boldsymbol{\theta} + \Delta\boldsymbol{\theta}$ from their supposed values $\boldsymbol{\theta}$.

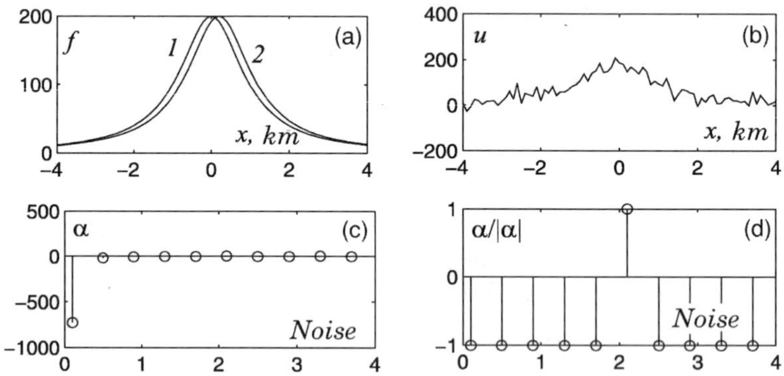

Fig. A.16 Qualitative interpretation. The model field $f_1(\boldsymbol{\theta})$ (*1*) and "real" field $f_1(\boldsymbol{\theta} + \Delta\boldsymbol{\theta})$ (*2*) (a), the model field with uncorrelated noise (b), dependence of α (c) and normalized α (d) as functions of the noise value.

An example of the similar numerical simulation for the field (A.26) is represented in Fig. A.16 for the next input data:

```
x=-4:0.1:4;  % observation points
m=1;  % intensity
h=1;  % depth
xi=0.0;  % location of the object
ddm=0.0;  % deviation of m
ddh=0.0;  % deviation of h
ddxi=0.2;  % deviation of xi
xm=0.0;  % expectation of the noise
p0=0.5;  % a priori probability of existence of the object
s=0.1;  % initial standard deviation of the noise
s1=s:0.4:4;  % the values of the standard deviation for the numerical simulation
```

A.3.6.1 *Exercises*

Using script `pOmg2.m` to implement the next exercises for the models (A.26) – (A.29):

(1) to implement the numerical simulation for a wide range of a noise/signal ratio (variable `s1`);
(2) to implement simulations for the different deviations of $\Delta\boldsymbol{\theta}$ between the "real" and model fields;
(3) to implement the simulation for a different number of the observation points (for example, 10, 20, 50, 100).

A.3.7 *Diffraction tomography*

Script `dtOmn1.m` is implemented a numerical simulation to restore the medium parameters by the diffraction tomography method (Secs. 10, 11).

The 1-D direct problem solution on the propagation of the elastic waves in uniform space, which contains a smooth perturbation of the parameters comparable in size with the wavelength, is solved by the finite difference method. The component of the wave field φ satisfies the partial equation (A.20) an the boundary conditions (A.21), (A.22). The restoration of the Lame parameters λ, μ and mass and mass density ρ is implemented by the diffraction tomography method using the next algorithm.

We introduce the tomography functionals p_ρ and p_E, which are connected with the values ρ and $E = \lambda + 2\mu$ correspondingly:

$$p_\rho(z, z_s, z_r, t) = -\int_0^\infty \varphi_{\text{out}}(z, z_r, t-\tau)\frac{\partial^2}{\partial t^2}\varphi_{\text{in}}(z, z_s, \tau)d\tau,$$

$$p_E(z, z_s, z_r, t) = -\int_0^\infty \frac{\partial}{\partial z}\varphi_{\text{out}}(z, z_r, t-\tau) \times \frac{\partial}{\partial z}\varphi_{\text{in}}(z, z_s, \tau)d\tau, \qquad \text{(A.35)}$$

then, under the Born approximation, the scattering (or *difference*) field $\Delta\varphi$ generated by the inhomogeneity, can be written as

$$\Delta\varphi \approx \int_L (p_E\delta E + p_\rho\delta\rho)dl, \qquad \text{(A.36)}$$

where z_s and z_r are the source and receiver locations; φ_{in} is the wave field generated by the source with its time dependence; φ_{out} is the wave field, generated by the "artificial" source with time dependence as $\delta(t)$-function, located in the observation point; L is the restoration region; δE, $\delta\rho$ are the desired parameters of the inhomogeneity (perturbation of E and ρ relatively reference medium).

If, for example, a value of $\delta\rho \approx 0$ and we shall find only δE, then the integral equation (A.36) after digitization can be presented as a system of the linear equations

$$P\boldsymbol{\delta}_E = \boldsymbol{\delta}_\varphi, \qquad \text{(A.37)}$$

where $\boldsymbol{\delta}_E$ is the desired vector and $\boldsymbol{\delta}_\varphi$ is a vector of the sampled difference field.

The simplest regularization of the solution of (A.37) consist of the solution of the next system of the linear equations:

$$(P'P + \varepsilon D'D)\boldsymbol{\delta}_E = P'\boldsymbol{\delta}_\varphi, \tag{A.38}$$

where ε is regularized parameter and D, for example, is an identity matrix.

Let us consider an example of the numerical simulation for the next model:

- $z_s = z_r = 0.003$ km;
- $\rho = 1$, $E = 1$, $v = \sqrt{E/\rho} = 1$;

the smooth (desired) inhomogeneity is given by the formula

$$\delta E(z) = \begin{cases} \delta E = 0, & \text{if } |z - \tilde{z}| > \Delta, \\ \delta E = E_m 0.5[1 + \cos(\pi(z - \tilde{z})/\Delta)], & \text{if } |z - \tilde{z}| < \Delta, \end{cases} \tag{A.39}$$

where $\tilde{z}$ is inhomogeneity location, Δ is its half-size, E_m is a maximum value of a perturbation of E.
$\tilde{z} = 0.3$ km, $\Delta = 0.03$ km, $E_m = 0.3$; Ricker wavelet

$$F(\omega) = \left(\frac{\omega}{\omega_0}\right)^2 \exp\left\{\left(\frac{\omega}{\omega_0}\right)^2\right\} \exp\{-2\pi i\omega/\omega_0\} \tag{A.40}$$

with the apparent frequency $\omega_0 = 25$ Hz; the interval L of restoration is $L = 0.21 - 0.39$ km; regularization parameter $\varepsilon = = 10^{-5}$.

Graphic presentation of the model and the results of the direct problem solution is represented in Fig. A.17 and Fig. A.18 respectively. The result of restoration is

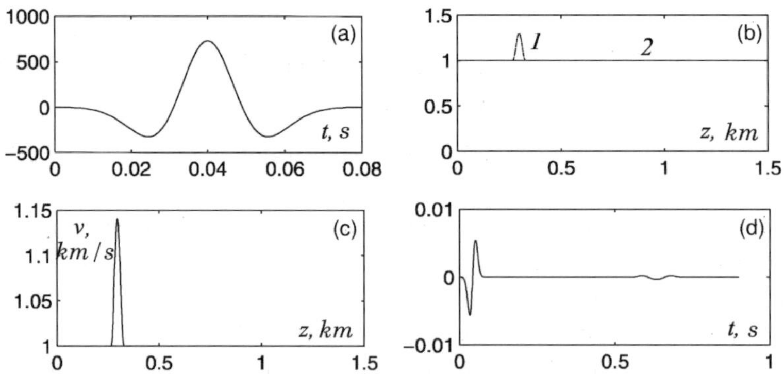

Fig. A.17 The data for simulation of the propagation process and the wave field in the first observation point. This representation is similar to Fig. A.11.

represented in Fig. A.19.

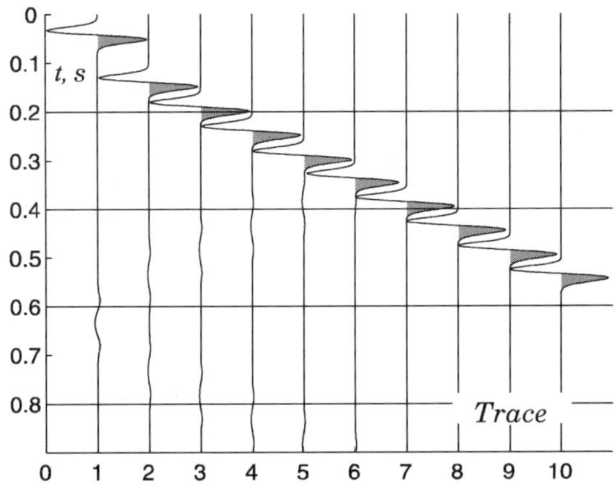

Fig. A.18 The seismogram for the observation points (km): 0.003, 0.1, 0.15, 0.2, 0.25, 0.3, 0.35, 0.4, 0.45, 0.5.

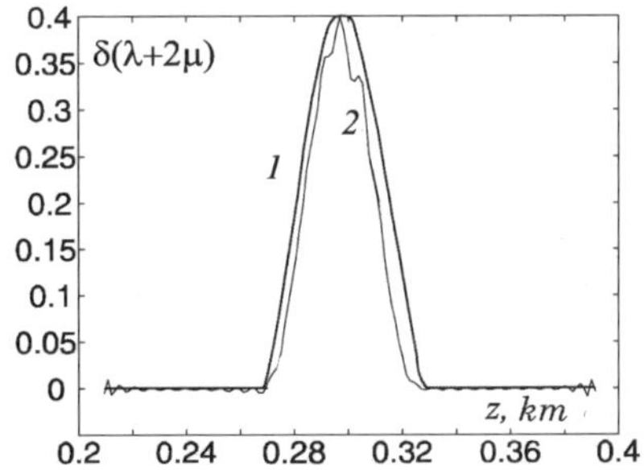

Fig. A.19 Restoration of the longitudinal velocity : *1* it is the model of inhomogeneity; *2* it is the restored perturbation of the velocity.

A.3.7.1 *Exercises*

(1) Using a script `dtOmn1`, to investigate an accuracy of the restoration of E by the diffraction tomography method in dependence on the contrast and size of inhomogeneity (for example, $E = 0.05, 0.1, 0.2, 0.5$ $\Delta = 0.01, 0.02, 0.03, 0.04$).

(2) To estimate accuracy of restoration in dependence on a noise value.

Appendix B

Tables

The tables which are represented below allow to implement the calculations with Laplace function, with density functions for χ^2 distribution and Student distribution.

Table B.1 The values of the function $\tilde{\Phi}(z) = 2(1-\Phi(z))$ (see Sec. 1.8.1, Eq. (1.42)).

z	$\tilde{\Phi}(z)$	z	$\tilde{\Phi}(z)$	z	$\tilde{\Phi}(z)$	z	$\tilde{\Phi}(z)$
0,0	1,00000	1,0	0,31732	2,0	0,04550	3,0	0,00270
0,1	0,92034	1,1	0,27134	2,1	0,03572	3,1	0,00194
0,2	0,84148	1,2	0,23014	2,2	0,02780	3,2	0,00138
0,3	0,76418	1,3	0,19360	2,3	0,02144	3,3	0,00096
0,4	0,68916	1,4	0,16152	2,4	0,01640	3,4	0,00068
0,5	0,61708	1,5	0,13362	2,5	0,01242	3,5	0,00087
0,6	0,54850	1,6	0,10960	2,6	0,00932	3,6	0,00061
0,7	0,48392	1,7	0,08914	2,7	0,00694	3,7	0,00042
0,8	0,42372	1,8	0,07186	2,8	0,00512	3,8	0,00029
0,9	0,36812	1,9	0,05714	2,9	0,00374	3,9	0,00020

Table B.2 The value x_P and probability $P(x > x_P)$ with degrees of freedom n (see Sec. 1.8.4, Eq. (1.45)).

	P				
n	0,99	0,90	0,50	0,10	0,01
1	0,0001157	0,0158	0,455	2,706	6,635
2	0,0201	0,211	1,386	4,605	9,210
3	0,115	0,584	2,366	6,251	11,345
4	0,297	1,064	3,357	7,779	13,277
5	0,554	1,610	4,351	9,236	15,086
10	2,558	4,865	9,342	15,987	23,209
15	5,229	8,547	14,399	22,307	30,578
20	8,260	12,444	19,337	28,412	37,566
25	11,524	16,473	24,337	34,382	44,314
30	14,953	20,599	29,336	40,256	50,892

Table B.3 The values t_P and probability P of the difference t from zero average more than t_P under a number of the degrees of freedom $\nu = n - 1$ (see Sec. 1.8.5, Eq. (1.47)).

	P				
ν	0,9	0,5	0,1	0,05	0,01
1	0,158	1,000	6,314	12,706	63,657
2	0,142	0,816	2,920	4,303	9.925
3	0,137	0,765	2,353	3,182	5,841
4	0,134	0,741	2,132	2,776	4,604
5	0,132	0,727	2,015	2,571	4,032
10	0,129	0,700	1,812	2,228	3,169
15	0,128	0,691	1,753	2,131	2,947
20	0,127	0,687	1,725	2,086	2,845
25	0,127	0,684	1,708	2,060	2,787
30	0,127	0,683	1,697	2,042	2,750
∞	0,1257	0,6745	1,6448	1,9590	2,5758

Bibliography

Ahlberg, J., Nilson, E. and Walsh, J. (1967). *The Theory of Splines and Their Applications* (Academic Press).

Akaike, H. (1974). A new look at the statistical model identification, *IEEE Trans. Autom. Control* **AC-19**, pp. 716–723.

Aki, K. and Richards, P. (2002). *Quantitative seismology.* (University Science Books).

Albert, A. (1972). *Regression and the Moore-Penrose Pseudoinverse* (Academic Press).

Alekseev, A., Lavrentev, M. and Preobrazhenskii, N. (1985). *Problems of reconstructive tomography (in russian)* (SO AN SSSR).

Alekseev, A., Mikhailenko, V. and Cheverda, V. (1983). *Numerical methods of solution of the inverse and direct problems in the theory of wave propagation (in russian)* (SO AN SSSR), pp. 165–172.

Anderson, N. (1974). On the calculation of filter coefficients for maximum entropy spectral analysis, *Geophysics* **39**, pp. 69–72.

Andrews, D. (1974). A robust method for multiple linear regression, *Technometrics* **16**, 4, pp. 523–531.

Aronov, V. (1990). *Methods of creation of maps of geological and geophysical characteristics and application of geometry for mapping of oil and gaz (in russian)* (Nedra, Moscow).

Babic, V. and Buldyrev, V. (1991). *Short-Wavelength Diffraction Theory. Asymptotic Methods* (Springer-Verlag).

Babic, V., Buldyrev, V. and Molotkov, I. (1999). *The Space–Time Ray Method* (Cambridge University Press).

Backus, G. and Gilbert, F. (1967). Numerical applications of a formalism for geophysical inverse problems, *Geophys. J. Roy. Astr. Soc.* **13**, pp. 247–276.

Backus, G. and Gilbert, F. (1968). The resolving power of gross earth data, *Geophys. J. Roy Astr. Soc.* **16**, pp. 169–205.

Bakushinski, A. and Goncharski, A. (1994). *Ill-posed problems: theory and applications* (Kluwer).

Belishev, M. (1990). Equations of the gelfand-levitan type in a multidimensional inverse problem for the wave equation, *Journal of Mathematical Sciences* **50**, 6, pp. 1940–1944.

Berkhout, A. (1984.). *Seismic migration — Imaging of acoustic energy by wave field extrapolation (practical aspects)* (Elsevier).

Berkhout, A. (1985). *Seismic migration — Imaging of acoustic energy by wave field extrapolation (theoretical aspects)* (Elsevier).

Beylkin, G. (1985). Imaging of discontinuities in the inverse scattering problem by inversion of a causal generalized radon transform, *J. Math. Phys.* **26**, 1, pp. 99–108.

Blagoveshchenskii, A. S. (1966). About inverse problem of seismic wave propagation (in russian), *Probl. Mat. Physici* , 1, pp. 68–81.
Blagoveshchenskii, A. S. (1978). Inverse problems of theory of elastic wave propagation (in russian), *Izv. AN SSSR. Fizika Zemli* , 12, pp. 45–54.
Blagoveshchenskii, A. S. (2005). An inverse problem of the theory of wave propagation in a random layered medium, *Differential Equations (Pleiades Publishing)* **41**, 10, pp. 1442–1448.
Bleistein, N., Cohen, J. and Stockwell, J. (2000). *Mathematics of Multidimensional Seismic Imaging, Migration and Inversion* (Springer).
Box, G. and Jenkins, G. (1976). *Time series analysis: forecasting and control* (Holden Day).
Brandt, S. (1998). *Statistical and Computational Methods in Data Analysis* (Springer).
Brekhovskikh, L. (1960). *Waves in layered media* (Academic Press).
Brekhovskikh, L. and Godin, O. (1998). *Acoustics of Layered Media I: Plane And Quasi-Plane Waves* (Springer-Verlag).
Brekhovskikh, L. and Lysanov, Y. (1991). *Fundamentals of Ocean Acoustics* (Springer-Verlag).
Bukhgeim, A. (2000). *Introduction to the Theory of Inverse Problems* (VSP Intl Science).
Burg, J. (1972). The relationship between maximum entropy spectra and maximum likelihood spectra, *Geophysics* **37**, pp. 375–376.
Claerbout, J. (1976). *Fundamentals of geophysical data processing* (Blackwell).
Claerbout, J. (1985). *Imaging the Earth's Interior* (Blackwell).
Cramer, H. (1946). *Mathematical methods of statistics* (Princeton University Press).
Ermakov, S. and Mikhailov, G. (1982). *Statistic simulation (in russian)* (Nauka, Moscow).
Ermakov, S. and Zhigljavsky, A. (1987). *Mathematical Theory of Optimal Experiment (in russian)* (Nauka, Moscow).
Faddeev, L. (1976). Inverse problems of quantum scattering theory, *J. Sov. Math.* , 5, pp. 334–396.
Fedorov, V. (1972). *Theory of Optimal Experiments* (Academic Press Inc.).
Fedotov, A. (1990). *Ill-Posed Problems with Random Errors in the Data (in russian)* (Nauka, Novosibirsk).
Goldin, S. (1984). Inverse kinematical problems of reflection seismology – i. classical problems in laterally inhomogeneous media, *Geophysical Journal of the Royal Astronomical Society* **79**, 1, pp. 317–325.
Goltsman, F. (1971). *Statistical models of interpretation (in russian)* (Nauka, Moscow).
Helgason, S. (1999). *The Radon Transform* (Birkhauser).
Herman, G. (1980). *Image Reconstruction from Projections* (Academic).
Holland, J. (1992). *Adaptation in Natural and Artificial Systems: An Introductory Analysis with Applications to Biology, Control and Artificial Intelligence* (MIT Press).
Huber, P. (2004). *Robust statistics* (Wiley).
Jensen, F. B., Kuperman, W. A., Porter, M. B. and Schmidt, H. (2000). *Computational ocean acoustics* (Springer-Verlag).
Johnson, N. and Lion, F. (1977). *Statistics and experimental design in engineering and in the physical sciences* (Wiley).
Karus, E., Kuznetsov, O. and Faizullin, I. (1986). *Interhole Sonic Survey (in russian)* (Nedra, Moscow).
Kennett, T. (1983). *Seismic wave propagation in stratified media* (Cambridge University Press).
Kireitov, V. and Pikalov, V. (1985). *Linear and nonlinear problems of computerized tomography (in russian)* (VTS SO AN SSSR, Novosibirsk).

Kiselev, Y. and Troyan, V. (1998). Numerical simulation in diffraction tomography (in russian), in *Voprosy geophiziki*, 35 (Saint Petersburg University).

Kolmogorov, A. (1956). *Foundations of the theory of probability* (Chelsea Publishing Company).

Kozlov, E. (1986). *Migration transform in seismic exploration (in russian)* (Nedra, Moscow).

Kravtsov, Y. (2005). *Geometrical Optics in Engineering Physics* (Alpha Science International).

Kravtsov, Y. and Apresyan, L. (1996). Radiative transfer: new aspects of the old theory, in *Progress in Optics*, Vol. XXXVI (North Holland), pp. 179–244.

Kullback, S. (1978). *Information theory and statistics* (Dover Publications).

Lavrentev, M. M. (1967). *Some improperly posed problems of mathematical physics (Springer Tracts Nat. Philos., vol. II)* (Springer-Verlag).

Lee, Y. (1960). *Statistical theory of communication* (John Wiley and Sons Inc.).

Levshin, A., Yanovskaya, T., Lander, A., Bukchin, B., Barmin, M., Ratnikova, L. and E.N., I. (1989). *Seismic surface waves in a laterally inhomogeneous Earth* (Kluwer Publ.).

Marchuk, G. (1982). *Methods of Numerical Mathematics* (Springer-Verlag).

Meshalkin, L. and Kurochkina, A. (1981). A new approach to the parametrization of regression dependences, *Journal of Mathematical Sciences* **17**, 6, pp. 2269–2275.

Morozov, V. (1993). *Regularization methods for ill-posed problems* (CRC Press).

Newton, R. (1981). Inversion reflection data for layered media: a review of exact methods, *Geophys. J. Roy. Astr. Soc* **65**, 1, pp. 191–215.

Nikitin, A. (1986). *Theoretical basis for processing of geophysical information (in russian)* (Nedra, Moscow).

Nolet, G. (1985). Solving or resolving inadequate and noisy tomographic systems, *J. Comp. Phys.* **61**, p. 463482.

Nolet, G., Trier, J. and Huisman, R. (1986). A formalism for nonlinear inversion of seismic surface waves, *Geophys. Res. Lett.* **13**, 1, pp. 26–29.

Penenko, V. (1981). *Mathematical methods of design of experiment (in russian)* (Nauka, Novosibirsk).

Petrashen, G. (1978). Basis of the mathematical theory of elastic wave propagation, in *Voprosy dinamicheskoi teorii rasprostraneniya seismicheskikh voln (in russian)*, 18 (Nauka, Leningrad).

Petrashen, G. (1980). *Wave propagation in anisotropic elastic media (in russian)* (Nauka, Leningrad).

Petrashen, G. and Kashtan, B. (1984). *Elementary of the theory of body wave propagation in anisotropic elastic media (in russian)* (Nauka, Leningrad).

Petrashen, G., Molotkov, L. and Krauklis, P. (1985). *Waves in homogeneous-layered elastic media (in russian)* (Nauka, Leningrad).

Petrashen, G. and Nakhamkin, S. (1973). *Continuation of the wave fields in problems of seismic exploration (in russian)* (Nauka, Leningrad).

Pikalov, V. V. and Preobrazhenski, N. G. (1983). Computer-aided tomography and physical experiment, *Sov. Phys. Usp.* **26**, 11, pp. 974–990.

Populis, A. and Pillai, S. (2002). *Probability, random variables and stochastic processes* (Mc-Graw Hil).

Pugachev, V. S. (1984). *Probability theory and mathematical statistics for engineers* (Pergamon).

Pytev, Y. (1990). *The methods of analysis and nterpretaion of the experiments (in russian)* (Vyssh. Shkola, Moscow).

Rao, K. (1972). *Linear statistical infernece and its applications* (Wiley).

Romanov, V. (1974). *Integral Geometry and Inverse Problems for Hyperbolic Equations. Springers Tracts in Natural Philosophy, vol. 26* (Springer-Verlag).

Romanov, V. (1987). *Inverse Problems of Mathematical Physics* (VNU Science Press).

Ryzhikov, G. (1976a). About covergence of the iterative methods of the solution of ill-posed problems (in russian), in *Problemy phiziki atmosfery*, 14 (Leningrad State University).

Ryzhikov, G. (1976b). Construction of algorithm of the solution of ill-posed problemswith the use of fructional-rational operator functions (in russian), in *Problemy phiziki atmosfery*, 17 (Leningrad State University).

Ryzhikov, G. and Troyan, V. (1985). Application of the methods of computational tomography to interpretative seismic problems (in russian), in *Vestnik Leningradskogo Universiteta*, 4 (Leningrad State University).

Ryzhikov, G. and Troyan, V. (1986a). Determination of the wave field and velocity in the media with the perturbation method (in russian), in *Vestnik Leningradskogo Universiteta*, 4 (Leningrad State University).

Ryzhikov, G. and Troyan, V. (1986b). Restoration of the velocity field in 3-d medium with the use of the elastic wave field functionals (in russian), *Izvestiya AN SSSR, Ser. Phizika Zemli* , 7, pp. 49–56.

Ryzhikov, G. and Troyan, V. (1986c). Two approaches to the problem of processing of the sonar data (in russian), Deponirovano VINITI 12.09.86, number 6649-86.

Ryzhikov, G. and Troyan, V. (1987). Linear estimation of the functionals of the elasic parameter fields and seismic thomographs (in russian), in *Voprosy dinamicheskoi teorii rasprostraneniya seismicheskikh voln*, 27 (Nauka, Leningrad).

Ryzhikov, G. and Troyan, V. (1989). Tomographic functionals and interpretative tasks of the elasitc wave sounding (in russian), in *Voprosy dinamicheskoi teorii rasprostraneniya seismicheskikh voln*, 28 (Nauka, Leningrad).

Ryzhikov, G. and Troyan, V. (1990). Back projection in diffraction tomography (in russian), in *Voprosy dinamicheskoi teorii rasprostraneniya seismicheskikh voln*, 29 (Nauka, Leningrad).

Ryzhikov, G. and Troyan, V. (1994). *Tomography and inverse sounding problems (in russian)* (Saint Petersburg Unibersity).

Ryzhikov, G., Troyan, V. and Yurin, A. (1989). Regualrization methods in 3-d seismic ray tomography problems (in russian), in *Voprosy dinamicheskoi teorii rasprostraneniya seismicheskikh voln*, 28 (Nauka, Leningrad).

Schumaker, L. (2007). *Spline Functions: Basic Theory* (Cambridge University Press).

Sheriff, R. and Geldart, L. (1995). *Exploration seismology* (Cambridge University Press).

Smylie, D., Clarke, C. and Ulrych, T. (1973). Analysis of irregularities in the earth's rotation, in S. F. B. Adler and B. A. Bolt (eds.), *Methods in Computational Physics*, Vol. 13 (Academic Press), pp. 391–430.

Späth, H. (1973). *Spline-Algorithmen zur Konstruktion glatter Kurven und Flächen* (R. Oldenbourg Verlag).

Stolt, R. and Weglein, A. (1985). Migration and inversion of seismic data, *Geophysics* **50**, 8, pp. 2458–2472.

Stratonovich, R. (1975). *The theory of information (in russian)* (Sov. Radio, Moscow).

Sudvarg, D. (1974). About using the polinomial multisections in description of the seismic interfaces (in russian), in *Application of the matematical statistics and computational mathematics for processing the seismic data* (Nauka, Novosibirsk).

Tarantola, A. (1984). Inversion of seismic reflection data in the acoustic approximation, *Geophysics* **49**, 8, pp. 1259–1266.

Tikhonov, A. and Arsenin, V. (1977). *Solutions of Ill-Posed Problems* (Wiley).

Tikhonov, A., Arsenin, V. and Timonov, A. (1987). *Mathematical Problems in Computer Tomography (in russian)* (Nauka, Moscow).

Timoshin, Y. (1978). *Impulse seismic holography (in russian)* (Nedra, Moscow).

Troyan, V. (1981a). Application of the spline functions for approxiamtion of geophysical information (in russian), in *Voprosy dinamicheskoi teorii rasprostraneniya seismicheskikh voln*, 20 (Nauka, Leningrad).

Troyan, V. (1981b). Application of the spline functions for approxiamtion of the geopysical information (in russian), in *Voprosy dinamicheskoi teorii rasprostraneniya seismicheskikh voln*, 20 (Nauka, Leningrad).

Troyan, V. (1981c). Statistacal methods for approxiamtion of the geophysical information, in *Annales Universitates Scientiarum Budapestinnensis de Rolando Eötvös Naminatae, Sectio Geologica*, XIII (Roland Eotvos University).

Troyan, V. (1982). *Statistical methods of processing of seismic information in in vestigations of the complex media (in russian)* (Nedra, Moscow).

Troyan, V. (1983). Solution of the interpretative seismic problems by the finite elements method (in russian), in *Vestnik Leningradskogo Universiteta*, 4 (Saint Ptersburg State University).

Troyan, V. (1984). Estimation of the seismic signal parameters by the correct moment method (in russian), in *Voprosy dinamicheskoi teorii rasprostraneniya seismicheskikh voln*, 24 (Nauka, Leningrad).

Troyan, V. (1985). The method of the local estimation of the seismic wave parameters (in russian), in *Vestnik Leningradskogo Universiteta*, 2 (Saint Petersburg State University).

Troyan, V. and Kiselev, Y. (2000). *Statistical methods of processing and interpretation of geophysical data (in russian)* (Saint Petersburg University).

Troyan, V. and Ryzhikov, G. (1994). 3-d diffraction seismic tomography. 1. tomography functional and reconstruction algorithm (in russian), *Geophisika* , 6, pp. 4–12.

Troyan, V. and Ryzhikov, G. (1995). 3-d diffraction seismic tomography. 2. resolution of t he method (in russian), *Geophisika* , 2, pp. 7–18.

Troyan, V. and Sokolov, Y. (1982). Approximation of the seismic horizon with taking into account borehole data (in russian), in *Voprosy dinamicheskoi teorii rasprostraneniya seismicheskikh voln*, 22 (Nauka, Leningrad).

Troyan, V. and Sokolov, Y. (1983). Determination of the bedded velocity by the area seismic data (in russian), in *Voprosy dinamicheskoi teorii rasprostraneniya seismicheskikh vol*, 23 (Nauka, Leningrad).

Troyan, V. and Sokolov, Y. (1989). *The methods of the geophysical data approximation with the use of computer (in russian)* (Nedra, Leningrad).

Troyan, V. and Windgassen, W. (1979). Approximation geophysikalischer daten unter berücksichtigung linearer bedingungen, *Geophysik und Geologie* **11**, 1, pp. 22–30.

Turchin, V., Kozlov, V. and Malkevich, M. (1971). The use of mathematical-statistical methods in the solution of incorrectly possed problems, *Sov. Phys. Uspechi* **13**, 6, pp. 681–703.

Ursin, B. (1983). Review of elastic and electromagnetic wave propagation in horizontally layered media, *Geophysics* **48**, pp. 1063–1081.

Ursin, B. and Berteussen, K.-A. (1986). Comparison of some inverse methods for wave propagation in layered media, *Proc., IEEE* **74**, pp. 389–400.

Vasilenko, G. and Taratorin, A. (1986). *Image Restoration (in russian)* (Radio i Svyaz', Moscow).

Voevodin, V. and Kuznetsov, Y. (1984). *Matrix and calculations (in russian)* (Nauka, Moscow).

Volkov, A. (1981). *Application of spline-function for geological and geophysical problems solution (in russian)* (Nauka, Moscow).

Yanovskaya, T. (1996). Ray tomography based on azimuthal anomalie, *PAGEOPH* **148**, 1/2, pp. 319–336.

Yanovskaya, T. and Porokhova, L. (2004). *Inverse Problems in. Geophysics (in russian)* (Saint Ptersburg University).

Zhdanov, M., Frenkel, M. A., Matusevich, V. I. and Urupov, A. K. (1988). *Seismic and electromagnetic migration (in russian)* (Nauka, Moscow).

Index

Absolute deviation (mean), 27
Algebra of events, 3
Amplitude spectrum, 374

Basis
  Hilbert space, 251
  Korhunen–Loeve, 282
Bayes
  formula, 11, 18
  strategy, 88, 280
Bias, 285
Bit, 62
Brightness, 132

Coefficient
  asymmetry (nondimensional), 28
  correlation, 274
  variation, 27
Complement, 3
Condition number, 259
Conditional
  mathematical expectation, 49
  variance, 49
Confidence
  bounds, 59
  interval, 191
  probability, 191
Correlation
  coefficient, 25
  function, 274
  Markovian, 108
  operator, 273
Covariance, 25
Cross-correlator, 274

Debye expansion, 136
Decile, 26
Decision rule, 84
  admissible, 88
Density function
  marginal, 17
  of random vector, 16
  standard normal, 46
Design matrix, 103
Design of experiment
  D-optimal, 290
  E-optimal, 291
  G-optimal, 291
  L-optimal, 291
  continuous, 290
  standardized, 290
Deviation
  mean-square, 24
  standard, 24
Dilatation, 119
Distance
  10 — 20 percent, 27
  interquartile, 27
Distribution
  a priori, 277
  of probabilities, 12
Distribution function
  cumulative, 12
  of random vector, 16
  standard normal, 46
Dynamic (Kalman) filtration, 382

Effective solution, 285
Eigenfunction, 253
Eigenvalue, 253
Ellipse
  of equiprobability, 50

of dispersion, 50
Entropy
differential, 66
mean conditional, 66
total, 66
Equation
characteristic system, 140
diffusion, 130
dispersion, 137
eikonal (general), 139
Euler, 254
Lame, 120
telegraph, 130
transport, 132, 306
wave, 126
scalar, 130
vector, 129
Wiener–Hopf, 390
Ergodic hypothesis, 76
Estimate
asymptotic efficient, 96
asymptotically bias, 93
efficient, 96
maximum a posteriori probability, 280
maximum a posteriori probability , 90
Event
certain, 3
elementary, 1, 2
impossible, 3
random, 2
Events
independent, 9
nonindependent, 9

Fisher matrix, 69
Flow of events, 41
ordinary, 42
Poisson, 42
stationary, 42
without a consequence, 41
Frechet derivative, 253
Function
autocorrelation, 74
correlation, 74
decision, 84
likelihood, 89, 164, 279
loss, 84
sufficient reception, 212
Functional, 251
convex, 268
Rayleigh, 254
stabilizing, 271
tomographic, 297
Functional gradient, 253
Fundamental sequence, 249

Generating function
central moments, 32
initial moments, 31
Geometrical optics, 135
Green's theorem
first, 257
second, 257

Independent random variables, 18
Inequality
Cauchy–Bunyakovskii, 250
Schwarz, 249
Information
a priori, 265
complete mutual, 65
Kullback, 209
partial, 65
Interpretation
combined, 88
interval-quantitative, 88
point-quantitative, 87
qualitative, 87
quantitative, 87
quantitative-qualitative, 88
Intersection, 2

Lame parameters, 119
Law
Coulomb, 127
Faraday, 127
Oersted, 128
of averages, 34

Maxwell hypothesis, 128
Median, 26
Method
likelihood ratio, 90
maximum likelihood, 89, 279
steepest descent, 267
Minimax procedure, 89
Mode, 27
Moment
central, 28
first, 73

initial, 28

Newton generalized scheme, 267
Norm
of function, 248
of linear operator, 252

Operator, 251
compact, 252
conjugate (Lagrange), 253
Fisher (information), 281
Hermite, 253
Hilbert–Schmidt, 252
Lame, 120
limited, 252
linear, 251
linear homogeneous, 77
linear regression, 286
non-negative, 255
positive, 255
positive defined, 255
self-conjugate, 253
unbounded, 263
unitary, 253
unlimited, 252

Pareto optimality, 291
Percentile, 26
Phase spectrum, 374
Poisson ratio, 119
Poynting's vector, 130
Principle
Fermat, 143
Huygens–Kirchhoff, 143
Probability
conditional, 9
elementary, 45
model, 4
space, 4
statistical, 5
Probability density
a posteriori, 278
conditional, 277
marginal, 277
Problem
ill-posed, 264
well-posed, 264
well-posed by Tikhonov, 264

Quartile, 26

Quasisolution, 272

Radon transform, 315
Random field (homogeneous), 274
Random fields (independent), 277
Random function
Markovian, 76
stationary, 74
wide sense, 75
Random variable
centered, 28
continuous, 14
discrete, 12
Range, 27
Rate of occurrence, 41
Refraction vector, 146
Regression model, 290
Regularization
n-th order, 270
Ivanov–Lavrentiev, 270
parameter , 271
Tikhonov, 270
Resolvability condition, 137
Risk
Bayes, 85
function, 85
Robust estimation, 98

Sample, 83
repeated, 83
size, 83
Shortwave asymptotic, 146
Significance level, 59
Singular numbers, 180
Slowness, 160
Slowness vector, 122
Space
Banach, 249
decision, 84
elementary events, 1
Hilbert, 249
metric, 249
complete, 249
observation, 84
outcomes, 1
parameter, 84
Spatially-time ray, 142
Spline functions, 217
Statistics, 83
Stress, 116

tenzor, 117
System of estimates
jointly asymptotic efficient, 96
jointly efficient, 96
System of normal equations, 168

Tomographic
experiment, 294

Uniform pressure modulus, 120
Union, 2

Vector potential, 129

Wavefront, 146

Young modulus, 119